Probability and Mathematical Statistics (Continued)

SEBER • Linear Regression Analysis
SEBER • Multivariate Observations
SEN • Sequential Nonparametrics: Invariance F
Inference
SERFLING • Approximation Theorems of Ma
TJUR • Probability Based on Radon Measures
WILLIAMS • Diffusions, Markov Processes, a
I: Foundations
ZACKS • Theory of Statistical Inference

Applied Probability and Statistics

ABRAHAM and LEDOLTER • Statistical Methods for Forecasting
AGRESTI • Analysis of Ordinal Categorical Data
AICKIN • Linear Statistical Analysis of Discrete Data
ANDERSON, AUQUIER, HAUCK, OAKES, VANDAELE, and WEISBERG • Statistical Methods for Comparative Studies
ARTHANARI and DODGE • Mathematical Programming in Statistics
BAILEY • The Elements of Stochastic Processes with Applications to the Natural Sciences
BAILEY • Mathematics, Statistics and Systems for Health
BARNETT • Interpreting Multivariate Data
BARNETT and LEWIS • Outliers in Statistical Data, *Second Edition*
BARTHOLOMEW • Stochastic Models for Social Processes, *Third Edition*
BARTHOLOMEW and FORBES • Statistical Techniques for Manpower Planning
BECK and ARNOLD • Parameter Estimation in Engineering and Science
BELSLEY, KUH, and WELSCH • Regression Diagnostics: Identifying Influential Data and Sources of Collinearity
BHAT • Elements of Applied Stochastic Processes, *Second Edition*
BLOOMFIELD • Fourier Analysis of Time Series: An Introduction
BOX • R. A. Fisher, The Life of a Scientist
BOX and DRAPER • Evolutionary Operation: A Statistical Method for Process Improvement
BOX, HUNTER, and HUNTER • Statistics for Experimenters: An Introduction to Design, Data Analysis, and Model Building
BROWN and HOLLANDER • Statistics: A Biomedical Introduction
BROWNLEE • Statistical Theory and Methodology in Science and Engineering, *Second Edition*
CHAMBERS • Computational Methods for Data Analysis
CHATTERJEE and PRICE • Regression Analysis by Example
CHOW • Analysis and Control of Dynamic Economic Systems
CHOW • Econometric Analysis by Control Methods
CLARKE and DISNEY • Probability and Random Processes: A First Course with Applications
COCHRAN • Sampling Techniques, *Third Edition*
COCHRAN and COX • Experimental Designs, *Second Edition*
CONOVER • Practical Nonparametric Statistics, *Second Edition*
CONOVER and IMAN • Introduction to Modern Business Statistics
CORNELL • Experiments with Mixtures: Designs, Models and The Analysis of Mixture Data
COX • Planning of Experiments
DANIEL • Biostatistics: A Foundation for Analysis in the Health Sciences, *Third Edition*
DANIEL • Applications of Statistics to Industrial Experimentation
DANIEL and WOOD • Fitting Equations to Data: Computer Analysis of Multifactor Data, *Second Edition*
DAVID • Order Statistics, *Second Edition*
DAVISON • Multidimensional Scaling
DEMING • Sample Design in Business Research
DILLON and GOLDSTEIN • Multivariate Analysis: Methods and Applications
DODGE and ROMIG • Sampling Inspection Tables, *Second Edition*
DOWDY and WEARDEN • Statistics for Research

continued on back

Spatial Data Analysis by Example

VOLUME I

Point Pattern and Quantitative Data

Spatial Data Analysis by Example

VOLUME I

Point Pattern and Quantitative Data

GRAHAM J. G. UPTON
Department of Mathematics,
University of Essex

BERNARD FINGLETON
Cambridge College of Arts and Technology

JOHN WILEY & SONS
Chichester · New York · Brisbane · Toronto · Singapore

Library of Congress Cataloging in Publication Data:

Upton, Graham J. G.
Spatial data analysis by example.

Bibliography: v. 1, p.
Includes indexes.
Contents: v. 1. Point pattern and quantitative data.
1. Spatial analysis (Statistics) I. Fingleton, B. (Bernard), 1949– . II. Title.
QA278.2.U68 1985 519.5 84–11806
ISBN 0 471 90542 9 (v. 1)

British Library Cataloguing in Publication Data:

Upton, Graham J. G.
Spatial data analysis by example.—(Wiley series in probability and statistics)
Vol. 1: Point pattern and quantitative data
1. Spatial analysis (Statistics)
I. Title II. Fingleton, Bernard
519.4 QA278.2

ISBN 0 471 90542 9

Photosetting by Thomson Press (India) Limited, New Delhi and Printed by Page Bros. (Norwich) Ltd.

G. J. G. U.: to Sue, Robin and Christopher

B. F.: to Eileen, Paddy, Ellen, Meave and James

Contents

Preface

The idea of producing a single volume 'user's guide' to the recent developments in spatial statistics arose from a sequence of phone-calls and conversations during the summer of 1981 which culminated in the production of a list of contents mildly resembling those of this volume and its successor. However, two failings of this dialogue were our underestimation of the likely length of the resulting compendium and also of the length of time required to complete it in its entirety. We had originally envisaged that the chapters would naturally fall into four groups, which themselves were arranged in pairs, and the change to two volumes therefore seems not unreasonable. At the least we hope that our readers will find that the two volumes make a convenient pair of bookends!

It is a pleasure to acknowledge the help and encouragement of the publishers in the various stages of preparation of this volume. Most especially we thank our families for their patience: the gestation period has been so long that they must often have wondered whether eventual delivery (deliverance?!) would ever come.

We are very grateful to Dr S. A. L. M. Kooijman for his assistance with the glasswort and anemone data and we gratefully acknowledge the following sources for permission to reproduce various figures and tables: The Association of American Geographers (Figure 1.30); Akademie-Verlag, Berlin (Figure 4.9); Blackwell Scientific Publications Ltd (Figures 3.14 and 4.14); *Biometrika* Trustees (Figure 1.39b, Appendices 3 and 4); Cambridge University Press (Figures 3, 1.7, 1.29 and 4.13); *Evolution* (Figures 3.6 a–d); the Geographical Association (Figures 1.34a and 1.34b); Gustav Fischer Verlag (Figure 1.11); Gordon and Breach Science Publications (Table 1.18); Harvard University Press (Figure 3.7); Institut National de la Recherche Agronomique Service des Publications, St. Cyr (Figure 2); the Institute of British Geographers (Figure 1.16); the Editors of the *Journal of the Royal Statistical Society, Series B* (Figure 4.3); dr W. Junk publishers (Figure 6); Regional Science Association (Figure 1.12).

Any typographical errors or errors of principle or practice that remain in this volume are of course our own responsibility and we would be grateful to be notified by readers if they detect such errors. We would also value any suggestions for additions or improvements to the present volume.

Introduction

This is the first volume of a two-volume work that deals with the analysis of spatial data. Spatial data comes in many varieties and it is not easy to arrive at a system of classification that is simultaneously exclusive, exhaustive, imaginative, and satisfying. We have identified four major clusters of techniques which appear, to us, to include the majority of the diverse examples of spatial data that we have encountered. We treat point pattern and quantitative data in this first volume and categorical and directional data in the subsequent volume. We do not claim that these volumes entirely account for spatial data in all its manifestations, nor that we have exhausted all aspects of the four clusters of techniques. We can, however, claim with confidence that what we have covered has certainly exhausted us!

Before giving a more detailed introduction to the subject-matter of the five chapters of this volume, we present an overview derived from the nearly 500 references included at the end of the book. We find lists fascinating and therefore present Table 1, which shows, in descending order, the five most commonly cited journals in our bibliography. Table 1 is somewhat misleading, however, since the apparent emphasis on biometrics is a side-effect of there being fewer journals in this field.

Figure 1 attempts to give a truer indication of the emphasis of the book. When we attempt to classify the references it quickly becomes apparent that there are roughly equal numbers falling in the four categories displayed in the diagram. Thus, roughly half the references have appeared in primarily theoretical statistical or biometrical journals and half in the more subject-oriented journals

Figure 1 deliberately situates the two applied fields side by side because we wish to emphasize the similarities of their statistical problems. We greatly hope

Table 1 The five most commonly cited journals (in descending order)

1. *Biometrics*
2. *Biometrika*
3. *Geographical Analysis*
4. *Journal of the Royal Statistical Society, Series B*
5. *Journal of Ecology*

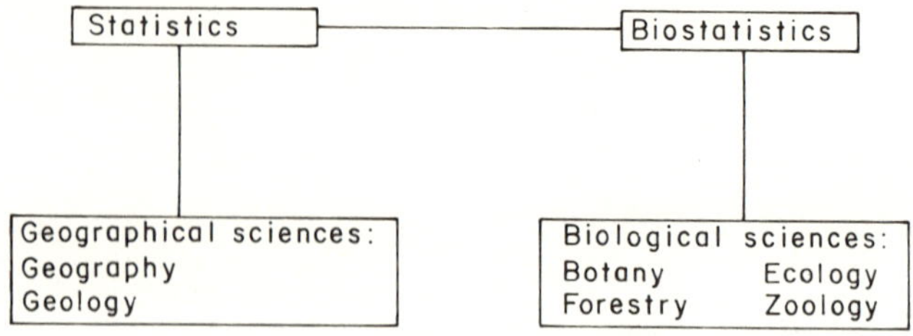

Figure 1. The contents of the book

that a side-effect of this work on spatial data analysis will be an increased cross-fertilization between these two subject-areas. The breadth of potential application of the methods that we describe is easily seen by glancing at the example index, which includes subjects as diverse as South-East-Asian birds, Southampton junior schools, desert shrubs, and forest trees! The common feature with all these examples is that individual observations can be located, more or less accurately, on the surface of the earth, so that a set of such observations is therefore capable of being displayed in the form of a map. This spatial attribute is, in many cases, a highly revealing piece of information concerning the causal processes and mechanisms operating amongst the data.

So that the reader may decide at the outset whether this work is likely to be suitable for his or her purposes, it seems sensible to set out here our guiding philosophy in the construction of this work. Throughout our writing we have had three precepts in mind:

(i) the practitioner seeking to *use* a statistical procedure does not want his or her understanding of that procedure obscured by immense numbers of theorems, lemmas, and other mathematical excrescences, but

(ii) it is manifestly unwise for the practitioner to use a statistical 'black-box' without having, at the least, a reasonable idea of its rationale, and

(iii) some readers will require more detail than we have provided and will welcome a reference to a source paper that provides that detail.

Our aim has therefore been to explain, clearly but succinctly, the ideas and principles underlying each statistical procedure without getting lost in too many mathematical formulae. Undoubtedly the idea of how many is 'too many' will differ from reader to reader and we shall not please everybody all of the time. However, we have *tried* to avoid proving results that could reasonably be stated and to avoid stating results if they were not helpful.

The degree of statistical and mathematical background that is required varies somewhat from chapter to chapter. As a rule, we presume that the reader is familiar with (or has met in the past) basic statistical ideas such as probability, probability distributions, variance, expectation, estimation, tests of significance, and the like. The chapter on regression (Chapter 5) anticipates rather more than this because of the impossibility of including a general introduction to the subject along with everything else. In that chapter we have assumed that the reader can stand (and understand) apparently complicated matrix expressions and is familiar with the general (non-spatial) methodology. Notice that in the second sentence of this paragraph we used the word 'ideas' rather than the word 'formulae'—the latter are supplied as required. Thus, in Chapter 1 we provide the formula for the Poisson distribution, though we certainly anticipate that most readers will have met that formula before.

The difference between point pattern and quantitative data is illustrated by the contrast between Figure 2 and Figure 3. Figure 2 is a map of 71 Swedish Pine saplings in a 10 m × 10 m square (see Strand, 1972, and Ripley, 1981). Since the saplings are presumably all roughly the same size, and if we regard their diameter as being negligible by comparison with the inter-sapling distances, then size variations can be ignored and the matter of interest may be regarded as being simply their existence at the points indicated on the map. This particular point pattern shows signs of being regularly spaced, and of not being representative of the sort of pattern to be expected if the saplings grew at random places within the plot. The various traditional methods for detecting departures from randomness (mostly based on quadrat counts) together with the very recently developed methods based on inter-item distances form the subject material for Chapter 1.

Figure 3 is an example of what we call quantitative data. Ostensibly it is merely

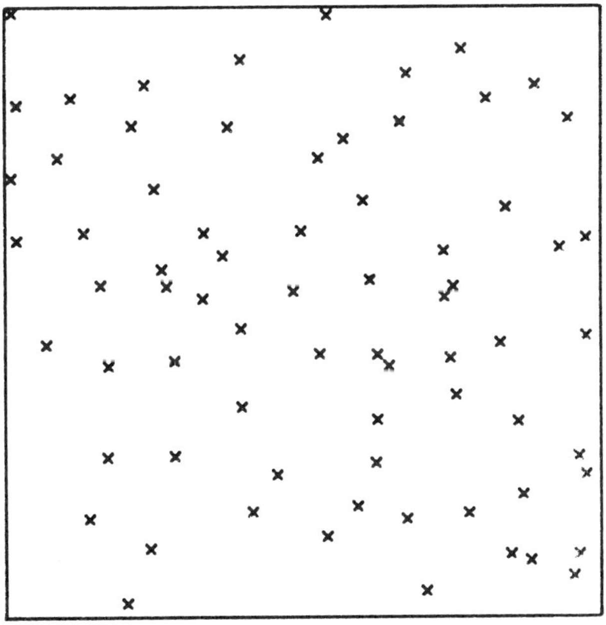

Figure 2. The distribution of 71 Swedish Pine saplings. (Redrawn from Strand, 1972. Reproduced by permission of INRA)

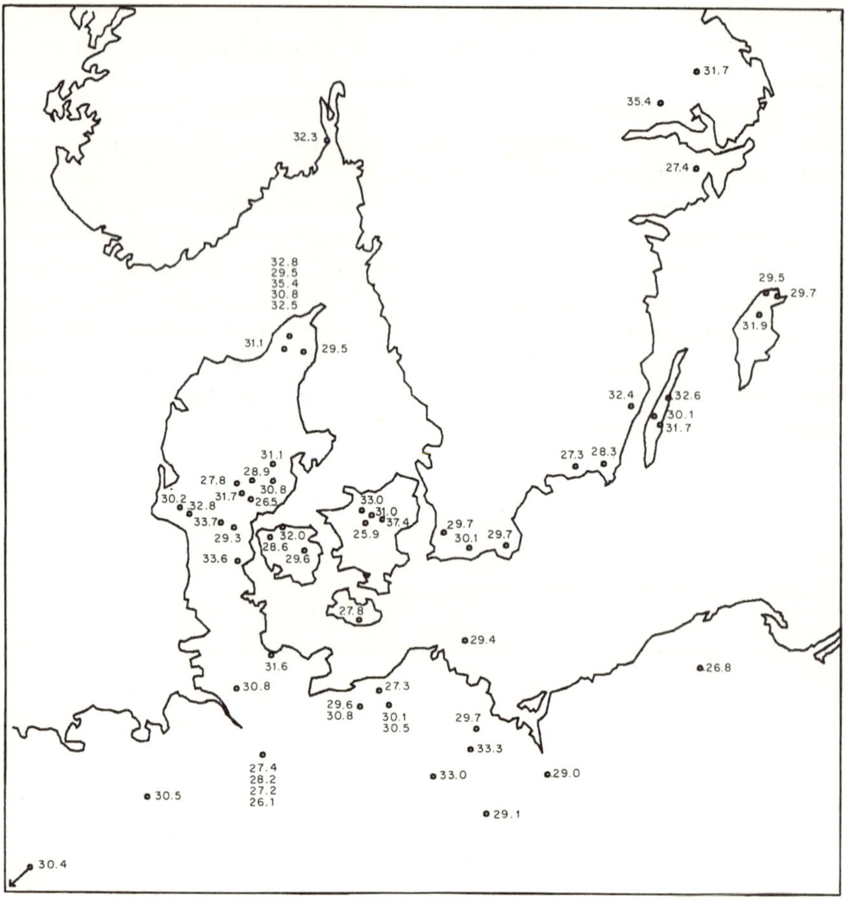

Figure 3. The distribution of the length/breadth index for the Bagterp spearheads. (Redrawn from Hodder and Orton, 1976. Reproduced by permission of Cambridge University Press)

another point pattern, but in this case the points possess values and, while the location of the points is not uninteresting, our principal concern is with the possibility of spatial variation in the values. The data displayed in Figure 3 relate to finds of so-called Bagterp spearheads which emanate from the early Bronze Age. A possible method of analysing the degree of localization of the manufacture of these spearheads (which have particular easily recognized characteristics) is based on the variation in the shape of the spearheads. A simple measure of shape is the ratio of the length of each spearhead to its breadth, and these are the values that are recorded in Figure 3, which is taken from Hodder and Orton (1976, p. 175) who plot data collated by Jacob-Friesen (1967).

Notice that the point locations in Figure 3 do not comprise the complete set of locations for these Bagterp spearheads, but rather reflect the locations of organized archaeological activity. Thus, the recorded locations are indicative but not definitive. By contrast, Figure 2 displays *all* the saplings in the region illustrated.

The arrangement of the book is that Chapters 1, 2, and 4 are principally concerned with point pattern data, while Chapters 3 and 5 are principally concerned with quantitative data. However, this division is not precise and there are many occasions on which quantitative methods are used to analyse point pattern data. An example is illustrated by Figure 4, which displays the numbers of bitterbrush (*Purshia tridentata*) plants in a 700 ft × 500 ft rectangular region to the north-east of Florence, Montana (Lyon, 1968). Here, the original point locations have been summarized as a series of quadrat counts. We shall see in Chapter 1 that the most sensitive insights into the pattern of objects are those obtained using functions of the distance between objects and their nearest neighbours, whereas in Chapter 2 we shall see that quadrat counts of this type provide the simplest means of obtaining a reliable estimate of the total number of plants in the larger region of interest.

As already intimated, in Chapter 1 we seek to identify the characteristics of a pattern, with the hope that these characteristics will reveal something helpful concerning the origins of that pattern, whereas in Chapter 2 the emphasis is on estimating the numbers present in a region by means of a detailed examination of a small part of that region. At first sight one might imagine that this latter task scarcely deserved a chapter to itself! However, the difficulties of the task are made apparent by considering some of the possible contexts:

(i) animals that run away from an approaching observer;

(ii) trees or bog plants in inhospitable terrain;

(iii) trees that have contracted a disease but may not yet show any outward signs of that disease.

While remote sensing, using satellite photographs or some other device, can ameliorate the problems caused by these types of data, such an approach will be expensive and may also be impracticable.

In Chapter 3 we turn to quantitative data, such as that exemplified in Figure 3. This chapter is concerned with the measurement of spatial autocorrelation: that

154	74	67	168	172	136	128
79	89	69	104	126	102	73
114	120	143	117	41	160	139
117	53	84	184	74	155	172
188	115	62	38	83	108	145

Figure 4. Numbers of bitterbrush (*Purshia tridentata*) plants in 35 100 ft × 100 ft quadrats (Lyon, 1968)

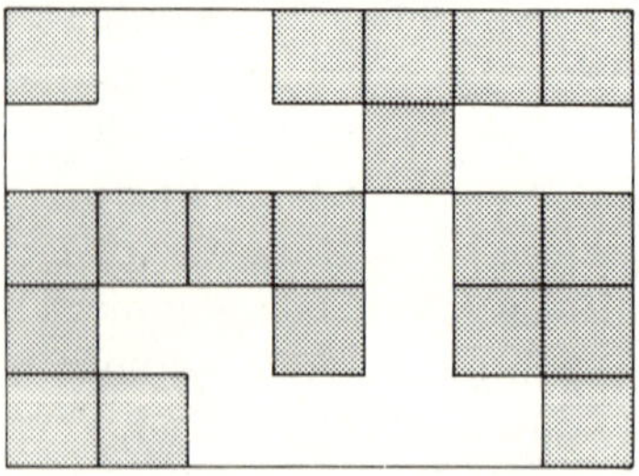

Figure 5. Black/white representation of bitterbrush counts

is, the pattern or organization in the values at fixed localities. We seek to discover whether these values are distributed haphazardly over the locations or whether there is some spatial order to their arrangement, so that, for example, all the high values are found to be grouped close together. The actual metric used may be the continuous metric of the length/breadth ratios of the Bagterp spearheads (this is the case with our subsequent analysis in Example 3.7 of variations of the wing length of a species of bird in the Far East) or it may be a much 'clumsier' metric such as 'lowest, lower, low, medium, high, higher, highest'. In the limit, the latter metric can be reduced to a two-case 'higher-than-average, lower-than-average' metric which we could refer to as a 'black and white' pattern.

Figure 5 illustrates how the detailed data illustrated in Figure 4 can be summarized using the black/white metric. It incidentally also illustrates that what was basically point pattern data can also be analysed in a relevant manner by methods recorded in Chapters 3 and 5. A major class of methods for measuring spatial autocorrelation hinge upon the numbers of occasions on which 'black' cells border on other 'black' cells. Figure 5 certainly provides us with a picture that suggests that there is spatial organization present in the location of the bitterbrush plants—there are few isolated cells.

Chapters 4 and 5 are concerned with bivariate and multivariate data, and therefore themes such as mutual attraction and repulsion, or the correlation between spatial distributions, come into prominence in place of the 'self-oriented' autocorrelation, inhibition, and clustering mechanisms which feature as the explanatory processes of the univariate spatial distributions of Chapters 1–3. Some of the procedures discussed in these chapters are still in the process of being developed and have methodological problems that are, as yet, unresolved. We have attempted to give the reader guidance and to provide some perspective on the available alternatives, including some illustration of the pitfalls that may arise from the uncritical application of non-spatial methods for spatial data.

Figure 6, taken from Erschbamer *et al.* (1983), is typical of the data discussed in Chapter 4. In the figure are recorded the point positions of several different types of plant which compete with one another for the available water supply in a

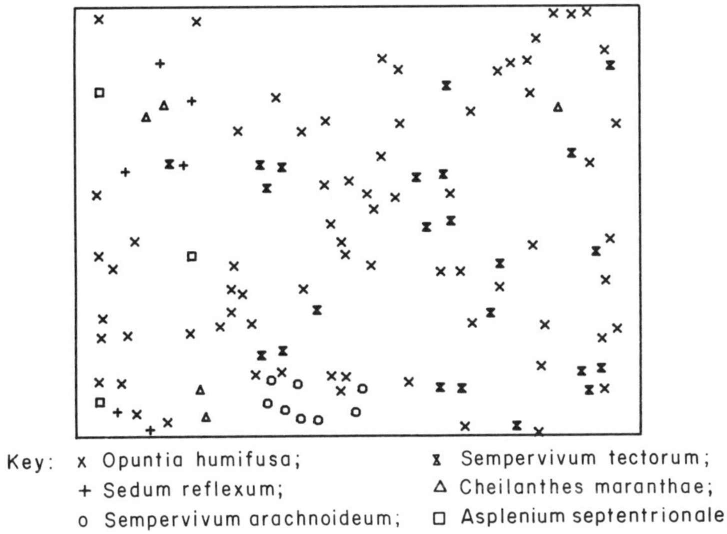

Figure 6. The distribution of six plant species in an Alpine valley. (Redrawn from Erschbamer *et al.*, 1983. Reproduced by permission of JUNK)

Central alpine valley in the South Tyrol. The ecologist is interested in the factors that determine their positions. Viewed as a point pattern we may use the methods described in Chapter 4 to determine the extent to which plants of different types are isolated from one another. Of course part of the isolation will be due not to physical repulsion on the part of the plants but rather to variations in the nature of the terrain. Erschbamer *et al.* note that there are several rocky outcrops within the study area, while the soil depth varies down to about 90 cm and also varies in its water content. Using the regression methods of Chapter 5 we could analyse this data from a rather different standpoint, comparing the spatial density of a particular species of plant with the (literally) underlying physical conditions: soil depth and available water.

An important consideration when applying regression procedures to spatial data is the presence of self-correlation (autoregression). As an instance, in Example 5.20 we examine a model that attempts to explain the variation in the number of different plant species among Californian localities, $\{Y_i\}$, using as explanatory factors the maximum altitude in the locality, the latitude of the locality, the area of the locality, and also an autoregressive component whereby Y_i is related to Y_j $(i \neq j)$, the number of species in the 'neighbouring' locality j. The methods for dealing with autoregressive spatial interaction have been the subject of some debate (cf. Ord, 1980 and Griffith, 1980, 1981).

At this point it is timely to make clear that the intention of our analyses is primarily to demonstrate the methodology and we make no attempt simultaneously to clinch some argument concerning the controlling mechanisms of Californian flora, or whatever. For this reason our analyses are presented 'warts and all' in order to give a true representation of our actual experiences with the methods and models that we are illustrating. Having said this, we do nevertheless hope that the actual results that we obtain will have intrinsic interest for their own sake.

Many of the procedures that we discuss require the use of a computer. After considerable hesitation we decided that it was not feasible to append the various computer programs that we wrote in order to achieve our results. Apart from a reluctance to reveal just how clumsy these programs were (!), there was the added consideration that some were written in FORTRAN, some in BASIC, some used the GENSTAT package, others the TILE4 package, and still others used subroutines from the NAG library. We concluded that not all our readers would know, or have access to, all these acronyms! Instead we have attempted to reduce the complexity of formulae wherever possible and to concentrate our efforts on describing the successful simpler techniques.

Obviously this book was not written in a void, and it is apposite to draw up a short list of major sources from which our ideas and inspiration spring. We have already presented, in Table 1, a list of the five most cited journals, though there are a further dozen that figure prominently in the list of references. To these journals must be added a quartet of books that can be warmly recommended if additional background reading is required. These are:

Cliff and Ord, *Spatial Processes* (1981), Haggett, Cliff, and Frey, *Locational Analysis in Human Geography* (1977), Pielou, *Mathematical Ecology* (1977), and Ripley, *Spatial Statistics* (1981). Further details are given in the list of references.

CHAPTER 1

The Identification of Pattern

1.1 ALTERNATIVE PATTERNS

A glance at the dictionary definition of 'pattern' reveals a number of alternative definitions that, while related to one another, vary subtly with context. In the context of this book the most appropriate definition would seem to be the following:

> Pattern is the zero-dimensional characteristic of a set of points which describes the location of these points in terms of the relative distances of one point to another (Hudson and Fowler, 1966).

As an example, consider the pattern given in Figure 1.1a. This is unaltered by either enlargement or rotation (Figures 1.1b, c), but becomes a different pattern when distorted (Figure 1.1d). In essence, a pattern is fixed by the angles between the lines that would join the points of the pattern.

The reason for the interest in a pattern of *points* is that, with an appropriate choice of scale, even huge objects may be best represented by a point. Thus, while London scarcely seems like a point when one is driving by car from Dover to Edinburgh, this becomes an entirely reasonable representation of its location for a schoolboy's map of the world. Evidently the reasonableness of the point approximation is a function of the scale at which we are working, and we must not lose sight of the fact that it is only an approximation. In a later section we shall briefly consider some of the implications of replacing the points by circles or discs of noticeable diameter.

In the previous paragraph we stressed that it was the scale of one's map that determined whether the subject of interest could reasonably be represented as a

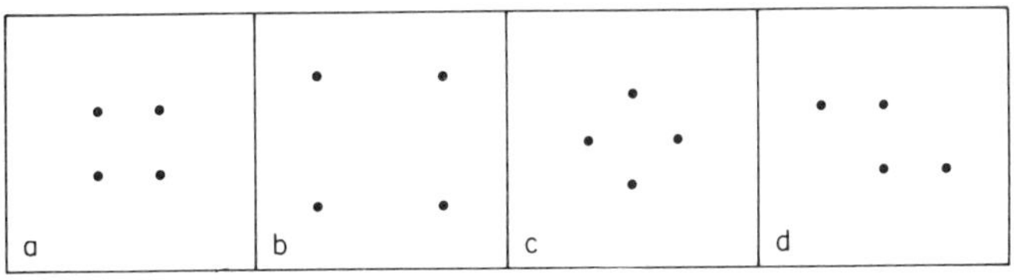

Figure 1.1 Three identical patterns (a–c) and a different pattern (d)

Table **1.1** *Types of objects analysed as point patterns*

Objects	Analysts
Neurones in bone	Aherne and Diggle (1978)
Magnetite crystals	Davis (1973)
Acorn barnacles	Kooijman (1979)
Liverfluke snails	Keuls (1963)
Bronze Age daggers	Bergmann (1970)
Sparrowhawk nests	Newton *et al.* (1977)
Atriplex bushes	Gulmon and Mooney (1977)
Drumlins	Hill (1973)
Towns	Glass and Tobler (1971)
Galaxies	Neyman and Scott (1958)

point. Given suitable scales, the actual physical sizes of objects that may be so represented is unbounded, as is shown by the examples in Table 1.1.

At one end of this list microscopes are required—at the other end it is telescopes that are needed! The range of disciplines (pathology, geology, marine biology, zoology, archaeology, ornithology, botany, forestry, physical and human geography, and astronomy) reflects the width of application of the techniques that will be discussed in this and the following chapter.

In the present section we are concerned with alternative patterns. It is natural to begin with the simpler patterns and so the first pattern to be considered is that which is the easiest to describe mathematically.

The random pattern

Suppose, for convenience, that the region that we are considering has area A (in some convenient units) and suppose also that there are a total of λA points scattered about at random within this region. The overall intensity of points is therefore $\lambda A/A = \lambda$ per unit area. We now imagine that the region as a whole is made up of a huge number, N, of extremely small sub-regions each having size A/N. Clearly, if we make N sufficiently large, then there will be a negligible chance of any sub-region containing more than one of the λA points. Now, a random pattern implies that the points could exist absolutely anywhere within the region, so that each of the N sub-regions must have an equal probability of actually containing a point. Since there are N sub-regions and λA points it follows that this common probability must be $\lambda A/N$. We now enquire what happens as N increases towards infinity. Since the preceding argument is precisely equivalent to the argument used in deriving the Poisson distribution as a limit of the binomial distribution, it follows that the probability of observing exactly r points in a unit area, when the intensity per unit area is λ, will be given by

$$p_r = \frac{\lambda^r e^{-\lambda}}{r!} \qquad (r = 0, 1, 2, \ldots). \tag{1.1}$$

An important property of the Poisson distribution (1.1) is that it has mean = variance = λ. The result (1.1) explains why a random distribution of points in space may be referred to as a (spatial) Poisson process: Keuls *et al.* (1963) describe the arrangement of points as a *Poisson forest*. An example of a Poisson forest is given in Figure 1.2.

A question of immediate interest is the following: 'Is it reasonable to expect a pattern of real data points to display randomness?' Naturally, the answer depends on what these points represent. As an example of a situation in which randomness might well be a reasonable description, consider the colonization of a newly ploughed field by plants that are propagated by airborne seeds. It is surely reasonable to suppose that these seeds land and germinate at haphazardly placed positions within the field so that the initial distribution of the newly germinated plants is entirely consistent with the hypothesis of randomness. Subsequently, however, two contradictory forces are liable to have an effect on the observed plant location. On the one hand, the competition between neighbouring plants is likely to result in some thinning out of close neighbours, so that the pattern becomes rather more regular in appearance. On the other hand, variations in the local ecology will result in some regions being more favourable for growth than are other regions. This will result in an apparent patchiness (or clustering) in the plant distribution. These two effects, and others such as inter-species competition, may well be sufficiently in balance that the resulting much-modified plant distribution may yet retain the appearance of randomness.

Greig-Smith (1979, p. 75) has remarked that 'One of the few generalizations that we can make about vegetation is that...its constituent species show...departure from randomness of distribution.' Nevertheless, this *is* only a

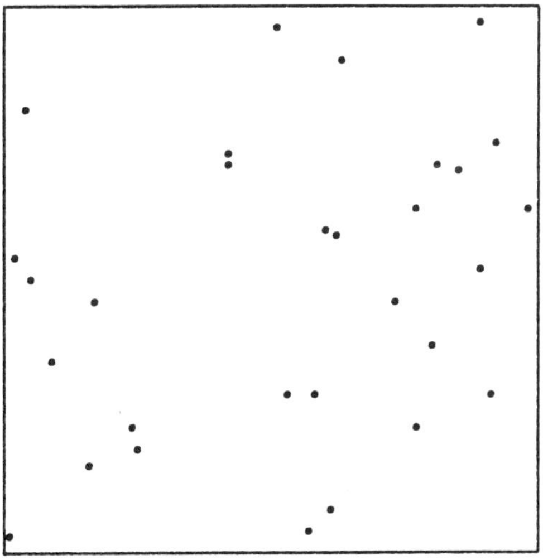

Figure 1.2 A realization of a Poisson forest

generalization. For example, Skellam (1952) notes that the effects of competition and irregular ecology on the distribution of the sedge *Carex arenaria* on sand dunes renders that distribution indistinguishable from a random distribution.

If competing contradictory forces can cause apparent randomness, then so can competing contradictory human decisions. For example, consider the influence of human decisions on the positioning in the countryside of human settlements (and, consequently, on the positioning of archaeological 'treasures'). Without doubt, the positioning of the settlements is not a matter of chance: a great many considerations must have been borne in mind before the final decision to 'pitch camp' was made. However, so far as the modern-day observer is concerned, the sheer number and complexity of the decisions that were involved, together with the fact that they were based on information that is no longer available to that modern observer, implies that 'every decision may [have been] optimal from a particular point of view, and yet the resulting actions as a whole appear random' (Curry, 1964, p. 138). Certainly it is the case that 'many of the observed archaeological patterns have a form which is similar to patterns produced by a random process' (Hodder and Orton, 1976, p. 9).

In view of the remarkable diversity of mechanisms which may lead to an apparently random pattern, it must be stressed that 'to say that a distribution is random ... is to say that the pattern has no discernible order and that its cause is undeterminable' (Dacey, 1964, p. 559): it is clear that 'a pattern is not necessarily indicative of the process which produced it' (Hudson and Fowler, 1966, p. 7).

However implausible the concept of randomness of distribution may appear on intuitive grounds, the above arguments suggest that it must be the first pattern that we should contemplate as an explanation for our own particular point distribution. Some tests for randomness will be the subject of a later section; we consider first the alternatives to a random pattern.

Clusters

Many spatial patterns display real or apparent clusters of points. An obvious situation in which we might expect to see clusters of individuals occurs with the distribution of the caterpillars of the Small White butterfly (*Pieris rapae*) in a cabbage patch—some leaves will have been chewed to pieces by an army of caterpillars while others remain untouched. Another familiar example of clustering is provided by the common strawberry (*Fragaria* × *ananassa*) which propagates itself by sending out surface runners, and, if left undisturbed for too long, will colonize an entire vegetable garden. Human settlements also display clustering, with the individual dwellings huddling around churches, fortifications, or some central meeting place.

In each of the above examples the cluster was only apparent because of the size of the region (cabbage patch, vegetable garden, or county) that was chosen for consideration. Thus, the regular pattern shown in Figure 1.3a might well be

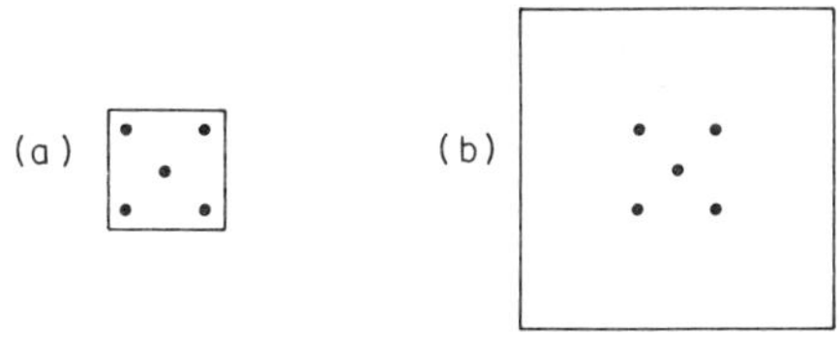

Figure 1.3 The effect of area size on the appearance of a pattern

regarded as a cluster if the region of enquiry were increased to that shown in Figure 1.3b.

That there is a cluster present in Figure 1.3b is essentially due to the impression that we get that there is a strikingly varying *intensity* of the points over the complete region being considered. The size of the region of inquiry therefore plays an important role, as Figure 1.4 emphasizes. For the larger regions shown, there are, respectively, clusters of regularly arranged points and a regular arrangement of clusters of points. For the smaller regions, however, only the regularity on the one hand, and a single cluster on the other hand, will be observed.

Of course the two example patterns given in Figure 1.4 are far too stylized to be realistic descriptions of patterns that one might expect to encounter. Nor would they be particularly easy to describe in mathematical terms. The alternatives to randomness that we shall consider are all derivatives of the simple Poisson process, varying only in the degree of extra complication that is introduced. Comprehensive summaries are to be found in Haggett *et al.* (1977, ch. 13) and Douglas (1975). There are two main classes of cluster models (though one is actually a special case of the other), namely the so-called generalized (or

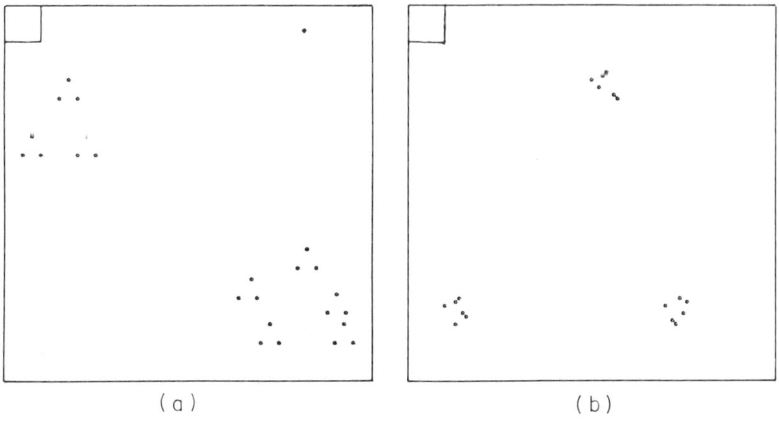

Figure 1.4(a) Two clusters of regularly arranged points. (b) A regular arrangement of clusters.

compound) Poisson distributions, and the Poisson cluster processes of Neyman and Scott (1952), Warren (1962), and Matern (1971).

Generalized Poisson distributions

In Figure 1.2 we presented one possible realization of a Poisson forest, in which each point represented a single individual. The difference with a generalized Poisson distribution is that each point represents a cluster of closely packed individuals, with the number of individuals in the cluster itself being an observation from some distribution. Examples of generalized Poisson distributions are given in Figures 1.5a and b.

The point pattern shown in Figure 1.5a was derived directly from that of Figure 1.2 by replacing the points in Figure 1.2 by clusters of points for which the number in a cluster was an observation from a Poisson distribution with mean 2. The pattern that results no longer looks random and does indeed suggest some sort of contagion between points, resulting in their clustering together. This particular combination of randomly located clusters with random cluster sizes was first suggested by Neyman (1939) in connection with the analysis of the distribution of corn borer larvae (*Pyrausta nubilalis*) and is known as the *Neyman Type A* distribution.

Although easy to describe, the Neyman Type A distribution has a rather revolting mathematical form. Possibly for this reason, a more favoured generalized Poisson distribution has been the *negative binomial* distribution, for which the probability of a randomly chosen unit of area containing exactly r

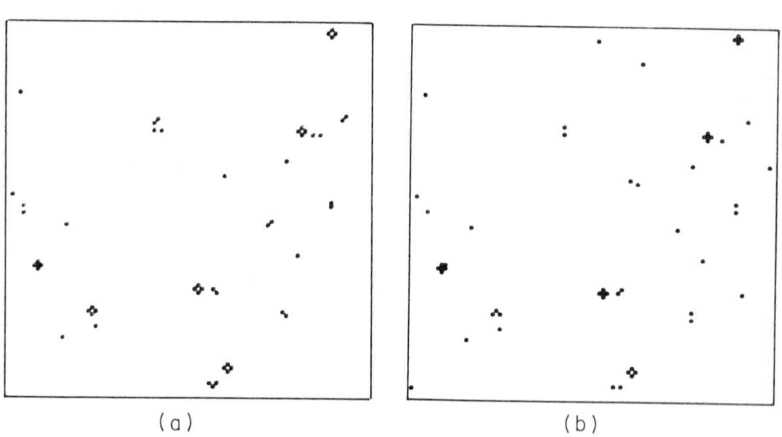

Figure 1.5 (a) A realization of a Neyman Type A distribution. (b) A realization of a negative binomial distribution

points is given by

$$p_r = \binom{r+k-1}{k-1} p^k (1-p)^r \qquad (r = 0, 1, 2, \ldots), \tag{1.2}$$

where p and k are two unknown parameters. This distribution occurs when the cluster size distribution is logarithmic:

$$p(\text{cluster size} = n) \propto (1-p)^n/n \qquad (n = 1, 2, \ldots).$$

Notice that the minimum value of n is 1 rather than 0 as was the case with the Poissonian clusters of the Neyman Type A distribution. The result is that visually there appear to be a lower proportion of clusters, though these are, on the whole, somewhat larger. An example pattern is given in Figure 1.5b, in which the same random numbers were used as in the generation of Figure 1.5a, so that a direct comparison between these two generalized Poisson distributions is possible.

Many other generalized Poisson models have been suggested and a comprehensive summary is provided by Ord (1972, p. 126). We mention here only two further models, the *Thomas* and *Short* models. Thomas (1949) suggested a simple modification of the Neyman Type A distribution, in which the size of the cluster was a Poisson observation *plus one*. An obvious interpretation is to regard the original locations as 'parents' and the additional Poisson clusters as 'children'. In the Thomas model the parents are effectively immortal! This would perhaps be a reasonable description of the pattern of spread of childhood diseases such as measles, in the sense that once one has had measles one can say (proudly) 'I have had measles' for the rest of one's life. On the other hand, it would not be appropriate for people who wished to say 'I have *got* measles' because it would be no longer true by the time that the disease was manifest in those that they had infected.

The second distribution of some interest is the so-called Short distribution (Cresswell and Froggart, 1963; Kemp, 1967). This distribution might be appropriate if there were areas of the region under scrutiny in which it was *impossible* for any clusters to exist—for example, if a region contains an island-less lake, then there will be no trees in the part of the region occupied by the lake. The Short distribution assumes that the cluster size is basically Poisson, but with a quantity of extra zeros reflecting the proportion of inhospitable terrain.

Further discussion of these generalized Poisson distributions is to be found in the next subsection dealing with compound Poisson distributions.

Compound Poisson distributions

The foundation of the generalized Poisson model lay in the Poisson distribution for cluster centres. It was supposed that over the entire region of interest these centres were distributed at a constant intensity λ, the parameter of the underlying Poisson process. Suppose instead that λ was not constant but varied with location, so that λ was much greater in some sub-regions than it was in others. What would be the result? Clearly, in regions of high λ there would be a far greater intensity of points than in other regions, with the result that the overall map of point locations would take on a somewhat mottled appearance. The resulting distribution of points per unit area is termed a *compound* Poisson distribution.

It is easy to imagine that compound and generalized Poisson distributions might appear very similar, and this is indeed the case. In fact Feller (1943) demonstrated that if there is a cluster size distribution leading to a particular generalized Poisson distribution, then there must exist a distribution to describe the spatial variation in λ that leads to an *identical* compound Poisson distribution. In other words, the negative binomial distribution is *simultaneously* an example of a generalized Poisson distribution *and* of a compound Poisson distribution. This problem of convergent evolution has come to be known as the problem of *equifinality*.

Pielou (1977, p. 123) concludes that it is therefore 'futile to try to reach conclusions about the mechanism underlying a particular pattern simply by observing the observed distribution of the number of individuals per unit', and she concludes by stating that: 'The fitting of theoretical frequency distributions to observational data can never by itself be sufficient to "explain" the pattern of a natural population.' Nevertheless, Cliff and Ord (1981a, pp. 94–96) suggest a technique that, for the negative binomial distribution, holds out some prospect of being able to distinguish between *true contagion* (the generalized distribution) and *apparent contagion* (the compound distribution), and this technique will be described in Section 1.3.

Poisson cluster processes

The compound and generalized Poisson models contain one particularly unrealistic constraint, namely that the 'offspring' are coincident with the 'parent'—in other words they suppose that the entire cluster is situated at a single point (thus Figure 1.5 is slightly misleading in that cluster members have been displaced in order to render them visible).

The consequence of this mathematical simplification is that extensive analysis of the structure of these patterns is possible. However, in some situations—depending on the physical scale of the investigation—a coincident cluster may be totally untenable as a working model, so that a third (displacement) component must be added to the previous models.

The general Poisson cluster process therefore consists of:

(1) parent events located at random and forming a Poisson forest;

(2) a number of offspring for each parent, these being independent observations from some specified distribution (e.g. Poisson, logarithmic); and

(3) the locations of each of the offspring relative to the position of its parent

being independent observations from some specified distribution (e.g. bivariate normal, bivariate uniform).

The third component simply specifies where the offspring are likely to be in relation to their parent. Relatively simple cases occur when the offspring are known to be at a fixed distance from the parent, and when the offspring are distributed at random over the region immediately surrounding the parent's location.

The initial work on this type of process is due to Neyman and Scott (1952, 1958) who were concerned with the pattern, on a two-dimensional photographic plate, of the images of the three-dimensional arrangement of the galaxies in the heavens. Warren (1962, 1971) considered the true two-dimensional case and referred to the model as the *centre–satellite* model. Further results have been obtained by Bartlett (1964, 1975) and Diggle (1975). Unfortunately, the added realism of these processes results in most intractable mathematics and none of their results seems likely to be of very great assistance for the data analyst.

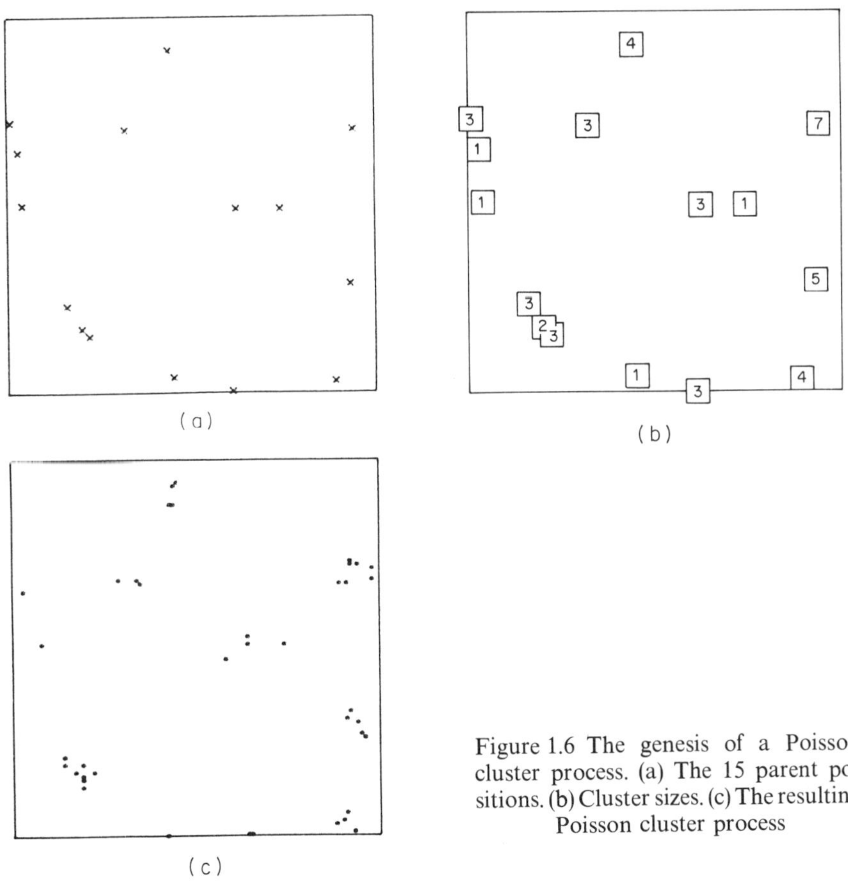

Figure 1.6 The genesis of a Poisson cluster process. (a) The 15 parent positions. (b) Cluster sizes. (c) The resulting Poisson cluster process

An example of the pattern produced by a Poisson cluster process is given in Figure 1.6c, with its genesis recorded in the preceding figures. Figure 1.6a records the positions of 15 randomly placed parents. Figure 1.6b shows square areas with these points as their centres, and with the numbers of offspring for each point (observations from a Poisson distribution with mean 3). Figure 1.6 c shows the effect of these offspring being randomly positioned within their own square area. Diggle (1979b) provides a useful review of these and other patterns, and observes that, as with the compound and generalized Poisson models (which are just degenerate forms of Poisson cluster models), there are two alternative mechanisms (true contagion and heterogeneity) that can lead to any given Poisson cluster pattern.

Inhibition models

The cluster models represented a move from the random pattern towards a pattern having a more uneven intensity of points, with local peaks at clusters. We now consider patterns that have a more even intensity. Such patterns are by no means infrequent, as is demonstrated by the distribution of Romano–British walled towns (Hodder and Hassall, 1971) reproduced in Figure 1.7.

The mechanism that led to the point distribution shown in Figure 1.7 was

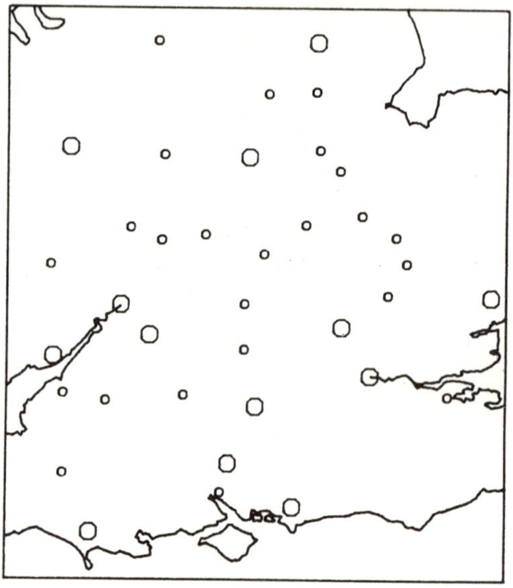

Figure 1.7 Walled town locations. (Redrawn from Hodder and Orton 1976. Reproduced by permission of Cambridge University Press)

clearly not a natural one, but was due to the design of man. We shall consider man-made patterns in the next sub-section; for the present we concentrate on modelling the regular patterns that arise from natural causes. Two natural mechanisms suggest themselves, one being static and the other dynamic. We commence by considering the static mechanism.

Imagine a flower-bed in which have been sprinkled, at random, the seeds of some choice annual flower. The flower-bed is kept free of weeds, the seeds germinate, and small plants begin to grow. However, quite apart from the effects of genetic variation, the plants do not grow equally well. Some plants quickly become sturdy bearers of masses of flowers while others languish: the difference being due to the differing amounts of inter-plant competition. Matern (1960) suggested that this type of situation could be modelled in the following way. We commence by generating a Poisson forest, and we then eliminate all pairs of points that are found to be less than a distance x apart. The quantity x is simply

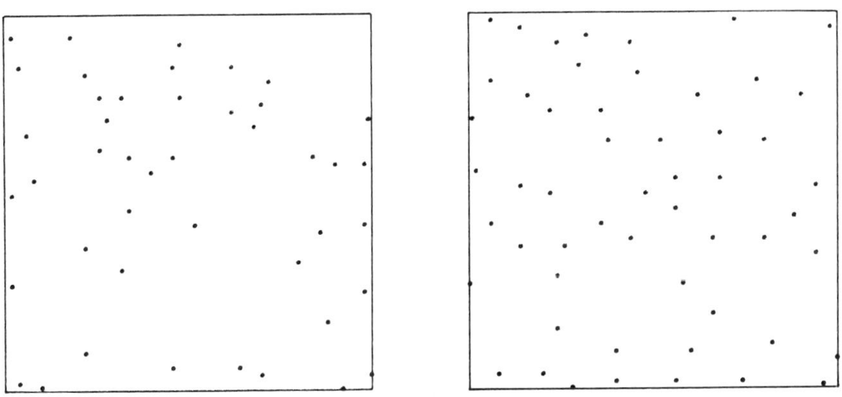

(a) (b)

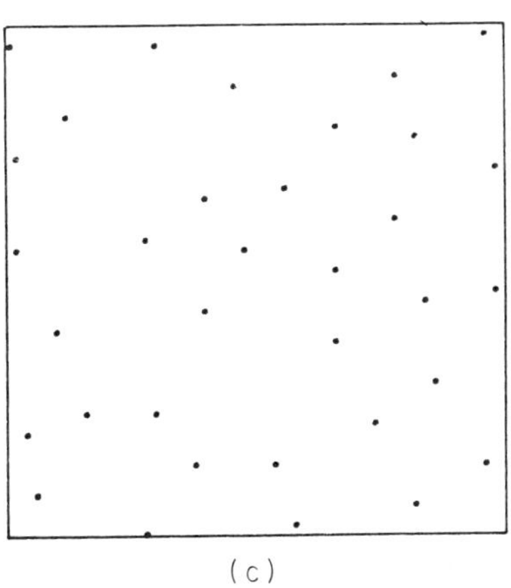

(c)

Figure 1.8 Realizations of some inhibition processes —(a) Matern's static process. (b) Matern's sequential process. (c) Diggle's sequential process

the diameter of the circle of influence of a mature plant. An example of the pattern that results is shown in Figure 1.8a.

The dynamic regular pattern arises as follows. Imagine a number of migrant birds arriving one after another in a wood in springtime. Each bird in turn selects a nest site from which it will repulse any rival bird of the same species. The territories so claimed form invisible circles (roughly) about the central nesting points, with these central points having been established sequentially rather than simultaneously. Matern (1960) varied his previous static model by suggesting that a Poisson forest be generated one point at a time, with points being discarded if they lay within a distance x of any previously generated point. Diggle *et al.* (1976) have suggested what appears to be a more natural variation to this procedure wherein points are only discarded if they lie within a distance x of any *retained* point. Typical patterns for these two variants are shown as Figures 1.8b and c.

From the practical point of view the most noticeable feature of the three patterns shown in Figure 1.8 is their similarity (apart from some incidental variation in point intensity). It must be doubtful as to whether it would be possible in practice to distinguish between these patterns. Paloheimo (1971) has established some properties for Matern's second process and Bartlett (1974) has used it as the basis for a description of the positioning of nests in gull colonies, although Ripley (1977) maintains that it would be more appropriate for botanical situations. Diggle's variant appears to be even less mathematically tractable than the other alternatives.

All three of these models are extreme in that they deny the possibility of any points occurring at distances of less than x apart. Ripley (1977) describes such processes as *hard-core* processes. Bartlett (1974) attempted a more realistic description by using a mixture of Matern processes each having its own value of x; his method was, however, essentially *ad hoc.*

Cormack (1979) notes that 'a hard-core inside which there is complete inhibition and outside which there is no interaction seems too extreme for any biological organism which displays plasticity of size and shape'. Cormack states that 'the total competitive effect experienced at a point in the plane is a function only of the number of neighbours it has': here an individual is described as a 'neighbour' of a point providing that it lies within a distance r of that point, with the appropriate value of r being dependent on what the points represent.

Strauss (1975) introduced a distribution which is characterized by Kelly and Ripley (1976) as having a joint density function for the point locations which is proportional to $\alpha^n e^{-\beta s}$, where α and β are constants, n is the number of points in the region of interest, and s is the number of neighbouring pairs of points (with 'neighbouring' used in the sense of the previous paragraph and with a predetermined value of r). Both α and β have non-negative values. If $\beta = 0$, then the process is simply a random Poisson process, since, in that case, each small area will be equally likely to contain a point. If $\beta = \infty$, then the process is a hard-core process with critical inter-point distance equal to r.

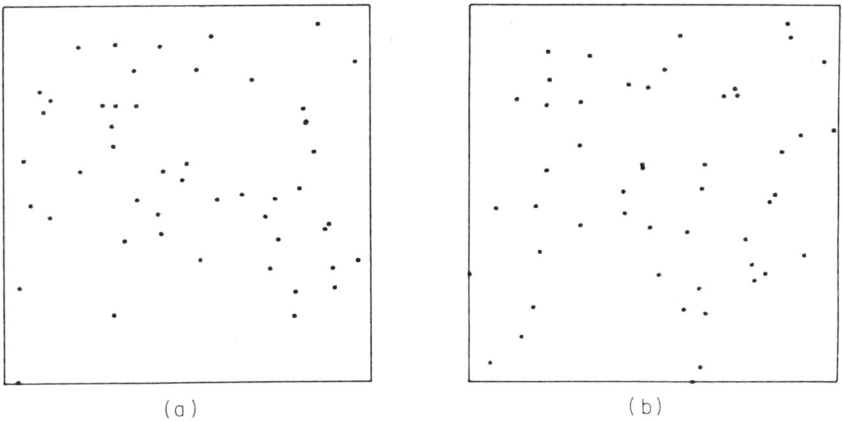

Figure 1.9 Two Strauss processes. (a) $\beta = 1.6$. (b) $\beta = 0.2$

Two examples of the patterns produced by realizations of the Strauss process are shown in Figure 1.9. These realizations were produced by means of the algorithm provided by Ripley (1979b). The range of pattern from random to regular is evident from these two figures. Ripley (1977) and Besag (1977a) have used a Strauss process as a model of the distribution of towns on the Spanish plain, using data previously analysed by Glass and Tobler (1971). Ripley (1981) also uses the Strauss process to model the distribution of Swedish pine trees in a wood and the distribution of magnetite crystals in a slab of rock. We shall address the problem of fitting this and other distributions in Section 1.5.

Central place models

The inhibitive models described in the previous subsection were derived principally with ecological applications in mind. However, as was demonstrated by Figure 1.7, regular patterns are not infrequent in human geography and archaeology. The workers in these fields have traditionally been concerned with deterministic rather than stochastic explanations of such patterns, being much influenced by the seminal work of Christaller (1933).

Consider the pattern in Figure 1.10 in which each *central place* supplies the functions and services demanded by the surrounding population. There is obviously a powerful inhibitory mechanism at work here, with central places going to the ultimate length to avoid each other. This is a very much more stringent prescription of regularity than any of the previous inhibitory models. As Figure 1.10 implies, the central places are supposed to be of varying importance (depending upon the goods and services supplied) and in Figure 1.10 there are three levels of central place being considered, with, as one would expect, the most

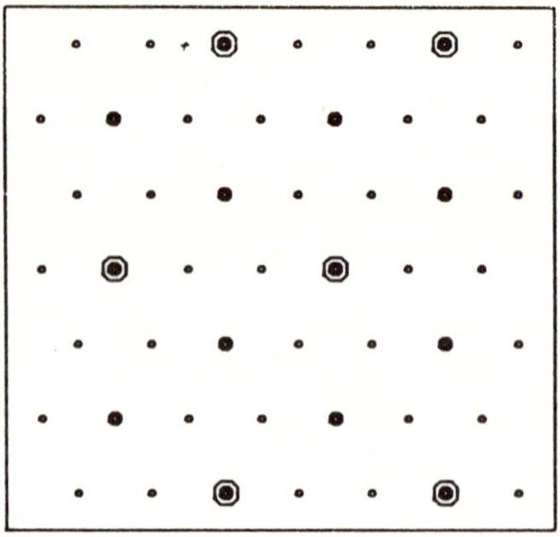

Figure 1.10 A central place pattern

important central places being the least frequent. Of course Figure 1.10 is an illustration of a deterministic pattern rather than a stochastic (mathematical) pattern and cannot hope to be more than an idealization of a real world that is subject to the vagaries of human behaviour.

In order to explain the regular settlement pattern of Figure 1.10, we commence with the location of the highest-order goods, the kind only sold in the most important central places. These goods are those that require the greatest demand before a supply is forthcoming—this is the so-called threshold or minimum range (Berry, 1967). Christaller (1933) makes several simplifying assumptions in order to establish the locations of these and lower-order goods. He assumes that the area to be settled is an isotropic surface (in other words, a featureless plain!) with equal transport opportunity in all directions. Demand, and thus by implication population density and purchasing power, is evenly and ubiquitously distributed, and this establishes 'the maximum distance over which people will travel to purchase a good or derive a service offered at a central place' (Carter, 1972, p. 71).

Given that the consumers are reasonable people, they will want to minimize travelling distance. To meet this desire on the part of the consumers, the highest-order centres therefore arrange themselves in a triangular lattice pattern, with each centre surrounded by its own market area over which it has a complete monopoly of the sale of the highest-order goods. The geometry of the centre positions implies that these market areas will be hexagons. Losch (1954), taking up Christaller's ideas, was able to show mathematically that hexagons enabled 'the best possible packing of trade areas in a region to the mutual advantage of both consumers (through distance minimization) and producers (because the largest number of independent enterprises is possible)' (Yeates and Garner, 1971, p. 149).

Initially these central places of major importance will be the sole sources of a variety of goods. However, to complete the story, we imagine that, with the

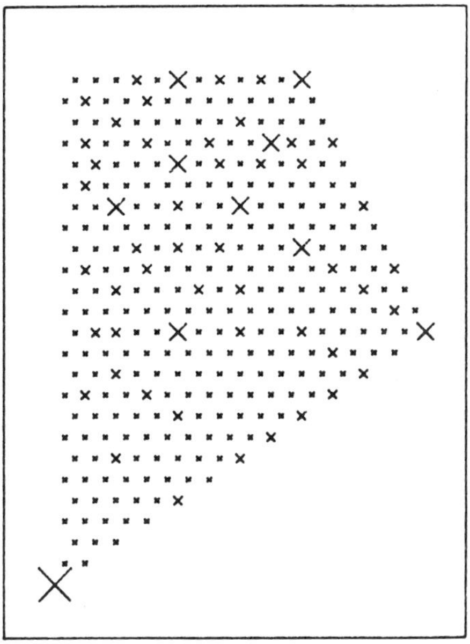

Figure 1.11 Centres of various sizes north-east of a metropolis. (Adapted from Losch, 1954. Reproduced by permission of Gustav Fischer Verlag)

passage of time, there may well be some goods whose minimum range decreases to the point where it is feasible to supply these goods from a completely new point of supply. Thus, while people may be prepared to travel 50 miles to buy a new computer, they would find it tiresome to have to travel that far for a loaf of bread. Therefore a subsidiary network is formed of second-order central places which infill between the highest-order places, and this process is continued until all demands are satisfied.

The above comprises only the briefest of outlines of one version of the central place model, this version being based on common-sense marketing principles. Alternative versions include the traffic principle which maximizes the number of central places on transport arteries connecting higher-order centres, and the administrative principle in which the goal is efficient administrative control achieved by higher-order centres attaining undivided command over subordinate centres. Losch (1954) has added extra mathematical rigour and generality to the model and has extended its potential applicability. Figure 1.11 shows the outcome of Losch's model applied to a sector of a region surrounding a metropolis.

Isard (1956) modified Losch's scheme to take full account of the impact of population density variations which increase the size and distort the shape of the central place market areas. An example of the effects of distortion is provided in

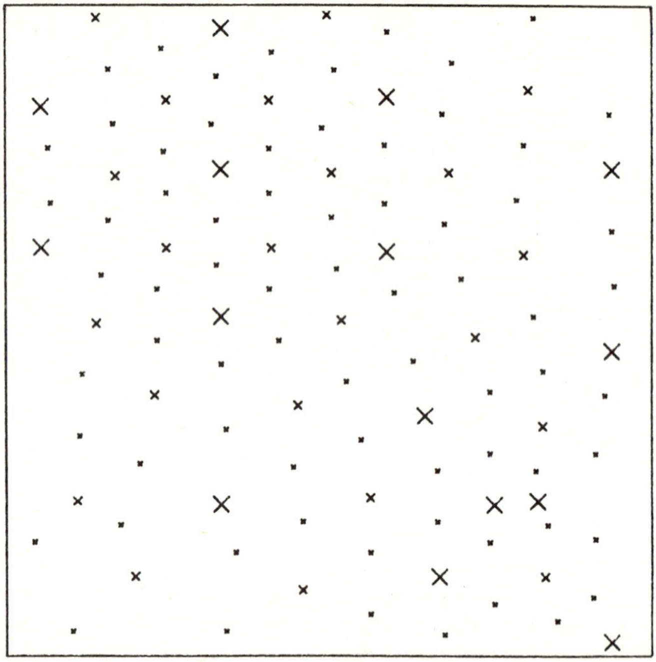

Figure 1.12 A distorted central-place pattern. (Adapted from Rushton, 1972. Reproduced by permission of Regional Science Association)

Figure 1.12 in which the central places have been located using a technique described by Rushton (1972) who contrasts his approach with those of Tobler (1963), Tornquist *et al.* (1972), and others.

A semi-stochastic approach is adopted by Dacey (1965) who systematically distorts the central place lattice by random displacements of the central places about their lattice positions. Other attempts to incorporate a probabilistic element in the model have been made by Dacey (1966a, b) and Curry (1964, 1967). White (1977) and Fingleton (1978) emphasize the effect of variable consumer behaviour, as well as factors such as an inhomogeneous demand surface, on the central place pattern. A new, integrated model has not yet evolved, although there has been some reinterpretation (Beavon, 1977) and the classical model has acted as the springboard for considerable research. Johnston and Wrigley (1981) give a useful summary of the literature that has emanated from the central place model.

So far in this subsection we have considered only permanent points of supply. However, impermanent ones, namely periodic markets, are very common phenomena, particularly in the less-developed world, and have been the subject of extensive investigation. A basic distinction in periodic market systems is between market-places and market-days (so that marketless days are possible at market-places).

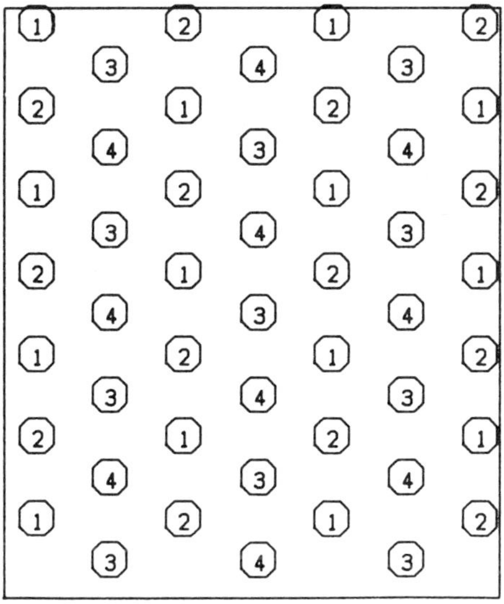

Figure 1.13 A periodic market point pattern

One hypothetical way in which the market-days might arrange themselves amongst the market-places is illustrated in Figure 1.13, where there is an obvious avoidance behaviour reminiscent of the earlier inhibitive spatial interaction. Obviously, any number of alternative rules might be devised to account for the pattern of market-days. Stine (1962) suggested that the lack of permanent markets reflected the fact that there was insufficient demand to warrant the goods being put on permanent offer; this view receives some support from remarks made by Berry (1967). Such marketing conditions will most often occur when the consumers are poor, relatively immobile, and live in more or less impassable terrain: the result is mobile entrepreneurs accumulating the requisite demand by moving around so-called market rings.

Tinkler (1973) provides an alternative rationale and also suggests a mechanism to generate the four-day market cycle illustrated in Figure 1.13. A premise of his approach is that no two contiguous market-places should have the same market-day since direct competition would be self-defeating, 'akin to running summer fetes in adjacent parishes on the same Saturday in June!' (Haggett *et al.*, 1977, p. 154). If market areas having the same market-day are given the same colour on the map, then it follows that a necessary requirement is that no two contiguous areas should be coloured in the same way. The problem of how few colours are required so as to colour a map with this restriction has long been a famous one. The conjecture that at least four colours are required was not finally verified until high-speed computers became available and the final proof is due to Appel and Haken (1977). The consequence of this requirement in the present context is that we can expect to observe four-day market cycles (since there may well be some virtue in having markets as frequently as possible subject to the contiguity restrictions). A comprehensive account of periodic market systems and some associated models is given by Bromley (1980) and by Haggett *et al.* (1977).

The county seat model

Dacey (1966a) suggested a model, which he termed the county seat model, that provides a bridge between the unrealistically deterministic central place model and some of the stochastic models discussed above, for it is based on the Poisson distribution and is aimed particularly to account for settlement distributions. Whilst not conflicting with the basic notions of central place theory, it holds out much more hope of reproducing the patterns of real data. The essential feature of the model is its ability to generate a pattern that is somewhat more regular than random without this regularity being as obtrusive as with the stylized central place model.

Dacey conceives of areas of territory being subdivided into 'counties', with each county containing not more than one highest-order central place termed the 'county seat'. In both Great Britain and the United States this makes obvious sense in the context of local administration. In addition to the (possible) county seat, each county may contain a number of other places distributed at random within the county.

Suppose that there are s ($\leq c$) county seats and n other places distributed at random amongst c counties subject to no county receiving more than one county seat. Let P_r and p_r be the respective probabilities of a county containing r places in all and r places that are not county seats. Let p_c be the probability of a county containing a county seat. Thus:

$$P_r = p_c p_{r-1} + (1 - p_c) p_r. \tag{1.3}$$

If $s = 0$ we have the Poisson process (1.1), whereas if $s = c$, then we have a Poisson process plus one. Obviously, if $s > 0$, then there are a greater number of counties containing one place than would have been expected from a purely random allocation of places. Dacey (1966b) extends the scope of the model by treating both p_c and the intensity of the Poisson process as random variables that vary from county to county: this results in a model that has a close affinity with the negative binomial distribution defined by equation (1.2). With this degree of flexibility it is perhaps not surprising that the model is found to provide a good description of 'the areal distribution of a wide variety of phenomena' (Dacey, 1966b, p. 179).

1.2 THE USE OF SCATTERED QUADRATS

Much of the previous section was concerned with ways to model the processes of pattern. A large number of different models were considered and examples of the patterns that result from these models were provided. For the remainder of this chapter we shall be working in the opposite direction—starting with an observed pattern and attempting to deduce the nature of the process that gave rise to that pattern.

There are two major approaches that have been suggested for the analysis of pattern, and both have a considerable history. One involves measuring the physical distances between points, the other involves analysing the variation in the numbers of points in selected sub-areas of the region under investigation. The distance methods are discussed in Sections 1.4 and 1.5, the counting methods are discussed in this and the next section.

All the counting methods rely on the use of the *quadrat*. Originally a quadrat was a square wooden frame that was placed on the ground to delimit a survey area, but the meaning of the word has broadened and now refers simply to a study area of fixed size and shape. The shape is usually square, but rectangular and circular quadrats are by no means uncommon.

The traditional method of sampling a small area using the wooden quadrats was to stand with one's back to the area and to toss the quadrats over one's shoulder. This was not entirely satisfactory because the restricted articulation of one's bones led to biases in the location of the quadrats: nowadays one uses tables of random numbers to locate the quadrats and this is the only practical method if the survey area is at all large. Unless restrictions are imposed on the quadrat positions, they may well overlap, as is demonstrated by the outcome illustrated in Figure 1.14.

Once the quadrats have been positioned, the next stage is to count the numbers of points that fall in each quadrat. Since, in practice, the 'points' actually represent

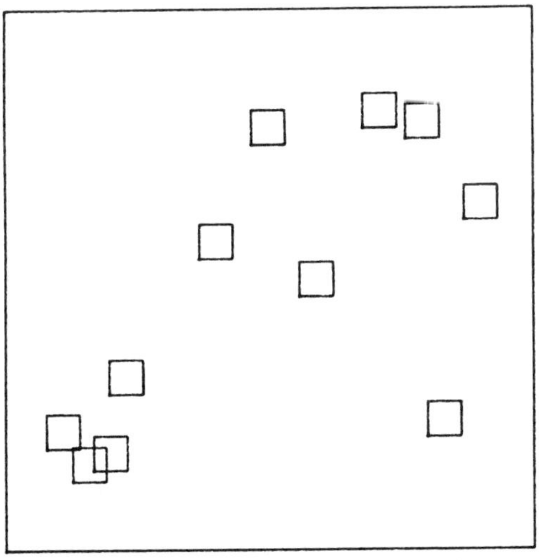

Figure 1.14 Randomly scattered quadrats

objects with a noticeable cross-section, it is necessary to specify some rules to cover the inclusion of objects that overlap the quadrat boundary.

The major use of quadrat counts is probably to provide an estimate of the total number of individuals in the study region and this use is discussed in Chapter 2. However, it is certainly also true that quadrat counts provide some limited information about the nature of the point pattern and that is our present concern. For example, suppose that ten randomly placed quadrats give counts of

1 0 1 1 0 0 1 1 2 99.

If the individuals were distributed at random, then we would expect quadrat counts varying according to a Poisson distribution. The observed set of counts is certainly not typical of such a distribution and we should have little hesitation in concluding that there was significant evidence that the pattern contained clusters of points.

Suppose, however, that the quadrat counts were

2 2 2 2 2 2 2 2 2 2.

These counts are all identical and the concept of a regular pattern such as that indicated in Figure 1.15a would naturally spring to mind. However, this pattern is by no means the only one that could give rise to these 'regular' counts, as Figure 1.15b makes clear. The important lesson to be drawn from these two figures is that counts from scattered quadrats are, at best, a very crude indicator of pattern because they take into account neither the relative positions of the points within the quadrats nor the relative positions of the quadrats themselves.

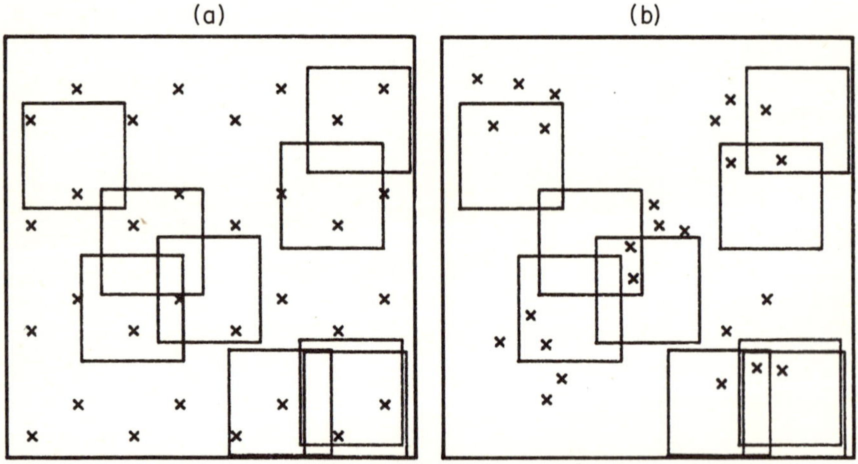

Figure 1.15 Equal quadrat counts. (a) With regularly spaced plants. (b) With clustered plants

However, it would be reasonable to argue that the point patterns of Figures 1.15a and b are sufficiently obvious that we would not have bothered to scatter quadrats in order to analyse the pattern. For the remainder of this section we shall assume that the hypothesis that the pattern forms a Poisson forest cannot be dismissed out of hand on purely visual grounds.

Indices

We have already remarked that the counts from equi-sized quadrats scattered over a Poisson forest will be observations from a Poisson distribution with parameter equal to the product of the intensity of the points per unit area and the area of the quadrats. A useful characteristic of the Poisson distribution is that this parameter is equal both to the mean and to the variance of the distribution. If the pattern is more regular than a Poisson forest, then the quadrat counts will be more uniform in size and will therefore have a relatively small variance (when compared with the size of the mean). On the other hand, if there are clusters, then some quadrats will have large counts so that the variance of the point counts will be relatively large. A natural test of a Poisson distribution is therefore provided by examining the value of the ratio sample variance/sample mean. This ratio was termed the *relative variance* by Clapham (1936) in the first published application of the ratio to the analysis of spatial data.

If we denote the observed counts in n quadrats by $x_1, x_2, \ldots, x_n$, then these counts have mean $\bar{x} = \sum x_i/n$ and variance $s^2 = \sum (x_i - \bar{x})^2/(n-1)$, where the summations are over the values of i from 1 to n. Hoel (1943) showed that

$$ID = (n-1)s^2/\bar{x}, \qquad (1.4)$$

the *index of dispersion*, has an approximate χ^2_{n-1} distribution under the assumption of randomness. This approximation is reasonable provided $n > 6$ and $\bar{x} > 1$ (Kathirgamatamby, 1953). Perry and Mead (1979) have examined the behaviour of the index of dispersion test and conclude that it is remarkably sensitive at detecting a lack of homogeneity in a point pattern.

A number of other indices have been suggested, principally for situations where it is thought that clusters may be present. All are sensitive to one or both of changes in the quadrat size and changes in the point intensity (as indeed is ID itself). Perry and Mead (1979) concluded that, in the case of ID, the behaviour of the test was principally dependent upon the size of the mean of the quadrat counts: the larger the mean the more likely was the ID test to recognize any heterogeneity that was present.

David and Moore (1954) suggested that the quantity

$$ICS = s^2/\bar{x} - 1 \qquad (1.5)$$

would provide an appropriate index of 'clumping' or 'contagiousness'. The notation ICS was suggested by Douglas (1975) as a mnemonic for the *index of cluster size*. For a Poisson forest ICS has mean 0 and is independent of the quadrat size. An interpretation of a positive value for ICS is as the number of other points intimately associated with a randomly chosen point. A negative value for ICS indicates some regularity in the positioning of the points (possibly due to the physical size of the objects that the points represent).

Waters (1959) suggested that if there were clustering according to the negative binomial distribution (equation (1.2)), then the parameter k of that distribution would provide a measure of aggregation. Douglas (1975) suggested that a suitable measure would be

$$ICF = \bar{x}/ICS, \tag{1.6}$$

and showed that, for the negative binomial, $ICF = k$. Douglas termed ICF the *index of cluster frequency*. Since $ICF \cdot ICS = \bar{x}$, which varies with the quadrat size, it is evident that, for a Poisson forest, ICF is proportional to the quadrat area, but is impervious to random thinning. If ICS is negative, then ICF has no useful interpretation.

Alternatives to the Douglas statistics have been suggested by Lloyd (1967). These are the *index of mean crowding*, IMC, and the *index of patchiness*, IP, defined by

$$IMC = \bar{x} + ICS, \qquad IP = IMC/\bar{x} = 1 + 1/ICF. \tag{1.7}$$

The index of mean crowding represents the average number of other points that are contained in the quadrat that contains a randomly chosen point and is obviously dependent on quadrat size. The index of patchiness is related to an earlier index, I_δ, introduced by Morisita (1959):

$$I_\delta = n\sum x_i(x_i - 1)/\{n\bar{x}(n\bar{x} - 1)\} = n\bar{x}\mathrm{IP}/(n\bar{x} - 1). \tag{1.8}$$

Both Lloyd and Morisita imagined that the population under scrutiny could be modelled as a mosaic of large patches of different densities, with the points being distributed within each patch in a random manner. Lloyd's intention in deriving IP was to find a measure that was independent of quadrat size. Morisita argued that the more clumped the distribution, the more likely were two of the sampled points, chosen at random from amongst the sample points, to belong to the same quadrat: I_δ is simply a scaled version of this probability.

Evidently both the mean and the variance of a set of quadrat counts are dependent on quadrat size and we can expect them to be related to one another in some way, even when the point pattern is not random. By collecting information from different sized quadrats (or by combining quadrat counts) we can obtain some idea of the nature of this relation. Iwao (1968), considering the nature of animal populations, suggested that the relation would take the form of IMC being a linear function of the mean. Taylor *et al.* (1978) note that this is equivalent to the suggestion of Bartlett (1936) that in general the relative variance was likely to be a linear function of the mean. Following a detailed study of quadrat counts for a huge variety of different species, Taylor *et al.* concluded that a better fit to the relation between variance and mean was provided by the multiplicative law suggested by Taylor (1961), namely that

$$s^2 = ax^b. \tag{1.9}$$

Taylor regarded the index b as being a characteristic of the distribution of a species, but it is difficult to see what would be its spatial interpretation.

Choice of quadrat size

A perennial problem in designing a study using quadrats is to establish what would be a suitable size for the quadrat. For example, suppose we choose a quadrat size that is so small that it can contain at most one sample point. Such a choice would hardly provide any useful information concerning the pattern! It would be equally absurd to divide the entire region into just two quadrats, with roughly half the points in each quadrat. Clearly we require some compromise between these two outrageous extremes.

Various suggestions have been made as to the optimal size. Bartlett (1948) recommended aiming for a quadrat with mean count about 1.6 (so that only about 20% of the quadrats have zero counts for a Poisson forest), Greig-Smith suggests a mean of 1.0 (about 40% zero counts) while Curtis and McIntosh (1950) suggest a mean of 4.0 (only 2% zero counts). Clearly it is not of crucial importance, but evidently to detect small-scale divergences from randomness one should aim for a quadrat mean in this region. However, as the example that follows indicates, quadrats with very much larger means may also be of some interest.

Example 1.1

Although there are numerous examples of the use of indices in the literature, the majority refer to the contiguous blocks of quadrats that will be discussed in the next section. The methods that are discussed there are likely to be more informative since they take some account of the relative positions of the quadrats.

A study that did use scattered quadrats is that of Dacey (1968a), and the context and quadrat locations are indicated in Figure 1.16. Dacey was concerned with the settlement of rural districts in Puerto Rico, and had available 352 maps of the entire area of Puerto Rico, these maps comprising 22 groups with 16 maps per group. Dacey selected at random one map from each group subject to the restrictions that it contained no towns of population 1000 or more, no coverage of the ocean, and no coverage of the Caribbean National Forest. One group contained no suitable map so that there were just 21 maps in the final collection. We shall regard these 21 (rectangular) maps as our 21 scattered quadrats: from a commonsense viewpoint it would be quite extraordinary if the rural dwellings in Puerto Rico were distributed at random and we can expect to detect strong

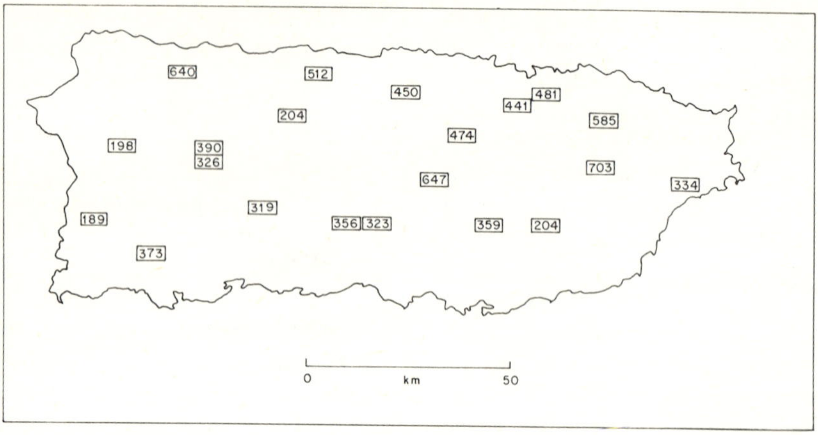

Figure 1.16 Randomly chosen quadrat maps of Puerto Rico. (Redrawn from Dacey, 1968a. Reproduced by permission of The Institute of British Geographers)

evidence of clusters (villages). Dacey's own analysis was conducted by dividing each map into smaller sub-quadrats and investigating hypotheses at this more detailed level of investigation, but this is irrelevant to our present purpose.

The 21 quadrat counts are as shown in the figure:

189	198	204	204	319	323	326
334	356	359	373	390	441	450
474	481	512	585	640	647	703.

For these counts $\sum x_i = 8508$ and $\sum x_i^2 = 3\,908\,730$ so that $\bar{x} = 405.1$ and $\sum(x_i - \bar{x})^2 = \sum x_i^2 - (\sum x_i)^2/n = 461\,774.6$. Hence, the sample variance $s^2 = 23\,088.7$ and is very much greater than the sample mean $\bar{x}$; indeed, the variance-to-mean ratio is 57 rather than the 1 that would have been expected if the houses had been scattered at random.

The values of the indices are $ICS = 56$, $ICF = 7.1$, $IMC = 460$, and $IP = I\delta = 1.14$. An apparent explanation of the variation in the quadrat counts is therefore that the quadrats contain an average of seven villages with the average village containing 57 houses. Whilst this seems a not unreasonable description of rural Puerto Rico, it must be recognized that it is, of course, a tremendous oversimplification of the reality.

1.3 THE USE OF CONTIGUOUS QUADRATS

In his presidential address to the British Ecological Society, Watt (1947) gave a description of the plant community as a working mechanism with the composition of the community depending on the combined processes exerted by competing plants and on the natural processes of plant regeneration. In his introduction (p. 1) Watt stated that:

> The ultimate parts of the community are the individual plants, but a description of it in terms of the characters of these units and their spatial relations to each other is impracticable at the individual level. It is, however, feasible in terms of the aggregates of individuals and of species which form different kinds of patches; these patches form a mosaic and together constitute the community. Recognition of the patch is fundamental to an understanding of structure as analysed here.

While we may disagree with the latter part of the first sentence (since this is precisely the concern of the next two sections of this chapter), it is the last sentence that forms the motivation for this present section, namely *recognition of the patch.*

The use of a square grid

One method of identifying the size of the mosaic patches (if they exist) was first suggested by Greig-Smith (1952) in a most influential paper. Greig-Smith proposed that instead of utilizing randomly scattered quadrats, which was often somewhat impracticable in the field, a square block of contiguous quadrats should be used. The advantage of such a block is that neighbouring quadrats can be combined so that we may obtain information about quadrats of more than one size.

Greig-Smith suggested that a 16 × 16 grid of quadrats should be used. Two methods of combining the individual quadrats so as to obtain blocks of sizes equal to 1, 2, 4, 8, 16, 32, 64, and 128 quadrats in area are shown in the two diagrams in Figure 1.17. The blocks of sizes 1, 4, 16, and 64 are square and are positioned in the same locations in each diagram (there will, for example, be a total of four blocks of size 64 in each case—the diagrams show only one of these four). The blocks of sizes 2, 8, 32, and 128 are rectangular and are all placed either

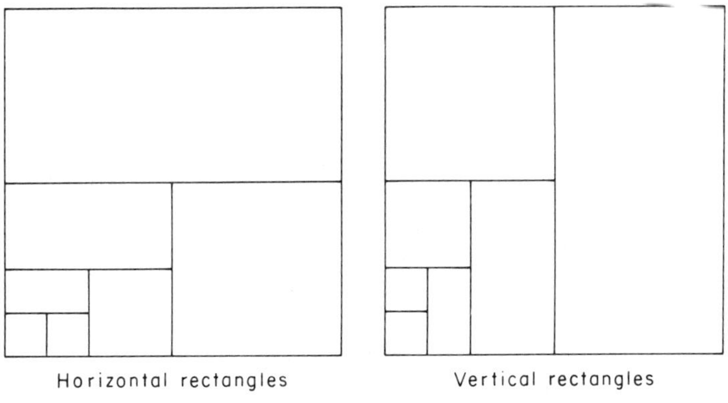

Figure 1.17 Two methods of combining quadrats in a 16 × 16 grid

'horizontally' or 'vertically'. Other arrangements might use a mixture of the two types of rectangle.

Greig-Smith's procedure is as follows. Denote by T_r the sum of squares at the block size $r(=1,2,\ldots,256)$. Next calculate

$$G_r = 2T_r - T_{2r}, \tag{1.10}$$

for $r = 1,2,\ldots,128$ and plot G_r against r. This description simplifies the arithmetic at the expense of obscuring the motivation! Essentially the procedure amounts to the construction of a hierarchical analysis of variance in which the total variation in the 256 separate quadrat counts is apportioned to the component variations at each of the block sizes. Thus, at the largest level of block size it is noted that the quadrat counts vary because this block of 128 quadrats has a different total point count to that block of 128 quadrats; similarly, *within* each of these blocks it is noted that the quadrat counts vary because this block of 64 quadrats has a different total point count to that block of 64 quadrats, and so on. The advantage of working with G_r is simply that it is an integer. Most examples of the method work with M_r rather than G_r, where $M_r = G_r/N$, and N is the total number of quadrats (256 in our case). The quantity M_r is an estimate of the variance of the counts from the blocks of size r and is the term denoted as the *mean square* in an analysis of variance.

If the points form a Poisson forest, then G_r (or M_r) will be more or less (given sampling fluctuations) constant, but if there is clumping, then G_r may be expected to reach a peak at a value of r indicating the clump size. This peak will be maintained for greater values of r provided that the clumps are not regularly spaced (Bartlett reported by Greig-Smith, 1952). Regularity in the point distribution would be indicated by a deep trough indicating the regular inter-point distance.

There are a number of disadvantages to Greig-Smith's procedure.

(a) The rectangular blocks may be arranged in different ways, with each resulting (in general) in a different value for G_r.

(b) The centres of gravity of two adjacent square blocks of size r lie the same distance apart (r) as the centres of gravity of two adjacent rectangular blocks of size $2r$. Thus, unless the clumps that exist are conveniently aligned rectangles, there are going to be some undesirable side-effects. Pielou (1977, p. 141) reports that the rectangular blocks are likely to give a saw-toothed appearance to the graph of G_r against r.

(c) The method only allows investigation of block sizes that are powers of 2. Thus, an apparent peak at, say, $r = 32$, can only be interpreted as indicating that clumps exist having some size between 16 and 64, and does not necessarily indicate that the clump size is 32.

(d) There is very little information available at the larger block sizes (e.g. there are only two blocks of size 128).

Of course the first two objections can be met by disregarding the intermediate rectangular blocks and working with $H_r = 4T_r - T_{4r}$ for $r = 1, 4, 16, 64$. This is the course adopted by Moellering and Tobler (1972) in their analysis of the classical data on the distribution of houses in the Tonami Plain in Japan, but naturally this course results in a reinforcement of the third objection.

In order to test whether the values of G_r provide significant evidence of the existence of clumps, Greig-Smith proposed that one used the value of the ratio G_r/G_1 (a variance ratio) and compared it with the tabulated significance points of an F distribution with $N/2r$ and $N/2$ degrees of freedom. However, Thompson (1958) has shown that if there is significant departure from randomness, then this test is only approximate.

Despite all these various objections the method is a popular one because of its simplicity, and when data presents itself in the appropriate form it does seem to be worth performing the relatively trivial calculations.

Example 1.2

Figure 1.18 shows quadrat counts for the distribution of the perennial shrub *Atriplex hymenelytra*, on a 40 m × 40 m square area of the floor of Death Valley, California. The counts were obtained by overlaying a 16 × 16 grid on Figure 1 of Gulmon and Mooney (1977). Choosing to use horizontal rectangles, we obtain combined counts for the various block sizes, as shown in Figure 1.19.

The sums of squares of the block counts are easily found to be $T_1 = 98$, $T_2 = 120$, $T_4 = 170$, $T_8 = 268$, $T_{16} = 492$, $T_{32} = 842$, $T_{64} = 1534$, $T_{128} = 2890$, and $T_{256} = 5776$. If we had used vertical rectangles we would have obtained $T_2 = 120$, $T_8 = 306$, $T_{32} = 866$, and $T_{128} = 3016$. A reasonable policy would seem to be to

0	0	0	0	1	0	0	0	0	0	0	0	0	0	0	0
0	0	0	0	0	0	1	0	0	1	0	0	0	0	0	0
0	0	1	0	0	1	0	1	0	0	0	0	0	0	0	0
1	1	0	0	0	0	0	1	1	1	2	0	0	0	0	0
0	0	0	0	0	1	0	1	0	0	0	0	0	2	0	0
1	0	0	2	2	1	1	0	0	1	0	0	0	0	0	0
0	1	1	0	1	1	1	0	1	1	1	0	0	0	0	0
0	0	1	0	2	1	0	0	0	0	0	1	0	1	0	0
0	1	0	1	0	0	0	0	0	0	0	0	0	0	0	0
0	0	0	1	2	0	2	0	0	1	0	1	0	0	1	1
0	1	0	2	1	0	0	0	0	0	0	1	0	0	0	0
0	0	1	0	0	0	0	0	0	0	1	0	0	0	0	0
0	0	1	0	1	0	0	0	1	2	0	1	1	0	0	0
0	0	0	0	0	2	0	0	1	0	0	0	0	0	0	0
0	0	0	1	0	1	0	0	1	0	0	0	0	0	0	0
0	0	0	0	0	1	1	0	2	0	0	1	0	1	0	0

Figure 1.18 Distribution of *Atriplex hymenelytra* in Death Valley, California. (After Gulmon and Mooney, 1977)

Block size 2

0	0	1	0	0	0	0	0
0	0	0	1	1	0	0	0
0	1	1	1	0	0	0	0
2	0	0	1	2	2	0	0
0	0	1	1	0	0	2	0
1	2	3	1	1	0	0	0
1	1	2	1	2	1	0	0
0	1	3	0	0	1	1	0
1	1	0	0	0	0	0	0
0	1	2	2	1	1	0	2
1	2	1	0	0	1	0	0
0	1	0	0	0	1	0	0
0	1	1	0	3	1	1	0
0	0	2	0	1	0	0	0
0	1	1	0	1	0	0	0
0	0	1	1	2	1	1	0

Block size 4

0	0	1	1	1	0	0	0
2	1	1	2	2	2	0	0
1	2	4	2	1	0	2	0
1	2	5	1	2	2	1	0
1	2	2	2	1	1	0	2
1	3	1	0	0	2	0	0
0	1	3	0	4	1	1	0
0	1	2	1	3	1	1	0

Block size 8

0	2	1	0
3	3	4	0
3	6	1	2
3	6	4	1
3	4	2	2
4	1	2	0
1	3	5	1
1	3	4	1

Block size 16

3	5	5	0
6	12	5	3
7	5	4	2
2	6	9	2

Block size 32

8	5
18	8
12	6
8	11

Block size 64

26	13
20	17

Block sizes 128,256

39
37

76

Figure 1.19 Successive block size counts for the data of Figure 1.18

take the average of the alternative rectangular block values so that we then get $G_1 = 76$, $G_2 = 70$, $G_4 = 53$, $G_8 = 82$, $G_{16} = 130$, $G_{32} = 174$, $G_{64} = 115$ and $G_{128} = 130$. Formally, the F-test gives no indication of significant departures from randomness at the 5% level, although, with horizontal rectangles, the value of G_{32} is just significant. There is, therefore, the suspicion of evidence that there may be some clumping with a clump size of around 80 m in diameter.

Randomly paired quadrats

Consider the two adjacent blocks of size 2 shown in Figure 1.20. The centres of gravity of the two individual quadrats at the top of the figure lie at a distance of 1.0 unit apart, whereas the diagonal distance shown has a length of about 1.4 units. Thus, the average inter-quadrat distance is 1.2 units, which is greater than the distance between the centres of gravity of the two rectangular blocks themselves. This is always the case for blocks of more than 1 quadrat. For example, for blocks of size 16, the distance between the centres of gravity is 4 units, whereas the average inter-quadrat distance is 4.3 and the individual inter-quadrat distances vary between 1 unit and 7.6 units. (Goodall, 1961 suggested that an appropriate summary measure of distance would be the geometric—rather than the arithmetic—mean of the inter-quadrat distances.)

In view of the restricted range of block sizes that can be examined using Greig-Smith's procedure, and because of the possibility of interpreting the blocking effects in terms of distances, Goodall (1974) suggested a procedure based on the random sampling of individual quadrats at a range of different distances. With his procedure the sampling unit is a single quadrat, so that the problems of block shape and orientation are no longer relevant. By sampling quadrats at the entire

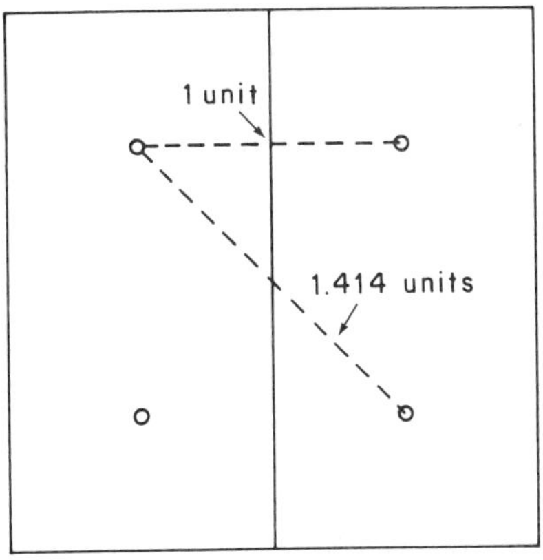

Figure 1.20 Two adjacent blocks of size 2

range of distances a more precise idea of the effect of distance is also obtained and it is possible to obtain rather more information concerning the greater distances.

A further major advantage of this method is that the lattice of sampled quadrats need not be square or even regular—the method could be used, with a little more difficulty, even with arrays that are incomplete and contain unsampled areas. These advantages are offset to some extent by the greater complexity of the random sampling procedure and by the reduction in precision at the shorter distances. Also, of course, one may well argue that the clumping effect is not entirely characterized by distance. The details of the method are explained in the following example.

Example 1.3

The so-called Middle Belt of Nigeria consists largely of savanna which is burnt annually. Hopkins (1965) examined the distribution of herb remains in a square of side 20 m in this savanna, using a 40 × 40 grid of quadrats with each quadrat having side 0.5 m. The quadrat 'counts' were the numbers of grams (in tens of grams) of dry weight remaining. Thus, a 'count' of 1 implies that the dry weight present amounted to at least 10 g but less than 20 g. The full array of quadrat counts is shown in Table 1.2.

As a preliminary analysis the Greig-Smith procedure was applied to the central 32 × 32 grid, giving the results shown in Table 1.3 (the T-values for rectangular blocks are averaged over the two orientations). The results show a steady rise in G_r as r increases, with a sharp peak at $r = 256$, suggesting clumping at a distance of $16 \times 0.5 = 8$ m.

There are a number of ways of conducting the random sampling of quadrats that is required by Goodall's paired-quadrat procedure. With a misshapen grid it

Table* 1.2 *Density diagram of the distribution of herb remains after burning savanna

0	0	0	0	1	1	2	0	0	0	0	1	0	0	2	1	0	2	1	1	1	1	0	0	0	1	3	0	2	0	0	0	0	0	0	0	0	0	0	0
0	2	1	1	0	1	0	0	0	0	3	1	0	0	0	0	3	2	1	0	0	0	2	5	0	0	0	0	0	2	5	0	0	2	1	1	2	0	0	0
0	1	1	3	0	0	1	0	1	0	2	0	1	1	2	1	1	2	0	1	1	0	2	3	1	0	0	0	0	0	2	0	0	0	0	0	0	0	1	1
0	0	2	1	0	0	0	0	0	1	2	0	0	0	0	0	1	2	2	5	2	2	2	1	1	0	0	2	1	0	0	0	0	2	1	0	2	0	1	2
0	0	1	0	0	0	2	1	0	2	1	1	0	0	0	0	0	0	1	2	0	1	0	1	3	0	0	0	1	0	0	1	0	2	1	4	0	1	0	1
0	0	1	0	0	0	0	3	1	1	1	0	1	1	1	4	0	1	0	0	0	0	0	2	0	0	1	1	0	1	2	3	1	1	1	0	0	0	0	0
0	0	1	0	0	0	1	0	0	1	1	1	1	6	0	0	0	2	1	0	0	0	0	1	3	2	1	1	1	0	0	1	2	1	2	0	2	1	0	1
0	0	0	1	0	0	0	2	0	0	0	0	0	1	0	1	0	0	0	0	2	3	1	0	1	2	2	1	1	1	1	2	4	1	0	2	1	1	1	0
0	0	1	0	0	0	0	0	0	0	0	2	0	1	1	0	0	0	0	0	0	0	0	0	0	0	0	0	0	4	0	0	1	0	0	3	1	0	1	0
1	0	1	0	0	0	0	0	0	0	2	0	2	2	0	0	0	1	0	0	0	0	0	1	0	0	3	5	0	1	4	2	2	1	0	0	0	1	1	1
0	1	2	0	0	0	0	1	0	0	0	1	0	0	0	2	2	0	0	1	0	0	0	2	2	0	2	1	0	0	7	2	2	6	12	0	0	6	2	0
0	0	2	0	0	0	0	0	0	0	0	0	1	0	1	0	0	0	2	0	0	1	2	0	1	0	2	1	1	1	1	0	0	2	14	4	2	0	0	1
0	0	0	0	0	0	0	0	0	0	1	0	0	0	0	0	0	1	1	0	1	0	0	3	0	0	1	3	0	2	0	0	1	0	0	3	4	3	0	1
0	0	1	0	1	0	0	0	1	0	0	1	0	1	0	2	0	0	0	0	0	0	0	0	0	1	0	0	0	0	1	1	1	2	1	0	0	1	1	0
1	0	0	0	0	0	0	0	0	0	0	1	0	2	0	0	0	0	0	0	0	1	0	4	1	0	0	0	0	0	0	0	0	2	0	2	0	0	0	1
0	0	0	0	0	0	0	2	0	0	1	0	0	0	0	1	0	0	0	0	0	0	1	2	1	0	0	1	0	0	2	1	0	0	3	0	1	0	0	2
0	0	0	0	0	0	0	0	0	0	0	0	1	0	2	0	0	2	1	0	0	0	0	0	0	0	2	0	0	0	0	0	0	3	2	3	0	1	0	0
0	1	0	0	0	3	1	0	0	2	0	0	1	1	1	0	0	1	2	0	1	0	0	0	0	3	1	0	1	0	0	2	1	1	0	0	0	2	1	0
1	0	0	0	0	0	0	1	0	0	2	0	0	0	0	0	0	0	0	0	0	1	0	0	0	0	1	1	0	0	2	1	2	0	0	0	0	1	1	1
1	2	2	3	0	0	1	0	0	0	0	1	1	0	0	3	0	1	0	0	0	0	1	2	0	2	1	0	2	0	0	2	0	0	0	0	1	0	1	0

1	0	2	0	0	0	0	1	1	1	0	0	0	0	0	1	3	1	0	0	1	1	0	0	0	1	0	0	0	0	3	1	2	1	1	2	1	0	0	0
0	2	1	0	0	0	3	0	0	0	0	0	0	1	2	5	3	0	0	0	0	0	0	0	0	0	3	1	1	0	0	1	0	0	0	1	1	1	0	1
1	1	1	1	0	4	2	0	1	0	0	1	1	0	0	0	2	2	0	0	0	0	0	2	1	1	1	1	0	1	0	0	0	0	1	2	1	0	1	0
1	2	1	1	0	2	2	0	0	1	1	0	0	1	1	0	0	0	0	0	0	0	0	1	0	0	0	0	0	0	0	2	1	0	1	0	1	0	1	3
0	1	0	2	0	3	1	1	0	1	0	0	1	0	2	0	1	1	2	1	0	0	1	0	0	0	0	0	4	0	1	1	0	4	1	0	0	0	2	1
2	1	3	1	0	4	1	1	1	0	0	0	0	0	0	3	0	1	1	0	0	1	1	2	1	0	0	1	9	0	0	0	0	0	0	0	2	2	3	1
1	1	1	0	0	2	2	0	1	0	0	0	0	1	0	0	0	0	0	0	0	0	0	1	0	3	0	0	0	0	0	2	0	2	0	0	0	1	0	0
1	1	1	1	2	1	3	1	0	0	1	0	1	0	0	0	0	0	0	0	0	1	0	0	0	0	0	0	3	0	0	0	0	0	1	1	0	0	1	1
0	0	0	2	1	1	1	0	0	0	0	0	2	0	0	0	0	2	0	0	0	0	0	0	0	4	0	0	1	0	0	1	1	1	2	0	0	4	0	0
1	0	0	1	2	2	2	1	0	1	0	0	0	0	2	1	0	0	0	0	0	0	2	0	0	0	0	0	1	0	2	0	0	0	0	0	2	1	0	0
1	0	2	0	3	0	0	6	0	1	0	1	0	0	0	0	0	2	0	0	0	0	1	0	2	1	0	0	0	0	0	1	0	0	0	2	0	4	0	1
0	0	0	0	1	2	1	0	0	1	4	1	0	1	0	2	2	1	1	0	0	0	0	0	0	0	0	0	0	0	0	0	0	0	0	1	0	0	0	0
0	0	1	1	0	0	1	0	0	0	0	1	1	0	0	0	0	1	0	1	0	0	0	0	0	0	0	0	0	0	0	0	0	0	1	0	0	0	0	0
0	2	1	0	0	2	3	0	1	2	0	2	0	0	1	0	0	1	0	0	2	1	0	0	0	0	0	0	0	0	0	0	0	0	0	0	1	0	0	1
0	0	3	0	0	0	1	0	0	1	0	4	1	1	0	0	0	0	1	0	0	0	0	0	0	0	0	0	0	0	0	0	0	0	1	0	0	0	0	0
3	1	4	2	0	2	0	2	0	0	3	3	2	1	4	2	0	1	0	1	0	0	1	0	0	0	0	0	0	0	0	0	0	0	0	0	1	8	1	0
1	1	1	1	0	0	0	3	0	0	5	1	1	3	0	0	1	1	0	1	1	0	0	0	0	0	0	0	0	0	0	0	0	0	0	1	0	1	0	1
1	0	2	2	0	0	0	0	0	0	0	0	0	1	4	2	4	0	1	2	1	0	0	0	0	0	2	1	4	3	0	0	0	1	0	0	0	0	0	0
0	0	0	0	1	0	0	0	3	0	0	0	0	0	0	1	4	2	0	0	1	2	1	0	1	0	0	0	1	0	0	0	0	0	0	3	0	0	0	0
1	0	2	2	0	0	0	2	1	1	2	0	0	0	1	1	4	2	1	0	1	2	0	1	0	1	0	0	0	1	0	0	0	2	1	0	0	0	1	0

Table **1.3** ***Results for Greig-Smith's procedure applied to the central*** **32 × 32** ***lattice of the savanna data of Table*** **1.2**

Block size (r)	Sum of squares (T_r)	Scaled variance (G_r)	Mean square (M_r)
1	1 607	901	0.88
2	2 313	1 051	1.03
4	3 575	1 539	1.50
8	5 611	1 689	1.65
16	9 533	1 741	1.70
32	17 325	2 121	2.07
64	32 529	5 664	5.53
128	59 394	6 458	6.31
256	112 330	23 083	22.54
512	201 577	3 730	3.64
1024	399 424		

would be simplest to label each quadrat with an identity number and then to choose random numbers from the corresponding range—this is the procedure that Goodall uses. However, with the square grid of this example the quadrats are automatically 'numbered' by their row and column positions. To choose a quadrat at random it suffices to generate a pair of random numbers, each being in the range 1–40. The quadrat so chosen then forms one half of the pair. The second quadrat is automatically specified by being at a distance d (the distance currently under investigation). Alternate 'second' quadrats are to the right or below the first quadrats. if this defines a quadrat that is outside the 40 × 40 grid, then the pair are rejected and a further pair of random numbers are chosen. To avoid correlated variance estimates it is also necessary to reject pairings that include quadrats that have already been sampled.

Suppose that we are considering quadrats that are separated by a distance $d = 21$, and suppose that the sequence of random numbers is 29 26 14 07 20 04.... The first pair of random numbers (29, 26) identify a quadrat having a count of 9. Suppose that it is the turn of a horizontal pairing, then since $d = 21$ this quadrat is required to be paired with the quadrat at position (50, 26). Since 50 is greater than 40 this quadrat is not in the grid and so the pairing is inadmissible, and we examine the next pair of random numbers. The procedure therefore selects the pair of quadrats (14, 07) and (35, 07) followed by the pair (20, 04) and (20, 05). Now, if the data represents a Poisson forest of counts, then each quadrat count has equal mean value and the difference between a pair of quadrat counts is an estimate of the variance of counts for quadrats at that distance apart. The two pairs of quadrats generated above therefore give individual estimates of variance of $(6 - 2)^2$ and $(5 - 1)^2$, respectively.

To illustrate the method in its entirety, ten pairs of quadrats were generated at each of the possible values of d from 1 to 39. With just ten pairs at each value the individual estimates are comparatively imprecise compared to most of those

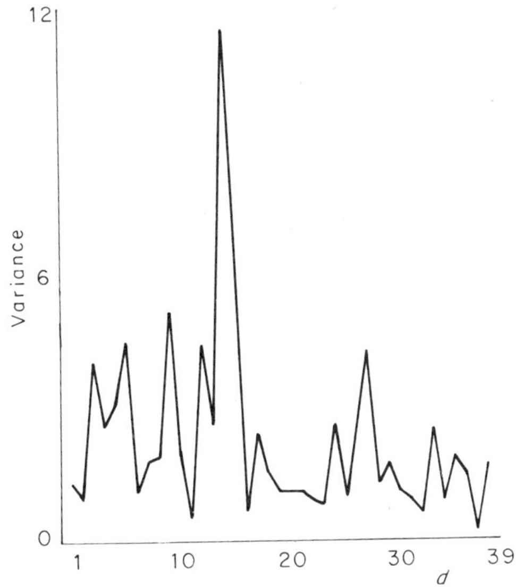

Figure 1.21 Graph of variation in variance estimates with change in d, the inter-quadrat distance

obtained using Greig-Smith's procedure. However, as can be seen from the plot of the results in Figure 1.21, the erratic nature of the graph fails to conceal evidence of some mild clumping at distances of 1.5–3 m (i.e. $d = 3$–6) and at distances of 12.5–15 m, while the clumping at the range previously indicated by the Greig-Smith analysis, 6.5–8 m, is very striking.

Hopkins (1965) used the linear version of the Greig-Smith analysis (discussed in the next subsection), and also detected variance peaks (a major one at 8 m and a minor one at 1 m). Drawing upon previous evidence concerning the nature of savanna fires, he was able to attribute the major peak to the natural physical consequences of a steady prevailing wind.

A transect of quadrats

Kershaw (1957, p. 291) remarked that Greig-Smith's method makes 'the implicit assumption that the areas of heterogeneity are isodiametric', whereas in ploughed fields, for example, 'areas of heterogeneity might be elongated in the direction of cultivation'. To cope with this type of heterogeneity, Kershaw suggested that Greig-Smith's method be adapted for a single row of contiguous quadrats: such a row is termed a *transect*. The benefit of this is that a transect of, say, 256 consecutive quadrats could be regarded as either 128 consecutive pairs of quadrats or as 64 consecutive quadruples, and so forth. A disadvantage of the transect is that it is inherently one dimensional, whereas the characteristic of pattern is that it is two dimensional—consequently, a one-dimensional transect may well fail to detect some important aspect of the actual pattern. Nevertheless, as with the original blocked version, the procedure is so easy to implement that it would appear to be a reasonable preliminary to a more detailed investigation. However, there are some subtle mathematical problems, as Example 1.4 makes clear.

Example 1.4

As a simple example, suppose that we consider an analysis of the counts in the first 32 quadrats of row 10 of Hopkins' data (given in Table 1.2). We must first form the successive block totals as shown in Table 1.4.

Calculating the sums of squares for these block totals, we find that $T_1 = 71$, $T_2 = 125$, $T_4 = 139$, $T_8 = 269$, $T_{16} = 353$, and $T_{32} = 625$, so that $G_1 = 17$, $G_2 = 111$, $G_4 = 9$, $G_8 = 185$, and $G_{16} = 81$. It would therefore appears that there is considerable evidence of pattern at the size (or distance) of 2 units. Notice that, with a transect, the distance between the centres of gravity of adjacent blocks is *always* equal to the average inter-quadrat distance and to the block size.

Suppose, however, that we consider the *central* 32 quadrats of row 10, thereby omitting the first four quadrat counts (1 0 1 0) included in the previous analysis, and adding a further four (2 1 0 0) at the other end of the original transect. The change would appear to be rather minor since 28 of the quadrats are common to the two data sets. The new sums of squares are $T_1 = 74$, $T_2 = 132$, $T_4 = 144$, $T_8 = 210$, $T_{16} = 410$, and $T_{32} = 676$, so that we obtain $G_1 = 16$, $G_2 = 120$, $G_4 = 78$, $G_8 = 10$, and $G_{16} = 144$. There is a peak at size 2 as before, but there is now a *trough* at size 8 where previously there was a peak! It is easy to see that the cause lies with the two quadrat counts of 5 and 4. In the original analysis these were combined for the first time in a block of size 8, but in the second analysis there combination is delayed to block size 16. Such an arbitrary decision seems to have been of disproportionate importance.

Detailed investigations of the causes of this type of shift in the location of the peak variance have been conducted by Usher (1969, 1975) and Errington (1973). Some possible methods for evading the problem are discussed in the next subsection.

***Table* 1.4 *Block calculations for the first* 32 *quadrats of row* 10 *of Table* 1.2**

Block size	Transect quadrat counts
1	1 0 1 0 0 0 0 0 0 0 2 0 2 2 0 0 0 1 0 0 0 0 0 1 0 0 3 5 0 1 4 2
2	1 1 0 0 0 2 4 0 1 0 0 1 0 8 1 6
4	2 0 2 4 1 1 8 7
8	2 6 2 15
16	8 17
32	25

Refinements to the use of the transect

Suppose that we have a transect of length n, where n is some power of 2 (e.g. $n = 32 = 2^5$). Denote the individual quadrat counts by $x_1, x_2, \ldots, x_{n-1}, x_n$. Kershaw's implementation of the blocked quadrat procedure requires us to calculate

$$T_1 = (x_1^2 + x_2^2 + \ldots + x_{n-1}^2 + x_n^2),$$
$$T_2 = (x_1 + x_2)^2 + \ldots + (x_{n-1} + x_n)^2,$$
$$T_4 = (x_1 + x_2 + x_3 + x_4)^2 + \ldots (x_{n-3} + x_{n-2} + x_{n-1} + x_n)^2,$$

etc.

Now

$$\begin{aligned} G_1 &= 2T_1 - T_2 \\ &= 2(x_1^2 + x_2^2) - (x_1 + x_2)^2 + \ldots + 2(x_{n-1}^2 + x_n^2) - (x_{n-1} + x_n)^2 \\ &= (x_1 - x_2)^2 + \ldots + (x_{n-1} - x_n)^2. \end{aligned} \tag{1.11}$$

Similarly, $G_2 = 2\mathrm{T}_2 - T_4$ can be shown to be given by

$$G_2 = \{(x_1 + x_2) - (x_3 + x_4)\}^2 + \ldots + \{(x_{n-3} + x_{n-2}) - (x_{n-1} + x_n)\}^2. \tag{1.12}$$

Thus, an alternative interpretation of the blocked quadrat G-values is as the sums of squares of differences of neighbouring block totals. The corresponding variance estimates are the values of G/n.

Suppose that the length of our quadrat is extended so that it now includes a further $m(<n)$ quadrats. There are therefore now $(m + 1)$ possible different sub-transects of length n. Usher (1975) suggested that a solution to the problem of which of the $(m + 1)$ should be used was to use them all and take the average of the values obtained. Thus, for example, in terms of the individual quadrat counts we would obtain:

$$(m + 1)\bar{G}_1 = \sum_{i=0}^{m} \{(x_{1+i} - x_{2+i})^2 + \ldots + (x_{n-1+i} + x_{n+i})^2\}, \tag{1.13}$$

and the corresponding variance estimates, termed by Ludwig and Goodall (1978) the *stepped-block quadrat variance* (SBQV) estimates are given by $\bar{G}_1/n$, etc.

However, this formulation places more emphasis on the central quadrats of the transect than it does on the peripheral quadrats. For example, the term $(x_1 - x_2)^2$ only appears when $i = 0$, whereas the term $(x_3 - x_4)^2$ appears both as the second term in the summation when $i = 0$ and as the first term in the summation when $i = 2$. A more even-handed procedure was suggested by Hill (1973), in which the variances V_1 and V_2, at block sizes 1 and 2, respectively, are given by

$$V_1 = \frac{1}{2(n + m - 1)}\{(x_1 - x_2)^2 + (x_2 - x_3)^2 + \ldots + (x_{n+m-1} - x_{n+m})^2\}, \tag{1.14}$$

$$V_2 = \frac{1}{4(n+m-3)}\{(x_1+x_2-x_3-x_4)^2+(x_2+x_3-x_4-x_5)^2+\cdots$$
$$+(x_{n+m-3}+x_{n+m-2}-x_{n+m-1}-x_{n+m})^2\}, \qquad (1.15)$$

with corresponding equations for larger block sizes (which need *not* be powers of 2) as far as the largest whole number less than or equal to $(m+n)/2$. The coefficient of the bracketed expression for V_r will be $1/2^r(n+m-2r+1)$. These variance estimates have been termed *two-term local-quadrat variance* (TTLQV) estimates.

Still another alternative has been suggested by Ludwig and Goodall (1978), this being a simple adaptation of the method of randomly paired quadrats discussed above. The authors suggest using *all possible pairs* of quadrats at *all possible* distances [1 to $(n+m-1)$], so as to make maximum use of all the information available. The separate variance estimates are, of course, highly correlated and the method cannot, therefore, be used for tests of significance. The variance estimates V_1^* and V_2^* for block sizes 1 and 2 are given by $V_1^* = V_1$ and

$$V_2^* = \frac{1}{2(n+m-2)}\{(x_1-x_3)^2+(x_2-x_4)^2+\ldots+(x_{n+m-2}-x_{n+m})^2\}, \qquad (1.16)$$

with corresponding equations for block sizes 3, 4, ..., $(n+m-1)$. The coefficient of the bracketed expression for V_r^* will be $1/2(n+m-r)$. These estimates will be referred to as the *paired quadrat variance* (PQV) estimates.

The estimate V^* will usually be considerably smaller than the blocked quadrat estimate, because V^* considers the effect of distance only, and is not concerned with block size. With the earlier methods we saw (Figure 1.20) that the individual quadrats in neighbouring blocks were not all equally spaced. This is equally true for transect blocks and consequently a comparison between blocks will involve the variation in counts due to their separation at a particular distance and *also* the variation due to the inequality of the separation distances. However, the most frequent separation distance for neighbouring blocks of size r, is also r, so that both techniques should be able to identify peaks and troughs of variance, though the block techniques would seem likely to produce a less peaked graph of variance against r than that produced by the distance technique. Ludwig and Goodall (1978) note that, in contrast to the smoothed variance pattern produced by the TTLQV method, the PQV method tends to produce an oscillating variance pattern when the underlying data have a regular periodicity (pattern at an interval r is also manifest at intervals $2r$, $3r$, etc.). Ludwig (1979) suggests that, because of the oscillatory nature of the PQV pattern, the TTLQV procedure is the more effective at revealing large-scale patterns when there are both large-scale and small-scale patterns present. The emphasis of the TTLQV procedure on larger-scale patterns is manifest in the results of Carpenter and Cheney (1983), who find that the PQV procedure performs well in a wide variety of situations and 'is a superior method for describing pattern'.

Example **1.5**

As an illustrative transect, consider the quadrat counts of row 10 of Hopkins' savanna burning data (Table 1.2). These were:

1010000000202200010000010035014221000111.

This is not ideal data because of the high proportion of zero counts (Ludwig and Goodall, 1978, p. 57), but serves as a convenient example of the problems that beset this type of analysis.

Since the transect contains 40 quadrats we have $n = 32$ (the highest feasible power of 2) and $m = 8$, so that there are nine possible block analyses that could be conducted (those starting at quadrats 1 to 9), and the results for these are summarized in Table 1.5.

Table 1.5 reveals the immense variation in the G estimates that can result from a change in the starting-point of a transect. The cyclic nature of the values is also apparent, particularly for G_8 which varies between 2 and 325. Since the individual values are so variable it would be unwise to be too impressed even by the average values, $\{\bar{G}_r\}$.

The variance estimates from the SBQV method ($\bar{G}_r/40$) are compared, in Figure 1.22, with those from the two exhaustive methods. As expected, the TTLQV method gives a smoothed 'continuous' representation of the SBQV results, while the PQV method gives much smaller estimates of variance at the larger values of r. The TTLQV graph shows peaks at 2, 7, and 16, while the PQV graph shows peaks at 2, 16, near 9, and near 24. Of course the data is so limited that it would be unwise to draw strong conclusions from the results, but it is interesting that the previous analysis of the entire 40 × 40 grid (Example 1.3) located a peak at 16 and also one at about 5.

Table **1.5** ***The results of nine separate block analyses of row* 10 *of Hopkins' savanna-burning data***

Cells in transect	G_2	G_4	G_8	G_{16}	G_{32}
1 to 32	17	111	9	185	81
2 to 33	58	12	6	250	144
3 to 34	17	65	27	325	121
4 to 35	58	18	54	130	144
5 to 36	16	120	78	10	144
6 to 37	58	12	130	2	144
7 to 38	17	65	165	25	169
8 to 39	58	22	50	74	196
9 to 40	17	121	5	97	169
Average values	35	61	58	122	146

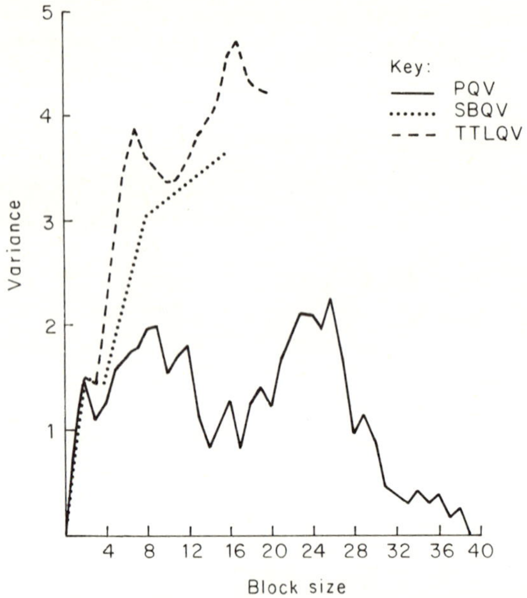

Figure 1.22 Pattern analysis of Hopkins' savanna-burning data

Tests of significance

It is natural to wish to test whether or not an apparent peak in the variance graph can be attributed to other than random variation. Thompson (1955, 1958) showed that the original F-test suggested by Greig-Smith (1952) was invalid whenever the data was not random. An alternative procedure has been suggested by Greig-Smith (1964, p. 86) and investigated by Zahl (1977) who states that it remains valid for non-random distributions. This procedure uses the test statistic

$$F_r = G_r/G_1, \tag{1.17}$$

which is to be compared with tables of the $F_{n/2r,n/2}$ distribution, where n is the total number of quadrats. An unusually large value of F_r provides evidence of a pattern.

Reference to Table 1.5 shows that this test is crucially dependent upon the starting position—alternate rows of Table 1.5 would give rise to quite different conclusions. The best procedure might be to use the compromise statistic

$$\bar{F}_r = \bar{G}_r/\bar{G}_1, \tag{1.18}$$

and to compare its value with the $F_{n/2r,n/2}$ distribution, as before. Zahl (1977) investigated the performance of $\bar{F}_r$ in the analysis of some artificial square arrays and concludes that it is reasonably efficient at detecting the artificial patterns that his arrays contained, and is in fact superior to the complex test that he had suggested earlier (Zahl, 1974). Zahl found that both his own test and the Greig-Smith test performed very much better than the test suggested by Goodall (1974) based on paired quadrats.

In the case of a transect there is one other test, which has been described by Cormack (1979, p. 163) as 'the only recommendable procedure' if 'a formal test is required of the existence of pattern': this is the '2 within 4' randomization test suggested by Mead (1974a).

Suppose that the consecutive counts in a transect (either quadrat counts or counts from blocks of quadrats) are $a,b,c,d,e,f,g,h,i,\ldots$. Mead's test procedure focuses attention on the separate quadruplets (a,b,c,d), $(e,f,g,h),\ldots$ and enquires, in essence, as to whether there is any repeating pattern in the arrangement of the four counts within the quadruplets. The basis of the test is the value of

$$S = |(a+b)-(c+d)| + |(e+f)-(g+h)| + \ldots, \tag{1.19}$$

where $|x|$ is the absolute value of x.

If the pattern of block counts is random, then the four counts (a,b,c,d) within the first quadruplet might have appeared in any order, so that the contribution to S from that quadruplet might equally well have been $|(a+c)-(b+d)|$ or $|(a+d)-(b+c)|$. By calculating the values of the three possible combinations that could have arisen from each of the quadruplets it is possible to calculate the exact distribution of S under randomness. However, there is a possibility that some quadruplets will dominate the distribution (Mead, 1974a, p. 301) and for this reason we replace the observed quadruplet contributions by coded contributions, as described below.

Let the three possible values of the contribution from a quadruplet have values denoted by w, x and y, with, for convenience, $w \leq x \leq y$. If $w = x = y$, the quadruplet is uniformative and is ignored. The remaining possibilities are $w = x < y$, $w < x < y$, and $w < x = y$. Mead suggests replacing the actual values (w, x, y) in these three cases by the coded values $(0, 0, 2)$, $(0, 1, 2)$, and $(0, 2, 2)$, respectively. Each quadruplet therefore contributes either 0, 1, or 2, with the total of the contributions being, T, say. Suppose that in our transect there are $n = (n_2 + n_3 + n_4)$ informative quadruplets, where n_2, n_3, and n_4 are the numbers of quadruplets coded as $(0, 0, 2)$, $(0, 1, 2)$, and $(0, 2, 2)$, respectively (with the subscripts 2, 3, and 4 corresponding to the totals of the coded values in each case). Mead shows that if the $\{n_i\}$ are large, then

$$Z = \frac{3T - (2n_2 + 3n_3 + 4n_4)}{(8n_2 + 6n_3 + 8n_4)^{1/2}} \tag{1.20}$$

has an approximate unit normal distribution, so that Z^2 has an approximate χ_1^2 distribution. Upton (1984) shows that the value of Z, like the previous blocked quadrat statistics, is dependent upon the starting-point. He suggests that the values of Z should be calculated for each of the four possible starting-points and that their sum should be compared with a χ_4^2 distribution. The same procedure, summing Z-values, could also be used to combine information from separate transects.

Example 1.6

Mead (1974a) analysed one of eight transects investigated by Kershaw (1957) and also analysed by Douglas (1975) and Ripley (1981). The observations were scores on a scale of 0–5 of the cover of *Nardus stricta*, a moorland grass. Mead found some evidence of pattern at block size 4, but none at larger block sizes. The 64 individual counts are given in Table 1.6, with the informative quadruplets for each of the four possible starting points being underlined.

Consider the first informative quadruplet for the transect starting at block 1, which is (0, 1, 6, 4). The observed value of $|(a+b)-(c+d)|$ is $|(0+1)-(6+4)| = 9$. The alternative values are $|(0+6)-(1+4)| = 1$ and $|(0+4)-(1+6)| = 3$, and so these three values, 9, 1, and 3, will be coded as 2, 0, and 1, respectively. The complete coded statistics for the transect starting at block 1 are given in Table 1.7.

It will be seen that a quadruplet (a,b,c,d) is coded as (0, 1, 2) if a, b, c, and d are all different, it will be coded as (0,0,2) if the two equal counts are either the two smallest counts or the two largest counts, and will be coded as (0,2,2) otherwise. The values of T, n_2, n_3, and n_4 for the four separate starting-points are therefore (11,3,4, and 0), (9,6,2, and 0), (8,6,1, and 0), and (7,3,3, and 0) so that the four values of Z^2 are $225/48 = 4.69$, $81/60 = 1.35$, $81/54 = 1.50$, and $35/42 = 0.86$. The combined test statistic therefore has the value $4.69 + 1.35 + 1.50 + 0.86 = 8.40$, which is less then 9.49, the upper 5% point of a χ_4^2 distribution, so that we conclude that there is not significant evidence, at this level, of a pattern at block size 4.

The reason for the disagreement with Mead's conclusion is simply that he considered only the transect starting at block 1, for which the value of Z^2 was 4.69—much the largest of the four possible values that he might have obtained.

Distinguishing between true and apparent contagion

In Section 1.1 we remarked that the negative binomial distribution was simultaneously an example of a generalized Poisson distribution and of a compound Poisson distribution. The generalized Poisson distribution arose when there was true contagion (clustering), while the compound Poisson distribution arose when there was apparent contagion (heterogeneity). In view of the equivalence of the end-products arising from these two processes, it may seem ambitious to hope to distinguish between them, but Cliff and Ord (1981, pp. 94–96) suggest a simple procedure which does hold out a slight prospect of success.

The Cliff–Ord procedure relies upon acceptance that the quadrat counts have

Table 1.6 The block totals of size 2 for Kershaw's Nardus stricta data

Block totals

0 0 0 2 2 0 0 0 0 0 0 0 0 0 1 6 4 5 6 1 0 0 0 2 0 0 5 4 5 3 1 8 2 0 0 0 1 8 9 3 0 0 0 2 1 0 0 0 0 0 0 0 0 0 0 0 1 0 0 0 0 0 0 0 0 1 4

Quadruplets starting at block 1

(0002), (2000), (0000), (0164), (5610), (0020), (0545), (3182), (0001), (8930), (0021), (0000), (0000), (0010), (0000), (0014)

Quadruplets starting at block 2

(0022), (0000), (0000), (1645), (6100), (0200), (5453), (1820), (0018), (9300), (0210), (0000), (0000), (0100), (0000)

Quadruplets starting at block 3

(0220), (0000), (0001), (6456), (1000), (2005), (4531), (8200), (0189), (3000), (2100), (0000), (0000), (1000), (0000)

Quadruplets starting at block 4

(2200), (0000), (0016), (4561). (0002), (0054), (5318), (2000), (1893). (0002). (1000), (0000), (0001), (0000), (0001)

Note Underlines denote informative quadruplets.

Table **1.7** *Informative quadruplets for the transect starting at block* **1** *for the Nardus stricta data given in Table* **1.6**

Quadruplet	(0,1,6,4)	(5,6,1,0)	(0,5,4,5)	(3,1,8,2)	(8,9,3,0)	(0,0,2,1)	(0,0,1,4)
Quadruplet type	(0,1,2)	(0,1,2)	(0,0,2)	(0,1,2)	(0,1,2)	(0,0,2)	(0,0,2)
Coded value	2	2	0	1	2	2	2

arisen from a negative binomial distribution. Now, while it is true that many data sets have quadrat counts for which the *hypothesis* of the negative binomial distribution is perfectly acceptable, it is rarely the case that we can be sure what mechanism gave rise to the data. The problem is that many other hypothetical distributions might also prove acceptable as descriptions of quadrat counts (see, for example, Haggett *et al.*, 1977, p. 430).

If we suppose that the negative binomial distribution *is* appropriate for the quadrat counts, then the basis of the Cliff–Ord procedure is simply that we determine the values of the parameters p and k of that distribution (see equation (1.2)), for the original quadrats, and for the counts obtained by combining blocks of quadrats. Cliff and Ord show that, subject to various limiting (and rather unlikely) conditions being true, the parameter p will remain constant in the case of true contagion, while the parameter k will remain constant in the case of apparent contagion.

Because of the presence of random variation, we cannot expect exactly constant parameter values and a certain variability is only to be expected. However, it seems likely that, in many cases, it will be clear that *neither p nor k* are constant for varying quadrat size. This could occur because the negative binomial has arisen as a result of a mixture of the two types of contagion, but it seems more likely that this arises because the original data did not have a negative binomial distribution at all, but had a distribution sufficiently similar that the goodness-of-fit test failed to detect significant differences from the negative binomial.

Example 1.7

The quadrat frequencies for the original 1600 cells of Hopkins' data (Table 1.2) are given in Table 1.8 in the column headed 'block size 1 × 1, Obs.'. Here 'Obs.' is short for observed frequency and 'Exp.' is short for expected frequency. In order to calculate the expected frequencies we should find the best fitting negative binomial distribution, using the method of maximum likelihood to obtain the parameter estimates (see Bliss and Fisher, 1953). This method is complicated, and it is very much easier to calculate parameter estimates from the method of moments. Denoting the observed mean and variance by m and v, respectively, the method of moments gives estimates $\tilde{p} = m/v$, $\tilde{k} = m\tilde{p}/(1-\tilde{p})$. The expected frequencies given in Table 1.8 are calculated using $\tilde{p}$ and $\tilde{k}$—expected frequencies

Table* 1.8 *Quadrat frequencies for various block sizes for Hopkins' savanna-burning data

	Block size					
	1 × 1		2 × 2		4 × 2	
Cell frequency	Obs.	Exp.	Obs.	Exp.	Obs.	Exp.
0	982	993	72	89	13	13
1	372	360	86	85	15	21
2	160	144	81	67	20	24
3	47	60	56	49	34	24
4	24	25	36	35	18	22
5	6	11	24	24	29	29
6	4	4	18	17	17	16
7	1		10	11	12	13
8	1		9	8	6	11
9	2		2	5	10	9
10	0	3	3	3	8	7
11	1		0		5	5
12	0		1		4	4
13	0		1	7	2	3
14 +	0		1		7	9
Total	1600		400		200	

calculated from the maximum likelihood estimates would be very slightly closer to the observed frequencies.

To test the hypothesis that the data may be considered to arise from a negative binomial distribution we use the Pearson *goodness-of-fit* test, and calculate

$$X^2 = \sum(\text{Obs.} - \text{Exp.})^2/\text{Exp.}, \tag{1.21}$$

where 'Obs.' and 'Exp.' are the corresponding observed and expected frequencies and the summation is over each of the possible quadrat counts. Thus, for the 1 × 1 block size, we calculate

$$\begin{aligned} X^2 &= (982-993)^2/993 + (372-360)^2/360 + \ldots + (4-4)^2/4 + (5-3)^2/3 \\ &= 7.8. \end{aligned}$$

The small tail frequencies corresponding to cell counts of 7 or more have been amalgamated because the distribution of X^2 becomes intractable if 'Exp.' is allowed to be very small. A value of 3 or more for 'Exp.' is perfectly reasonable. Provided this precaution is taken, the distribution of X^2, when the null hypothesis (that the data accords with a negative binomial distribution) is true, is approximately χ^2 with d degrees of freedom. The value of d is equal to the number of different observed values, less the number of parameters estimated from the data, less 1 for the equalization of the overall totals of the observed and expected cell frequencies. Thus, in the case of the block size 1×1 data, the number of observed values is 8 (0, 1, 2, 3, 4, 5, 6 and 7 +), the number of parameters estimated from the data is 2 (p and k) and therefore the value of d is $8 - 2 - 1 = 5$. Since the value of X^2, 7.8, is not an unusually large observation when compared to the distribution of values that might be obtained from a χ_5^2 distribution, we can happily accept that the block size 1×1 data is in accord with the hypothesis of a negative binomial distribution.

The values of X^2 for the 2×2 and 4×2 cases are, respectively, 11.6 and 19.8 with the values of d being 9 and 12. The value 11.6 is typical of what we would expect from a χ_9^2 distribution, while the value 19.8, which is somewhat on the large side, does not exceed the upper 5% value (21.0) of a χ_{12}^2 distribution. Therefore, for each set of quadrat counts the negative binomial seems to be an acceptable explanation.

The method of moments estimates of the parameters p and k, for various block sizes, are given in Table 1.9 (maximum likelihood estimates would be very similar). It is obvious that neither parameter is constant—in fact, $\tilde{k}$ appears to increase as the square root of the number of quadrats, q, that are being amalgamated. The successive values of $\tilde{k}/\sqrt{q}$ are 0.82, 0.75, 0.82, 0.82, 0.74, and 0.78. These values show impressive consistency, but the authors, unfortunately, are unable to interpret this as evidence of any specific alternative distribution. We can, however, assert that it now seems rather unlikely that the negative binomial is an appropriate description, despite the evidence of the X^2 tests. The true distribution must, however, be very similar in form and it may be as well to recall that the data are not genuine counts but are artificial scores.

Summary of the use of quadrats

The use of quadrat counts as indicators of pattern is very popular. However, as we have seen, all the methods that have been suggested are beset by problems. Not one of the methods appears to take proper account of the spatial nature of the data—which is, of course, paradoxical. All the methods based on transects are limited by being one dimensional, whereas pattern is two dimensional. The effect of the starting-point (for transects) and of the block orientation (for grids) has been seen to be of alarming importance. Virtually all the significance tests are suspect, even if we can be sure of what they are testing. Nevertheless, the

Table 1.9 Estimates of p and k for various block sizes for Hopkins' savanna-burning data

Block size	1×1	2×2	4×2	4×4	8×4	8×8
p	0.56	0.37	0.31	0.24	0.17	0.13
k	0.82	1.50	2.32	3.24	4.20	6.22

popularity of the simpler methods is justified by the fact that, against the odds, the plots of variance against block size do seem to convey interpretable (and credible) information. There is clearly scope for the introduction of some more efficient, two-dimensional test procedures.

At present, the best advice that we can give is, wherever possible, to use several different test procedures, choosing procedures that analyse the data in different ways. The combination of Mead's test, the paired quadrat method, and the TTLQV method suggest themselves as forming a suitable battery of tests for transect data. For a grid of counts, we suggest the use of Zahl's test for the Greig-Smith procedure, together with transect procedures applied separately to randomly chosen rows and to randomly chosen columns.

1.4 THE USE OF SAMPLED POINT LOCATIONS

In the previous sections that were concerned with quadrat counts the locational information was very limited and imprecise. Our ability to recognize the existence of a pattern was critically dependent upon the fortuitous choice of an appropriate size and location of a quadrat.

The obvious method of obtaining precise information about the locations of points is to record their positions using, for example, x- and y-coordinates. For a number of reasons this is not quite as easy as it sounds: establishing physical locations in the (three-dimensional) real world presents problems for surveyors. There is also the small detail that the objects in question may have a considerable size (e.g. galaxies!); if we have to represent such objects by a single point then there may be some disagreement over the location of that point. However, this is largely a question of scale: if we are concerned with the location of trees in a wood, then it will be reasonable to represent each tree by a point on the map of the wood, whereas if we are concerned with a suburban front garden, then a tree would probably be better represented as an irregular circle on the plan of the garden.

In this section we are concerned with sample information concerning point positions. It is often impracticable to locate every object in the region of interest and sampling methods have a long history in forestry (from which many of the examples and methods discussed in this section have been derived). All these methods are based on the distances between neighbouring points in the pattern or between an arbitrary location in the region of interest and the nearest point in the pattern. In forestry, the methods were developed in the first instance in order to estimate the total volume of wood in a forest. However, investigations showed that the efficiency of the estimation procedure was affected by the pattern of the trees and it was this that led foresters to be concerned with methods for describing pattern and for detecting non-random patterns.

We commence our discussion by describing the various measurements of distance that might be made. In order to distinguish points in the pattern from random sampling locations, for the remainder of this and the next section we shall suppose that we are concerned with a population of plants so that the word 'plant' refers to a point in the pattern while the word 'point' refers to a random sampling location. The 'plants' could, of course, be churches, grocery shops, or anything else of interest!

Distances between points and plants

The plant that is nearest to a randomly chosen plant is called its *nearest neighbour*, the next nearest plant is the second nearest neighbour, and so forth. To see that nearest-neighbour distances provide information about pattern, imagine a commercial orchard in which there are regimented rows of apple trees, the trunk of each tree being 15 m from the next tree in that row and with the parallel rows being 20 m apart. Evidently, whichever tree is randomly selected, its nearest neighbour will be at a distance of 15 m—the 'distribution' of nearest-neighbour distances will be pathological. On the other hand, with randomly scattered plants, some will be close to others while a few will be isolated, so that there will be a well-defined distribution of possible values for the nearest-neighbour distance.

We define nearest-neighbour distance in terms of the distance between a *randomly chosen plant* and its nearest neighbour. However, we cannot choose a plant at random without having first enumerated every plant in the region. Methods appropriate to the case where *all* the plant locations are known are considered in Section 1.5; in the present section we are concerned with methods for samples, so that we are unlikely to have information about the locations of the entire plant population. An alternative procedure is as follows: instead of choosing a *plant* at random, we choose a *point* at random and then work from that point. No bias arises, since random 'plants' and random 'points' can be regarded simply as different types of random 'spots'! To distinguish between the different types of distance we denote a *plant–plant* distance by W and a *point–plant* distance by X.

Suppose that our study area is square and has sides of unit length. We can obtain a random point in the study area by the simple expedient of obtaining two random numbers in the range $(0, 1)$ by using random number tables or a random number generator. If the study area is not a square, then it is simple to describe a circumscribing square that includes the study area and to proceed as before, ignoring any generated points that lie outside the study area.

A natural extension of the use of the randomly chosen point is to consider the neighbouring plant as having been randomly chosen (by proxy) and to consider the distance from this plant to its nearest neighbour. This *point–plant–plant* distance is denoted by Y in Figure 1.23. However, the distribution of the distance Y is *not* the same as that for the plant–plant distance, W, for the following reason. The location of the plant D, the nearest neighbour to the randomly chosen plant C, is completely unrestricted, whereas the location of the plant B, the nearest neighbour to the plant A, is subject to the restriction that it cannot lie within the

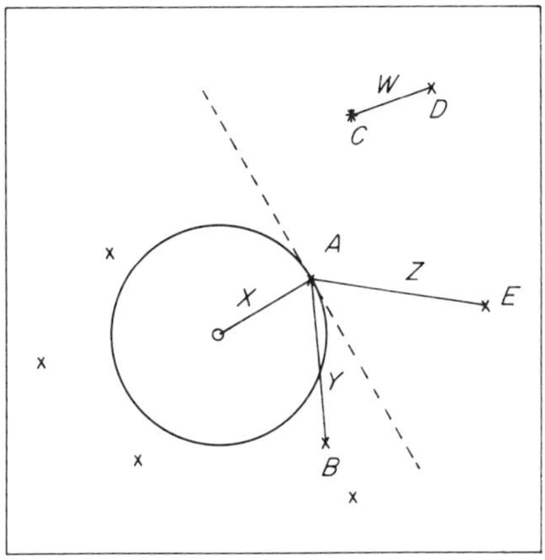

Key * A randomly chosen plant
x Other plants
o A randomly chosen point
A, B, C, D, E. Plant locations
W, X, Y, Z. Nearest-neighbour distances

Figure 1.23 Various nearest-neighbour measurements

circle of radius X centred on O, the randomly chosen point. This is because the plant A must, by definition, be nearer to O than any other plant. Kendall and Moran (1963, p. 38) have shown that the effect of this restriction is to increase the average value of Y, so that, in the case of randomly distributed plants, $E(Y) \simeq 1.19\, E(W)$. Further results concerning the joint distribution of X and Y are given by Cox and Lewis (1976) and Cormack (1977).

Suppose that we divide the entire two-dimensional plane into two half-planes, with the tangent at A to the circle with centre O providing the dividing line between these two half-planes. The restriction on the location of the plant B can then be seen to refer only to the half-plane containing the point O. Besag and Gleaves (1973) suggest that attention be restricted to the other half-plane, and that we measure the distance Z between the plant A and its nearest neighbour, E, *in this half-plane* (see Figure 1.23). In the case of randomly distributed plants it is then the case that X, W, and $Z/\sqrt{2}$ have identical distributions.

Some distributional results

Suppose that the mean intensity of plants per unit area is ρ and that the plants are distributed at random. As before, we define X to be the distance between a randomly chosen point and the nearest plant. Now, if X is greater than some value x, say, then this implies that there are no plants whatsoever in a region of area πx^2, for which the mean number of plants would be expected to be $\rho\pi x^2$.

Since the distribution of plants is random, the probability of no plants in the region of area πx^2 is simply the probability of an observation of 0 from the Poisson distribution (1.1) with $\lambda = \rho\pi x^2$. Thus, we have

$$P(X > x) = e^{-\rho\pi x^2}, \tag{1.22}$$

so that, on differentiating, we obtain the probability density function of X as

$$f(x) = 2\rho\pi x e^{-\rho\pi x^2}. \tag{1.23}$$

If we write $U = X^2$, then after some arithmetic juggling it is found that

$$f(u) = \rho\pi e^{-\rho\pi u}, \tag{1.24}$$

which is a scaled version of the density function of a chi-squared distribution with two degrees of freedom. This result was obtained independently by Skellam (1952), Moore (1954), and Morisita (1954).

Because of the additivity of chi-squared distributions, it follows that if $X_1, X_2, \ldots, X_n$ are independent point–plant distances, and $U_i = X_i^2\ (i = 1, \ldots, n)$, then the distribution of $2\pi\rho \sum U_i$ will be chi-squared with $2n$ degrees of freedom.

Thompson (1956) considered the joint distributions of $X, X_{(2)}, \ldots, X_{(k)}$, where $X_{(r)}$ is the distance to the rth nearest neighbour of a randomly chosen plant. Writing $U_{(r)} = X_{(r)}^2$, Thompson's result was that the joint distribution of U, $U_{(2)}, \ldots, U_{(k)}$, was

$$f(u, u_{(2)}, \ldots, u_{(k)}) = (\rho\pi)^k e^{-\rho\pi u_{(k)}}. \tag{1.25}$$

Integrating over $U, U_{(2)}, \ldots, U_{(k-1)}$, Thompson obtained the result that $2\pi\rho U_{(k)}$ has a chi-squared distribution with $2k$ degrees of freedom. It follows that

$$E(X_{(k)}) = k(2k!)/\rho^{1/2}(2^k \cdot k!)^2, \tag{1.26}$$

and

$$E(X_{(k)}^2) = k/\rho\pi. \tag{1.27}$$

Eberhardt (1967) gives a somewhat different derivation of the result (1.25) and notes that the result

$$E(X_{(k)}^{-2}) = \rho\pi/(k-1), \tag{1.28}$$

for $k \geq 2$, applies to a variety of underlying plant distributions.

All the preceding results have been couched in terms of X, the distance from a randomly chosen point to a neighbouring plant, but are equally appropriate for W, the plant–plant distance. The results also apply to the T-square distance, Z, subject to adjustment by a factor of 1/2 (for Z^2) or $1/\sqrt{2}$ (for Z) to take account of the restriction of this measure to the *half*-plane.

It is convenient at this point to recall that if A and B are independent random variables having chi-squared distributions with a and b degrees of freedom, respectively, then bA/aB has Fisher's F distribution with a and b degrees of freedom. This is usually written as the $F_{a,b}$ distribution. Furthermore, the ratio $A/(A+B)$ will have a beta distribution with parameters a and b. This distribution is usually written as $B(a/2, b/2)$. The upper-tail significance points of the beta distribution are directly related to those of the F distribution. In particular, for the case where $b = a$, if F_α is the upper α point of the $F_{a,a}$ distribution, then the upper α point of the corresponding $B(a/2, a/2)$ distribution is $F_\alpha/(1+F_\alpha)$. Because of this correspondence, tests based on the F distribution are equivalent to those based on the beta distribution and, because tables of percentage points of the F distribution are the more widespread, we shall concentrate on that distribution. The only reason for mentioning the beta distribution at this point was that many of the source papers discuss tests in their Beta form. These results form the basis for a number of the tests of randomness that have been proposed.

Cox and Lewis (1976) examined the joint distribution of X and Y and noted that two distinct cases arose: the case where $Y \leq 2X$ and where $Y > 2X$. The relevant diagrams are shown in Figure 1.24, where O refers to the random point, A to the neighbouring plant, and B to its neighbouring plant. The point C is a point of intersection of the circles centred on O and A.

Cormack (1977) provides an elegant argument to show that when $Y > 2X$, the distribution of $4X^2/Y^2$ is uniform in (0, 1), no matter what is the pattern of the plants whose locations are being sampled. This ratio is therefore totally uninformative concerning the nature of that pattern. Anticipating this result, Cox and Lewis (1976) focused attention on the case $Y \leq 2X$. Since $AC = Y$ and $AO = X$, elementary trigonometry tells us that

$$\sin(\beta/2) = \tfrac{1}{2}Y/X. \tag{1.29}$$

Cox and Lewis found it easier to calculate the joint distribution of X and β, rather than X and Y, and they found that this distribution involves the expression

$$\theta = 2\pi + \sin\beta - (\pi + \beta)\cos\beta. \tag{1.30}$$

In the case where the plants are distributed at random, they showed that

$$R = \tfrac{4}{3}(1 - \pi/\theta) \tag{1.31}$$

is distributed uniformly in (0, 1), and that this is not the case for either a clustered pattern of plants or a regular pattern of plants. Hines and Hines (1979) noted that, to a very close approximation,

$$R \simeq \tfrac{3}{2}Y^2/(2X^2 + Y^2). \tag{1.32}$$

Satyamurthi (1979) has suggested that the areas of the circles shown in Figure 1.24 are themselves informative. The circle centred on O has area

$$A_x = \pi X^2. \tag{1.33}$$

For the circle centred on A, we consider only the area that excludes the circle

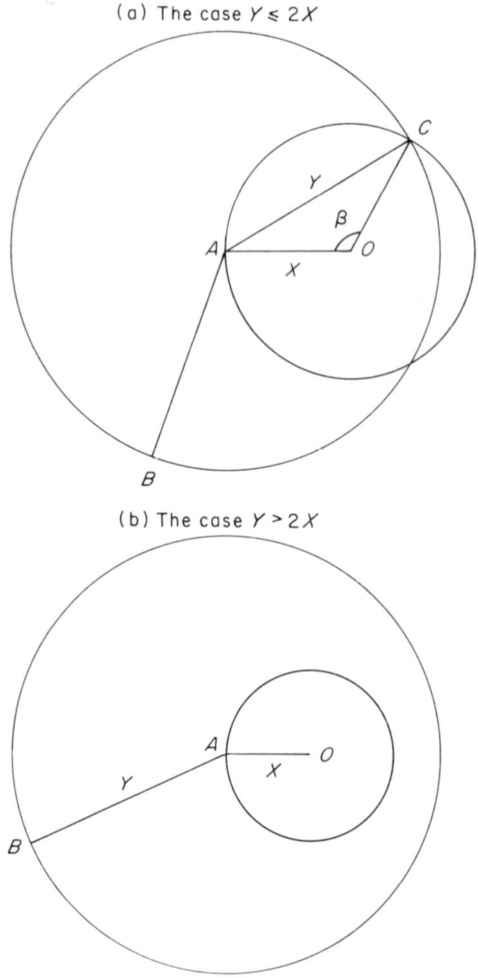

Figure 1.24 The possible configurations in point–plant–plant calculations.
(a) The case $Y < 2X$. (b) The case $Y > 2X$

centred on O. This is given as (Persson, 1964, p. 21)

$$A_y = \pi(Y^2 - X^2), \qquad \text{for } Y \geq 2X; \tag{1.34a}$$

$$A_y = Y^2(\pi + \beta)/2 - \beta X^2 + XY\cos(\beta/2), \quad \text{for } Y < 2X, \tag{1.34b}$$

where β is given by equation (1.29). For a collection of n point locations, let $\bar{A}_x$ and

$\bar{A}_y$ be the average values of these areas, and let $\bar{A}$ be given by

$$\bar{A} = (\bar{A}_x + \bar{A}_y)/2. \tag{1.35}$$

Then, for reasonably large n, under the null hypothesis of a random pattern of plants, $\bar{A}_x$, $\bar{A}_y$, and $\bar{A}$ are all normally distributed, with the same mean and with variances $(\bar{A}_x^2/n)$, $(\bar{A}_y^2/n)$, and $(\bar{A}^2/2n)$.

Testing for randomness

When the plants are distributed at random the distributions of X^2, Y^2, W^2, and Z^2 are all known, as we have seen. It is therefore natural to consider testing for departures from randomness by calculating the observed values of these statistics and assessing whether or not these observed values are sufficiently extreme as to cast significant doubt on the null hypothesis of a random plant distribution. The problem therefore becomes one of deciding which are the most appropriate test statistics. In the last thirty years a huge number of alternative test statistics have been suggested, the majority of which are given in Table 1.10.

A drawback of the apparently simple statistics S and Q is that they require the evaluation of an estimate, $\hat{\rho}$, of plant intensity. Mountford (1961) has shown that Q has a normal distribution with a variance of the form $1/n + g(\hat{\rho})$. If $\hat{\rho}$ is based upon a great many plant observations, then $g(\hat{\rho})$ will be close to zero and we may then regard $\sqrt{n}(Q-1)$ as having an approximate unit normal distribution. As we have seen, the distribution of S is chi-squared with $2n$ degrees of freedom, but this test does not seem very powerful (Brown and Holgate, 1974).

The statistic E was first introduced by Eberhardt (1967), is pleasantly simple, and appears to be sensitive to the presence of clusters (Hines and Hines, 1979). A table of significance points, reproduced from Hines and Hines (1979), is given in Appendix D.

The statistic P was introduced as a possible estimator of intensity, assuming randomness (when it has a chi-squared distribution with $(n-1)$ degrees of freedom), and does not perform well in the comparisons made by Hines and Hines.

The statistics L_N and L_F, introduced by Holgate (1965), involve the distance $X_{(2)}$ to the second-nearest neighbour. Both Diggle *et al.* (1976) and Byth and Ripley (1980) find that these statistics are very ineffective. The motivation behind Holgate's tests was that they avoided the calculation of W^2: the difficulty with the use of W^2 being simply that it involves the selection of plants at random. This selection problem appears to imply the complete enumeration of all the plants in the study area—a task at best tedious, but often entirely impracticable. The practicality of W^2 tests has been greatly increased by the suggestion by Byth and Ripley (1980) of the use of a semi-systematic sampling scheme which is illustrated in Figure 1.25.

Byth and Ripley suggest that a regular grid of study points should be

Table **1.10** ***Alternative distance tests for randomness***

Basic Measurements	Test statistic	Asymptotic distribution	Author
W	$S = 2\pi\hat{\rho}(\Sigma W^2)$	χ^2_{2n}	Skellam (1952), Moore (1954)
	$Q = \pi\hat{\rho}(\Sigma W^2)/n$	$N(1, 1/n)$	Pielou (1959)
X	$E = n(\Sigma X^2)/(\Sigma X)^2$	See Appendix D	Eberhardt (1967)
	$P = 12n\{n\log(\Sigma X^2/n) - \Sigma(\log X^2)\}/(7n+1)$	χ^2_{n-1}	Pollard (1971)
$X, X_{(2)}$	$L_N = \{\Sigma(X^2/X^2_{(2)})\}/n$	$N(1/2, 1/12n)$	Holgate (1965)
	$L_F = (\Sigma X^2)/(\Sigma X^2_{(2)})$	$F_{2n,2n}$	Holgate (1965)
X, W	$H_N = \{\Sigma(X^2/(X^2 + W^2)\}/n$	$N(1/2, 1/12n)$	Byth and Ripley (1980)
	$H_F = (\Sigma X^2)/(\Sigma W^2)$	$F_{2n,2n}$	Hopkins (1954), Moore (1954)
X, Z	$T_E = 2n\Sigma(2X^2 + Z^2)/\{\Sigma(X\sqrt{2} + Z)\}^2$	See Appendix D	Hines and Hines (1979)
	$T_P = 48n[n\log\{\Sigma(2X^2 + Z^2)/n\} - \Sigma\log(2X^2 + Z^2)]/(13n+1)$	χ^2_{n-1}	Diggle (1977a)
	$T_N = \Sigma\{2X^2/(2X^2 + Z^2)\}/n$	$N(1/2, 1/12n)$	Besag and Gleaves (1973)
	$T^*_N = \Sigma 2[\{\min(2X^2, Z^2)\}/(2X^2 + Z^2)]/n$	$N(1/2, 1/12n)$	Diggle *et al.* (1976)
	$T_F = 2(\Sigma X^2)/(\Sigma Z^2)$	$F_{2n,2n}$	Besag and Gleaves (1973)
	$T_C = -\Sigma[\log\{2X^2/(2X^2 + Z^2)\}]/2$	χ^2_{2n}	Cormack (1979)
X, Y	$M = \min(R_1, R_2, \ldots, R_n)$	$\text{Beta}(1, n')$	Cox and Lewis (1976)
	$R = (\Sigma R)/n'$	$N(1/2, 1/12n')$	Cox and Lewis (1976)
	$D = \bar{A}_X - \bar{A}_Y$	$N(0, (\bar{A}^2_X + \bar{A}^2_Y)/n)$	Satyamurthi (1979)

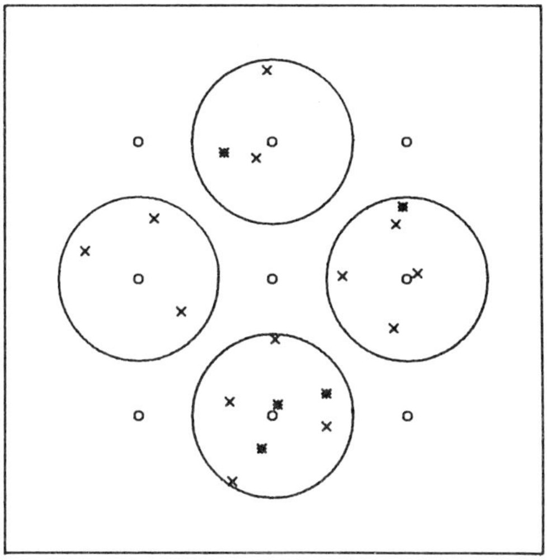

Figure 1.25 A semi-systematic sampling scheme

superimposed on the study area. Alternate study points are used as the points for calculating point–plant distances (for X^2). Around each of the remaining study points is prescribed a small sample area within which all the plants are enumerated. Byth and Ripley suggest that these sample areas should be such that they contain, on average, about five plants. The selection of random plants, for calculating plant–plant distances (for W^2), is based on the population of plants within the collection of sample areas. The reduction in power due to this restricted sampling procedure appears to be quite slight (Byth and Ripley, 1980) and the results confirm the findings of Diggle *et al.* (1976) and earlier authors that the Hopkins tests are efficient detectors of non-randomness. Diggle and Cox (1983) note that the Byth–Ripley sampling scheme may be costly to implement and may not always be practicable.

The subscripts used in Table 1.10 give an indication of the relevant probability distributions: H_N has a normal distribution with mean 1/2 and variance $1/12n$, while H_F has an $F_{2n,2n}$ distribution. Byth and Ripley (1980) find that H_N is particularly sensitive to the presence of regularity while H_F is more sensitive to the presence of clusters.

The statistics T_E, T_P, T_N, and T_F are T-square versions of E, P, H_N, and H_F, respectively, and have identical sampling distributions. The statistic T_N^*, which has the same asymptotic normal distribution as T_N, was suggested as being particularly suitable for lattice clusters (a regular arrangement of clusters), but this claim was not borne out by the results of Hines and Hines (1979). There appears

Table 1.11 *Locations and diameters of the beadlet anemones illustrated in Figure* 1.26

x	y	d	x	y	d	x	y	d	x	y	d	x	y	d	x	y	d
27	7	6	54	5	4	127	1	2	129	5	6	160	3	5	185	2	5
197	5	4	211	2	6	239	6	6	252	1	3	266	3	7	47	16	4
74	15	4	82	14	5	97	17	6	147	14	4	178	18	6	201	20	5
214	18	6	228	18	6	257	19	3	11	22	5	38	25	3	79	26	5
121	22	3	160	25	4	238	28	4	269	30	3	22	32	6	37	37	5
51	31	3	68	31	5	75	40	3	107	40	6	130	32	5	152	40	4
201	36	5	211	36	3	220	30	7	231	37	3	26	47	4	65	43	6
82	50	3	87	42	3	98	45	6	118	46	7	139	41	6	143	48	4
176	47	5	193	44	4	205	45	3	242	42	5	254	43	4	279	50	4
75	55	4	96	56	5	115	54	4	137	56	3	158	51	5	198	56	5
209	55	4	223	60	6	239	56	2	250	52	3	263	52	4	11	69	3
20	61	6	33	69	4	48	62	7	71	67	4	89	68	5	108	62	5
124	70	5	133	66	4	172	62	4	179	70	4	203	64	3	242	68	4
271	62	3	26	75	3	47	77	5	59	78	5	92	77	6	100	71	3
149	75	4	250	74	4	268	74	4	276	74	4	15	85	3	38	83	4
70	80	7	85	87	4	126	82	4	146	84	4	151	88	3	160	85	4
172	87	4	186	82	6	177	83	4	220	82	6	235	86	6	256	85	4
12	93	6	42	97	5	58	99	6	92	95	5	103	90	4	108	96	3
114	99	6	118	94	4	127	94	4	167	91	3	199	99	6	210	98	4
226	97	5	235	100	3	246	92	3	254	95	3	271	99	4	8	105	5

45	103	4	50	101	6	93	109	3	106	103	3	172	102	4	189	109	5
244	108	4	260	102	5	272	109	3	77	110	5	105	117	6	123	114	3
136	111	5	147	113	3	183	115	3	221	111	3	231	117	3	239	117	3
242	116	3	249	115	3	273	120	6	10	124	6	37	129	5	52	122	5
54	130	3	87	122	6	96	130	7	115	122	5	129	125	3	153	124	3
166	128	6	180	124	5	187	128	3	197	121	5	226	126	4	233	128	4
246	125	4	249	124	3	257	126	7	269	130	4	15	139	6	54	137	5
69	134	5	79	134	3	89	139	3	121	134	4	128	133	5	156	139	6
168	136	4	237	140	2	251	140	2	263	140	3	39	146	5	56	145	2
68	149	4	80	141	5	102	143	4	121	143	4	133	147	4	159	147	5
178	141	5	202	146	7	211	144	3	221	149	4	230	144	5	249	150	2
268	146	3	2	152	4	13	156	5	33	151	4	45	152	4	71	157	3
91	150	3	116	155	4	124	156	2	130	159	4	141	158	3	144	151	2
164	156	5	188	154	4	230	154	3	236	154	3	272	155	3	28	160	3
27	165	3	40	160	3	47	161	4	60	162	4	88	162	4	120	167	3
151	164	4	174	169	3	248	164	4	269	167	7	12	178	4	22	178	7
35	179	5	47	174	4	67	174	3	98	176	4	112	173	4	134	171	5
145	180	8	161	171	3	166	173	3	189	173	6	203	173	4	230	172	4
246	178	5	258	171	6	211	102	3									

Note: One unit corresponds to about 0.475cm.

to be nothing known about the behaviour of T_C, except that its distribution under the null hypothesis is chi-squared with $2n$ degrees of freedom. Hines and Hines demonstrate that T_E is very effective at detecting clusters and also at detecting regularity, although the results of Byth and Ripley (1980) suggest that the semi-systematic version of the Hopkins tests, H_N and H_F, are somewhat superior both to T_E and to T_N and T_F.

Hines and Hines (1979) also investigated the performance of the Cox–Lewis statistics M and $\bar{R}$. They give the asymptotic distribution of M as being $B(1,n')$ and their results suggest that both statistics are effective as detectors of regularity. Notice that the parameter in the beta distribution and the divisor in the expression for $\bar{R}$ is n' $(\leq n)$, the number of randomly sampled cases for which $Y \leq 2X$.

Finally, we note that, according to Satyamurthi (1979), under the hypothesis of spatial randomness, the statistic $D = \bar{A}_X - \bar{A}_Y$ has an asymptotic normal distribution, with mean 0 and variance estimated by $(\bar{A}_X + \bar{A}_Y)^2/2n$. The power of this test is not known.

Example 1.8

Kooijman (1979) has analysed data concerning the locations of 231 beadlet anemones (*Actinia equina*), and we are most grateful to Dr Kooijman for supplying us with the coordinates of the positions of these anemones, together with their diameters. This information is reproduced in Table 1.11, in which 1 unit corresponds to about 0.475 cm. The location of the sample plot, which measured 97.5×145 cm, was the north face of a boulder well above low-tide level at Quiberon in May 1976. Beside the anemones, star barnacles (*Chthamalus stellatus*) were abundant.

Kooijman (1979) showed that there was clear evidence of negative autocorrelation (see Chapter 3) at distances of up to 5 cm. This reflects the physical size of the anemones which had diameters of between 1 and 4 cm. There would also be some extra separation since anemones would avoid having their tentacles touching when these are extended during high tide. The positions and sizes of the anemones are indicated in Figure 1.26.

Despite the locations of all 231 anemones having been recorded, we shall use the semi-systematic sampling scheme in order to demonstrate its effectiveness. The size of the plot and the intensity of 'plants' suggest using a 4×6 grid of equi-spaced locations, and this results in the information available being that illustrated in Figure 1.27. The 12 grid squares contain a total of 69 'plants' indicated by X, with the 12 randomly selected 'plants' being further indicated by *. The 12 grid points are indicated by dots while the lines indicate the relevant point–plant and plant–plant distances, which are summarized by $\sum X^2 = 455$, $\sum W^2 = 1125$, and $\sum\{X^2/(X^2 + W^2)\} = 3.456$. In order to calculate this last statistic the 12 values of X_i are paired at random with the 12 values of W_i.

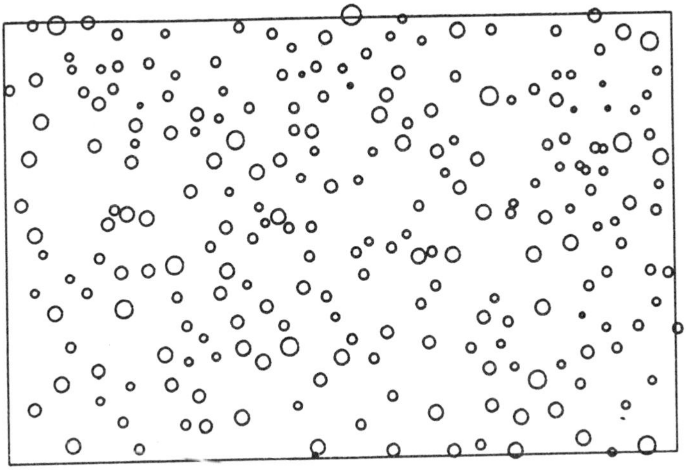

Figure 1.26 The positions of sea anemones on a rock. (After Kooijman, 1979)

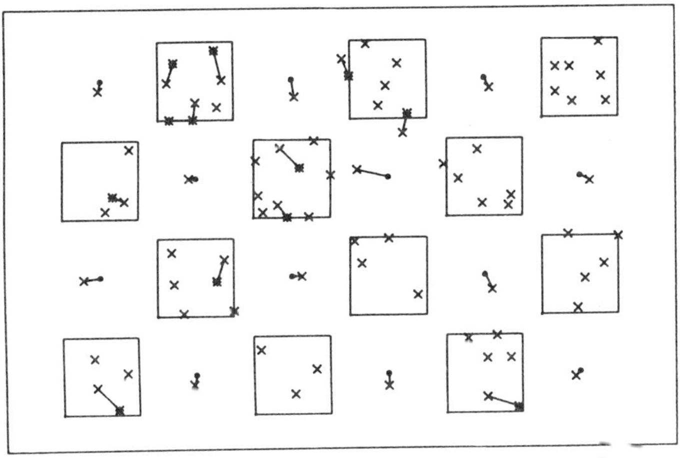

Figure 1.27 Semi-systematic sampling of the anemone data

Using the definitions given in Table 1.10, we find that $H_F = 455/1125 = 1/2.473$. The value 2.473 exceeds the two-sided 2.5% point (2.27) of an $F_{24, 24}$ distribution. Consequently the value of H_F provides strong evidence of regularity (because $H_F < 1$; clustering would be indicated by $H_F > 1$). Stronger evidence of regularity is provided by the second Hopkins statistic $H_N = 3.456/12 = 0.288$. This is to be compared with the normal distribution with mean 0.5 and variance $1/(12 \times 12)$, so that it is convenient to calculate $z = 12|0.288 - 0.5| = 2.544$ which corresponds to a two-sided tail probability of about 1%. Since the observed value of H_N is *less* that the mean of 0.5, this value provides significant evidence of regularity as opposed to clustering. The greater significance provided by the H_N test is in accord with the findings of Byth and Ripley (1980).

We next calculate the various T-square statistics, using the definitions given in Table 1.10. The 12 point–plant–plant distances involved in these calculations are illustrated in Figure 1.28. The dotted line in this figure indicates the inner region within which the points have been randomly sampled. It is unwise to utilize the full sample plot, since a point near the edge might have, as its closest plant, a plant *outside* the actual sample plot. By sampling within an inner frame we avoid this difficulty.

The values of the four T statistics, T_F, T_N, T_P, and T_E, are respectively 1/1.592, 0.406, 5.88, and 1.158. It is immediately clear that the T-square sampling procedure has led to a marked reduction in the significance of the values of T_F and T_N as opposed to those of H_F and H_N—in fact none of T_F, T_N, or T_P is significant at the two-sided 10% level. The statistic T_E fares rather better, since its value only just exceeds the lower-tail 5% point (1.151) of its distribution under the hypothesis of randomness (see table in Appendix A).

In view of the relative success of T_E, it is of interest to consider its free sampling version, E, which is a function only of point–plant distances. Since all the alternative statistics have involved the calculation of a total of 24 distances, we augment the 12 sample points used for T by an additional 12 randomly sampled points from within the inner region of Figure 1.28. The values that result are $\sum X^2 = 1080$ and $\sum X = 150.35$ so that $E = 24 \cdot (150.35)^2/1080 = 1.147$. For a random distribution of points, the lower 5% value of the distribution of E is 1.151 (see table in Appendix A); consequently, since our observed value is more

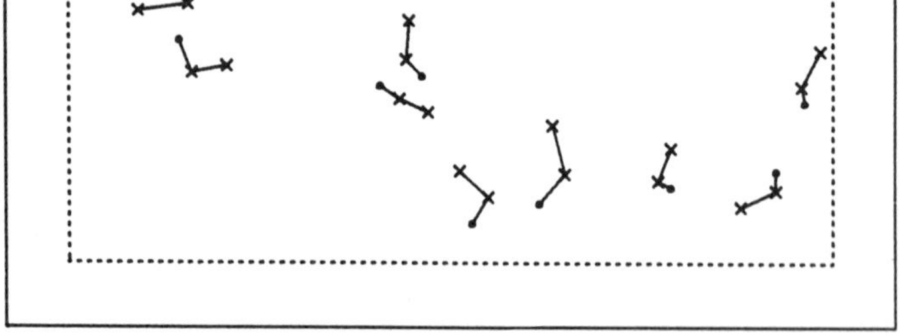

Figure 1.28. T-square sampling of the anemone data

extreme than this, the statistic has succeeded in giving us a mild indication of the regularity that is undoubtedly present.

The remaining statistics of interest are $\bar{R}$, M, $\bar{A}_x$, and $\bar{A}_y$. Of the 12 plant–plant distances in Figure 1.28, four are more than twice the size of the corresponding point–plant distance, and the effective sample size is reduced from 12 to 8. The exact value of $\bar{R}$ is difficult to calculate, but the approximation given by equation (1.32) greatly eases matters. Using this approximation we have $\bar{R} = 0.7055$ so that $z = \sqrt{96}|0.7055 - 0.5| = 2.013$, providing significant evidence at the two-sided 5% level of a departure from randomness. The value of M was 0.5252. The values of $\bar{A}_x$ and $\bar{A}_y$ for this sample of 12 point–plant distances were 170.693 and 244.580, respectively, so that the test statistic is $z = \sqrt{24}(-73.887)/415.273 = -0.872$, which is nowhere near significant.

Example 1.9

As an example of a pattern that appears to show clumping of 'plants', Figure 1.29 illustrates the distribution of archaeological sites containing funnel-necked beakers in a small area of the south Polish loess zone. Figure 1.29 has been redrawn from Hodder and Orton (1976, p. 90) who report the earlier analysis by Hodder (1975) of the data that was first published by Kruk (1973). Hodder and Orton conclude that while groupings of sites do occur, the pattern suggests apparent rather than true contagion (i.e. heterogeneity rather than clusters). They argue that the intensity of the underlying Poisson distribution may vary with the position of rivers and the pattern of site retrieval.

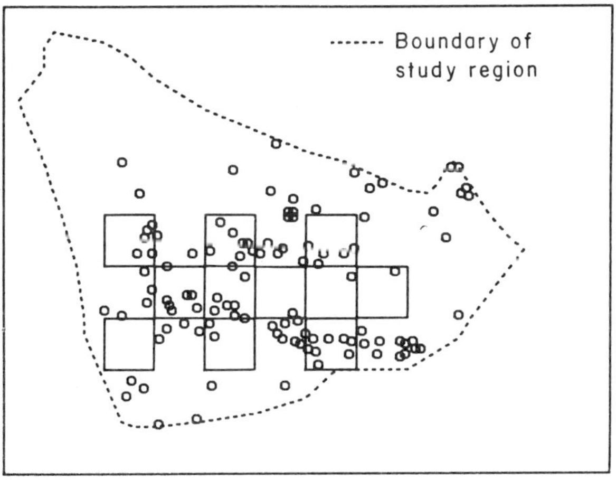

Figure 1.29 Distribution of beakers in southern Poland. (Redrawn from Hodder and Orton, 1976. Reproduced by permission of Cambridge University Press)

The dotted line in Figure 1.29 indicates the boundary of the study region, but clearly the sites are not distributed at random throughout this region and we concentrate on the central rectangle indicated by the nine squares shown in the figure. These are the squares used for semi-systematic sampling and have been arbitrarily located in the region where sites are plentiful (and not, for example, in the north-west corner). The semi-systematic sampling leads to $X^2 = 11\,225$ and $W^2 = 5250$ so that $H_F = 2.14$, which is a little less than the critical 5% point (2.24) of the $F_{18,18}$ distribution. The value of H_N is 0.588, giving a z-value of 0.913 which is clearly not significant. This is in agreement with Byth and Ripley's (1980) findings that H_F is the better detector of aggregation.

The values of the remaining statistics are $T_F = 1.02$, $T_N = 0.57$, $T_P = 14.70$, $T_E = 1.351$, $E = 1.284$, $\bar{R} = 0.348$, and $M = 0.092$ (with $n' = 7$). Also, $\bar{A}_x =$ 4310.963 and $\bar{A}_y = 1886.76$, so that $\sqrt{2n}(\bar{A}_x - \bar{A}_y)/(\bar{A}_x + \bar{A}_y) = 1.189$, which is not significantly large when compared to the values to be expected from a unit normal distribution. Indeed, the only test statistic that approaches significance is T_P, which is very close to 15.51, the upper 5% point of a χ_8^2 distribution. In view of Hodder and Orton's conclusions concerning the possible presence of heterogeneity in the distribution of these sites, it is of particular interest that Diggle (1977a) suggested the use of T_P especially as a means of detecting heterogeneity as opposed to clusters (for which he recommended T_N).

Of course the failure of any of the statistics to determine a significant departure from a random distribution of 'plants' is simply a reflection of the very small sample size that has been used. The eye detects the river-valley effects on the site locations as a result of observing the locations of all 109 sites—our samples have looked at just 18 positions. If we double the sample size we obtain $T_N = 0.509$ (clearly not significant) and $T_P = 54.22$ which far exceeds the upper 0.1% point of a χ_{17}^2 distribution. There is now overwhelming evidence in favour of heterogeneity rather than clusters being the appropriate description of the pattern.

Example 1.10

As a final example, data has been extracted from the paper by Hill (1973) concerning the locations of drumlins (material deposited by glaciers) in Northern Ireland. Hill considers an area of some 2500 km in County Down. From his paper we have extracted the data referring to a square of side 8 km situated in the northern end of the Upper Ards peninsular. The positions of the 235 drumlins in this region are indicated in Figure 1.30. Hill conducts a trend surface analysis of his data and concludes that there is some evidence that the drumlins occur in north-east to south-west bands. A portion of what may be such a band is suggested by the distribution of points in the figure.

As with the previous analyses we commence by calculating the Hopkins statistics, using in this case a 4×4 grid, giving eight sampling squares. The resulting values are $H_F = 1/1.639$ and $H_N = 0.344$, both of which provide non-significant indications of regularity. All the other statistics confirm this im-

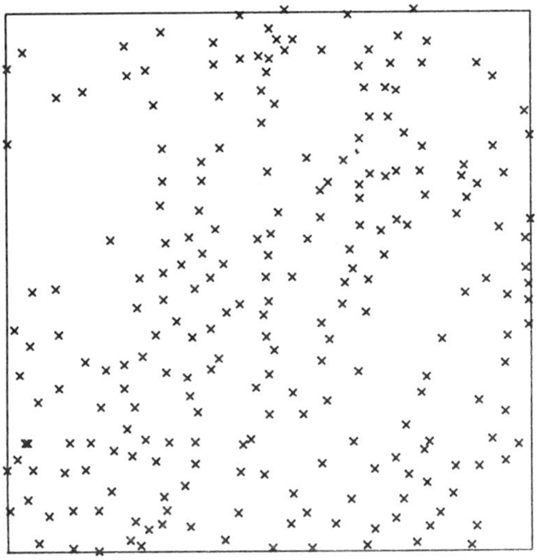

Figure 1.30 Drumlins in a 8 km square of the Upper Ards peninsular, Co. Down. (After Hill, 1973. Reproduced by permission of Association of American Geographers from the Annals of the Association of American Geographers, Vol. 63, 1973, pp. 229–241)

pression. For example, T_P, which we might expect to provide evidence of heterogeneity in view of Hill's own findings, takes the value 10.0 for a sample of 16 point–plant–plant measurements, which is certainly not a significantly large value to have observed by chance from a χ^2_{15} distribution. We must conclude that, for this restricted region, there is no evidence of other than a random distribution of 'plants'.

Recommendations

The results of our three analyses serve to confirm the earlier empirical studies: the H_N and H_T tests are powerful tools for detecting departures from randomness. Our experience suggests that it would be wise to use rather larger samples than we have used, although a sample of size 20 should be ample to reveal any pattern. If it is desired subsequently to estimate plant intensity (see Chapter 2), then much larger samples could be used, since intensity estimation with only 20 sample points is scarcely worth while.

If we have already calculated the Hopkins statistics, then there seems to be relatively little extra information that can be gained from the other statistics, although there is clearly an advantage, for clustered data, in also calculating the values of T_P and T_N as a possible means of distinguishing between true and apparent contagion. Nevertheless, probably the best advice is to calculate as many as possible of the statistics, remembering that while the tests are not entirely independent, each examines the data in a subtly different fashion.

It is also possible to construct tests based on other facets of the data and these tests (as Marriott, 1981, observes) may be both simpler and more powerful against specific alternatives. Marriott observes that, for hard-core models, the best estimator of the core diameter is the minimum plant–plant distance, so that, for a test of randomness against a hard-core alternative, it suffices to count the number of point–plant distances which are less than this minimum plant–plant distance. The significance of the result could be tested using the Monte Carlo approach.

1.5 THE USE OF MAPPED PLANT LOCATIONS

In this section we examine various techniques that have been developed specifically for the situation in which the entire 'plant' map is available for inspection. Of course, if there are a great many plants in the population, then the sample techniques discussed in the previous section will be very efficient at detecting departures from randomness, and it would also be possible to use the quadrat counting techniques discussed in Sections 1.2 and 1.3. However, none of these techniques would be very successful for a small population of plants, partly because a sample from a small population does not amount to much (!), but also because with a small population the assumption that sampled plant–plant distances are independent observations ceases to be even approximately correct (because of the shortage of plants).

We shall consider three broad classes of techniques: those that utilize plant–plant or point–plant distances, those that involve the counting of plants in various sizes of plot, and those that involve a head-on attempt at the study of pattern via the description of the shape and size of the areas surrounding individual plants.

In Section 1.1 we discussed various more-or-less realistic models of plant distribution as alternatives to the rather unbelievable null hypothesis of randomness. When the complete plant map is available it becomes feasible to consider going a stage beyond significance testing and considering the fitting of alternative models. This work is still very much in its infancy, but clearly holds out the promise of our being able to draw more far-reaching inferences from a map than were possible hitherto.

Mapping Methods

The ease with which objects can be mapped depends largely upon the size of their images as perceived by the observer. Thus, both the positions of deformed blood cells on a glass slide as seen through a microscope and the (two-dimensional) positions of stars as seen through a telescope are relatively easy to record, whereas the positions of trees in a forest as recorded by an observer walking through that forest will be liable to considerable error—a case of not being able to see the wood for the trees! In the latter situation a possible solution may be to use an aerial photograph in place of the ground-based observer, though this will not suffice if it is the understorey which is of interest.

The most common procedure for ground-based mapping involves triangulation which, in a forest, involves considerable effort, and is not outstandingly accurate. Mosby (1959) states that the distance accuracy to be expected, using reconnaissance-type mapping techniques, is only 1 part in 80 (4.5°) for a hand-held compass and 1 part in 300 (1.2°) for a staff-held forester's compass, whereas distances measured using a tape can be expected to be accurate to 1 part in 5000.

The increased accuracy obtained by the direct measurement of distance suggested to Rohlf and Archie (1978) that a more accurate alternative to triangulation would be provided by using the geometric construction illustrated in Figure 1.31. In the figure, A and B represent two points whose coordinates (x_a, y_a) and (x_b, y_b) have been previously determined, while P is the point of interest. The distances (d_{ab}, d_{ap}, d_{bp}) between the three points are measured, and from these distances the angle θ is determined using equation (1.36):

$$\theta = \tan^{-1}\{(y_b - y_a)/(x_b - x_a)\} - \cos^{-1}\{(d_{ab}^2 + d_{ap}^2 - d_{bp}^2)/2d_{ab}d_{ap}\}. \quad (1.36)$$

Notice that the formula is not symmetric in the a and b subscripts, and is written for the case where the point A is on the left-hand side of an observer stationed at P. With this proviso the angle θ is in the range $(-90°, 90°)$. The coordinates (x_p, y_p) of the new point P are then given by

$$x_p = x_a \pm d_{ap}\cos\theta, \qquad y_p = y_a \pm d_{ap}\sin\theta, \quad (1.37)$$

with the minus sign being relevant when θ itself is negative.

One problem with this procedure is that errors are likely to accumulate if the points A and B are themselves points whose locations have been determined by calculation. The best solution would seem to be to refer new points, wherever possible, to the same two fixed central marker points. When these become too

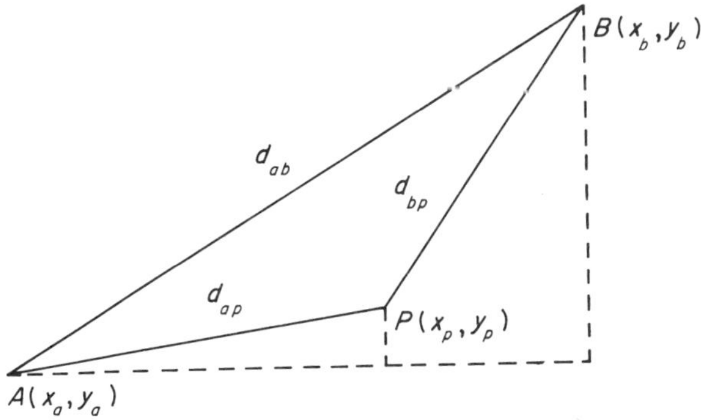

Figure 1.31 Determination of coordinates from distance measurements

remote, a local 'centre' (with calculated coordinates) should then be used, so that the error accumulation is minimized.

Rohlf and Archie (1978) suggest using *three* fixed marker points, so that the location of each point can be calculated three times over from the three possible triangles, with the result that the average of the three estimates may be expected to be still more precise. They also give details of a somewhat complex error minimization procedure, based on the principle of least squares, which seems to fix locations with improved accuracy.

Example 1.11

Suppose that the points A and B are fixed marker points with coordinates (1, 0) and (0, 1), respectively, and that the true but unknown position of the point P is given by the coordinates (0.63, 1.19). The distance between A and B is $\sqrt{2}$ by construction, but we presume that the other distances are only correct to 2 decimal places. This is a rather pessimistic presumption by comparison with Mosby's (1959) expectation of an accuracy of 1 part in 5000. The true values of d_{ap} and d_{bp} are 1.2462... and 0.6580.... So we take $d_{ap} = 1.25$ and $d_{bp} = 0.66$. Using equation (1.36) we have

$$\theta = \tan^{-1}(-1) - \cos^{-1}(3.1269/3.5355) \simeq -72.82°,$$

so that

$$x_p = 1 - 1.25\cos(-72.82°) = 0.6308,$$

and

$$y_p = 0 - 1.25\sin(-72.82°) = 1.1942,$$

which are evidently very close to the true values, (0.63, 1.19).

If we introduce a third marker point, C, with coordinates (0, 0) then we have $d_{cp} = 1.35$, compared to the true value of 1.3465.... We now have two extra triangles from which to estimate (x_p, y_p). Using the triangle ACP we obtain (0.6300, 1.1940), while from the triangle BCP we obtain (0.6310, 1.1935), leading to an 'average' position of (0.6306, 1.1939). It is noteworthy that the three estimates do *not* bracket the true value, as might have been expected.

Edge effects and interdependence

In the previous section we were examining only a relatively small proportion of a large population and it was therefore reasonable to assume that the resulting plant–plant and point–plant distances were all independent of one another. When the population is very small this will not be the case, however. In a careful examination of the problem, De Vos (1973) cites as a counter-example the pathological case of a population comprising just two plants, A and B, for which the distance from A to its nearest neighbour (B) will be identical to the distance from B to its nearest neighbour (A). The consequence of this interdependence is that all test statistics will have a somewhat greater variance than would otherwise have been the case.

The major difficulty with all the nearest-neighbour techniques is the influence of the boundary of the study region (cf. Haggett, 1980). The expected nearest-neighbour distance for a plant near the boundary will be greater than for a point well inside the region, because the former is denied the possibility of neighbours on the outside of the boundary. In the examples at the end of the previous section we attempted to circumvent the problem by restricting attention to an inner region of the study area, which was surrounded by a 'buffer zone'. Plants in the buffer zone were not selected by the randomization procedures, but were allowed to be the nearest neighbours of selected plants. In this way we hope to have located correctly all the relevant nearest neighbours, but at the expense of wasting a lot of information—since the buffer zone may well contain a considerable proportion of the total number of plants.

There is one simple solution to the edge-effect problem which can be employed whenever the study region is rectangular (or square). This solution effectively involves the invention of an edgeless representation of the study region by imagining that the study area was wrapped around a toroid, which is an object shaped like a lifebelt or a serviette ring. With this representation we imagine that, simultaneously, the top and bottom of the rectangle are joined together and so are the two sides, so that the nearest neighbour of a plant in the north-east corner of the rectangle may be a plant in the south-west corner!

An equivalent effect is obtained by reproducing the observed pattern a further eight times and then surrounding the original pattern with the eight copies to form a 3×3 array, as shown in Figure 1.32. The idea of this toroidal edge correction dates back at least to Fraser and van den Driessche (1972), while

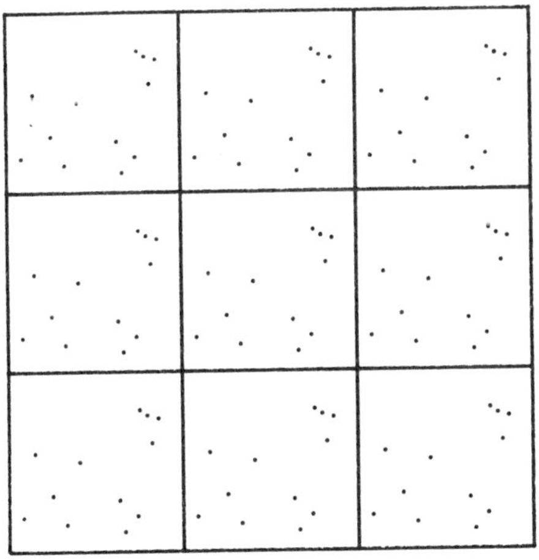

Figure 1.32 A representation of a toroidal edge correction for a rectangular region

Ripley (1979a) shows that it performs well. The representation given in Figure 1.32 suggests the interpretation that, since the observed rectangle was a random sample of all the rectangles that might have been observed, our best guess at the appearance of any other rectangle will be that it looks identical to our sampled rectangle.

Tests based on distance measures

All the tests in the previous section that dealt with sampled point patterns were based upon the distribution of *squared* point–plant or plant–plant distances, whereas the most widely used (and misused) of all the test for randomness is based directly upon the distribution of W, the plant–plant nearest-neighbour distance. This test is due to Clark and Evans (1954), who derived the test in the following fashion. The probability density function for W, by analogy with that given by equation (1.23) for the point–plant distance, X, is given by

$$g(w) = 2\rho\pi w \exp(-\rho\pi w^2) \qquad (1.38)$$

where ρ is the mean intensity of plants per unit area. By straightforward integration we obtain:

$$\mathrm{E}(W) = 1/2\sqrt{\rho} \quad \text{and} \quad \mathrm{E}(W^2) = 1/\rho\pi, \qquad (1.39)$$

so that

$$\mathrm{var}(W) = (4-\pi)/4\rho\pi. \qquad (1.40)$$

Applying the central limit theorem, Clark and Evans then argued that $\bar{w}$, the sample value of the nearest-neighbour distance, could be regarded as being an observation from a normal distribution with mean $1/2\sqrt{\rho}$ and variance $(4-\pi)/4N\rho\pi$, where N is the number of observations in the sample.

Since the test requires knowledge of ρ, which is best estimated by N/A, where A is the total area of the study region, the Clark–Evans procedure has generally been applied to the complete population of plants in the study area. Because of the combined influences of the edge effects and interdependence, this is not a valid procedure. Clark and Evans recognized the edge-effect problem and recommended the use of a buffer zone, but this has not always been utilized in published applications of the technique. A number of suggestions have been made as to possible remedies, culminating in the work of Matern (1972) and Donnelly (1978). Donnelly used a combination of numerical integration and simulation techniques on a variety of regular and irregular study areas and suggested that

$$CE = \{\bar{w} - \mathrm{E}(\bar{W})\}/\mathrm{var}(\bar{W}) \qquad (1.41)$$

could be considered to be an observation from a unit normal distribution, providing that the plant distribution was random and that the expectation and variance were calculated from

$$\mathrm{E}(\bar{W}) = 0.5\sqrt{A/N} + 0.0514P/N + 0.041P/N^{3/2} \qquad (1.42)$$

and

$$\operatorname{var}(\bar{W}) = 0.0703A/N^2 + 0.037P\sqrt{A/N^5}, \tag{1.43}$$

where P is the length of the perimeter of the study region.

Pielou (1977) notes that the Clark–Evans test measures the extent to which the intensity varies from place to place, rather than providing information concerning the sizes of the patches of different intensity. Cliff and Ord (1981a), referring to the uncorrected version of the test, state that it is not very powerful, whereas the separate studies by Ripley (1979a), using the corrected version of the test, and by Diggle (1979a), using a Monte Carlo procedure, both suggest that the procedure is pleasingly powerful at detecting either regular or clustered distributions.

Brown (1975) and Brown and Rothery (1978) considered three test statistics based on squared nearest-neighbour distances, and these are included in the comparative power study conducted by Ripley (1979a). Ripley's results suggested that the Clark–Evans test is generally superior to all three, except in the case of regular alternatives, where Brown's B statistic can prove to provide an extremely powerful test. In the form used by Ripley, the B statistic is

$$B = N\left(\prod_i w_i^{*2}\right)^{1/N} \Big/ \left(\sum_i w_i^{*2}\right). \tag{1.44}$$

This statistic is evidently the ratio of the geometric mean of the quantities $\{w_i^{*2}\}$ to their arithmetic mean. To understand the definition of w_i^* it is first necessary to observe that the value of B would be unaffected by replacing w_i^{*2} by πw_i^{*2} in both the numerator and denominator of equation (1.44). Now πw_i^{*2} looks like the area of a circle and, for most i, that is exactly what it is—in fact for all interior plants $\pi w_i^{*2} = \pi w_i^2$, where w_i is the distance from the ith plant to its nearest neighbour. For an edge plant, however, the circle centred on that plant (plant j, say), with radius w_j, will overlap the edge of the region, as illustrated in Figure 1.33, and when this occurs πw_j^{*2} is the area of that part of the circle that lies within the region, so that w_j^* is the *effective* nearest-neighbour distance. The statistic B was first proposed by Moran (1951) in the context of renewal processes.

No adequate approximation to the sampling distribution of any of these statistics is known. Consequently, it is necessary to use a Monte Carlo approach. The value of B is calculated for each of 99 separate realizations of the random distribution of N plants within the study region. The actual observed value of B is then compared with these and, if the observed value is sufficiently extreme, then the hypothesis of randomness is rejected. If a significantly high value of B is obtained, then we have evidence (Brown, 1975) of *local* regularity. The data could, for example, be located in widely spaced regular clusters (as in Figure 1.4a).

A quite different style of distance test has been proposed by Ripley and Silverman (1978), based on the use of the smallest distances between pairs of objects, and is appropriate to the case where the alternative to randomness is that the points display the type of regularity manifest by hard-core or Strauss

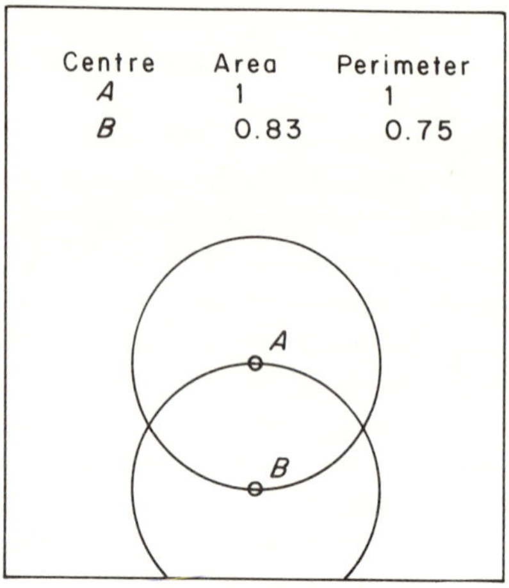

Figure 1.33 The effect of the edge on a peripheral point

processes—compare the theoretical patterns of Figure 1.9 with the schools pattern in Figure 1.34 below.

Silverman and Brown (1978) showed that under the null hypothesis of randomness for the arrangement of n objects over a region of area A, the statistic

$$S_i = n(n-1)\pi d_i^2/A \tag{1.45}$$

will have an approximate χ^2_{2i} distribution. Here d_1 is the smallest of the distances between objects, d_2 is the second smallest, and so forth. Ripley and Silverman (1978) suggest that attention be confined principally to d_1 and to d_3. Their findings are confirmed by the work of Saunders *et al.* (1982). Upper and lower bounds to the distribution of S_i in the case of a hard-core model are given by Westcott (1982).

If the alternative to randomness is the model of non-overlapping discs of radius r, then an estimate of r is provided by

$$\hat{r} = \{d_1^2 - 2A/n(n-1)\pi\}^{1/2}, \tag{1.46}$$

though this is not a very precise estimate. The approximate 95% confidence bounds, r_L and r_U are given by

$$r_L = \{d_1^2 - 7.4A/n(n-1)\pi\}^{1/2}, \qquad r_U = \{d_1^2 - 0.05A/n(n-1)\pi\}^{1/2}. \tag{1.47}$$

***Example* 1.12**

In order to illustrate the application of the test based on distance measures we reanalyse two data sets previously studied by Pinder and Witherick (1972). The data refer to the positions of tors (hills) on the Bodmin Moor granitic boss and, by way of contrast, to the locations of junior schools in Southampton. The data are illustrated in Figures 1.34a and 1.34b, which have been redrawn from those given in the source paper. The calculations that follow are based on the illustrated data, rather than on the original data.

The dotted lines in the figures represent idealized versions of the boundaries of the study areas. In both cases the true boundary (see Pinder and Witherick, 1972) has a jagged edge, whereas the idealized boundary forms a convex polygon (a figure with straight sides and with no vertex indented). The choice of a convex polygon was motivated by the desire to use the exceedingly powerful computer program, TILE4, developed by Green and Sibson (1978). A brief description of the capabilities of this program is given in Appendix B. The vast majority of the calculations and geometrical constructions that are included in the remainder of this chapter were performed with the aid of this program.

It is not clear what sort of pattern we should expect to see displayed by the Bodmin tors since a case could be made for either clustering or regularity (they actually appear to be clustered in the figure): consequently, any test of the null

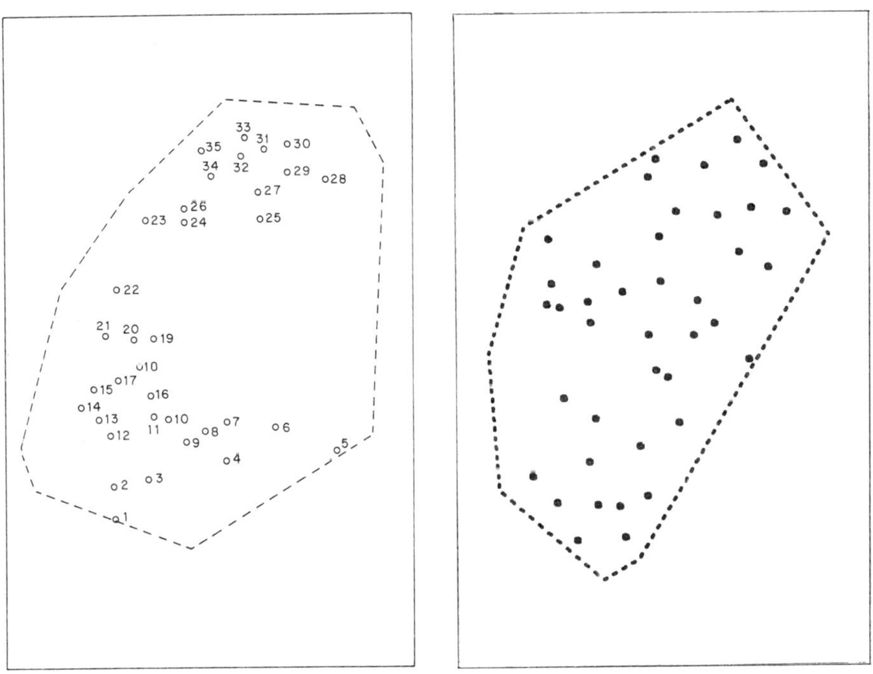

(a) (b)

Figure 1.34 (a) Tors on Bodmin Moor. (Redrawn from Pinder and Witherick, 1972.) (b) Locations of Southampton junior schools. (Redrawn from Pinder and Witherick, 1972.) (Reproduced by permission of the Geographical Association)

hypothesis of randomness must pay equal attention to both tails of the test statistic. For the junior schools, on the other hand, clustering is inconceivable—we strongly expect the regularity that is apparent in the figure, and consequently only one tail of the distribution of the test statistic is relevant.

The area, A, of the convex polygon enclosing the Bodmin data, is calculated by TILE 4 as being 194.45 km, with the perimeter, P, being of length 52.96 km. Using equations (1.41)–(1.43) we find that $\mathrm{E}(\bar{W}) = 1.2668$ km compared to the observed value of $\bar{w} = 1.0914$ km, while var $(\bar{W}) = 0.01493$ km, so that the test statistic, CE, has the value -1.435. The negative value indicates a departure from randomness towards uniformity, but this is not a very significant departure since the tables show that the probability of a unit normal random variable having an absolute value greater than 1.435 is about 15%. The absolute value is relevant because, being a two-tailed test, we must study both tails of the distribution.

For the schools data, we find that $A = 51.44$ km and $P = 28.55$ km, so that $\mathrm{E}(\bar{W}) = 0.6083$ km and var$(\bar{W}) = 0.00301$ km. The observed value of $\bar{w}$ is 0.7244 km, which, as expected, is considerably greater than the value that would be expected under the hypothesis of randomness and this leads to $CE = 2.117$. The probability of a unit normal random variable exceeding 2.117 is only about 1.7% (one-tailed test), so that we are indeed able to conclude that there is strong evidence of regularity in the data.

We turn now to the values of Brown's B statistic. The calculation of the modified areas πw^{*2} is performed automatically by the TILE4 program and we obtain values for B of 0.783 for the tors data and of 0.826 for the schools data. To interpret these values we require knowledge of their probability distributions under the null hypothesis of a random arrangement of 'plants'. These distributions are summarized, in cumulative form, in Table 1.12. It will be seen that both values are considerably greater than the means of their respective distributions, indicating some *local* regularity in each case. In the case of the tors this might, if it were significant, be explained by recognizing that, within those areas of the Bodmin granitic boss in which tors occur, there will be some local inhibition. As it happens, the value obtained for the tors is not significant at the 5% level, since it was exceeded by 7 of the 99 arrangements and is therefore estimated as

Table* 1.12 *The cumulative distributions of B for the two data sets based on* 99 *random arrangements in each case

Value of B	0.30	0.35	0.40	0.45	0.50	0.55	0.60	0.65	0.70	0.75	0.78
Freq. $\geq B$	99	96	93	86	72	59	44	33	16	15	7

(b) Schools

Value of B	0.30	0.35	0.40	0.45	0.50	0.55	0.60	0.65	0.70	0.75	0.80
Freq. $\geq B$	99	97	95	86	80	65	44	20	7	1	0

being in the top 8% of the distribution. The value for the schools data, however, was the most extreme value calculated and is clearly significant.

The results for these two data sets confirm Ripley's (1979a) findings concerning the usefulness of B as a detector of regularity. They also serve to stress the fact that different test statistics put the emphasis on different characteristics of any departure from randomness. For the tors data, the Clark–Evans statistic emphasized the clumping of the tors, whereas Brown's B statistic emphasized the regularity within those clumps.

Using the Ripley–Silverman (1978) 'quick test' we find, for the tors data, $d_1^2 = 0.2788\,\text{km}^2$ and $d_3^2 = 0.4711\,\text{km}^2$ giving $S_1 = 5.36$ (which is less than 5.99, the upper 5% point of χ_2^2) and $S_3 = 9.06$ (which is less than 10.64, the upper 10% point of χ_6^2), so that we have no reason to reject the hypothesis of randomness. For the schools data, we find $d_1^2 = 0.0946\,\text{km}^2$ and $d_3^2 = 0.1988\,\text{km}^2$ giving $S_1 = 9.02$ (which is greater than 7.38, the upper 2.5% point of χ_2^2) and $S_3 = 18.94$ (which is greater than 18.55, the upper 0.5% point of χ_6^2), so that, in this case, we must certainly reject the hypothesis of randomness in favour of that of regularity.

Using equations (1.46) and (1.47) we find that an estimate of r, the radius of the discs in the non-overlapping disc model, is 0.24 km with approximate 95% confidence bounds being 0.13 and 0.30 km. We could interpret this as implying that schools were located so that children should not live more than about one-quarter of a kilometre from their local junior school.

Refined nearest-neighbour analysis

The two distance measures discussed in the previous subsection considered the distribution of nearest-neighbour distances in a somewhat superficial fashion. Diggle (1979b) has suggested that a more penetrating test would be obtained by looking at the complete *distribution function* of either W, the nearest-neighbour distance, or X the distance from a sampling point to the nearest plant: Diggle refers to these techniques as constituting a *refined* nearest-neighbour analysis. The suggestion to base a test on the distribution of W had been made earlier by Roder (1975). We commence with a description of the test based upon W.

Suppose that we have n plants distributed within a closed region R. If the plants occur at random, then the probability density of the nearest-neighbour distances is that given by equation (1.35) and the corresponding cumulative distribution function is

$$P(W \leq w) = G(w) = 1 - \exp(-\rho\pi w^2). \tag{1.48}$$

Diggle suggested that a suitable test statistic would be

$$d_w = \sup |G(w) - \widehat{G(w)}|, \tag{1.49}$$

where sup $(\cdot)$ means the greatest value attained (as w varies) and $\widehat{G(w)}$ is the observed distribution function for the sample points, calculated in the fashion described below. To perform a test based on d_w, it is necessary to generate a sequence of 99 (say) random configurations of n plants within the region R, and to compare the value of d_w actually obtained with the values of the 99 alternatives. If the observed value is unusually large, then this is evidence that the data shows some departure from non-randomness at a distance equal to the corresponding value of w. If $\widehat{G(w)} > G(w)$, then this is an indication of clustering, whereas if $\widehat{G(w)} < G(w)$, then this indicates regularity. Variants on the d_w statistic have been suggested by Saunders *et al.* (1982).

A complementary test statistic to d_w is given by the corresponding statistic for the distribution function of X:

$$d_x = \sup |F(x) - \widehat{F(x)}|, \tag{1.50}$$

where, for a random pattern, by analogy with $G(w)$, $F(x) = 1 - \exp(\rho\pi x^2)$. Here again, an unusually large value of d_x signifies that the observed pattern departs significantly from randomness.

Diggle (1979a) has investigated the power of these two statistics and finds that both perform well and can distinguish from randomness patterns that display either clustering or regularity. His results suggest that d_x is a particularly successful detector of clusters, while d_w is particularly successful as a detector of regularity.

Calculation of $\widehat{G(w)}$ or $\widehat{F(x)}$ is not as straightforward as one would suppose because of the necessity to allow for edge effects. An appropriate estimate of $G(w)$ is given by

$$\widehat{G(w)} = \frac{\#(b_i > w \geq w_i)}{\#(b_i > w)}, \tag{1.51}$$

where '#' should be read as 'the number of plants for which', and b_i and w_i are respectively, the distances from plant i to its nearest boundary and to its nearest neighbour. A corresponding formula holds for the calculation of $\widehat{F(x)}$.

The references to the distance to the boundary that occur in equation (1.51) have the effect that plants (or, in the case of $\widehat{F(x)}$, points) adjacent to the boundary have their w- (or x-) values ignored in the subsequent calculations. A convenient procedure for calculating $\widehat{G(w)}$ is as follows.

(i) Compute the values of w and b for each plant, and suppose that $w > b$ for k of the n plants. Ignoring these k w-values we have a total of $(2n - k)$ values, of which $(n - k)$ are w-values, $(n - k)$ are b-values, and the remaining k are b-values for which the w-value has been discarded. It is useful to distinguish these, so we refer to them subsequently as c-values.

(ii) Arrange the $(2n - k)$ values in increasing order of magnitude as the first column of a table having eight columns whose headings are value (v), value label (b,c, or w), $\#(b \leq v)$, $\#(c \leq v)$, $\#(w \leq v)$, $\widehat{G(v)}$, $G(v)$, and $|\widehat{G(v)} - G(v)|$. In the column headed $\#(b \leq v)$ insert 1 opposite to the smallest v which is labelled b, 2 opposite to the next smallest b-value, and so forth, repeating the procedure in an analogous fashion for $\#(c \leq v)$ and $\#(w \leq v)$.

(iii) Compute $G(v)$ using equation (1.45) and $\widehat{G(v)}$ using

$$\widehat{G(v)} = \frac{\#(w \leq v) - \#(b \leq v)}{n - \#(b \leq v) - \#(c \leq v)}. \tag{1.52}$$

(iv) Compute the values of $|\widehat{G(v)} - G(v)|$ and locate the maximum, which is then the observed value of d_w.

Calculation of $\widehat{F(x)}$, and hence of d_x, is performed in an exactly analogous fashion with the point–plant and point–boundary distances being calculated for each of the m sample points. Two questions remain to be answered, so far as $\widehat{F(x)}$ is concerned: How large should m be? How should the sample points be arranged within the study region? The answers to these questions are not cut and dried. Diggle (1979b) favours a regular grid of sample points, and this seems much the simplest solution, though Ripley (1981) advocates a stratified arrangement. So far as the number of points is concerned, Diggle (1979a, 1979b) has chosen $m = n$ and this choice is ratified by the recent theoretical study of Diggle and Matern (1980). On the other hand, Ripley (1978) has suggested that m should be at least 100 and preferably equal to $10n$; later (Ripley, 1981) he has suggested that m be chosen to have a value between $10n$ and 1000. The pragmatic answer must be to choose a convenient value of m, at least as large as n and preferably greater than 100, but consistent with a balance between the cost of the computer time and the value of the results.

The distribution function $F(x)$ has been defined as the probability of there being one or more plants within a distance x from a randomly chosen point. Suppose that there are no plants within a distance x of a point P. This implies that if we were to draw circles of radius X around each plant, then not one of those circles would overlap P: this situation is illustrated in Figure 1.35.

The inner region of Figure 1.35 is surrounded by a guard area of width x (so that the guard area varies with x). The proportion of the inner area that is shaded provides an unbiased estimate of $F(x)$—although this is a somewhat impractical estimation procedure. However, this view of $F(x)$ provides an insight into its behaviour. For clustered points it is obvious that the shaded circles will have a huge amount of overlap so that $\widehat{F(x)}$ will be much less than expected, whereas with regularly spaced points $\widehat{F(x)}$ will be greater than expected. To summarize:

Regularity	*Clustering*
$\widehat{F(x)} > F(x)$	$\widehat{F(x)} < F(x)$
$\widehat{G(w)} < G(w)$	$\widehat{G(w)} > G(w)$

In cases where regularity is present it will be useful to plot $\rho\pi x^2 - \widehat{F(x)}$ against x. This is because if N shaded circles of radius x do not overlap, then the total shaded area will be $N\pi x^2$ and the shaded proportion of the total area, A, will be $N\pi x^2/A = \rho\pi x^2$, which is therefore the maximum value that $\widehat{F(x)}$ can achieve. For a hard-core situation (see Section 1.1) with non-overlapping discs of radius r we would have $\widehat{F(x)} = \rho\pi x^2$ for all $x \leq r$, with $\widehat{F(x)} < \rho\pi x^2$ thereafter.

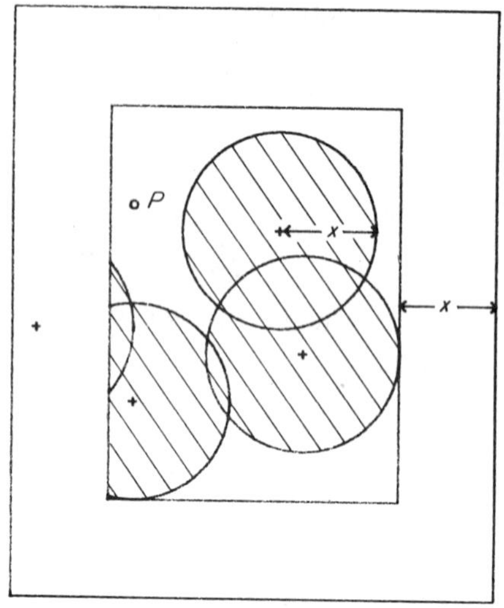

Figure 1.35 An alternative interpretation of $F(x)$

***Example* 1.13**

We now apply these refined nearest-neighbour methods to the two Pinder and Witherick data sets introduced in Example 1.12. There are six tors which are closer to the boundary region than they are to their neighbouring tor (these are the tors labelled 1,2,5,22,23, and 35 in Figure 1.34a), which leaves 64 values to be placed in order. The full results are shown in Table 1.13, from which it can be seen that $d_w = 0.243$. Since no less than 13 (see Table 1.14a) of the 99 random plant arrangements have values of d_w greater than 0.243, this test fails to distinguish any departure from randomness.

In order to calculate d_x the tors map was overlaid with a grid of equi-spaced points, of which 130 fell in the study region. The value obtained for d_x was 0.218, which is appreciably greater than any of the values obtained from the 99 different random plant arrangements. The largest of these values was 0.189 and the full set is summarized in Table 1.14b. A plot of $F(x) - \hat{F}(x)$ against x is shown in Figure 1.36a. From this figure it is clear that this statistic has detected clustering for values of x corresponding to $F(x) \simeq 0.9$, in other words for $x \simeq 2$ km. There is also a very slight suggestion of some regularity for smaller x (near 0.8 km).

Table 1.14c and d summarize the empirical distribution functions for $G(w)$ and $F(x)$ for the Southampton schools data, with the data for $F(x)$ being based on 113 grid points. For this data the observed values were $d_w = 0.265$ and $d_x = 0.199$, so that both statistics find significant evidence of non-randomness. Notice that d_x is highly significant and yet the data is clearly regular rather than clustered. This confirms that both statistics form quite general tests of non-randomness.

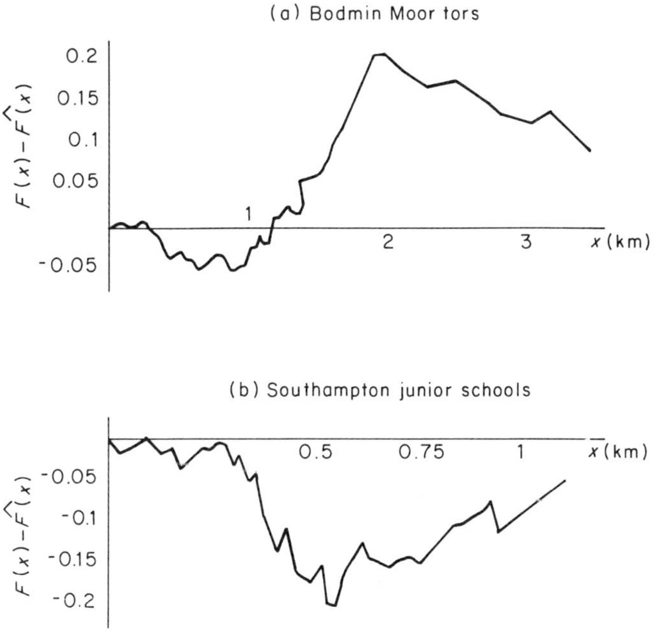

Figure 1.36 Plots of $F(x)-\widehat{F(x)}$ against x for two data sets

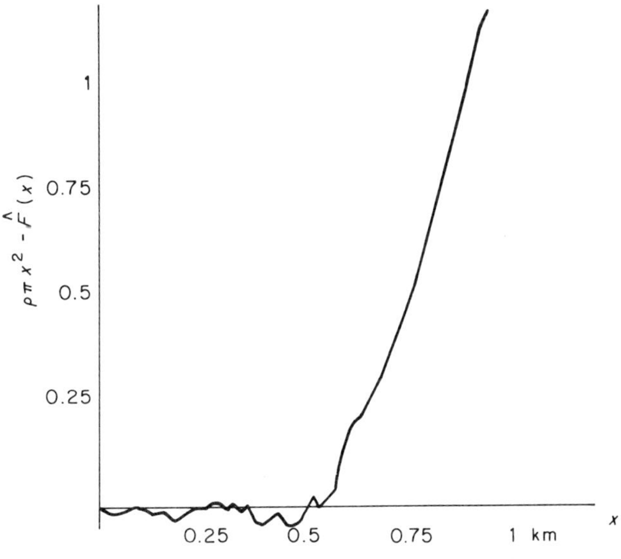

Figure 1.37 Examination of a hard-core model for the schools data

Table* 1.13 *Calculation of $\hat{G}(v)$ for the tors data

Value (v)	Value label	$\#(b \le v)$	$\#(c \le v)$	$\#(w \le v)$	$\hat{G}(v)$	$G(v)$	$\lvert\hat{G}(v) - G(v)\rvert$
0.06	c	0	1	0	0/34 = 0.000	0.002	0.002
0.24	c	0	2	0	0/33 = 0.000	0.032	0.032
0.52	$w(\times 2)$	0	2	2	2/33 = 0.061	0.146	0.085
0.60	$w(\times 2)$	0	2	4	4/33 = 0.121	0.187	0.066
0.68	$w(\times 2)$	0	2	6	6/33 = 0.182	0.234	0.052
0.78	$w(\times 2)$	0	2	8	8/33 = 0.242	0.295	0.053
0.79	c	0	3	8	8/32 = 0.250	0.302	0.052
0.82	$w(\times 2)$	0	3	10	10/32 = 0.313	0.320	0.007
0.83	$w(\times 2)$	0	3	12	12/32 = 0.375	0.329	0.046
0.84	w	0	3	13	13/32 = 0.406	0.334	0.072
0.86	$w(\times 2)$	0	3	15	15/32 = 0.469	0.345	0.124
0.92	w	0	3	16	16/32 = 0.500	0.384	0.116
0.95	w	0	3	17	17/32 = 0.531	0.405	0.126
0.97	w	0	3	18	18/32 = 0.563	0.418	0.145
1.04	w	0	3	19	19/32 = 0.594	0.461	0.133
1.06	w	0	3	20	20/32 = 0.625	0.477	0.148
1.07	$w(\times 2)$	0	3	22	22/32 = 0.688	0.484	0.204

***Table* 1.13** (*Contd.*)

Value (v)	Value label	$\#(b \leq v)$	$\#(c \leq v)$	$\#(w \leq v)$	$\hat{G}(v)$	$G(v)$	$\lvert \hat{G}(v) - G(v) \rvert$
1.08	w	0	3	23	23/32 = 0.719	0.488	0.231
1.14	w	0	3	24	24/32 = 0.750	0.528	0.222
1.20	w	0	3	25	25/32 = 0.781	0.561	0.220
1.21	c	0	4	25	25/31 = 0.806	0.566	0.240
1.25	c	0	5	25	25/30 = 0.833	0.590	0.243
1.44	b	1	5	25	24/29 = 0.828	0.699	0.129
1.46	w	1	5	26	25/29 = 0.862	0.707	0.155
1.48	w	1	5	27	26/29 = 0.897	0.716	0.181
1.57	w	1	5	28	27/29 = 0.931	0.759	0.172
1.61	b	2	5	28	26/28 = 0.929	0.775	0.154
1.80	b	3	5	28	25/27 = 0.926	0.843	0.083
1.86	c	3	6	28	25/26 = 0.962	0.864	0.098
1.87	b	4	6	28	24/25 = 0.960	0.865	0.095
1.99	w	4	6	29	25/25 = 1.000	0.897	0.103
2.00	b	5	6	29	24/24 = 1.000	0.900	0.100
⋮	⋮	⋮	⋮	⋮	⋮	⋮	⋮
4.65	b	29	6	29	1.000	1.000	0.000

See text for explanation of the symbols b, c, and w.

Table 1.14 *The cumulative distribution of d_w and d_x for the two data sets, based on* 99 *random arrangements in each case*

(a) Tors—d_w

Value of d_w	0.08	0.10	0.12	0.14	0.16	0.18	0.20	0.22	0.24	0.26	0.28	0.30
Freq. $\geq d_w$	99	94	80	69	47	38	29	19	13	10	9	3

(b) Tors—d_x

Value of d_x	0.04	0.06	0.08	0.10	0.12	0.14	0.16	0.18	0.20
Freq. $\geq d_x$	99	89	58	27	12	7	2	1	0

(c) Schools—d_w

Value of d_w	0.08	0.10	0.12	0.14	0.16	0.18	0.20	0.22	0.24	0.26	0.28	0.30
Freq. $\geq d_w$	99	91	88	71	58	48	34	24	17	4	2	1

(d) Schools—d_x

Value of d_x	0.02	0.04	0.06	0.08	0.10	0.12	0.14	0.16	0.18
Freq. $\geq d_x$	99	97	90	66	38	21	10	4	0

Figure 1.36b shows the graph of $F(x) - \widehat{F(x)}$ plotted against x. The regularity of the pattern is manifest in the consistently negative values of $F(x) - \widehat{F(x)}$. The maximum deviation occurs for x between 0.5 and 0.6 km. To obtain further insight we turn to the plot of $\rho\pi x^2 - \widehat{F(x)}$ against x, which is shown in Figure 1.37. Recall that $\rho\pi x^2$ is the maximum value attainable by $\widehat{F(x)}$ with non-overlapping circles so that negative values of $\rho\pi x^2 - \widehat{F(x)}$ reflect errors of calculation or measurement. The graph suggests that the hard-core model is (unsurprisingly) a rather good description of the observed pattern. At about $x = 0.6$ km the value of $\rho\pi x^2 - \widehat{F(x)}$ 'takes off'! This implies that the radius of the non-overlapping discs is 0.6 km or, equivalently, that there is a general policy of spacing schools by at least 1.2 km (although, of course, the two closest schools are only 0.3 km apart).

Combined count and distance analysis

Ripley (1976, 1977) introduced a statistic, $K(t)$, that combines quadrat counting with distance measurement. This 'second-order' statistic is difficult to interpret on its own, but has a simple interpretation when multiplied by the intensity parameter ρ:

$\rho K(t) = \text{E}$(number of plants within a distance t of a randomly chosen plant).

The area that lies within a distance t of a randomly chosen plant is πt^2, and hence the expected number of plants that lie within the area, for a Poisson forest with intensity ρ, will simply be $\rho\pi t^2$. Consequently, for a Poisson forest, we have

$$K(t) = \pi t^2, \tag{1.53}$$

or, equivalently, that

$$L(t) = t - \{K(t)/\pi\}^{1/2} = 0. \tag{1.54}$$

The use of the square-root transformation is due to a suggestion by Besag in the discussion of Ripley (1977).

Diggle (1979b) has shown that, when plants are distributed according to a Poisson cluster process (see Section 1.1), then the form of $K(t)$ is given by

$$K(t) = \pi t^2 + \text{E}[N(N-1)]F(t)/\rho\mu, \tag{1.55}$$

where μ is the mean of N, the number of offspring per parent, and $F(t)$ is the distribution function for the distance between two offspring of the same parent.

One definition of clustering would be that there were more plants at short distances from one another than one would otherwise have expected. Thus, for small t, the observed value of $K(t)$ would be greater than πt^2. For the Poisson cluster process we can see that, as t increases, so $F(t) \to 1$, and thus $\{\pi t^2 - K(t)\}$ will tend to a fixed 'deficit' of points of magnitude

$$\text{E}[N(N-1)]/\rho\mu.$$

Without some further assumptions, however, we shall still not be able to identify the precise process involved.

With a process that displays regularity, such as one of the inhibition processes, there will be a shortage of small inter-plant distances, and hence $L(t)$ will be positive for small t. Consequently, it is natural to consider a plot of $L(t)$ against t as a means of identifying the characteristics of the plant population. Notice that $L(t)$ and $K(t)$ involve *all* plant–plant distances and not merely the nearest-neighbour distances, and consequently we might expect to obtain rather greater insight than heretofore.

In considering the estimation of $K(t)$ we must again take account of the effect of the edge, which has a more dramatic effect on the perimeter of a would-be circle than it has on the area, as was indicated in Figure 1.33. The appropriate estimate is

$$\widehat{K(t)} = A \sum_{r \neq s} \sum k(x_r, x_s)/N^2, \tag{1.56}$$

with

$$\widehat{L(t)} = t - \{\widehat{K(t)}/\pi\}^{1/2}, \tag{1.57}$$

where, in $\widehat{K(t)}$, the sum is over all plants whose inter-plant distance is t or less. The function $k(x_r, x_s)$ is defined through the inverse function $1/k(x_r, x_s)$ which is the proportion of the perimeter of the circle centred on the plant x_r and passing through the plant x_s which lies within the region of interest. The double summation will include both $k(x_r, x_s)$ and $k(x_s, x_r)$ and, as Figure 1.33 indicates, these need not be the same.

Both Diggle (1979a, 1979b, 1981a) and Ripley (1978, 1981) have used $\widehat{K(t)}$ extensively in order to investigate the plausibility of various alternatives to randomness. Only in the case of the Poisson cluster process is the explicit form of $K(t)$ known and even here the distribution of $\widehat{K(t)}$ is unknown (as, indeed, it is with the Poisson forest itself). Consequently, all the tests based on $\widehat{K(t)}$ have been based on a comparison of the observed value of $\widehat{K(t)}$ with those values obtained from samples randomly generated from the hypothesized distribution. In cases where the Poisson forest is not an acceptable model of the plant distribution, the plot of $\widehat{L(t)}$ against t provides some guide as to the appropriate alternative. However, if the values of the parameters of the alternative distribution are adduced from the sample data, then validation is very difficult unless further independent data is available. Byth (1981) has considered statistics equivalent to $K(t)$ for testing the null hypothesis of a clustered plant distribution, while Galiano (1982) proposes calculating huge numbers of inter-plant distances and examining the changes in apparent plant intensity in circles of varying radius. As yet little is known concerning the properties of these statistics.

An obvious choice of statistic for testing the null hypothesis of a random plant distribution is

$$L = \sup |t - \{\widehat{K(t)}/\pi\}^{1/2}|. \tag{1.58}$$

Since the plant pattern supplies increasingly limited information as t increases, Ripley suggests that the range of inquiry should be limited to values of t which are less than $\sqrt{A/2}$, where A is the area of the study region.

It is best to use the Monte Carlo procedure to investigate the significance of an observed value. However, an approximate guide is given by Ripley (1978, 1979a). He suggests that approximate 5% and 1% significance points for L are $1.42\sqrt{A/N}$ and $1.68\sqrt{A/N}$, respectively.

Example 1.14

We continue our analysis of the Pinder–Witherick data sets. Figure 1.38a shows the plot of $\widehat{L}(t)$ against t for the tors data. An interpretation of this figure is that the tors pattern combines aspects of inhibition and clustering. The positive values for $\widehat{L}(t)$ for t less than 0.8 km suggest that it is very unusual for tors to occur extremely close together, while the increasingly negative values of $\widehat{L}(t)$ for larger t suggest clustering. Indeed, it can be seen from the figure that $\widehat{L}(t)$ appears to stabilize at a value of around -0.65 km, suggesting that, for all but the smallest values of t, a Poisson cluster process might be appropriate. The observed value of

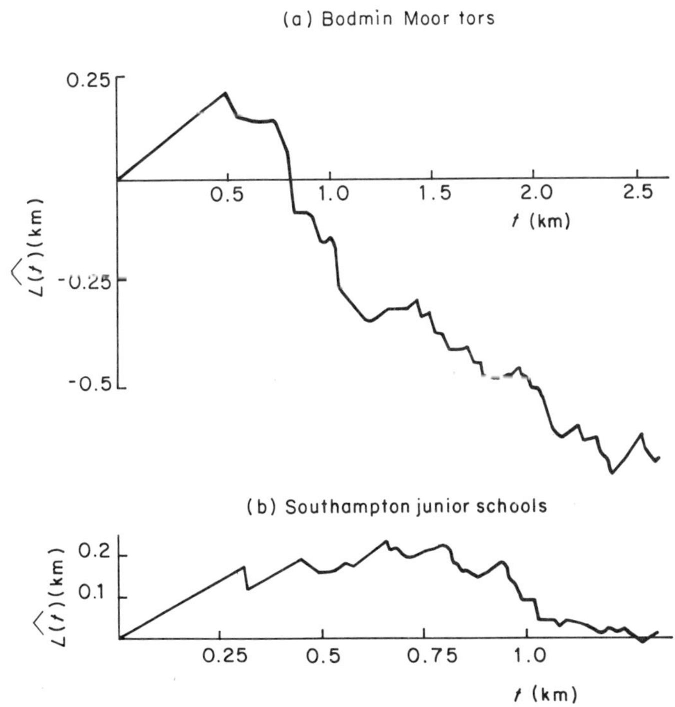

Figure 1.38 Plots of $\widehat{L}(t)$ against t for two data sets

the test statistic, L, is $(-)0.71$ which is more extreme than Ripley's approximation to the 1% value (0.67). This is confirmed by the simulation results which are summarised in Table 1.15a, for which the largest value obtained has a magnitude of 0.56.

The dependence of $\widehat{L(t)}$ on t for the schools data is illustrated in Figure 1.38b. Here it can be seen that the natural regularity of the spacing of the schools affects all inter-school distances up to a range of about 1 km, but that, thereafter, the distances appear random. This is the type of effect that we would expect to observe with one of the sequential inhibition models of Section 1.1 and confirms the conclusions of the analysis based on $F(x)$. The maximum value of L is 0.23 which does not exceed Ripley's approximate 5% value of 0.255. The simulation results given in Table 1.15b suggest that Ripley's value may be a little pessimistic.

For both these data sets it can be seen that the ability to study a range of inter-plant distances provides the useful bonus that one can detect the simultaneous presence of two or more pattern mechanisms. The method is considerably more demanding on computer time, but the results are proportionately more detailed.

Generalized quadrat analysis

A natural way to discover whether the plant intensity varies from region to region of the study area is to divide that area into its component regions and to consider the variation between the counts in these separate regions—this was the simple idea that underpinned the original David–Moore index of clumping (see equation (1.5)). Two defects with the David–Moore statistic were its dependence on the size of quadrat and on the choice of location for the quadrat grid. Recent work by Liebetrau (1977, and 1978), Liebetrau and Rothman (1977), and, independently, by Jolivet (1978, 1980) has been effectively aimed at producing a generalized form of the David–Moore statistic. Liebetrau and Jolivet have suggested slightly different, though related, statistics, and the results of Ripley (1979a) demonstrate that there is little to choose between the two alternatives. A full discussion of the underlying theory connecting these two statistics with Ripley's $K(t)$ statistic is

Table* 1.15 *The cumulative distributions of L for the two data sets, based on* 99 *random arrangements in each case

(a) Tors

Value of L	0.120	0.168	0.216	0.264	0.312	0.360	0.408	0.456	0.504	0.552
Freq. $\geq L$	99	97	89	74	49	29	14	8	5	1

(b) Schools

Value of L	0.055	0.082	0.109	0.136	0.164	0.191	0.128	0.245
Freq. $\geq L$	99	94	76	49	25	12	5	1

given by Ripley (1981). Our discussion will be restricted to the statistic first given by Liebetrau and Rothman (1977).

The essential function that emerges from the analysis of Liebetrau and Rothman is the sample estimate of the average count in a quadrat of size $r \times c$ placed at random in the study area. This average count is given by

$$M(r,c) = \sum_i \sum_j (r - |x_i - x_j|)_+ (c - |y_i - y_j|)_+, \tag{1.59}$$

where (x_i, y_i) and (x_j, y_j) are the locations of two different plants in the study area. The notation $(z)_+$ means that the value of the bracketed term is the larger of z and 0, so that there are no negative contributions to $M(r, c)$.

Of course, the value of $M(r, c)$ depends upon our choice of values for r and for c. If we had a prior expectation that the data points would cluster in approximately rectangular clumps of size $r_0 \times c_0$, then the choice of values $r = r_0$ and $c = c_0$ would be of obvious special interest. Usually, however, we shall not have any such preconception and therefore it will be simplest to let $r = c = t$, say, and to examine the way in which the value of $M(t, t)$ changes with t. By being able to vary t at will, we have overcome the second limitation of the David–Moore statistic.

Liebetrau intended his statistic for use with rectangular (or square) study areas—if the study area is irregular in shape then the test would be applied to an interior regular region. As usual it is necessary to take account of edge effects and Ripley suggests using the toroidal correction. To implement the toroidal correction, suppose that the study region is of size $R \times C$, and consider the distance between points in the x-dimension. For points i and j, this will be the smaller of $|x_i - x_j|$, $|x_i - x_j - R|$, and $|x_i - x_j + R|$, with a similar triple of possibilities, involving C, for distances in the y-dimension. Using this toroidal correction, the maximum distances between two points in the x- and y-directions are, respectively, $R/2$ and $C/2$. Consequently, using rectangular quadrats, the maximum interesting value of t will be the smaller of $R/2$ and $C/2$.

To perform a test (of the null hypothesis of randomness) based on $M(t, t)$ we make use of the normal approximation given by Liebetrau and Rothman (1977), and calculate the value of

$$z(t) = \frac{3\sqrt{A}\{2M(t,t)A - n(n-1)t^4\}}{2\sqrt{2}nt^3(A - t^2)}, \tag{1.60}$$

which has a unit normal distribution. Here A is the area of the study region (i.e. $A = RC$) and n is the number of plants in that region.

If we use a computer, it is a simple matter of calculate the values of $z(t)$ for a range of t between 0 and the smaller of $R/2$ and $C/2$. Of course, the values obtained for $z(t)$ will not be independent of one another as t varies, but the locations of any peaks or troughs in $z(t)$ will indicate the approximate sizes of clusters or of inhibitory discs.

***Example* 1.15**

The variable quadrat tests require a rectangular region. Whilst we could reanalyse the Pinder–Witherick data sets using an internal rectangle, the reader would doubtless prefer a change of topic and so we turn instead to two tree distributions analysed by Diggle (1979b). Figure 1.39a shows the locations, within a square plot of side 5.7 m, of 65 Japanese black pine saplings. The data set is due to Numata (1961) and has also been analysed by Bartlett (1964) and by Besag and Diggle (1977). All these analysts report finding the data to be compatible with the hypothesis of complete spatial randomness. By way of contrast, Figure 1.39b shows the locations within a square plot of side 75 ft, of 62 redwood seedlings which display an obviously clustered pattern. The redwood data set is due to Strauss (1975) and has been analysed subsequently by Ripley (1977) and by others in the discussion of his paper. Strauss gives details of the plot, which apparently contained the stumps of former trees. The locations of these stumps, which might be expected to be the cluster centres, were not recorded. Diggle (1979b) suggests that the data set is consistent with the hypothesis of a Matern process (see Section 1.1) having 27 parents.

For both sets of data, $z(t)$ was evaluated at equi-spaced intervals of t between 0 and the maximum possible (2.85 m and 37.5 ft, respectively), with the results being summarized in Figure 1.40. Figure 1.40a illustrates the variation in $z(t)$ with t for the black pine data (solid line) contrasted with the extremes of the variation in $z(t)$ noted for 24 groups of 65 *randomly* distributed observations. The complete plots for the 25 groups are not shown because the resulting diagram would be indecipherable! Instead, only the most extreme values (indicated by the dotted

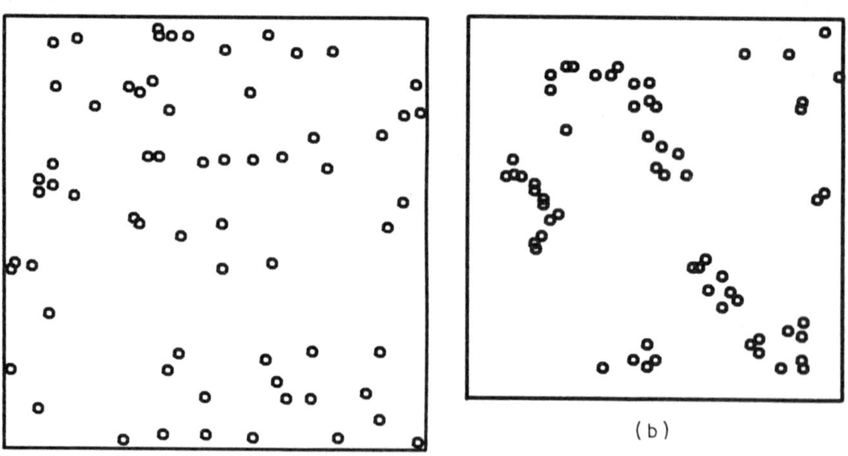

(a) (b)

Figure 1.39 Two sets of trees. (a) 65 Japanese black pine saplings. (Numata, 1961; redrawn from Diggle, 1979b.) (b) Locations of 62 redwood seedlings. (Strauss, 1975; redrawn from Diggle, 1979b. Reproduced by permission of *Biometrika* Trustees)

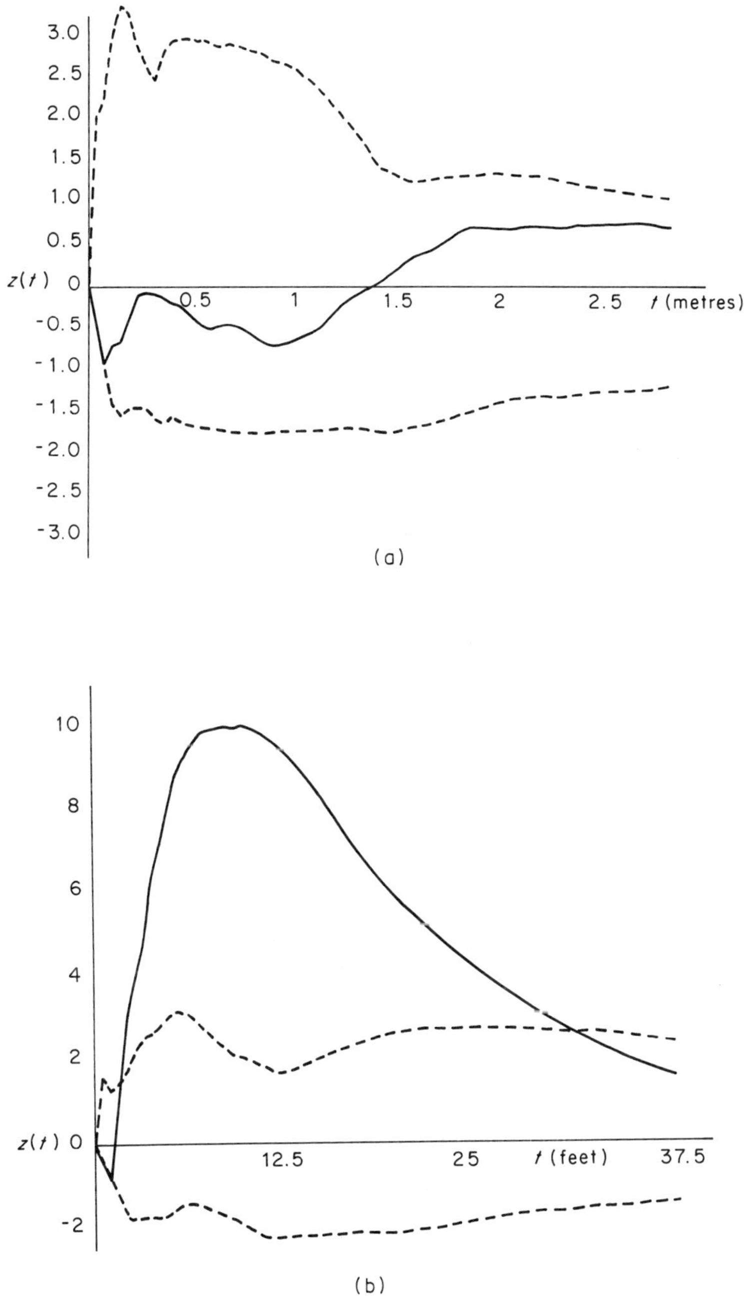

Figure 1.40 Envelopes of simulations of $z(t)$ for the trees data. (a) Black pines. (b) Redwood seedlings

lines) are given. These define the 'envelope' of the variation of $z(t)$ with t to be expected from random arrangements of points. Since the observed graph of $z(t)$ lies well within the envelope for this data, it is apparent that the black pine data is indeed consistent with the hypothesis of randomness. The maximum deviation is a value of 0.944 which corresponds to a two-tail probability of about 35%—which is certainly not surprising.

Figure 1.40b reveals quite a different story! The graph of $z(t)$ exhibits a sharp peak in the vicinity of $t = 10$, which implies that the average physical size of the clusters (whose presence cannot be doubted) is about 100 ft^2. The envelopes for both data sets show the inevitable convergence on zero as t approaches 0 and there is also some indication of tapering for large t. It would be wise therefore not to pay too slavish respect to the alleged unit normal distribution of $z(t)$: that should be regarded as being no more than an approximation.

Reflexive nearest neighbours

One possible indication of a clustered arrangement of plants would be an unusually small number of reflexive nearest neighbours (Dacey, 1969). The plants A and B are defined to be reflexive nearest neighbours if, simultaneously, A is the nearest neighbour of B and B is the nearest neighbour of A. As a generalization of this concept, A and C would be nth-order reflexive nearest neighbours if A were the nth nearest neighbour to C and vice versa.

Clark and Evans (1955) showed that, for a Poisson forest, the expected proportion of reflexive nearest neighbours was $6\pi/(3\sqrt{3} + 8\pi) = 0.6215$, while Clark (1956) showed that, for points distributed at random along a line, the proportion of nth-order neighbours is $(2/3)^n$. The exact results for a Poisson forest are given by Cox (1981) and other properties are derived by Pickard (1982). A recent example of the application of Cox's results is provided by Askew (1983).

An early advocate of the reflexive nearest-neighbour approach was Dacey (1960), but the counter-example of Porter (1960) has led to widespread doubt as to the usefulness of this technique (see Haggett *et al.*, 1977, pp. 442–445).

1.6 CIRCLES—NOT POINTS

During the previous sections we have frequently referred to models of inhibition, involving non-overlapping discs, whilst nevertheless continuing to regard the data as an array of *points* rather than discs. We might argue, with some justification, that it is the positions of the disc centres that is crucial and that, with the more sophisticated techniques, we can identify the presence of discs.

Suppose, however, that we actually *know* that we have discs and, moreover, we have detailed knowledge concerning the diameters of these discs. It must surely be helpful to feed in this knowledge as part of our data in order to obtain a more precise pattern analysis. Since this could be true for both sampled data and mapped data, our brief comments are contained in this separate section.

Simberloff (1979) considers the moments of the plant–plant distance, W, for the case of circles of radius r. He shows that the original results of Clark and Evans (1954), given as equations (1.39) and (1.40), should be modified to become

$$E(W) = r + e^{-\rho\pi r^2}\{1 - \Phi(\sqrt{2\rho\pi r})\}/\sqrt{\rho} \tag{1.61}$$

and

$$E(W^2) = r^2 + 1/\rho\pi, \tag{1.62}$$

for the case of randomly distributed circles. The function $\Phi(\cdot)$ in equation (1.61) is the unit normal distribution function. These formulae should be reasonably accurate approximations for small ρ and r. No allowance is made here for edge effects and it will be wise to use a buffer zone when computing the analogue of any of the previous test statistics. When ρ is large, the influence of 'packing problems' will be detected and the previous approximations will no longer be valid.

Simberloff also considers the distribution of W^2. We showed above (equation (1.24)) that, for n randomly distributed points, W^2 has a χ^2_{2n} distribution. Simberloff shows that, for n sparsely distributed randomly placed circles,

$$2\rho\pi(W^2 - nr^2) \sim \chi^2_{2n}, \tag{1.63}$$

which suggests that the tests of Hopkins and others should be based on $(W^2 - nr^2)$ rather than on W^2.

The point–plant distance distribution is not affected by the representation of plants as circles; consequently, provided that the distance is measured to the circle centre, we have $2\rho\pi X^2 \sim \chi^2_{2n}$, as before. Combining this result with that of equation (1.63) we obtain as an estimate of the circle radius:

$$\tilde{r} = \{(W^2 - X^2)/n\}^{1/2}. \tag{1.64}$$

***Example* 1.16**

As an example of the modified tests that result from the incorporation of knowledge concerning disc radius, we reconsider the beadlet anemone data of Example 1.8. The anemones are represented as circles in Figure 1.26 and their diameters are recorded (in units of about 0.475 cm) in Table 1.11. Inconveniently for our analysis, the anemones are not all the same size! However, the variation in radius is not very great and it is an easy matter to obtain an estimate of the average radius and to compute the adjusted values of the most effective of the earlier tests, namely the tests based on Hopkins' statistics with semi-systematic sampling.

From our previous results for the sampled data we have $\sum X^2 = 455, \sum W^2 - 12r^2 = 1075$, and $\sum\{X^2/(X^2 + W^2 - r^2)\} = 3.328$. Consequently, the revised values of the test statistics are $H_F = 1/2.36$ and $H_N = 0.277$. The changes from the original values ($H_F = 1/2.48$ and $H_N = 0.288$) are not very great, and our conclusions concerning the apparent regularity of the anemone distribution are unaltered. Since, however, we *are* now allowing for the physical size of the anemones, it is evident that size alone did not account for this regularity. Using equation (1.64) we get the estimate $r = 0.475 \times 7.5 = 3.56$ cm, which is much greater than the largest observed radius (about 2 cm). Since these anemones are aggressive beasts, we might interpret this as suggesting that each anemone occupies a territory some 12 times its own area.

1.7 SPATIAL TESSELLATIONS

The description of pattern in terms of the number of individuals in a given area or in terms of the distribution function for distances between individuals would appear to entirely miss the spatial nature of the pattern. Intuitively one expects to need to characterize a pattern by describing the *positions* of each individual relative to the positions of its immediate neighbours.

If we consider identifying 'neighbours' by joining one to another with a line, then we have commenced upon the business of subdividing the entire region of interest into sub-regions or *tiles*, with each tile being linked to one or more points. The mosaic of tiles that results is called a *tessellation*.

Unfortunately, as we shall see, the extra sophistication of the tessellation brings with it mathematical complexity so that relatively little is known of the properties of the tessellations that we shall be describing.

The Dirichlet tessellation

This tessellation is due to Dirichlet (1850), but has been rediscovered in a number of contexts and goes by a corresponding variety of names: thus 'Voronoi polygons' (Voronoi, 1909), 'Thiessen polygons' (Thiessen, 1911) 'Wigner–Seitz cells' (Wigner and Seitz 1933), 'the cell model' (Meijering, 1953), and 'the S-mosaic' (Pielou, 1977) all refer to the same tessellation. A comprehensive bibliography is provided by Boots and Murdoch (1983).

Figure 1.41 illustrates the Dirichlet tessellation for a simple case. Each individual is contained within a polygon which has the property that every point within that polygon is nearer to that individual than it is to any other individual. It is natural to regard each polygon as being the territory or service area for its own individual. Crain (1972) characterized the tessellation as corresponding to the case in which '"crystals" (two dimensional) are allowed to grow uniformly until mutual contacts prevent further growth'. In the figure, the point A has five neighbours (the points B, C, D, E, and F). Meijering (1953) observes that there are two types of neighbours, *direct* neighbours (points B, C, D, and E) and *indirect* neighbours (the point F). A neighbour is labelled as indirect if the line joining it to the point A does not cross the tile boundary at right angles. Notice that the tile boundaries are all segments of the perpendicular bisectors of the lines joining A to its neighbours.

There are three obvious properties of Dirichlet tiles which might serve as the

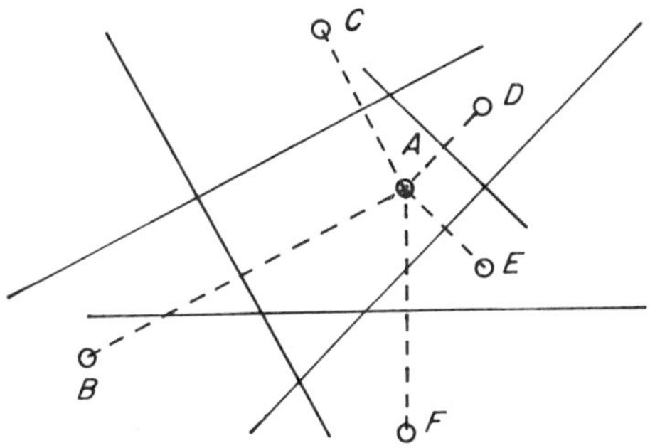

Figure 1.41 The Dirichlet tessellation

basis for a test of significance. These are the number of sides (N), the length of the perimeter (S), and the area (A), and these have, respectively, zero, one, and two dimensions. Unfortunately, the underlying mathematics for these tiles is horrendous and little is known of their properties. Meijering (1953) has shown that, for a Poisson forest, $E(N) = 6$, $E(A) = 1/\rho$, and $E(S) = 4/\sqrt{\rho}$, where ρ is the intensity of the plants. Gilbert (1962) obtained the further result that $E(A^2) = 1.28/\rho^2$, but these are the only theoretical results the have been obtained.

The result $E(N) = 6$ can easily be shown to be true for *any* point process, as follows. Connect together all neighbouring points so as to form a tessellation of triangles (the Delaunay triangles). The average interior angle of a triangle is obviously $\pi/3$ since a triangle always contains three angles that sum to π. Suppose that each point has an average of N^* neighbours. There will therefore be an average of N^* triangles meeting at each point. Now these triangles will use up the entire 2π angle available at that point and consequently the average angle per triangle will be $2\pi/N^*$. Since this must equal $\pi/3$, we have the result $2\pi/N^* = \pi/3$, which gives us $N^* = 6$. A more rigorous version of this argument was given by Collins (1972). The important feature is that it applies *whatever* the arrangement of points.

Consequently, knowledge that the average number of contacts is six is uninformative and does not, for example, suggests the hexagon-based central place theory (cf. the discussion in Boots, 1975a).

Naturally, the paucity of theoretical results has led to much empirical work. Results for 7000 random polygons are given by Vincent *et al.* (1976) for 11 000 random polygons by Crain (1972), and for a further 46 000 by Crain (1978). Pride of place goes to Hinde and Miles (1980) who report results for two *million* random polygons, none of which had more than 14 sides. Their empirical distribution for N is given in Table 1.16.

The few published studies to make use of these results have relied upon standard statistical techniques to compare the empirical results from the theoretical distributions with those observed for the given data. Vincent *et al.*

***Table 1.16** Empirical probability distribution for N, the number of sides of a Dirichlet tile for the Poisson process* (**Hinde and Miles, 1980**)

N	3	4	5	6	7	8
Frequency	22 628	214 246	518 251	588 812	398 266	180 322
Proportion	0.0113	0.1071	0.2591	0.2944	0.1991	0.0902

N	9	10	11	12	13	14	15+
Frequency	59 062	14 858	2984	493	68	10	0
Proportion	0.0295	0.0074	0.0015	0.0002			

(1976) apply the Pearson goodness-of-fit test to compare distributions of N. Boots (1973) calculates moments estimates for alternative models and compares these with the sample estimates from his data using the Kruskal–Wallis test, and later (Boots, 1975a) using factor analysis.

The concentration upon numbers of neighbours leads not unnaturally to the question as to the average number of neighbours of all polygons that border polygons having n neighbours! This might be described as a count at one stage removed.... Suggestions as to formulae for this quantity have been made by Aboav (1970, 1980), Weaire (1974), and Boots (1980), but these formulae are of doubtful merit (Boots, 1982) and we are forcefully reminded of the lament (in a different context) by Pielou (1975): 'How we might interpret observed departures from expectation is not clear.' Boots and Murdoch (1983) investigate the relations between neighbouring polygons in the case of a random plant pattern.

An alternative approach, which takes more account of the actual *shape* of the Dirichlet tiles, was described by Mead (1966). This involves the calculation of the eccicircularity and abcentricity of each tile. Mead defines *eccicircularity* as 'the extent to which the polygon is elliptical rather than circular in general shape', while *abcentricity* is a measure of the distance between the location of the plant and the location of the centroid (centre of gravity) of the tile that surrounds that plant. These statistics are calculated automatically by the TILE4 program described by Green and Sibson (1978), as are two other useful shape parameters: the radius of the smallest circumscribing circle and the radius of the largest inscribing circle. Unfortunately, nothing seems to be known concerning the distributions of any of these statistics. The TILE4 program is part of a powerful suite of spatial routines devised by Professor Sibson and available (for a small fee) on application to him at the University of Bath, U.K. A more compact algorithm, limited to Dirichlet polygons, is listed by Boots and Murdoch (1983): this algorithm consists of some 300 lines of FORTRAN IV.

The results of Hinde and Miles (1980) reveal that the distribution of tile perimeters, S, for randomly generated triangles is virtually identical to that of a normal distribution with mean $4/\sqrt{\rho}$ and variance $0.9472/\rho$, where ρ is the plant intensity. A feasible statistic upon which a test of randomness could be based would therefore be

$$U_k = \hat{\rho} \sum_{i=1}^{k} (S_i - 4/\sqrt{\hat{\rho}})^2/0.9472. \tag{1.65}$$

Here, $S_1, S_2, \ldots, S_k$ are the perimeters of a sample of k of the n tiles that subdivide the study region whose total area is A and $\hat{\rho} = n/A$ is the sample estimate of plant intensity. With a careful choice of tiles, the distribution of U_k under randomness will be approximately χ_k^2.

The choice of the k tiles is obviously restricted to tiles away from the border of the region to avoid edge effects: this precludes both tiles that abut the edge and, in some cases, internal tiles that lie close to the edge. Since every tile abuts other tiles, every edge length is part of the perimeter of *two* tiles. To avoid including any edge more than once in the calculation of U (since this would increase the variance above the theoretical value) it is necessary to select tiles that do not touch one another. A feasible scheme is to number the plants in some logical order and then to choose the tile associated with every mth plant subject to its not bordering any previously chosen tile.

An entirely different investigative procedure was suggested by Chapman (1970). Chapman observed that the distribution of the areas of the Dirichlet tiles provided information concerning the nature (i.e. clustered or random or regular) of the pattern. He suggested summarizing this information using the quantity $\sum A_i \log(A_i)$, which is simply a scaled version of the usual measure of entropy. According to Chapman, the value obtained for this quantity provides insights into the amount of organization that is present in the underlying pattern.

Example 1.17

Figures 1.42a and b show the Dirichlet tessellations for the two data sets introduced in Example 1.15. The more variable size of the redwood tiles, brought about by the clustering of the redwood seedlings, is immediately apparent.

Although there are 65 black pine sapplings and 62 redwood seedlings, the edge is close to a rather high proportion of the 'plants' and there are only 27 and 28 tiles that are unaffected by the possible presence of neighbouring trees outside the sample plot. The distributions of N for these two restricted data sets are given in Table 1.17, together with the 'expected' frequencies obtained from the proportions given in Table 1.16. Using the Pearson goodness-of-fit test we get values of $X^2(=\sum(O-E)^2/E)$ of 2.19 for the black pines and 7.48 for the redwoods, neither of which exceeds the upper 5% point (7.81) of a χ_3^2 distribution. The test is evidently not very sensitive, but the results of Vincent *et al.* (1976) and of Boots (1977, 1982) suggest that regularity will be indicated by an excessive proportion of

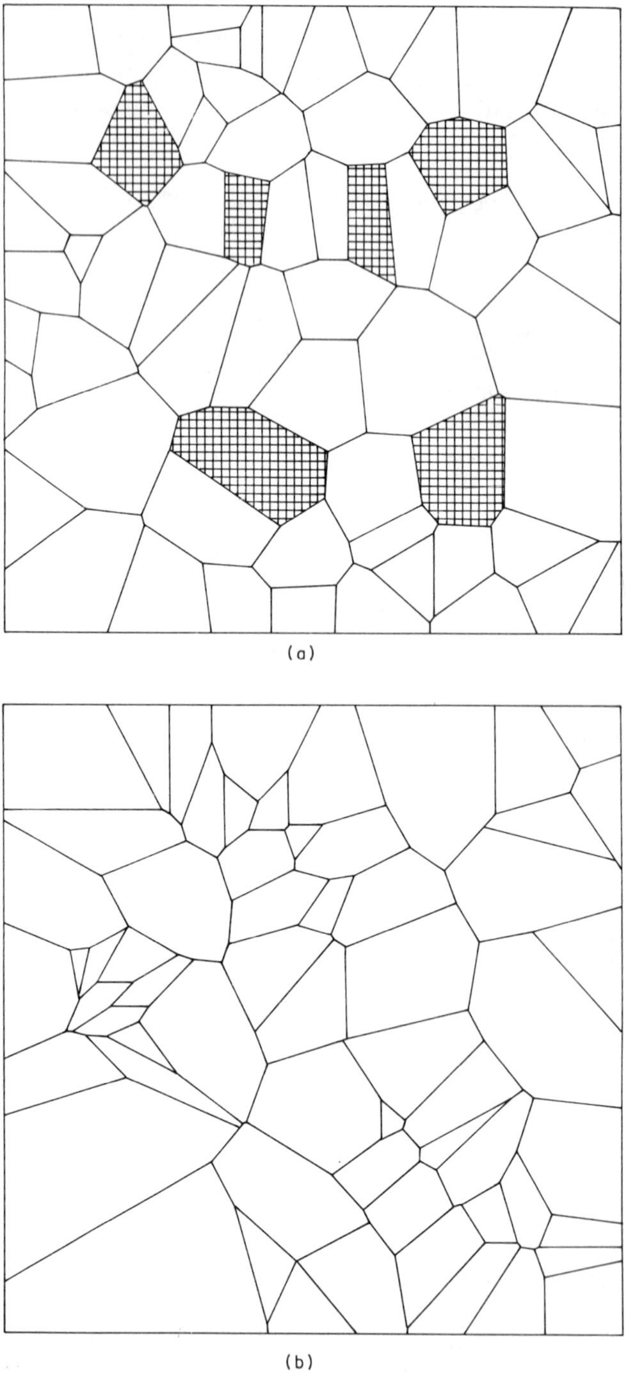

Figure 1.42 Dirichlet tessellations, (a) Black pines. (b) Redwood seedlings

Table 1.17 Numbers of sides of interior tiles for two data sets

Number of sides		≤ 4	5	6	≥ 7
Redwood	observed	8	6	7	7
seedlings	expected	3.32	7.25	8.24	9.18
Japanese	observed	2	5	11	9
black pines	expected	3.20	7.00	7.95	8.85

hexagons and clusters by an excessive proportion of triangles.

We consider next the test based on U_k (equation (1.61)). With the limited number of tiles available, we chose $m = 3$, which led to $k = 6$ for the black pines and to $k = 8$ for the redwoods. In the case of the black pines, the tiles selected are indicated in Figure 1.42a. The TILE4 program calculates individual tile perimeters automatically and thus calculation of U_k was a simple matter. We find that, for our tiles, $U_6 = 4.17$ for the black pines and $U_8 = 21.09$ for the redwoods. The first of these figures is an entirely typical value for an observation from a χ_6^2 distribution, but the latter exceeds the upper 1% point (20.09) of a χ_8^2 distribution, suggesting that this test may be a powerful method of detecting departures from randomness.

The Delaunay tessellation

We have already invoked the Delaunay tessellation in connection with the demonstration that $E(N) = 6$ for the Dirichlet tessellation. If all the neighbouring points of the Dirichlet tessellation are joined, then we have a tessellation of triangles, as illustrated in Figures 1.43a and b for the two trees data sets.

The Delaunay tessellation is in fact the dual of the Dirichlet tessellation: Miles (1970) shows that, if a region contains n_i interior polygons and n_b boundary polygons, then the Delaunay tessellation will consist of $(2n_i + n_b - 2)$ triangles.

For the case of a Poisson forest, rather more is known about the theoretical properties of this tessellation than is known about the corresponding Dirichlet tessellation, largely due to the impressive paper by Miles (1970), who examines the circumradius R, the inradius I, the triangle perimeter S, the area A, and the length L, of a randomly chosen side. His results include expressions for the correlation between any pair of these statistics, as well as for the mean and variance of each statistic—all in terms of the plant intensity, ρ. Perhaps more useful than these various moments are the joint density function $f(\alpha, \beta)$ of two randomly chosen angles from the triangle and its marginal form $f(\alpha)$, the density function of a single randomly chosen angle. These are

$$f(\alpha, \beta) = (8/3\pi)\sin\alpha\sin\beta\sin(\alpha+\beta) \qquad (\alpha, \beta > 0, \alpha+\beta < \pi), \tag{1.66}$$

$$f(\alpha) = 4[(\pi-\alpha)\cos\alpha + \sin\alpha]\sin\alpha/3 \qquad (0 < \alpha < \pi). \tag{1.67}$$

Boots (1974, 1975b) examines a 'plant' distribution by forming the corresponding Delaunay tessellation, choosing an angle at random from each triangle, and

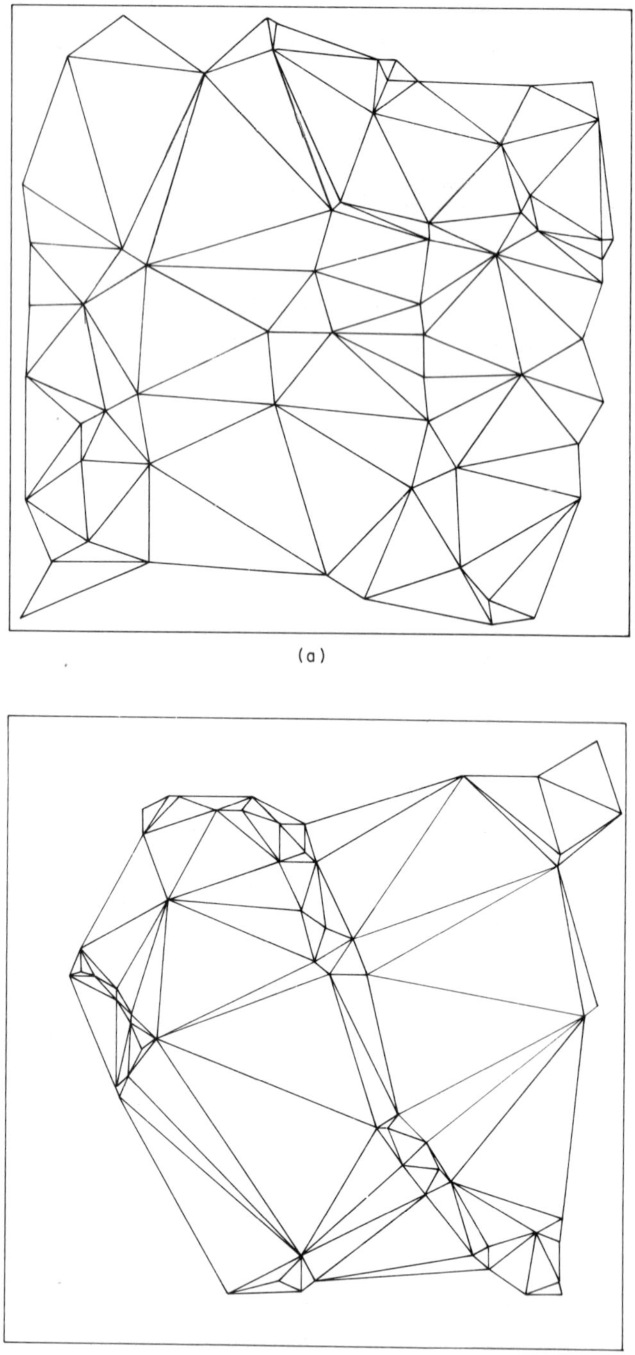

(a)

(b)

Figure 1.43 Delaunay tessellations. (a) Black pines. (b) Redwood seedlings

comparing the observed cumulative distribution with that obtained by integrating $f(\alpha)$:

$$F(\alpha) = \tfrac{1}{3}[2\sin^2\alpha + (1/\pi)(\alpha\cos 2\alpha - \tfrac{3}{2}\sin 2\alpha + 2\alpha)]. \quad (1.68)$$

Mardia *et al.* (1977) propose using the von Mises distribution, $f(\theta) \propto \exp(\kappa\cos\theta)$ as an approximation to $f(\alpha)$, and they construct a test based on the estimated value of κ. There have been no comparisons of the efficiency of these tests as detectors of departures from randomness, but the necessity (and difficulty) of choosing just one angle from each triangle is a discouraging feature.

The Delaunay tessellation is not the only triangle-based tessellation that has been suggested. Fraser and van den Driessche (1972) discuss a so-called 'least diagonal network' that also results in triangles. Since this tessellation was developed for intensity estimation, rather than for pattern analysis, we postpone further discussion until Chapter 2.

The line tessellation

One final tessellation involving straight-edged tiles is the tessellation obtained by throwing lines at random on the plane. This tessellation, illustrated in Figure 1.44, was called the *L*-mosaic by Pielou (1977). Various theoretical properties have been calculated (Goudsmit, 1945; Miles, 1964a): thus $E(N) = 4$, $E(S) = 4\pi/\rho$, $E(L) = \pi/\rho$, $E(A) = \pi/\rho^2$, and $\text{var}(A) = \pi^2(\pi^2 - 2)/\rho^4$. The lines are obtained by generating points at random, with intensity ρ, and then drawing a line through each point at a random angle. A convenient summary of the known moments is provided by Crain and Miles (1976), who also provide the distributional results obtained from a study of some 2 000 000 tiles. Of these

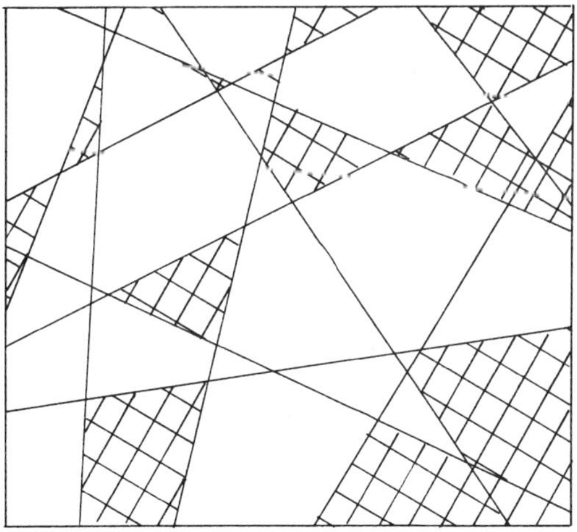

Figure 1.44 The line tessellation

***Table* 1.18 *Empirical probability distribution of N, the number of sides of polygons determined by random lines in a plane* (*Crain and Miles*, 1976. *Reproduced by permission of Gordon and Breach Science Publishers*)**

N	3	4	5	6	7	8	9	10
Frequency	37 228	39 561	20 082	6200	1380	205	26	4
Proportion	0.3556	0.3779	0.1918	0.0592	0.0132	0.0020		

empirical results the most interesting may be the distribution of N, the number of polygon sides: this is reproduced in Table 1.18.

Exact theoretical expressions have, in fact, been obtained for the proportions of triangles (Miles, 1964a) and quadrilaterals (Tanner, 1983). The theoretical proportions are (to four decimal places) 0.3551 and 0.3815 which provide close agreement with the figures obtained by Crain and Miles (1976).

With the exception of some work on paper fibres (Miles, 1964b), the model does not seem to have found much application.

Curved-boundary tessellations

The Dirichlet cells may be considered to be the end-product of the growth, at a constant rate, of scattered nuclei—the constant growth rate for each nucleus leads to the straight edges for the tiles. If, however, we allow the nuclei to grow at different rates, or we allow the nuclei to be born at different times, then we obtain a quite different tessellation, first investigated by Johnson and Mehl (1939). This tessellation has been the subject of recent investigations in a geographic context by Boots (1972, 1973) with somewhat inconclusive results. Very little is known about the properties of the model, however. It is known that cells will have, on average, less than six neighbours and a few properties have been calculated, in terms of growth rate and density, by Evans (1945). Boots (1981) developed the links between the Johnson–Mehl and Dirichlet tessellations and suggests yet another variant based upon differential weighting of the individual regions of space.

CHAPTER 2

The Estimation of Spatial Intensity

2.1 INTRODUCTION

In Chapter 1 we introduced two literal conventions involving the terms 'intensity' and 'plant' which will be maintained in the present chapter. We have used *intensity* as an alternative to density and *plant* as a general term for any object of interest. Thus, the intensity of a sample of plant locations is simply the average number of objects of interest per unit area. Later in this chapter we shall substitute the global term 'animal' in place of 'plant' for situations where this seems more appropriate. note that our use of the word 'intensity' is different to that employed by Pielou (1977, p. 155), who uses 'intensity' to describe the *variation* in plant density.

The estimation of spatial intensity is directly linked to the identification of pattern: they are, as it were, two sides of the same coin. The information collected in order to perform an analysis of pattern can also be used to provide an estimate of plant intensity. As a consequence the layout of this chapter follows that of Chapter 1. We begin by considering the advantages and disadvantages of the use of quadrats. Subsequently, we consider in turn, distance-based methods and methods that involve the calculation of the areas of the tiles comprising various tessellations.

2.2 QUADRAT METHODS

Suppose that the plants are distributed over the region of interest with an average intensity ρ. The quadrat counts obtained from n quadrats each of unit area are denoted by $x_1, x_2, \ldots, x_n$. An obvious and unbiased estimate of ρ is provided by

$$\hat{\rho}_Q = \frac{1}{n}\sum x_i, \tag{2.1}$$

and, if the total region studied has area A, then an unbiased estimator of the total number of plants in that region, T, is given by

$$\hat{T} = A\,\hat{\rho}_Q\,. \tag{2.2}$$

If we suppose that the plants are distributed at random, then the observations

$\{x_i\}$ come from a Poisson distribution with parameter ρ, and therefore

$$\mathrm{var}(\hat{\rho}_Q) = \rho/n, \qquad \mathrm{var}(\hat{T}) = A^2\rho/n. \tag{2.3}$$

However, as we noted in Chapter 1, a set of plants displaying a Poisson distribution is a comparatively rare phenomenon. If the Poisson hypothesis is not tenable, then the best estimate of variance available is s^2, the sample variance of the $\{x_i\}$. Thus, the variance of $\hat{T}$ is given approximately by A^2s^2/n.

Practical problems

There are a number of rather obvious practical problems associated with the use of quadrats. The first problem arises when the quadrat itself has an appreciable physical size, such as is the case in the sampling of forests. The problem here is that, in a forest, it may be well-nigh impossible to accurately set out the quadrat boundaries. Indeed, this is likely to be the case with any rugged or treacherous terrain.

Suppose that the quadrat has been accurately delineated. It is now necessary to make an accurate count of the individuals within the quadrat. This is often very time-consuming and may be physically demanding. Also, even with stationary objects, it may be quite difficult to achieve an accurate count. Thus, Ferns and Mudge (1981) found that about one-sixth of the nests in a gull colony were not counted in an initial count.

Table 2.1 illustrates a more unexpected aspect of the problems involved in estimating T from quadrat counts. Batcheler (1975b) was interested in censusing a small population of red deer (*Cervus elaphus*). Since the deer themselves do not wait around to be counted, censusing takes the form of counting the number of groups of pellets, where a 'group' has a strict definition, but it is up to the observer to make a judgement about what constitute distinct groups. Table 2.1 reports the values of $\hat{T}$ for an area of 1 acre which was censused using quadrats of four different sizes. It is apparent that the larger the size of quadrat that is used, the smaller is the resulting estimate of T. An identical phenomenon is reported by Ghent (1963) in connection with stands of trees.

***Table 2.1** Estimates of pellet group intensity derived from quadrats of varying sizes in varying terrain (Batcheler, 1975a, p. 646)*

	Quadrat size (acres)			
Terrain	0.0001	0.0005	0.0010	0.0100
Tall grass	5900	3020	1875	1229
Short grass	5533	2707	2006	1443
Bush	4433	1947	1487	655

A partial explanation is, doubtless, that with larger quadrats it is easier to overlook individuals—with the result that the larger quadrats give depressed estimates of T. However, Batcheler's study shows clearly that this is not a complete explanation. It would appear that a major factor is that the 'individuals' being studied are not singletons but members of groups (of pellets or samplings) although the observer may have difficulty in recognizing this fact (cf. Figures 1.4a and b). With very small quadrats it is rather difficult to distinguish the presence of a group of about the quadrat size (cf. Figure 1.3). Batcheler concludes that the estimates of T that are based on the smallest (0.0001 acre) quadrats are certainly over-estimates of T, but that, since the larger quadrats become prone to underestimation due to missed individuals, it is not clear which of the remaining estimates is the most accurate.

The data of Table 2.1 also illustrates another problem of quadrat estimation, namely that the overall plant distribution is unlikely to be entirely homogeneous over the entire region of interest. If there are well-defined distinct sub-areas, then it is clear that separate estimates of T should be obtained for each sub-area, and that the estimate of T for the entire region should be obtained as a weighted sum of the separate estimates. Thus, if the deer region consists of tall grass: short grass: bush in the proportions 1:1:2, then the small-sized quadrats give a pooled estimate of $(5900 + 5533 + 2 \times 4433)/4 = 5075$ pellet groups per acre.

However, there 'are many kinds of populations... which occur in almost unstratifiable and usually intricately interspersed [intensity] phases... Consequently, on practical grounds, the sample units have to be made small, and this entails a high chance of many plots containing no members within the lower [intensity] phases' (Warren and Batcheler, 1979). In other words, with this type of heterogeneous plant population, whilst $\hat{T}$ will remain unbiased with any reasonable quadrat allocation procedure, the variance of $\hat{T}$ may be unacceptably large, and therefore we turn to alternative procedures for intensity estimation.

2.3 THE USE OF LINE TRANSECTS

In Chapter 1 the transect was defined as 'a single row of contiguous quadrats', but a better definition might be that a transect is a section through a study area. Eberhardt (1978) identifies three general categories of methods based on the transect. He describes these categories by the terms 'strip transect', 'line intercept' and 'line transect'. Eberhardt's strip transect is simply our row of contiguous quadrats and, for the purposes of intensity estimation, there are no relevant differences between a randomly placed strip transect and a series of scattered quadrats—although the independence of the scattered quadrats is likely to give a more robust estimate of the overall intensity.

The line intercept method is a procedure to estimate plant cover in a given region. One walks along a line and notes the proportion of one's walk for which the plant in question covers the transect line. Eberhardt gives some details of the limited application of this procedure to the estimation of intensity instead of cover.

The line transect is probably the simplest and least expensive method of surveying a wildlife population. It consists simply of an observer traversing a straight path of known length, L, through the region of interest. The observer records the position of each animal detected. The problem is then to translate these records into an estimate of population intensity.

If it were true that all members of the population were certain to be visible to the observer, then there would, of course, be no problem. In practice, however, it will generally be the case that the more distant animals have less chance of being detected than have the closest animals (and the same would be true for plants in cases where the nearby plants mask those further away). The statistical problem therefore arises from the need to include the differential sighting probabilities within the estimation procedure. A comprehensive discussion both of the practical problems and of the recent theoretical developments is provided by the excellent monograph of Burnham *et al.* (1980).

The impact of detection on the estimate of intensity

In order to achieve an estimate of the average intensity, ρ, of the animal population in a region R, it is necessary to make a number of assumptions (Burnham and Anderson, 1976; Ramsey, 1979):

(i) the animals are distributed at random within the region with mean intensity ρ;

(ii) the locations of different animals are independent of one another and the detections of different animals are independent events;

(iii) measurements are made without error;

(iv) the animals may be regarded as occupying fixed positions during the survey period so that no animals are counted more than once:

(v) the probability that an animal at a distance r from the line transect will be detected is given by the function $g(r)$, which is such that

(vi) $g(0) = 1$—which implies that an animal actually on the line is certain to be detected.

Of course assumption (i) is not likely to be exactly true and, if there are obvious differences of terrain, then it will be sensible to calculate separate intensity estimates for each terrain type.

If animals appear in tight clusters, then it will be sensible to regard each cluster as a separate unit and to estimate cluster intensity rather than animal intensity. If it seems likely that detection varies with cluster size, then it will be sensible to estimate intensity separately for each size of cluster and to obtain the overall intensity estimate by an appropriate pooling of the separate results. The problems of estimating intensity when animals move before detection or are only intermittently visible (as in the case of blowing whales) are addressed by Smith (1979) and Schweder (1977).

Clearly, the form of the detection function, $g(r)$, plays a crucial role in the determination of any estimate of ρ. A very simple idea (Kelker, 1945) is to assume that

$$g(r) = 1, \quad \text{for } 0 \leq r \leq w. \tag{2.4}$$

This simply states that all animals within a distance w of the line transect are certain to be detected. If we ignore any animals that we detect at a distance greater than w, then we are in effect merely counting the number of animals, $n(w)$, in a transect of area $2wL$, so that an estimate of intensity is

$$\hat{\rho}_k = n(w)/2wL, \tag{2.5}$$

which has variance estimated by $\hat{\rho}_k/2wL$.

A variation on Kelker's procedure has been advocated by Eberhardt (1978). This is the 'jack-knifed' estimator formed by considering the subdivision of the range $(0, w)$ into the two components $(0, \frac{1}{2}w)$ and $(\frac{1}{2}w, w)$:

$$\hat{\rho}_J = \{4n(\tfrac{1}{2}w) - n(w)\}/2wL, \tag{2.6}$$

which has variance estimated by $5\hat{\rho}_J/8wL$. Of course there is an arbitrary element in the value of w chosen for these two estimators. If too large a value of w is chosen, then these estimators will underestimate the value of ρ if any animals have been overlooked, while too small a value of w will be needlessly wasteful of the data referring to sightings at distances greater than w. Eberhardt (1979) suggests that an idea of the validity of the method may be obtained by studying how it behaves as w varies. He suggests that w be chosen so that between 60 and 90% of the observations are utilized. Eberhardt also produces a generalization of equation (2.6) which allows for the unequal subdivision of the range $(0, w)$—although there seems little reason to depart from the bisection specified by equation (2.6).

Before we consider alternative specifications of $g(r)$ to that provided by equation (2.4), it is useful to consider the connection between the detection function $g(r)$, and $f(r)$, the probability density function of the distances from the central line of those animals that are actually detected.

Imagine that the line transect bisects a strip of finite width $2W$. Since the animals are assumed to be uniformly randomly distributed over the study area, the actual density function of the distance from the central line for those animals situated within the strip is simply W^{-1}. However, when we take account of the detection function $g(r)$ we find that the probability that an animal within the strip is detected is given by

$$p = W^{-1} \int_0^W g(r)\mathrm{d}r, \tag{2.7}$$

and, writing

$$\mu = \int_0^W g(r)\mathrm{d}r, \tag{2.8}$$

we have that $p = \mu/W$.

Suppose that there are actually N animals within our strip of area $2WL$ and that we observe n of these N. The value n is clearly an observation on a random variable with mean Np, while N is itself a (Poisson) random variable with mean $2WL\rho$, where ρ is the average intensity of the animals. In other words

$$\mathrm{E}(n) = 2WL\rho \cdot p = 2WL\rho \cdot (\mu/W) = 2L\mu\rho. \tag{2.9}$$

It follows from equation (2.9) that an estimate of ρ is given by

$$\hat{\rho} = \frac{n}{2L}\left(\widehat{\frac{1}{\mu}}\right), \tag{2.10}$$

and that we now require an estimate of μ^{-1}. Such an estimate is provided by the link between $f(r)$ and $g(r)$ which is simply that

$$f(r) = g(r)/\mu. \tag{2.11}$$

The particularly useful application of equation (2.11) arises because of our assumption (vi) which was that $g(0) = 1$. Setting $r = 0$ in equation (2.11) and combining with equation (2.10), we have that

$$\hat{\rho} = \frac{n}{2L} \cdot \widehat{f(0)}, \tag{2.12}$$

so that our problem becomes one of estimating the probability density at 0.

Note that $\mathrm{E}(n)$ is independent of W so that there is no harm in replacing W by ∞ in the definition of μ. Ramsey (1979) refers to μ, so-defined, as the effective half-width of the survey. Note also that an implication of equation (2.11) is that $f(r)$ and $g(r)$ must have the same general form.

Parametric descriptions of the detection function

Various alternatives to Kelker's truncated uniform $g(r)$ have been suggested in the literature although, with the exception of the work by Hayes and Buckland (1983), there has been little theoretical consideration of the probable form of $g(r)$. Thus, Eberhardt (1968) suggested the family of truncated detection functions specified by

$$g(r) = 1 - (r/k)^{\alpha}, \quad \text{for } 0 \leq r \leq k, \tag{2.13}$$

while Gates *et al.* (1968) and Gates (1969) have considered the exponential function

$$g(r) = \exp(-\lambda r), \quad \text{for } 0 \leq r, \tag{2.14}$$

for which the minimum variance unbiased estimate of ρ is

$$\hat{\rho}_G = n(n-1)/2L\sum r_i. \tag{2.15}$$

Another suggestion is that of Hemingway (1971) who notes that a more realistically shaped detection function is provided by the half-normal function:

$$g(r) = \exp(-\lambda^2 r^2), \quad \text{for } 0 \leq r, \tag{2.16}$$

for which an approximate estimate of intensity is provided by

$$\hat{\rho}_H = L^{-1}(n^3/2\pi\sum r_i^2)^{1/2}. \tag{2.17}$$

Pollock (1978) notes that the exponential and half-normal functions are both members of the Weibull family for which

$$f(r) = \lambda \exp(-\lambda^k r^k)/\Gamma(1 + 1/k), \tag{2.18}$$

where r, k, and λ are positive constants and $\Gamma(\cdot)$ is the gamma function. Pollock notes that members of this family satisfy the relation

$$1\text{n}\{f(r)\} = \ln\{f(0)\} - \lambda^k r^k, \tag{2.19}$$

which is of the general form $y = \alpha + \beta x^k$. If the data is grouped together, then an estimate of the probability density corresponding to the mid-point of a group is provided by the ratio of the number of observations in the group to the product of the group-width and the total number of observations. These pairs of values can be input into a standard computer package for non-linear regression and the values of α and β (and hence of $f(0)$) can be estimated. This procedure generalizes the method initiated by Anderson and Pospahala (1970).

Pollock gives an expression for the variance of the density estimates arising from the assumption of the Weibull family, but this is too complex to be worth reproducing here. Instead, we note the advice of Burnham and Anderson (1976) who suggested that the most useful estimates of variance are those obtained empirically as a consequence of using several independent transects. Such estimates do not rely on possibly inappropriate distributional assumptions.

Hayes and Buckland (1983) note that all the previously mentioned detection functions are entirely heuristic, without theoretical justification. They developed a hazard-rate approach to the problem. From either of two alternative sets of assumptions they arrived at the detection function

$$g(r) = 1 - \exp(-ay^{-b}), \tag{2.20}$$

where a and b (> 1) are positive constants that are to be estimated from the data. Whatever the values of a and b, this detection function has a shoulder at $r = 0$ like that of the half-normal distribution and unlike that of the exponential distribution. Despite their development of this detection function, Hayes and Buckland recommend the use of non-parametric techniques, which we now consider.

Non-parametric procedures

Although the assumption of a parametric form for $g(r)$ and $f(r)$ may make the ensuing calculations a little easier, if the assumption is wildly inaccurate then the corresponding density estimate will be distorted.

Burnham and Anderson (1976) enumerate a number of alternative non-parametric strategies. If we work with grouped data, then we can draw a histogram of the observed distribution of distances and can then fit a curve through this histogram. Alternatively, we could fit a smooth curve to the observed cumulative distribution function, and then differentiate that curve. In either case we get an estimate of $f(0)$. A similar procedure is to estimate the density function directly using a Fourier series model (Crain *et al.*, 1979; Burnham *et al.*, 1980; Buckland, 1982). With this approach the probability density function $f(r)$ is approximated by

$$f(r) = 1/w^* + \sum_{k=1}^{m} a_k \cos(k\pi r/w^*), \tag{2.21}$$

where $m \geq 1$ and w^* is a perpendicular distance beyond which all observations are discarded in the analysis. The resulting density estimate is

$$\hat{\rho}_c = \frac{n^*}{2Lw^*}\left\{1 + \frac{2}{n^*}\sum_{k=1}^{m}\sum_{i=1}^{n^*}\cos\left(\frac{k\pi r_i}{w^*}\right)\right\}, \tag{2.22}$$

where n^* is the number of retained observations. A problem with this approach is that both m and w^* are arbitrary.

A further simple suggestion made by Burnham and Anderson (1976) was that, for some i, we could estimate $f(0)$ by

$$f(0) = i/nr_{[i]}, \tag{2.23}$$

where $r_{[i]}$ is the ith *order statistic* of the observed distribution of right-angle distances $\{r_i\}$—in other words $r_{[i]}$ is the ith largest of the n observations on r. The problem here, of course, is to decide an appropriate value of i. Burnham and Anderson followed Barlow *et al.* (1972) in suggesting that one should choose that value which maximizes the estimate $f(0)$. However, this procedure is likely to be very sensitive to any inaccuracies of measurement associated with the smallest values of r. Patil *et al.* (1979, 1982) have investigated the properties of $f(0)$ and show that the value of i which provides the minimum mean square error has the form

$$i = c(f, \nu)n^{2\nu/(2\nu+1)}, \tag{2.24}$$

Table 2.2 Recommended values of c and v (equation (2.24)) for various detection functions (Patil et al., 1982)

Detection function			Parameter	
			c	ν
Truncated (2.13)	$\alpha=1$	(triangular)	2.00	1
	$\alpha=2$		1.63	2
Weibull (2.18)	$k=1$	(exponential)	1.26	1
	$k=2$	(half-normal)	1.14	2
	$k=3$		1.27	3

where ν is a positive integer and $c(\cdot)$ is a constant function dependent on the value of ν and on the form of the probability density $f(r)$. Patil *et al.* (1982) suggest that ν is unlikely to take values other than 1 or 2 and the values of ν and c that they recommend for a number of detection functions are summarized in Table 2.2.

In limited investigations of alternative detection functions with $\nu=1$, Patil *et al.* report that $c \leq 2$ in each case and that c was generally near 1. This is encouraging since it suggests that c is relatively insensitive to the form of $f(\cdot)$. On the basis of this consistency in the value of c, and since the shapes of the triangular and exponential detection functions appear less plausible than that of the half-normal detection function, Patil *et al.* suggest the general purpose estimator

$$\hat{\rho}_P = (n^{4/5} - 1)/2Lr_{[n^{4/5}]}, \tag{2.25}$$

where the -1 in the numerator is introduced to reduce bias. This estimator has the approximate variance $\hat{\rho}_P^2/n^{4/5}$. Of course, $n^{4/5}$ is unlikely to be an integer, and Patil *et al.* recommend that, so far as $r_{[n^{4/5}]}$, is concerned, the value taken should be the integer part of the number obtained. Thus, when $n=100$, $n^{4/5}=39.8$ and we therefore use the value of the 39th largest of the 100 observations on r. The stability of this type of estimate to the choice of power of n may be judged by examining the variation with k of the ratio $(k-1)/r_{[k]}$. It will be found that the value of the estimate is remarkably insensitive to the value of k, provided that k is not too near to 0 or n.

The use of sighting distances

As originally used, the line transect method did not involve the measurement of the right-angle distances $\{r_i\}$ which we have so far considered. Instead, the observer walking along the transect line merely recorded the 'sighting' distances $\{s_i\}$. These are the radial distances from the observer at which the animals were 'flushed'. Given the sighting angles $\{\theta_i\}$ (see Figure 2.1) we can, of course, convert sighting distances to right-angle distances, or vice versa, using the relation $r_i = s_i \sin\theta_i$.

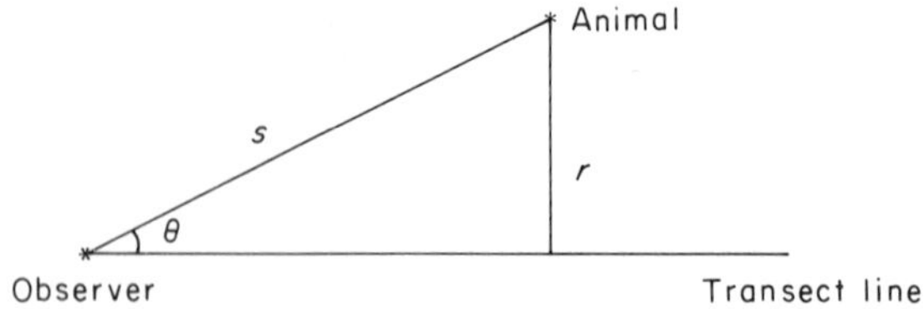

Figure 2.1 The relation between right-angle distance (r), sighting distance (s), and sighting angle (θ)

An advantage gained by using sighting distances as opposed to right-angle distances is that the former may be measured more accurately. This is most obvious in the case where s_i is very large and θ_i is very small: a rather small error in the measurement of θ_i will cause a relatively large error in r_i. This is one of the reasons why Burnham *et al.* (1980) advocate working with grouped right-angle distances (cf. Hayes and Buckland, 1983, p. 40). Clearly, there are other instances (e.g. with prominent immobile prickly pear cacti!) where the right-angle distance is both easier to measure and more accurate than the radial distance.

The traditional estimator of intensity based on sighting distances is that proposed by Hayne (1949):

$$\hat{\rho}_s = \frac{1}{2L} \sum_{i=1}^{n} s_i^{-1}. \tag{2.26}$$

The basis of Hayne's estimator is that animals are not visible until they move, which they do as soon as the observer reaches the critical distance, s_0, and that when they move they are immediately visible to the observer. From the point of view of the animal this implies that there is a circular 'no-go' area that surrounds it. The transect line is laid down at random and the animal will be observed if, and only if, the line intersects the circle.

Some straightforward calculations concerning the intersection of random lines and circles reveals that $\sin\theta$ has a uniform distribution in (0, 1), and hence that the average sighting angle under this detection model will be about 33°. However, studies by Robinette *et al.* (1974) and by Burnham *et al.* (1980) of various real-life data sets, suggest that Hayne's model is too simple and that $\hat{\rho}_s$ will be biased.

Generalizations of Hayne's estimator have been suggested by Burnham and Anderson (1976) and by Burnham (1979). The latter author proposed that the animal would be more likely to flush in cases where the observer was walking directly towards it, than in cases where the observer's course would pass the animal to one side. This gives rise to the idea of an elliptic 'no-go' area, with the major axis of the ellipse being parallel to the transect line. Hayes and Buckland (1983) note that 'none of [these] models appears to be founded upon realistic assumptions'. Following a detailed analytical investigation they conclude that a right-angle distance model should be preferred to the sighting distance approach

Table 2.3 *Line transect data* (s, θ, r) *for groups of Grant's Gazelle sampled in Tanzania, Africa, in April* 1972 (*Burnham*, 1979) (*Distances are given in metres*)

s	θ	r	s	θ	r	s	θ	r	s	θ	r
400	15	104	200	18	62	50	52	39	300	0	0
600	46	432	250	7	30	120	61	105	500	90	500
130	31	67	90	15	23	40	50	31	500	28	235
120	57	101	350	63	312	200	54	162	500	68	464
150	43	102	250	3	13	150	3	8	200	43	136
60	9	9	150	61	131	400	48	297	600	68	556
250	48	186	150	61	131	350	33	191	700	57	587
200	45	141	300	29	145	500	16	138	300	47	219
150	52	118	300	31	155	350	9	55	500	17	146
450	28	211	200	46	144	200	73	191	500	17	146
100	42	67	450	46	324	400	68	371	300	38	185
350	80	345	500	15	129	450	18	139	800	72	761
120	9	19	700	48	520	200	47	146	400	73	383
150	12	31	700	38	431	200	0	0	120	42	80
350	22	131	500	63	446	200	46	144	100	39	63

unless there are 'very good reasons [for choosing the latter, in which case] the assumptions of that model should be closely examined.'

Example 2.1

Burnham (1979) analyses some data which refer to the sighting distances and sighting angles for a line transect survey of Grant's Gazelle (*Gazelle granti* Brooke) conducted in April 1972, in Tanzania, Africa, by P. Hemingway. Methodological details of the survey are given by Hemingway (1971). The data, which are given in Table 2.3, refer to groups of gazelle rather than to individual animals, and we therefore seek to estimate the group intensity.

It is obvious from Table 2.3 that the original sighting distances were recorded to the nearest 10, 50, or 100 m, and the sighting angles to the nearest degree. The right-angle distances have been calculated to the nearest integer based on the given s- and θ-values, but, especially for large values of s, the r-values may be rather inaccurate.

A visual impression of the sightings is provided by Figure 2.2. For the purposes of this diagram it has been assumed that all the observations were made in front of, and on the left-hand side of, the observer. If Hayne's model were correct, then we should expect this diagram to show as an arc of a circle, while, if Burnham's elliptical flushing model were correct, then we should expect an elliptical outline with the long axis along the transect line. Evidently these data are not in accord with either of these models—if the outline is elliptical then it is the short axis that bestrides the transect line. This would be appropriate if the animals were more

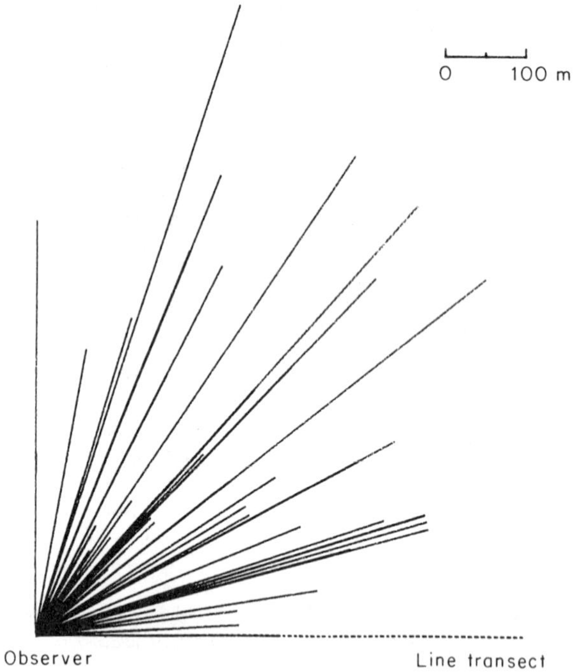

Figure 2.2 Sightings of Grant's Gazelle in Tanzania

likely to 'freeze' in position if they were being approached directly.

Figure 2.3 displays the histogram of right-angle distances. It is immediately clear that the 'natural' assumption of a detection function decreasing monotonically with distance would appear not to be the case—indeed, there is a pronounced peak at about 110 m and none of the density functions included in Table 2.2 has non-zero modes.

The probable explanation of this apparently odd result is that the animals move away from the transect line stealthily without being detected. It is worthy of note that, provided that the lengths of the pre-detection moves are not too great and provided that these movements only apply to animals that were initially in the immediate vicinity of the transect line, then these movements will have little effect on the estimates of Patil *et al.*, since these estimates are based on the values of observations in the central region of the cumulative distribution of distances.

For this data we find that $\sum r_i = 11\,538$, $\sum r_i^2 = 3\,932\,910$ and $\sum s_i^{-1} = 0.3033$, while the length of the transect was 135.1 km. From these results we have $\hat{\rho}_G = 1.135$, $\hat{\rho}_H = 0.692$, and $\hat{\rho}_s = 1.122$.

If we take $w = 300$ m, then Kelker's estimate is $\hat{\rho}_k = 0.568 \pm 0.164$, while the jack-knifed version is $\hat{\rho}_J = 0.605 \pm 0.189$. For these estimates the confidence bounds quoted are approximate 95% bounds obtained by calculating

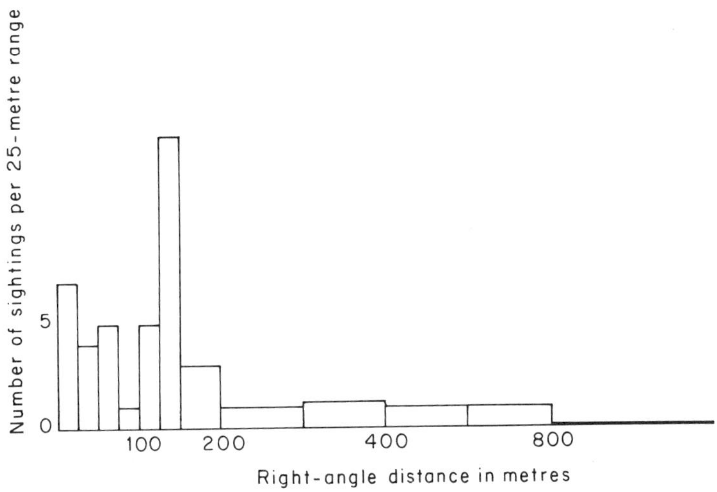

Figure 2.3 Histogram of right-angle distances for sightings of Grant's Gazelle

$1.96(\text{var}(\rho_{k,J}))^{1/2}$. The confidence bounds are correct only if the model assumed is also correct.

Turning next to the estimators of Patil *et al.* (1979, 1982), we find that $60^{2/3} = 15.33$, $60^{4/5} = 26.46$, and $60^{6/7} = 33.43$. If we use their suggested estimator, $\hat{\rho}_P$, given by equation (2.26), then we get an intensity estimate (with an approximate 95% confidence interval) of 0.719 ± 0.274. By way of comparison, the five values of c and v given in Table 2.2 give rise to the estimates (reading from top to bottom of the table) 0.778 ± 0.390, 0.500 ± 0.191, 0.664 ± 0.332, 0.765 ± 0.292, and 0.517 ± 0.175. These results seem reasonably consistent with one another, which suggests that this procedure is reasonably robust with respect to variations in the choice of parameters.

Burnham (1979) obtained the estimate 0.812 ± 0.357 for this data using a formula based on his assumption of an elliptical 'no-go' area. It is perhaps reassuring to find so similar a set of intensity estimates arising from such disparate (and implausible) assumptions. There have been few rigorous attempts at a comparison of the performance of the various estimators (an exception being that of Robinette *et al.*, 1974) and firm recommendations must await the results of further trials such as that on which Example 2.2 is based.

***Example* 2.2**

Burnham *et al.* (1980) have published data taken from the 1979 master's thesis of Laake, who conducted an investigation of line transect sampling by placing a known number of wooden stakes in a sagebrush meadow east of Logan, Utah, in

Table 2.4 ***Perpendicular distances (in cm) from a transect length 1 km, for a sample of wooden stakes with known intensity 37.5 stakes per hectare***

9	20	40	40	41	45	53	92	100	123	127	131	147	161
167	202	244	290	295	305	305	312	317	317	325	340	347	359
361	379	386	396	408	440	441	453	485	489	497	566	605	608
625	642	647	650	660	710	767	768	793	816	827	849	904	910
933	1005	1040	1071	1159	1182	1372	1423	1524	1815	1816	1860		

the United States. An example of a set of data collected by Laake is given in Table 2.4.

For these data $\sum r_i = 38\,536$ and $\sum r_i^2 = 35\,733\,370$ so that $\hat{\rho}_G = 59.1$ and $\rho^H = 37.4$, compared with the true intensity, ρ, which was known to be 37.5 stakes per hectare. The Kelker estimates are $\hat{\rho}_K = 31.9 \pm 8.7$ and $\hat{\rho}_J = 48.1 \pm 10.9$, where here the 95% confidence intervals are based on the assumption of the underlying model being correct. Burnham *et al.* (1980) used the Fourier series approach and obtained the estimate 39.0 which is remarkably close to Patil's non-parametric $\hat{\rho}_P = 39.1 \pm 14.2$.

Once again the Patil approach appears both accurate and appealingly simple and it is clear that in the face of the present evidence this must be our recommended estimation procedure.

2.4 DISTANCE METHODS

In Chapter 1 we saw that the most sophisticated measures of pattern were those derived from distance measurements. The distances concerned were either those between a randomly chosen plant and its nearest neighbour, $\{W_i\}$, a randomly chosen sampling point and the neighbouring plant, $\{X_i\}$, or the distance between that plant and its nearest neighbour—either unrestricted $\{Y_i\}$ or restricted to the half-plane opposite to the sampling point $\{Z_i\}$ (see Figure 1.24).

It is obviously convenient and efficient if the measurements collected in order to analyse the plant pattern can also be used to obtain an estimate of the plant intensity. Since distance measures are especially informative concerning plant pattern it is natural to hope that they will be equally useful for the present purpose. In Section 2.1 we remarked that the estimation of intensity using the counts from randomly placed quadrats provided a theoretically robust method of intensity estimation—although there were a number of interesting practical problems. We must now remark that those very properties of distance measurements that make them useful for pattern identification make them correspondingly less useful for the estimation of overall intensity. However, in recent years there have been considerable efforts put into the construction of robust distance-based intensity measures and, in this section, we consider these new estimators roughly in order of increasing complexity.

Point–plant methods

Most early methods of estimating plant intensity were based on the distance from randomly selected sampling points to one or more of the nearest-neighbouring plants. The basic results are directly related to the earlier equations (1.24)–(1.28) and we shall use the same notation.

Let $X_{(r)i}$ be the distance from the ith sampling point to the rth nearest-neighbouring plant. We showed in Chapter 1 that, for a Poisson forest, $X^2_{(r)i}$ may be regarded as an observation from a chi-squared distribution with $2r$ degrees of freedom. This result led Moore (1954) to consider the maximum likelihood estimator

$$\hat{\rho} = n \Big/ \pi \sum_i X^2_{(1)i}, \tag{2.27}$$

where n is the number of sample points. Moore showed that this estimator was biased and suggested using the unbiased estimator $(n-1)\hat{\rho}/n$, for which he obtained an expression for the variance. A generalization of Moore's results is provided by Pollard (1971) who showed that there is a family of unbiased estimators summarized by

$$\hat{\rho}_{(k)} = (nk-1) \Big/ \pi \sum_i X^2_{(k)i}, \tag{2.28}$$

for which

$$\text{var}\,(\hat{\rho}_{(k)}) = \rho^2/(nk-2). \tag{2.29}$$

However, as Pollard remarked, these estimators are sensitive to the underlying assumption of a random distribution and become grossly in error if there is a severe deviation from randomness.

An early suggestion that attempted to counteract the problems of the heterogeneity of ρ across the plant pattern was that of Morisita (1957). He suggested the estimate

$$\hat{\rho}_M = 2 \sum_i \{X^{-2}_{(3)i}\}/n\pi. \tag{2.30}$$

Byth (1982) reports that this estimator is not as robust as some more complex estimators that we shall consider below. However, her investigations suggest that, for highly clustered plant distributions, the estimator

$$\hat{\rho}_B = 3n/\pi \sum X^2_{(3)i} + (1 - 3/\mu) \tag{2.31}$$

works well. In this expression μ is the average number of plants in a cluster and the expression is only recommended for cases where $\mu \geq 3$. For $\mu = 3$, the expression is essentially that of equation (2.28). It is of interest to note that both Byth and Morisita connect the distance to the *third* nearest plant with the presence of a clustered pattern. For the cases that Byth considers the variance of $\hat{\rho}_B$ would appear to be about $\hat{\rho}_B^2/5n$, although this may not always be the case.

In the previous section, which was concerned with intensity estimation based on line transect data, we discussed the procedure developed by Patil *et al.* (1979, 1982). This procedure was developed initially for the present situation, with the distances being the $\{X_{(1)i}\}$-values. Patil *et al.* (1982) recommend the estimator

$$\hat{\rho}_P = (n^{2/3} - 1)/n\pi\{X_{(1)[j]}\}^2, \tag{2.32}$$

where $X_{(1)[j]}$ is the jth largest of the n measured values of $X_{(1)}$ and j is the largest integer less than $n^{2/3}$. The choice of 2/3 as the power of n in this expression is somewhat arbitrary, but, in practice, the estimator is remarkably insensitive to the power chosen, provided that j is not too close to either 0 or n. The one proviso with Patil's procedure is that it should not be used for cases where 'tight clusters' occur. The estimator has approximate variance $\hat{\rho}_P^2/n^{2/3}$ which, it should be noted, is appreciably larger than that of any of the other estimators that we shall consider.

A somewhat more complex estimator based on the distances from a random sampling point to both the nearest and the second-nearest-neighbouring plants has been suggested by Lewis (1975). This

$$\hat{\rho}_L = 16n^2/\{6\sum X_{(2)i} - \sum X_{(1)i}\}^2. \tag{2.33}$$

Lewis's theoretical results suggest that this estimator should be robust against deviations towards regularity. The estimator appears to have a variance of about $\hat{\rho}_L^2/n$.

Combined point–plant and plant–plant calculations

The most informative of the procedures for the investigation of plant patterns are those procedures which involve both point–plant and plant–plant calculations. This is because departures from randomness (towards either clustering or regularity) affect the two types of distance in opposite fashions.

When the plants are clustered, most nearest-neighbour plant–plant distances will be small. Our estimator of intensity, $\hat{\rho}_W$ (however calculated), based on these distances alone, will surely over-estimate the *average* intensity for the entire area because it will, in effect, assume that the intensity *within* clusters applies to the region as a whole. On the other hand, the estimator of intensity, $\hat{\rho}_X$ (however calculated), based on point–plant distances alone, will surely underestimate the average intensity by assuming that the intensity *between* clusters applies to the region as a whole.

Satyamurthi (1979) argues that, for a clustered process, one should be content to report the pair of values $\hat{\rho}_X$ and $\hat{\rho}_W$, but this would not help us to estimate T, the total number of plants. A better policy might be to regard each cluster of plants as being represented by a 'super-plant' situated at the centre of gravity of its cluster. One then calculates the point–super–plant distance $\{X_i'\}$, and an intensity estimator $\hat{\rho}_{X'}$. In addition, note is made of the number of plants per super-plant, $\{n_i\}$, which have mean μ. Provided that the *clusters* of plants are distributed at random, we have the *overall* intensity estimator

$$\hat{\rho}_U = \mu \rho_{X'}. \tag{2.34}$$

If we work with $\hat{\rho}_X$ and $\hat{\rho}_W$, however, then it is natural to consider as estimates of overall intensity either their arithmetic mean $(\hat{\rho}_X + \hat{\rho}_W)/2$ or their geometric mean $(\hat{\rho}_X \hat{\rho}_W)^{1/2}$. This is the approach adopted by Byth (1982), who considered the three possible intensity estimators $kn/\pi \sum d^2_{(k)i}$, $an/(\sum d_{(k)i})^2$ and $b/\pi d^{*2}_{(k)}$. Here a and b are appropriate constants depending on k, $d^*_{(k)}$ is the median of $d_{(k)1}, d_{(k)2}, \ldots, d_{(k)n}$, and $d_{(k)i}$ itself might be any of $X_{(k)i}$, $Y_{(k)i}$, or $Z_{(k)i}/\sqrt{2}$, where $Z_{(k)i}$ is the distance from the ith random sampling point to its kth nearest neighbour using T-square sampling (see Figure 1.28).

Altogether Byth (1982) considered a total of 40 different combined estimators of 'average' intensity, including those estimators that had been recommended earlier by Diggle (1975, 1977b). She found that the most robust of these composite estimators was

$$\hat{\rho}_E = n^2 \Big/ \left\{ 2\sqrt{2} \left(\sum_i X_{(1)i} \right) \left(\sum_i Z_{(1)i} \right) \right\}. \tag{2.35}$$

Byth considered a variety of artificially constructed heterogeneous or clustered patterns (but no inhibitory or hard-core patterns): her results suggest that $\mathrm{var}(\hat{\rho}_E) \simeq \hat{\rho}_E^2/n$. Note that, for strongly clustered processes, Byth found the estimator $\hat{\rho}_B$ (given earlier by equation (2.31)) performed in a superior fashion.

Cox's procedures

Cox (1976) has proposed two rather complex intensity estimators, of which the simpler is defined by

$$\hat{\rho}_0 = \frac{4}{n\pi} \sum_i{}' (Y_i^{*\,-2}). \tag{2.36}$$

In this definition $Y_i^* = \max(Y_i, \varepsilon)$, where ε is some arbitrary small positive number and the summation is over only those values of Y_i which satisfy $Y_i > 2X_i$ (see Figure 1.24). The purpose of the introduction of the mysterious ε is that, if it were not present, then there would be a distinct danger of the estimate being totally dominated by a single small value of Y_i^*. Indeed, the complex expression for $\mathrm{var}(\hat{\rho}_0)$ that is provided by Cox shows that the variance is infinite if $\varepsilon = 0$!

Batcheler's procedure

Virtually all the procedures that have been discussed during this section have been developed on theoretical grounds and have been tested by recourse to published or simulated data sets. The estimation procedure that we now consider, on the other hand, is dominated by practical considerations. Batcheler's aim has been to produce a procedure for intensity estimation which is both robust and practicable. A summary of the development of his procedure is provided by Warren and Batcheler (1979) and we shall confine our remarks to its salient features.

The twin cornerstones of Batcheler's work (Batcheler and Bell, 1970; Batcheler, 1971) are the following practical observations.

(1) There is an upper limit, R, to the distance from some nominated point or plant over which it is practicable to search for the nearest neighbour.

(2) While both point–plant and plant–plant distances are useful in estimating the intensity of a non-random plant population, it is impracticable to collect these as separate operations.

Figure 2.4 illustrates the measuring procedure that Batcheler has devised. This procedure is the direct extension to three measurements of the procedures of Cox and Lewis (1976) illustrated in Figure 1.24.

The three measured distances are $OA = X$, $AB = Y$, and $BC = V$. Measurements commence from the randomly located sampling point O and are subject to the practical limit on searching distances. Thus, all of X, Y, and V have values less than the upper limit R, and the presence of this upper limit is indicated in Figure 2.4 by the circles of radius R centred on O, A, and B. Notice that, since all measurements are made to the *nearest* plant, we have dropped the suffix (1) in order to simplify the notation (i.e. here $X_i \equiv X_{(1)i}$).

The locations of the plants A, B, and C are defined as follows. The plant A is the

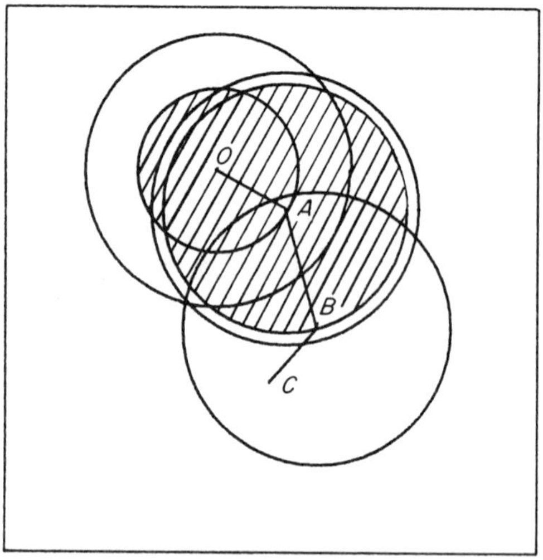

Figure 2.4 Batcheler's joint point–plant–plant–plant measurements

nearest plant to the point O. Plant B is the nearest neighbour to plant A, and plant C is the nearest neighbour (discounting plant A) to plant B.

In Figure 2.4 the shaded area indicates the treeless zone within which it is clear that plant C does not lie. Thus, while BC is a plant–plant distance, it is subject to even more stringent geometrical constraints than was the distance AB, and the distance V will not have the same distribution as the general plant–plant distance W.

Of course it will not always be the case that all of X, Y, and V will be less than R. If one of these measurements is greater than R, then there will be no information concerning the succeeding measurements. Thus, for a given sampling point, i, we may obtain no distance values at all, or just X_i, or both of X_i and Y_i, or all of X_i, Y_i, and V_i. Consequently, if we choose n sample points and denote by n_X, n_Y, and n_V the numbers of observations on X, Y, and V, respectively, then we have $n_V \leq n_Y \leq n_X \leq n$.

The essence of Batcheler's approach has been to take $\hat{\rho}_R$, the maximum likelihood estimator of ρ, for the case of a random plant distribution, and to multiplicatively adjust it to compensate for any deviations from randomness that calculations reveal. Batcheler (1971) notes that, given the restriction on the searching distance, the maximum likelihood estimator is given by

$$\hat{\rho}_R = n_X/\pi\left\{(n - n_X)R^2 + \sum_{i=1}^{n_X} X_i^2\right\}, \tag{2.37}$$

where, for simplicity, the subscripts 1 to n are attributed to the n_X X-values which, being less than R, are actually recorded exactly. If R is infinite, then we obtain the simple expression for $\hat{\rho}$ given by equation (2.27).

Batcheler (1973) investigated a large number of artificial populations and concluded that a key quantity was CV, the coefficient of variation of the point–plant distances:

$$CV = \mathrm{var}(X)/\bar{X}, \tag{2.38}$$

where $\bar{X}$ is the sample mean of the $\{X_i\}$-values $(= \sum X_i/n_X)$, and $\mathrm{var}(X)$ is the sample variance of the $\{X_i\}$-values. For a Poisson distribution we have, when $R = \infty$, $\mathrm{E}(CV) = \mathrm{E}\{\mathrm{var}(X)\}/\mathrm{E}(\bar{X})$ and so, from equations (1.39) and (1.40):

$$\mathrm{E}(CV) = \sqrt{(4 - \pi)/\pi}. \tag{2.39}$$

When the upper-limit restriction, R, is imposed, the result (2.39) no longer holds and an exact result is not available. Batcheler (1973) reports that simulations suggest that a good polynomial approximation to $\mathrm{E}(CV)$ is then obtained by using

$$\mathrm{E}(CV) = -1.0319 + 0.4892 f^2 - 0.7182 f^4 + 0.6095 f^6, \tag{2.40}$$

although it is notable that this does not provide the correct value for $f = 1$. In this expression f is the fraction of sample points for which the distance X could be calculated, that is

$$f = n_X/n. \tag{2.41}$$

Batcheler (1973) produces a number of alternative estimators of spatial intensity which are all more or less dependent on the quantities

$$A_{Y,R} = \frac{1}{\mathrm{E}(CV)}\left\{\frac{nn_Y \operatorname{var}(X)}{n_X^2 \bar{X}\bar{Y}}\right\}^{1/2}, \tag{2.42a}$$

$$A_{V,R} = \frac{1}{\mathrm{E}(CV)}\left\{\frac{nn_V \operatorname{var}(X)}{n_X n_Y \bar{X}\bar{V}}\right\}^{1/2}. \tag{2.42b}$$

When $R = \infty$, these expressions simplify to

$$A_Y = A_{Y,\infty} = \{\pi \operatorname{var}(X)/(4-\pi)\bar{X}\bar{Y}\}^{1/2}, \tag{2.43a}$$

$$A_V = A_{V,\infty} = \{\pi \operatorname{var}(X)/(4-\pi)\bar{X}\bar{V}\}^{1/2}. \tag{2.43b}$$

Batcheler recommends that, in cases where $A_{Y,R} < 0.5$ (which occurs with regular plant patterns), one should use the estimator

$$\hat{\rho}_{A,R} = \hat{\rho}_R/(1.0065 + 0.3401f + 0.4235f^2), \tag{2.44}$$

where $\hat{\rho}_R$ was given by equation (2.37) and f was defined by equation (2.41). The recommendation is, of course, based almost entirely on empirical grounds. In the case where $f = 1$, we obtain the simplification

$$\hat{\rho}_{A,\infty} = 0.5649\hat{\rho}, \tag{2.45}$$

where $\hat{\rho}$ was given by equation (2.27).

In cases where $A_{Y,R} \geq 0.5$, Batcheler (1975a) recommends the use of

$$\hat{\rho}_V = \hat{\rho}_R(b^{A_{Y,R}} + b^{A_{V,R}})2a, \tag{2.46}$$

where a and b are given by

$$a = (1 + 2.473f); \qquad b = (1 + 2.717f). \tag{2.47}$$

Batcheler gives the approximate variance of $\hat{\rho}_V$ as $\bar{A}^2\hat{\rho}_V^2/n$, where $\bar{A}$ is the average of $A_{Y,R}$ and $A_{V,R}$. Note that, for values of $\bar{A}$ less than 1, the implication of this variance formula is that Batcheler is claiming a greater precision for his estimator than has been attributed to any of the preceding estimators.

Despite his stringent rules, Batcheler (1975b), without comment, uses a different estimator, namely

$$\hat{\rho}_{Y,R} = \hat{\rho}_R b^{A_{Y,R}}/a. \tag{2.48}$$

Clearly, $\hat{\rho}_{Y,R}$ is to be preferred to $\hat{\rho}_V$ in cases where the upper-limit restriction, R, has the effect of severely limiting the number of observations on V. This may be the case with the estimation of deer pellet intensity that is the subject of Batcheler (1975b), but it cannot be the case with the tree survey conducted by Batcheler and

Hodder (1975). Nevertheless, these authors use $\hat{\rho}_Y$ in preference to $\hat{\rho}_V$, despite taking R as ∞. This results in the simplified expression

$$\hat{\rho}_{Y,\infty} = n(3.717)^{A_Y}/3.473\,\pi\sum X_i^2, \tag{2.49}$$

with approximate variance $A_Y^2\hat{\rho}_{Y,\infty}^2/n$.

Of course, from the practical point of view, $\hat{\rho}_{Y,R}$ is much preferred to $\hat{\rho}_V$, since it dispenses entirely with the third set of measurements $\{V_i\}$. It is indeed difficult to see—as Figures 1.24 and 2.6 suggest—how the extra distances can give any information of a distinctive character beyond that provided by the $\{Y_i\}$. On practical grounds it would seem that $\hat{\rho}_{Y,\infty}$ (or $\hat{\rho}_{Y,R}$, if $R < \infty$ is a practical necessity) is the preferred form of the estimator for cases where $A_{Y,R} > 0.5$, with $\hat{\rho}_{A,\infty}$ (or $\hat{\rho}_{A,R}$) being used otherwise.

Precision of the estimates

The variances quoted for $\hat{\rho}_{(k)}$, $\hat{\rho}_L$, and $\hat{\rho}_E$, based upon the assumption of a random point distribution, all take the form $\hat{\rho}^2/n$, which implies that one standard deviation corresponds to 10% of the value of $\hat{\rho}$ for the case $n = 100$ and to 1% of $\hat{\rho}$ for the case $n = 10\,000$.

For Batcheler's procedures the variance estimate depends upon the value of $A_{Y,\infty}$ (or $\bar{A}$). For a random point process $A_{Y,\infty} \simeq 1$ and thus the sample sizes are as shown above. For a clustered process $A_{Y,\infty} > 1$, and the required sample sizes become $100A_{Y,\infty}^2$ and $10\,000A_{Y,\infty}^2$, respectively. For a regular point process $A_{Y,\infty} < 1$ and the required sample sizes are again $100A_{Y,\infty}^2$ and $10\,000A_{Y,\infty}^2$, which may be as small as 25 and 2500, respectively.

To achieve the same accuracy with the very simple $\hat{\rho}_P$, which makes no assumptions about the nature of the plant distribution, we need—as one would expect—very much greater sample sizes. In fact, for a 10% standard deviation we require $n = 1000$, and for 1% we need $n = 1\,000\,000$! These large sample sizes are the price that we must pay in order to obtain a simple but robust estimation procedure.

Example 2.3

In order to gain some insight into how these various intensity estimators behave with different types of data, we shall examine three data sets: one with a near random pattern, one with a markedly clustered pattern, and one with a noticeably regular pattern.

Figure 2.5 reproduces a section of the recordings made by Dr Kooijman of the position of glasswort (*Salicornia europaea*) plants at 'De Kwade Hoek', which is a salt-marsh to the west of Stellendam in the Netherlands. We are most grateful to Dr Kooijman for making the data available to us. Kooijman (1979) has shown that, over the entire plot (only one-quarter of which is shown in Figure 2.5), there

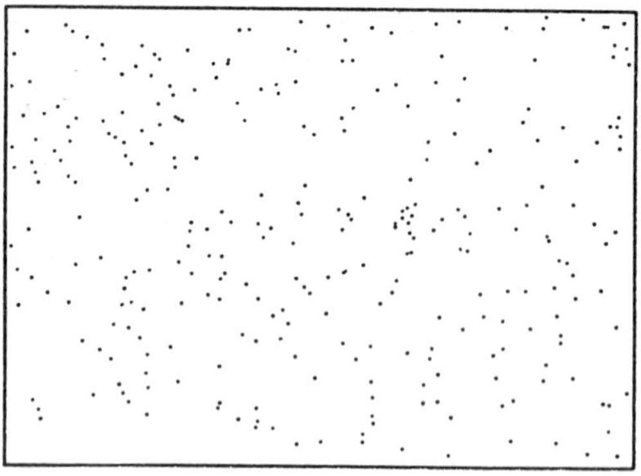

Figure 2.5 Locations of 296 Glasswort (*Salicornia europaea*) plants in a salt marsh near Stellendam. (Redrawn from Kooijman, 1979)

is evidence of a change in the intensity of the plants which is associated with the direction of the sea. Kooijman also finds some evidence of clustering (vegetative?) at short distances. For our purposes, however, we may regard this plant distribution as being as near to an example of a purely random pattern as one can ever expect to see!

By way of contrast, Figure 2.6 reproduces the 481 distinct recorded positions of maples in Lansing woods, Clinton County, Michigan: the data have been

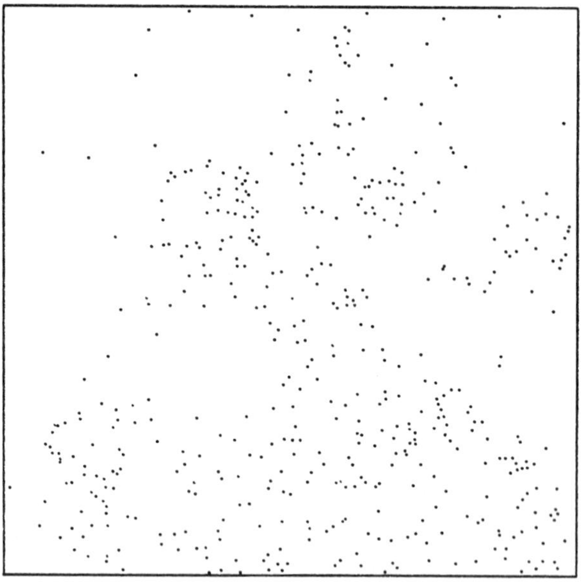

Figure 2.6 Locations of 481 maples in Lansing Woods. (Redrawn from Diggle, 1979b)

redrawn from the figure given by Diggle (1979b). These maples are quite clearly *not* scattered at random! Diggle also provides maps of the locations of oaks and of hickories for the same site. It is evident, from a comparison of these maps, that, at this location, hickories and maples are complementary trees—the gaps in the hickories map correspond to clusters of maples, while the gaps in the maples map correspond to aggregations of hickories.

As an example of a set of objects that have a noticeably regular pattern, we use the data on beadlet anemones that were given earlier in Table 1.11. These data were analysed in Examples 1.8 and 1.16 and were illustrated in Figure 1.26. The conclusion of that pattern analysis was that, even when the anemones were treated as discs rather than points, there was significant evidence of a regularity of position which suggested that each anemone was surrounded by its own 'territory'.

Because the estimation procedures vary in complexity from the simple $\hat{\rho}_{(1)}$, which uses only the distance from the randomly chosen point to the nearest plant, to $\hat{\rho}_V$, which uses point plant–plant distances, it is difficult to compare the efficiency of alternative estimators in any fair manner. It is clear that it is easier to take a single observation on X, than a triple of observations on X, Y, and V. However, if the major expense of the operation lies in reaching the original sampling point, then the marginal cost of the Y and V observations may be of little significance. On the other hand, if the measurement of accurate point–plant or plant–plant distances is the major cost, then the triple of (X, Y, V) observations will be nearly three times as expensive to obtain as was the single X observation. Evidently the costs involved vary according to the subject-matter and are strongly influenced by the physical scale of the sampling frame. It is no coincidence that $\hat{\rho}_V$ was developed for sampling in forests, where random sampling points may be located at considerable distances from one another.

In Table 2.5 we present the various summary statistics that we obtained using randomly sampled points for each of the three data sets. In order to avoid edge effects, the sampling has been performed, in each case, from within a central area which amounted to between one-third and two-thirds of the entire area available.

The sampling procedure involved, in each case, the generation of 30 random sampling points from which a wide variety of point–plant and point–plant–plant distances were calculated. For a further 15 random plants, a more restricted set of statistics was calculated, and for yet a further 45 random points only the distance to the nearest plant was calculated. These various sets of measurements were then amalgamated to give results, of varying detail, for samples of sizes 30, 45, and 90, respectively. The idea here was that each statistic calculated involved the same number of individual distances (in other words we are assuming near-zero relocation costs). In considering estimators of the $\hat{\rho}_{(k)}$ type we have assumed that it would be necessary to make k separate measurements. This may not be entirely fair to these estimators, but, as we shall see, it does not affect our recommendations.

The figures given in Table 2.5 are expressed in terms of the units used in the

Table 2.5 Summary statistics required for various intensity estimates

Statistic	Data set: Glassworts	Maples	Anemones
Plot dimensions	475 × 333 cm	280 × 280 m	145 × 97.5 cm
Central area dimensions	380 × 238 cm	225 × 225 m	100 × 50 cm
Basic unit of distance	0.475 cm	28 cm	0.475 cm
One measurement ($n = 90$)			
$\sum X^2_{(1)i}$	63 192.03	137 232.27	4824.432
$X_{(1)[20]}$	13.567	12.962	4.557
One or two measurements ($n = 45$)			
$\sum X^2_{(1)i}$	28 764.64	69 038.11	2290.123
$\sum X^2_{(2)i}$	71 628.01	152 638.04	5427.971
$X_{(1)[12]}$	15.396	12.344	4.557
$\sum X_{(1)i}$	1011.31	1403.37	299.698
$\sum X_{(2)i}$	1654.58	2234.97	481.014
$\sum Z_{(1)i}$	1365.93	1843.10	499.173
$\mathrm{var}(X)$	137.657	574.378	6.685
$\bar{X}$	22.474	31.186	6.660
$\sum Y_{(1)i}$	1251.13	1799.22	475.805
$\bar{Y}$	27.803	39.983	10.573
$A_{Y,\infty}$	0.898	1.298	0.589
One, two, or three measurements ($n = 30$)			
$\sum X^2_{(1)i}$	21 207.91	50 439.63	1404.63
$\sum X^2_{(2)i}$	49 680.22	104 577.11	3143.93
$\sum X^2_{(3)i}$	63 390.54	153 124.77	5226.74
$X_{(1)[9]}$	16.841	12.962	5.254
$\sum X_{(1)i}$	704.21	977.91	190.46
$\sum X_{(2)i}$	1121.92	1495.47	298.44
$\sum Z_{(1)i}$	1066.62	1248.30	315.398
$\mathrm{var}(X)$	161.294	640.093	6.738
$\bar{X}$	23.474	32.597	6.349
$\sum Y_{(1)i}$	829.39	1258.92	306.418
$\bar{Y}$	27.646	41.964	10.214
$A_{Y,\infty}$	0.954	1.309	0.617
$\sum V_{(1)i}$	889.71	1344.18	228.411
$\bar{V}$	29.657	44.806	7.614
$A_{V,\infty}$	0.921	1.266	0.714
$\bar{A}$	0.937	1.288	0.665

digitization process that led to Figures 2.5, 2.6, and 1.26. For the two Dutch data sets the digitization unit corresponded to approximately 0.475 cm, while for the maples one unit corresponds to about 28 cm. Thus, the values of $\bar{Y}$, for the case $n = 45$, correspond to inter-plant distances of about 13 cm for the glassworts, 11 m for the maples, and 5 cm for the sea anemones.

Because of the nested nature of the results presented in Table 2.5, it is easy to obtain alternative sub-sample results. Thus, while $\sum x_{(1)i}^2 = 28\,784.64$ for the first 45 sampled points, it equals $(63\,192.03 - 28\,784.64) = 34\,407.39$ for the second 45 sampled points. The order given for the summary statistics may seem peculiar at first sight, but this order actually corresponds to the order in which these quantities are required in the calculation of the various test statistics that were introduced earlier in this section.

In real life, of course, when estimating plant intensity we would not have the entire mapped plant pattern available for our scrutiny. From prior knowledge of the characteristics of the 'plants' that we were considering, we would doubtless have some idea as to whether the pattern was likely to be clustered or regular. However, a preliminary to the estimation of plant intensity should surely be an examination of the pattern (using sampling points) to test for deviations from a random pattern. We saw, in Chapter 1, that the most powerful tests were those of Hopkins, which involved random plant–plant distances $\{W_i\}$. None of the intensity estimators discussed in this chapter has used these distances, since they imply an enumeration of the plants which would render estimation superfluous. If the semi-systematic sampling scheme of Byth and Ripley (1980) were practicable, and a grid of equi-sized quadrats was laid down (cf. Figure 1.25), then the quadrat-based measures of intensity that were discussed in Section 2.2 would probably be preferable to the distance-based estimators of the present section.

Discounting the Hopkins' procedures, any of the remaining test statistics listed in Table 1.10 could be used to test for departures from randomness. It is not our purpose here to spend much time on pattern tests for these data sets and we content ourselves with reporting the values of Eberhardt's E statistic for each of the three data sets. For the case $n = 30$, only the maples data displays an undoubtedly significant departure from randomness. For this data set $E = 30 \times 50\,439.63/977.91^2 = 1.582$, which far exceeds the upper 0.5% point of the null distribution of E(1.4897, see Table A.1 of Appendix A). The value of E for the anemones data, with $n = 30$, is 1.162, which is just less than the lower 5% point of the null distribution of E. On increasing the sample size for this data set to $n = 45$, we get $E = 1.147$ and this now displays significant evidence of a departure from randomness (towards regularity) at the 1% level.

Even at the largest sample size ($n = 90$) the glasswort data shows no significant deviation from a random pattern. The value of $\sum x_{(1)i}$ is 2121.29 (this is not given in Table 2.5) which leads to $E = 1.264$, which is very close indeed to the mode of the null distribution of E.

From these preliminary investigations of pattern we can hazard some shrewd guesses concerning the nature of the results that we shall obtain. In particular, we can expect that the $\hat{\rho}_{(k)}$ estimators and $\hat{\rho}_L$ will be badly affected by the clustered nature of the maples data and will produce severe underestimates of ρ.

Using the summary statistics of Table 2.5 and the relevant formulae given above, we can easily obtain the values of the various estimates of intensity. Thus, for the glassworts, with $n = 90$, $\hat{\rho}_{(1)} = 89/\{63\,192.03\pi\}$ (using equation (2.28)). The value that results, 4.483×10^{-4}, is difficult to interpret, since it refers to the digitized units rather than to the actual distances. The object of intensity estimation is that, having obtained an estimate of intensity, $\hat{\rho}$, we can then multiply that estimate by A, the area of the region studied, and so obtain an estimate ($A\hat{\rho}$) of the total number of plants in the region, and this suggests that the value of $A\hat{\rho}$ would be the more interesting statistic to report.

In order to avoid edge effects we have confined our random points to a central area within the region of interest. If the plant intensity within this central area were greatly different from that of the region as a whole, then this might produce distorted estimates of plant intensity and hence of plant numbers. For this reason, since the true counts for both the central regions and the entire areas are known for all three data sets, these counts are given (together with the corresponding plant intensities) in Table 2.6, together with separate estimates ($A\hat{\rho}$) of the numbers in the two regions, for each of a number of sample-size/estimation-procedure combinations. As it happens, however, the differences in the observed intensities in the two regions are not very great and do not appear to have biased the estimators in any way. Our subsequent discussion will therefore concentrate on the final three columns of Table 2.6, since these refer to the estimates for the total regions of interest.

For the case $n = 90$, we had, for the glasswort data, that $\hat{\rho}_{(1)} = 4.483 \times 10^{-4}$. The units of $\hat{\rho}_{(1)}$ were plants per squared digitizing unit. Since a digitizing unit is equivalent, in this case, to about 0.475 cm, it follows that the estimate of intensity expressed in terms of plants per square metre is obtained by multiplying the previous result by $10^4/(0.475)^2$, which leads to the value 19.87. The central region for the glasswort data had an area of $A = 3.8 \times 2.38 = 9.044\,\text{m}^2$, and hence the estimated number of glasswort plants within this area is $19.87 \times 9.044 \simeq 179$ (allowing for various round-off errors). Similarly, the area of the entire plot is $15.82\,\text{m}^2$ and hence the estimated number of plants within the entire plot is 314. These figures, 179 and 314, are the figures recorded in Table 2.6.

Using equation (2.29) we can easily deduce that, if $A\hat{\rho}_{(k)}$ is the estimated number of plants in an area of size A, using the estimator $\hat{\rho}_{(k)}$, then, for randomly distributed plants, this estimator has standard deviation estimated by $A\hat{\rho}_{(k)}/\sqrt{(nk-2)}$. If an estimated value is more than two (estimated) standard deviations different from the true value, then it is recorded in italics in Table 2.6 as an indication of its poor performance. On the other hand, if an estimated value is within one (estimated) standard deviation of the true value, then this is indicated by the presence of an asterisk in Table 2.6.

Table 2.6 Alternative estimates of numbers of plants in three data sets

Method	Central area			Entire plot		
	Glassworts	Maples	Anemones	Glassworts	Maples	Anemones
One measurement ($n = 90$)						
$A\hat{\rho}_{(1)}$	179*	*132*	117	314*	*206*	*296*
$A\hat{\rho}_P$	147*	257*	*65*	257*	402*	164
One or two measurements ($n = 45$)						
$A\hat{\rho}_{(1)}$	195*	*130*	122	341*	*203*	308
$A\hat{\rho}_{(2)}$	159	*119*	104	278*	*186*	263*
$A\hat{\rho}_P$	139*	346*	79*	243*	541*	200*
$A\hat{\rho}_L$	163*	*144*	97*	285*	*225*	244*
$A\hat{\rho}_E$	207*	*177*	96*	363	*277*	241*
$A\hat{\rho}_{Y,\infty}$	186*	210	78	326*	*328*	198
[Average, $A\bar{\rho}$	175	188	96	306	293	242]
One, two, or three measurements ($n = 30$)						
$A\hat{\rho}_{(1)}$	174*	*117*	131	305*	*183*	331
$A\hat{\rho}_{(2)}$	151	*115*	119	265*	*180*	301
$A\hat{\rho}_{(3)}$	179*	*118*	108*	313*	*185*	273*
$A\hat{\rho}_P$	130	350*	67	227*	547*	168
$A\hat{\rho}_L$	159*	*144*	112*	277*	*225*	286
$A\hat{\rho}_E$	169*	*167*	106*	297*	*261*	267*
$A\hat{\rho}_{Y,\infty}$	181*	195	88*	318*	*304*	222*
$A\hat{\rho}_V$	178*	189	94*	311*	*296*	237*
[Average, $A\bar{\rho}$	165	174	103	289	273	261]
True value	176	264	97	296	481	231
Intensity (plants m^{-2})	19.5	5.2×10^{-3}	194	18.7	6.1×10^{-3}	163

Notes:
1. Estimates in italics differ by *more than* **2** *s.d.* from the true value.
2. Starred estimates differ by *less than* **1** *s.d.* from the true value.

Whilst the maples data is evidently clustered, as opposed to random, the clusters could not be described as 'tight' clusters, and the concept of a mean number per cluster centre does not seem relevant. For this reason, none of the estimates $\hat{\rho}_V$, $\hat{\rho}_M$, or $\hat{\rho}_B$ has been calculated. Also omitted from the table are $\hat{\rho}_0$ (because of the difficulty in specifying a 'natural' value for ε) and $\hat{\rho}_A$ (because $A_{Y,\infty} > 0.5$ for all three data sets).

We commence our discussion of the results by considering the near-random glassworts data. For this data, all eight estimators appear to behave very well: almost all the estimates given lie within one estimated standard deviation of the true value. Of course, for small sample sizes, one standard deviation is a not inconsiderable amount: for $n = 30$ it amounts to about 55 plants for each estimator except $\hat{\rho}_P$ (96 plants) and the Batcheler estimates $\hat{\rho}_{Y,\infty}$ and $\hat{\rho}_V$ (about 52 plants). When $n = 90$ the standard deviation of $\hat{\rho}_{(1)}$ is still about 33 plants, so that a realistic range for the most precise of the individual estimates given would be 314 ± 66 plants, which may not seem 'precise' at all!

With near-random data, a reasonable procedure with k independent estimators of intensity, each based on n observations, would be to take their average, $\bar{\rho}$, which would have the greatly decreased standard deviation $\bar{\rho}/\sqrt{nk}$. This is, of course, just as expensive in terms of observations as taking nk observations with any single estimator, and no more precise. However, the variety of estimators may give a more robust estimate by 'spreading the risk' amongst different types of procedure—rather like a portfolio of shares. For the data of Table 2.6, however, the estimators are *not* independent, since the same randomly chosen points were used as the starting points for each estimator. In this case 'inconvenient' points *could* produce a bias in each estimator, though this seems unlikely. One would expect that ρ would have a smaller variance than *any* of the individual estimators—provided, of course, that these estimators were valid (i.e. that the data were near random). Certainly, for the glasswort data the values of $\bar{\rho}$ for $n = 30$ and for $n = 45$ appear reassuringly close to the true values.

By contrast with the glasswort results, the estimates for the strongly clustered maples data are disastrous. As was anticipated, the $\{\hat{\rho}_{(k)}\}$ estimators and $\hat{\rho}_L$ all give gross underestimates of the true intensity. Sadly, $\hat{\rho}_E$ and Batcheler's complex procedures did very little better (though 30 *is* a small sample size for these purposes). Only $\hat{\rho}_P$ emerges with any laurels. However, even here, one should not be too confident. Although the full sample of 90 observations gives the reasonable estimate 402 ± 179 (a two standard deviation range), the three separate subsamples of size 30 gave 547 ± 352, 582 ± 375, and 158 ± 102. The final estimate here being a further disastrous underestimate.

We consider finally the anemones data which has been shown to display mild regularity. Here we would expect the $\hat{\rho}_{(k)}$ estimates to slightly over-estimate ρ (as is the case). Table 2.6 reveals also that neither $\hat{\rho}_L$ nor $\hat{\rho}_{Y.\infty}$ perform quite as well as one would have hoped, while $\hat{\rho}_P$ is unimpressive. However, none of the estimates is as disgraceful as those achieved with the maples data!

Recommendations

One must be wary of making recommendations, or drawing too firm a set of inferences, from the limited results furnished by the previous example. Nevertheless, certain basic precepts appear to have been established.

(1) It is important to establish the nature of the plant pattern (clustered, near-random, or regular) *before* deciding upon the intensity estimation procedure.

(2) The $\hat{\rho}_{(k)}$ family should only be considered for use with near-random patterns.

(3) With non-random patterns, accurate estimates will require a very considerable sampling effort. Whilst quite small samples suffice to detect the existence of a non-random pattern (say $n = 20$ random points), if such a pattern exists then a

reliable intensity estimate may require ten times as many sample points.

(4) The fact that alternative estimators give similar estimates cannot be taken as any indication that their common value is close to the true value.

None of the estimators has put on a convincing display with our (limited) example analyses. If the selection of random sampling points is very cheap (as with the glassworts or anemones), then the best procedure may be to use Eberhardt's E statistic to test for randomness, followed by $\hat{\rho}_P$ as an intensity estimator. Alternatively, if the selection of a new sampling point incurs noticeable cost, then the T-square methods appear useful, since T_E is a promising alternative to E and the combination of the T_P and T_N statistics were suggested as a method for distinguishing between heterogeneity and clustering. These tests could be followed by the calculation of $\hat{\rho}_E$. Finally, if sample points are expensive, then it would seem best to use the combination of $\bar{R}$ and E to test for randomness, followed by $\hat{\rho}_{Y,\infty}$ or $\hat{\rho}_{Y,V}$. A summary of these recommendations is displayed in Table 2.7.

Table 2.7 Recommendations for intensity estimation

Cost of choosing a new sampling point	Recommended pattern test(s) (see Table 1.10)	Recommended intensity estimator(s)	
Negligible	E	$\hat{\rho}_{(1)}$	for near-random patterns
		$\hat{\rho}_E$	otherwise
Low	T_P, T_N, T_E		
High	$\bar{R}, E$	$\hat{\rho}_{Y,\infty}$ or $\hat{\rho}_V$	

2.5 AREAL METHODS

The most obvious way to estimate plant intensity is to count the number of plants in some sub-area of size a and then use $\hat{\rho} = n/a$ as one's intensity estimate. In this section, however, we consider the estimation of ρ via its inverse, $1/\rho$. Since ρ is the number of plants per unit area, it follows that $1/\rho$ is the amount of area per plant.

All the estimates considered in this section rely upon the conceptual tessellation of the region of interest using one of the tessellations described in Chapter 1. Sample points are then chosen and the areas $\{a_i\}$ of the tiles associated with these sample points are determined. The average tile area, $\bar{a}$, is then an estimate of $1/\rho$ and hence we could use $1/\bar{a}$ as an estimate of ρ (Brown, 1965). Alternatively, since $1/a$ is an estimate of ρ, we can use $(\sum a_i^{-1})/n$ as an estimate of ρ (Ord, 1978). Note that while Ord's estimator is certainly an unbiased estimator of ρ, Brown's estimator is only asymptotically unbiased.

Since these types of methods are likely to be of particular interest to the forester, whose prime concern is not tree intensity but total tree volume, we shall include in our discussion some related methods which are designed with the forester in mind.

The Dirichlet tessellation

An example of the construction of the Dirichlet tessellation was provide in Figure 1.41. For any given plant *A*, one constructs the perpendicular bisectors of the lines *AB*, *AC*,..., where *B*, *C*,... are neighbouring plants. The intersections of these bisectors then form a convex polygon surrounding *A*; this polygon is the Dirichlet tile associated with *A*.

In the present application of Dirichlet tiles, we are concerned solely with the calculation of the *area* of the tile. There are many mathematical procedures that could be used to achieve this end. For example, a polygon with *n* sides could be subdivided into *n* component triangles each with vertex *A* (rather like the segments of an orange), and the area of each triangle could be found by measuring the sides of the triangle. For the entire polygon this would involve a total of $2n$ measurements.

A very simple method for use in the present context of intensity estimation was suggested by Brown (1965). His method is as follows. On ordinary graph paper, record the positons of the plant of interest, *A*, and its neighbours *B*, *C*,.... Using a transparent right-angle triangle, or its equivalent, draw the lines through *B*, *C*, etc. which are perpendicular to *AB*, *AC*, etc., respectively, as shown in Figure 2.7. This has the effect of avoiding the need to bisect *AB*, *AC*, etc. and simultaneously doubles the scale of the plot along both axes. The enclosed area is then estimated by counting the number of small squares of the graph paper which lie inside the constructed figure. To allow for the change of scale, this number should be divided by 4 to give the number of squares inside the Dirichlet tile and the corresponding physical area may then be calculated.

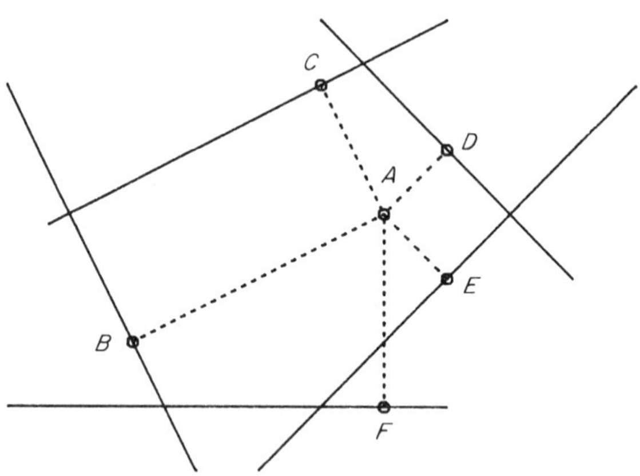

Figure 2.7 Calculating the area of a Dirichlet tile

A simple sampling procedure is to choose points at random within a central region, and to choose as the foci of the Dirichlet tiles those plants which were nearest to each sample point. Alternatively, one might first subdivide the entire region of interest into sub-areas and then sample points independently from each sub-area.

Brown evidently rediscovered the Dirichlet tile for himself, and was unaware of its various names, since he referred to the polygons as demonstrating the *area potentially available* (APA) for each plant. Brown's interest lay with estimating the intensity of trees and he gives various field results that he obtained with his graphical procedure in connection with stands of pine trees in New Zealand. Ord (1978) described the Dirichlet tiles as forming the basis of an 'ideal' estimator (his quotation marks) but Ord then dismissed the procedure as being impracticable in view of the labour involved in measuring the areas. Certainly it is the case that comparatively distant plants can affect the shape of the Dirichlet tile and therefore it may be necessary to make a rather large number of measurements. More easily obtained tessellations will be considered subsequently.

Example 2.4

As an example of the Dirichlet tile procedure, we reanalyse the three sets of data examined in Example 2.3. All calculations of area were done by hand, using Brown's graphical procedure. A major difficulty in the accurate construction of Dirichlet tiles is that of including relatively distant plants whose contribution to the polygon is the removal of a corner. While this may have an important bearing on the number of sides possessed by the polygon, it has only a minor impact upon the area of a tile. For this reason, while the hand-constructed tiles may not be of exactly the correct shape, any discrepancy in area is likely to be less than that implied by the method of counting squares.

Since the average number of sides of a Dirichlet tile is six, an average of six locational coordinates will need to be recorded. In practice one can expect to take more measurements, since it will not always be clear which are the relevant neighbouring plants. The results given in Table 2.8 therefore refer to just ten tiles, since, in practice, we found that we used only about two-thirds of the positions recorded on the graph paper. Thus, ten tiles corresponds to some 90 plant positions and this therefore places the amount of measurements on an approximately equal footing with the results presented in Table 2.6.

We consider first the procedure suggested by Brown (1965). Suppose that we take n sample points and determine n areas, $a_1, a_2, \ldots, a_n$. These have mean $\bar{a}$ and variance estimated by

$$s^2 = \{\sum a_i^2 - (\sum a_i)^2/n\}/(n-1). \tag{2.50}$$

The estimate of intensity is $1/\bar{a}$ and the estimated number of plants in the region of interest is $A/\bar{a}$, where A is the total area of the region. The two standard deviation

Table 2.8 Intensity estimation using the areas of randomly chosen Dirichlet tiles

Statistic	Data set: Glassworts (areas in cm^2)	Maples (areas in m^2)	Anemones (areas in cm^2)
Areas of ten randomly chosen tiles	300, 355, 400, 490, 525, 590, 655, 830, 1110, 1135	30, 45, 90, 125, 145, 155, 410, 740, 875, 1750	25, 48, 49, 54, 57, 57, 61, 64, 65, 66
Brown's procedure:			
Average tile area	639.0	436.5	54.6
Estimated no. of plants in plot	248	180	259
2σ range for no. of plants	191 to 351	100 to 876	227 to 301
Ord's procedure:			
Average reciprocal tile area	0.00188	0.00935	0.01974
Estimated no. of plants in plot	297	733	279
2σ range for no. of plants	216 to 378	207 to 1260	213 to 345
True number of plants in plot	296	481	231

range for the estimate of the mean of the area distribution is $(\bar{a} - s\sqrt{n}, \bar{a} + s/\sqrt{n})$ and hence the estimated two standard deviation range for the number of plants in the region of interest is $(A/\{\bar{a} + s/\sqrt{n}\}, A/\{\bar{a} - s/\sqrt{n}\})$, which is asymmetrical about the point estimate $A/\bar{a}$.

With the alternative procedure of Ord (1978), we estimate ρ by using $(\sum a_i^{-1})/n$, and calculate the variance of the reciprocals as

$$s^{*2} = \{(\sum a_i^{-2}) - (\sum a_i^{-1})^2/n\}/(n-1). \qquad (2.51)$$

With this procedure we then obtain the symmetrical two standard deviation range as $(\sum a_i^{-1})/n \pm s^*/\sqrt{n}$.

The degree of precision in the estimate of the mean area reflects the value of s^2 (or s^{*2}). With a clustered plant distribution there will be many very small tile areas and a few very large tile areas, leading to relatively large values of s^2 and s^{*2} and hence to wide confidence intervals. With a regular plant distribution, all the tiles will be of comparable area and hence s^2 and s^{*2} will be small, leading to narrow confidence intervals.

With the exception of Batcheler's procedures, the nominal accuracy of the distance estimators discussed in the previous section was independent of the actual plant pattern (or, rather, that pattern was presumed to be random for all except $\hat{\rho}_P$). These areal estimators, on the other hand, have an accuracy that can be estimated and which directly reflects the pattern in hand. This may well seem to be a desirable feature, since no prior assumptions need be made and it is quickly apparent when the pattern is clustered that more observations will be required to achieve tolerable accuracy.

The results of Table 2.8 reveal that all six estimates lay within two standard deviations of the correct values—although only two lay within one standard deviation. For the 'random' glassworts, the ranges $351 - 191 = 160$ and $378 - 216 = 162$ are comparable to those obtained with the distance estimators (e.g. with $n = 90$, $\hat{\rho}_{(1)}$ gave 314 ± 66 plants). For the clustered maples, the ranges $876 - 100 = 776$ and $1260 - 207 = 1053$ far exceed all the ranges of the distance estimators, but, conversely, the ranges for the regular anemones ($301 - 227 = 74$ and $345 - 213 = 132$) are rather tighter than those ranges previously obtained.

On the basis of these very limited results it appears that there is little to choose between these two areal procedures. Any slight bias in Brown's procedure appears to be amply compensated for by the much tighter confidence intervals that result from working directly with the areas rather than with their reciprocals.

Weighted polygons and weighted triangles

The dual of the polygonal Dirichlet tessellation was seen, in Chapter 1, to be the tessellation of Delaunay triangles (see Figures 1.43a and b). Delaunay triangles do not lend themselves directly to the estimation of density because each triangle is associated with *three* plants. However, it is easy to partition the area of a triangle into three components, proportional to the three plants that act as vertices of the triangle.

Figure 2.8 displays a triangle whose vertices are the plants B, C, and D. These plants have (in general) different sizes, measured in terms of diameter, or volume, or whatever we consider to be appropriate in the particular context. On the diagram these sizes are indicated by w_b, w_c, and w_d, which are the relative *weights* assigned to the three plants. These weights are scaled versions of the plant sizes:

$$w_b + w_c + w_d = 1. \tag{2.51}$$

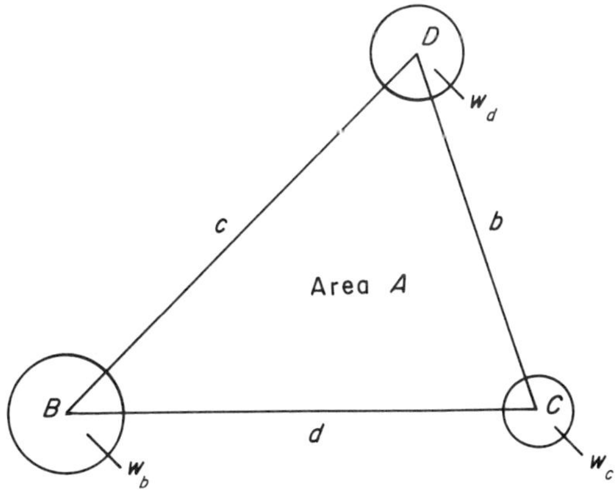

Figure 2.8 Weighting a Delaunay triangle

The idea is that T, the total area of the triangle, should be divided into three components of sizes Tw_b, Tw_c, and Tw_d (Fraser, 1977). Since every plant will be associated with four or more of these triangles, the total area assigned to a plant will be the sum over all the relevant Delaunay triangles.

In order to calculate the area of a triangle, it is perfectly feasible to draw the triangle on graph paper and, as before, to count the number of squares on the graph paper that lie within the triangle. There is, however, for a triangle a much simpler arithmetic procedure. Let the lengths of the sides opposite to B, C, and D be b, c, and d, respectively and let s be the semi-perimeter of the triangle:

$$s = (b + c + d)/2. \tag{2.53}$$

The area of the triangle may then be calculated using

$$T = \{s(s-b)(s-c)(s-d)\}^{1/2}. \tag{2.54}$$

Of course, with this procedure there is no need to determine the actual boundaries of the polygonal area associated with any given plant. It is sufficient to determine the triangle tessellation together with the vertex weights. Nevertheless, one cannot help being curious as to the nature of the associated polygons. Since there is no part of the weighting procedure that describes how the triangle should be partitioned, there is no unique subdivision, and hence no unique set of polygons. However, one possible procedure, illustrated by Figure 2.9, is as follows. Choose points P (on CD), Q (on DB), and R (on BC) in such a way that the lines BP, CQ, and DR intersect at X. With a correct choice of the locations of P, Q, and R (not easily done) we could then get the required subdivision into the component areas Tw_b, Tw_c, and Tw_d, as shown.

If it is felt desirable to determine the (weighted) polygons associated with each plant—perhaps for comparison with the Dirichlet polygons—then the difficulty in correctly locating P, Q, and R poses a problem. One solution (Fraser, 1977) is to use a revised system of weights. As before, w_b, w_c, and w_d reflect the sizes of the plants at B, C, and D, respectively. Suppose we now use the weights w_b^*, w_c^*, and

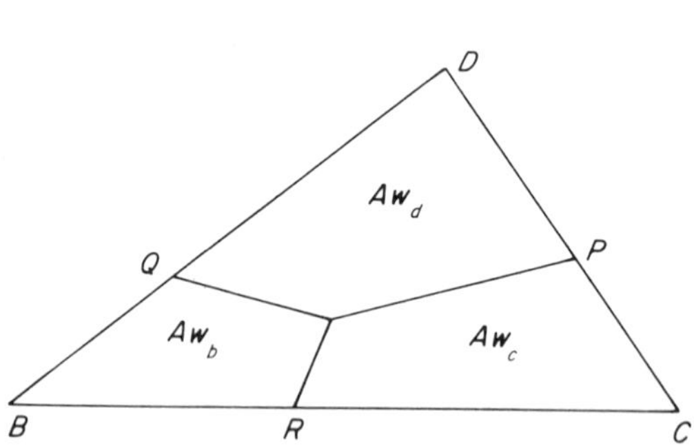

Figure 2.9 The weighted subdivision of a triangle

w_d^*, where, for example,

$$w_b^* = \frac{w_b^2(w_b w_c + 2w_c w_d + w_b w_d)}{w(w_b + w_c)(w_b + w_d)}, \tag{2.55}$$

where

$$w = (w_b w_c + w_b w_d + w_c w_d). \tag{2.56}$$

Simple algebra suffices to show that $w_b^* + w_c^* + w_d^* = 1$, as required. Why should we use such complex weights? The answer is that they correspond to a very simple choice of locations for P, Q, and R. In fact, we have that

$$CP/CD = w_c/(w_c + w_d), \qquad DQ/DB = w_d/(w_b + w_d), \qquad BR/BC = w_b/(w_b + w_c). \tag{2.57}$$

Thus, if a polygonal diagram is required, then it is easier to use the revised $\{w^*\}$ weights, whereas if only polygonal areas are required, then it is easier to use the simple $\{w\}$ weights.

Fraser (1977) presents details of an extended investigation of the choice of possible weights using a variety of data sets relating to stands of trees. In this context the characteristic of interest is not plant intensity but volume of wood. Fraser found that the most consistent estimates were obtained by using the $\{w^*\}$ estimates given by equation (2.54) with the $\{w\}$ weights themselves being calculated from the fractions of the basal areas of the trees that lay within the triangle in question.

As an example of Fraser's preferred weights, suppose that the basal diameters of trees B, C, and D were D_b, D_c, and D_d, respectively. Thus, the basal areas are in the ratio $D_b^2 : D_c^2 : D_d^2$. Now, the angle $\widehat{DBC}$ may be calculated from

$$\sin \widehat{DBC} = 2T/cd, \tag{2.58}$$

where T is the area of the triangle whose sides have lengths b, c, and d. Note that in determining $\widehat{DBC}$ one must be careful to distinguish between obtuse and acute angles. The weights recommended by Fraser are therefore in the proportions:

$$D_b^2 \sin^{-1}(2A/cd) : D_c^2 \sin^{-1}(2A/bd) : D_d^2 \sin^{-1}(2A/bc). \tag{2.59}$$

Notice that, when all the weights are equal, the point of intersection, X (see Figure 2.9), becomes the centroid of the triangle. This is the centre of gravity of a uniform triangle. Changing the weights relocates X to the centre of gravity of the non-uniform triangle. Even with equal weights, the polygon formed around any vertex will not be a Dirichlet polygon (except in the case of a network of equilateral triangles). This is because the Dirichlet construction uses *perpendicular* bisectors of the sides and these intersect at the circumcentre of the triangle and not at the centroid.

For interest, a section of a polygon corresponding to Fraser's suggested angle weighting as specified by equation (2.59) is illustrated in Figure 2.10. It will be

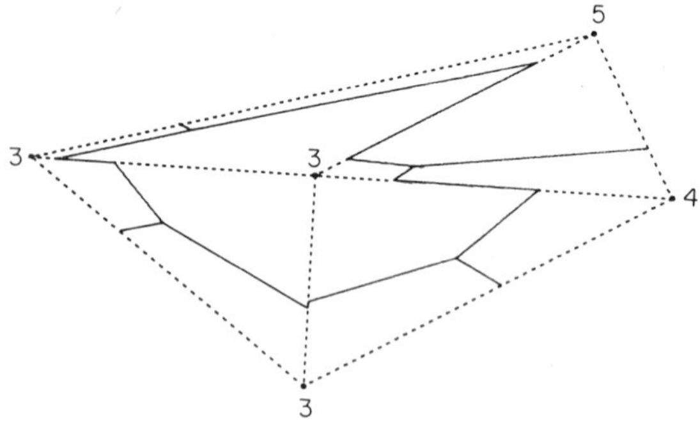

Figure 2.10 Spiky polygons resulting from angle-area weights

seen that the polygons that correspond to this procedure are startlingly jagged!

Suppose that the polygonal area associated with the ith tree is denoted by a_i, and let the estimated volume of that tree be V_i. Then an unbiased estimate of the total volume of timber is

$$\hat{V} = (A/n)\sum(V_i/a_i), \tag{2.60}$$

where A is the area of the total region of interest, and n is the number of trees that are sampled. In the same way, if we are concerned with the ground cover of plants, and we denote the diameter of the ith plant by D_i, then we have an overall estimate given by

$$\hat{G} = (A/n)\sum(\pi D_i^2/a_i). \tag{2.61}$$

Finally, if we are interested in plant intensity, then we use

$$\hat{\rho} = (A/n)\sum(1/a_i), \tag{2.62}$$

which is simply the estimate ascribed to Ord in the previous subsection.

All the discussion in this subsection has implicitly assumed that the basic triangles used for the calculation of these various weighted polygons were derived from the Delaunay tessellation. Since the Delaunay tessellation is itself derived from the polygonal Dirichlet tessellation, it follows that we are using an extremely roundabout route to arrive at our final weighted polygons. This procedure is scarcely one that we would wish to use in the field.

A practical procedure that leads directly to a tessellation of triangles, as opposed to polygons, was proposed by Fraser and van den Driessche (1972). Their suggestion involved locating all pairs of points which were *least diagonal neighbours* (LDN) of each other. A pair of plants are deemed to be least diagonal neighbours provided that no other plants occur on the straight line joining the pair, and provided that the said line between the pair cannot be intersected by a

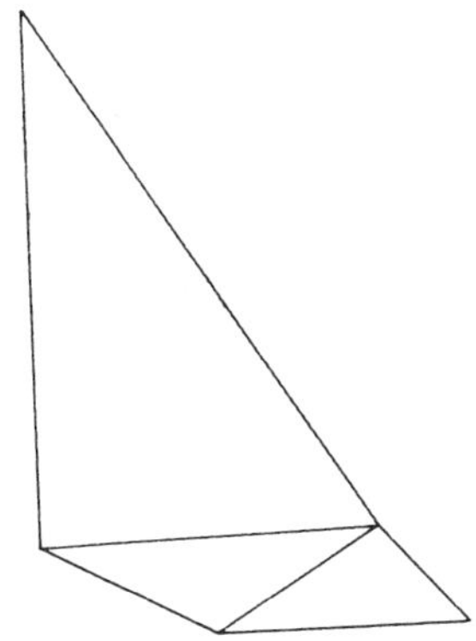

Figure 2.11 The least-diagonal-neighbour tessellation.

shorter straight line between any other pair of individuals. An example is shown in Figure 2.11.

In practice, the triangles formed by joining all least diagonal neighbours will all be almost identical to the triangles of the Delaunay tessellation. However, as Figure 2.11 illustrates, there will be a small proportion of quadrilaterals $ABCD$ which are divided into the pair of LDN triangles ABC and DBC instead of the Delaunay triangles ABD and ACD.

Fraser (1977) recommended his weighted polygons for use in connection with the LDN triangles. Since most Delaunay triangles are also LDN triangles, and since the latter are much easier to discover, there seems every reason to prefer the LDN triangles in the context of intensity estimation.

Example 2.5

Figure 2.12 shows the Delaunay tessellation (dotted lines) for a portion of the anemone data, together with the dual Dirichlet tessellation (unbroken lines). For those wishing to reconstruct these tessellations, the coordinates of the anemone centres were given in Table 1.11, and those shown in Figure 2.12 are all contained in the rectangle defined by $110 \leq x \leq 175$ and $50 \leq y \leq 130$.

By way of contrast, Figure 2.13 shows the result of using Fraser's procedure to construct weighted polygons based on the Delaunay triangles. In this case the weights were the diameters of the anemones as given in Table 1.11 and the simple ratios specified by equation (2.56) were used to construct the polygons. Thus, for example, a line connecting the centres of anemones of diameters 3 units and 4 units is intersected by the weighted polygon at a point $3/(3+4) = 3/7$ of the way along its length.

Although there are 21 anemones located in Figures 2.12 and 2.13, complete polygons are available for just eight of these, and consequently we could not expect great accuracy from any estimator. However, for the sake of completeness,

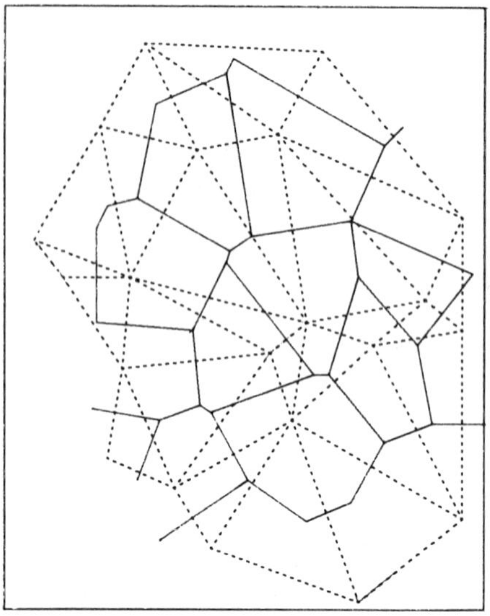

Figure 2.12 Delaunay and Dirichlet tessellations for part of the anemone data

Table 2.9 reports the values of $\hat{G}$ and of $A\hat{\rho}$ calculated using equations (2.61) and (2.62), together with the exact values of these quantities.

The eight anemones for which results are reported in Table 2.9 are those shown in the figures, reading from left to right and from top to bottom of the figures. Notice that with the Dirichlet polygons there is little evidence of any correlation

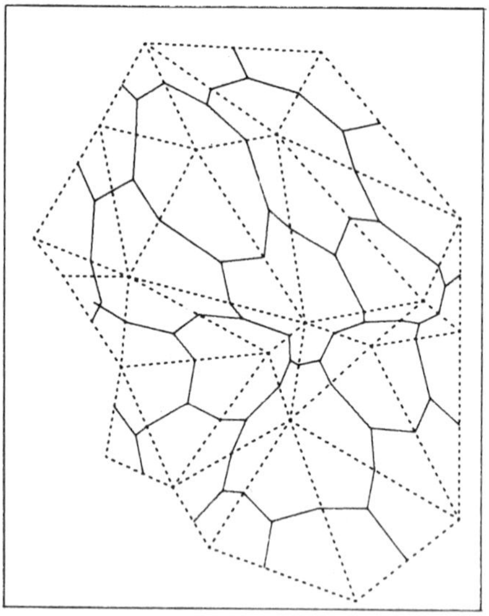

Figure 2.13 Delaunay and tree-weighted tessellations for part of the anemone data

Table 2.9 Results of using various polygonal tessellations in connection with part of the anemone data

	Individual anemones								$\hat{G}$	$A\hat{\rho}$
Anemone diameter D (units)	5	3	4	3	3	4	4	4		
Dirichlet polygon area (cm^2)	55	69	52.5	50.5	29	40.5	39	63	3053.4	304
Weighted polygons										
D and $\{w^*\}$ weights (2.55)	68	61	55.	47	32.5	35.5	36.5	66	3013.8	304
D and $\{w\}$ weights	65	62	54.5	49	33	35.5	36	65.5	3030.3	304
D^2 and angle weights (2.59)	92	53	49	38	32.5	35	47	55	2969.1	310
True values									3273.3	231

between polygon area and anemone diameter. The various weighted polygons naturally enhance this correlation, but it is interesting to note that the estimates of ground cover (G) and of the total number of anemones ($A\rho$) are relatively unaffected by the choice of polygon.

As a postscript to this very limited investigation of the anemone data, Figure 2.14 shows the alternative triangular tessellation that involves the *least diagonal* construction. It can be seen to be very similar to the Delaunay tessellation given in Figures 2.12 and 2.13.

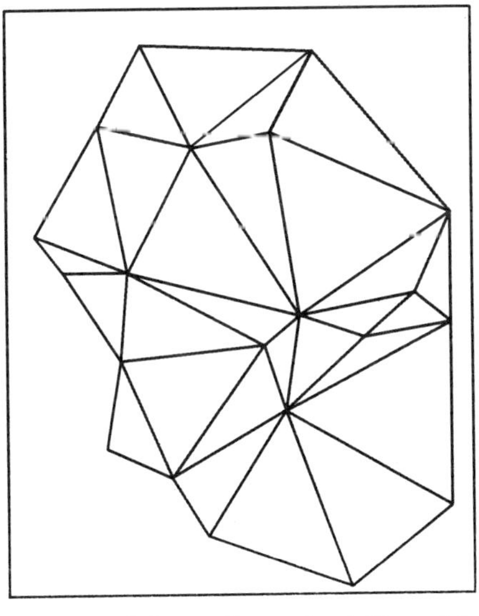

Figure 2.14 Least-diagonal-neighbour triangles for part of the anemone data

2.6 THE ANGLE-COUNT METHOD

The angle-count method was introduced by Bitterlich (1948) as a method of estimating the proportion of the area of a forest that is actually covered by the bases of trees. To use the method, an observer goes to each of a number of points in the forest, chosen either randomly or systematically, and counts the number of trees that subtend, at that point, an angle greater than or equal to some predetermined fixed angle 2α. To accomplish this task, foresters make use of an instrument called a *relascope*. Despite its name, the angle-count method could be considered to be a direct alternative to the areal methods of the previous subsection.

The geometry of the angle-count procedure is illustrated in Figure 2.15. In this figure, the observer's position is given as the point O, while the tree (whose diameter is D) is centred on a point which is at a distance $D/2\sin\alpha$ from O. At this distance, the tree subtends an angle of exactly 2α at O. If the centre of the tree were any further from O, then the angle subtended at O would be less than 2α, and the tree would not be counted.

It follows from this geometrical argument that all trees of diameter D that lie within the circle with centre O and radius $D/2\sin\alpha$ will be counted. This circle has area $\pi D^2/4\sin^2\alpha$. Suppose that the number of trees of diameter D that are counted by this procedure, when using n sampling points, is denoted by $\#D$. The estimated intensity of trees of this diameter will therefore be $4\sin^2\alpha(\#D)/n\pi D^2$. Similarly, if we consider trees of some other diameter, D^*, say, then we shall have the analogous result that the trees of this diameter have the estimated intensity $4\sin^2\alpha(\#D^*)/n\pi(D^*)^2$.

Suppose that we now consider all possible values of D. The overall estimated intensity will be

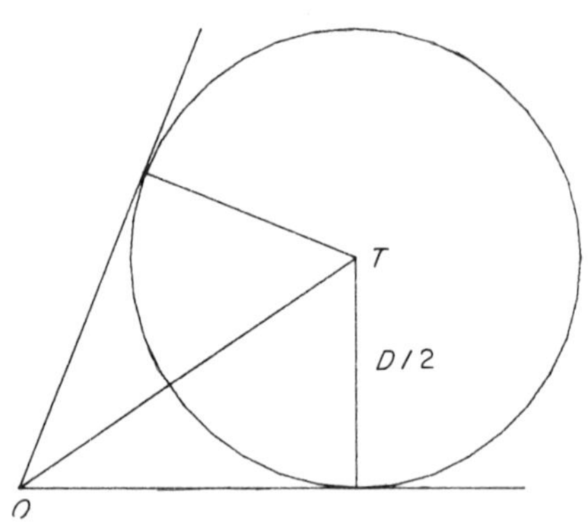

Figure 2.15 The geometry of the angle-count method

$$\hat{\rho}_B = \frac{4\sin^2\alpha(\#D)}{n\pi D^2} + \frac{4\sin^2\alpha(\#D^*)}{n\pi(D^*)^2} + \ldots = \frac{4\sin^2\alpha}{n\pi} \mathrm{m} \sum_{i=1}^{m} (D_i)^{-2}, \tag{2.63}$$

where m is the total number of counted trees and D_i is the observed diameter of the ith counted tree. If we now consider Bitterlich's original problem—that of estimating the proportion of the forest covered by the bases of the trees—we shall discover the reason for the name of this method. A tree of diameter D_i has basal area $\pi D_i^2/4$. Thus, the total basal area of all trees of diameter D is $(\#D)\cdot\pi D^2/4$, and hence the total proportion of the forest area that is taken up by trees of this diameter is

$$\{(\#D)\cdot\pi D^2/4\}/\{\pi D^2/4\sin^2\alpha\} = (\#D)\sin^2\alpha. \tag{2.64}$$

Consequently, when we consider all possible diameters, since the bases of the trees do not overlap, the overall proportion of the forest area that is taken up by the bases of trees is

$$(\#D)\sin^2\alpha + (\#D^*)\sin^2\alpha + \ldots = m\sin^2\alpha, \tag{2.65}$$

which is a simple scaling of m, the number of 'angle-counted' trees.

Ord (1978) seems to have been the first to suggest using $\hat{\rho}_B$ to estimate the spatial intensity of trees. He gives various theoretical results for the variance of $\hat{\rho}_B$ for regular and for random patterns and presents results for data referring to the trees of Lansing woods.

A crucial decision with regard to the angle-count method concerns the choice of α. Holgate (1967) presents a detailed theoretical study. His conclusions are basically that one should use as small a value of α as is consistent with accurate measurement. This is essentially because the smaller the value of α, the greater will be the value of m and this naturally leads to increased precision. The disadvantage of a small value of α is that a small error of measurement can then lead to a large error in the area of the resulting 'acceptance' circle surrounding the observer. Additionally, there is the problem that, as α is decreased, so the chance of missing trees (because near trees obscure ones view of far trees) increases, and the value of m may therefore be underestimated.

According to Holgate, the choice of $\alpha = 104°\ 18'$ was a common early choice, whereas Ord, writing ten years later, states that $1°\ 44'$ is popular! We would suggest that anyone using this method should choose their α-value according to the nature of the woodland. There is no need always to use the same value of α. Indeed, it may be thought worth while to use a range of values within a single forest in order to gain some reassurance about individual values of α.

It is evident from the construction of equation (2.63) that, like the estimators of the previous section, $\hat{\rho}_B$ is essentially the average of the reciprocal of a set of areas. There are, however, two major differences from the previous estimators. The first is that the actual value of m will rarely equal n, the number of sample points—

whereas in the previous section each sample point generated precisely one tile area. The second difference is that the area 'generated' will often overlap one another—whereas with the tessellations that was not possible. The net effect of these differences is that, once again, the more clustered the pattern of trees, the more observations are required in order to get a reasonably precise estimate.

2.7 MOSAICS

Previously in this chapter, and in Chapter 1, we have been concerned almost exclusively with individuals that could be regarded as points. However, this simply reflects the scale of the abstract maps that we have been using. In many situations it may be far more reasonable to associate an individual item with an area. Examples would include buttercups on a lawn, heathers on a small area of moorland, and conurbations as represented on an Ordnance Survey map. Notice that we are dealing here with 'concrete' areas, and not with the abstractions represented by the tessellations that we considered above.

There are at least three properties of mosaics that may be of interest to us: the plant intensity, the total plant area, and the boundary length per unit area.

Plant intensity

Using quadrats, there are no mathematical problems in extrapolating from the small quadrat area to the overall area. There are, however, some practical problems. Because of their physical size, a mosaic may well overlap the boundary of a quadrat—a decision must therefore be made as to whether either the 'centre' or the bulk of the mosaic area lies within the quadrat. Another problem that will arise with any method of estimating plant intensity for mosaics is that it may be difficult to distinguish between neighbouring plants since neighbouring mosaics may coalesce. This is just as true of conurbations as it is of heathers and buttercups.

More interesting, from the mathematical point of view, are the problems posed by using line transects: this is the so-called *line-intercept* method. The procedure is illustrated in Figure 2.16, which shows a rectangular study area of width W and length L, which is traversed by a line transect oriented parallel to one of the sides of the study area. With an irregularly shaped study area some degree of compromise would obviously be required.

The complications that arise are caused by the unequal sizes of the individual plants. It is obvious that a large plant is much more likely to be detected than a small plant. Furthermore, a long thin plant is more likely to be detected if it is oriented at right-angles to the transect line. Clearly, it is this width, w_i, which determines the chance of an individual being detected.

Denote by N the unknown number of plants in the study area. Clearly, our

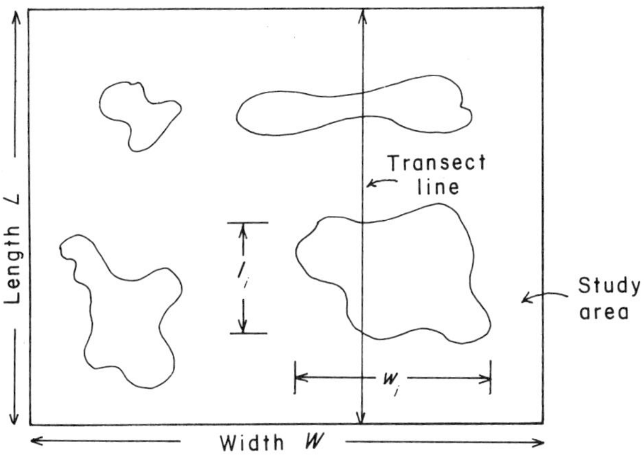

Figure 2.16 Estimating mosaic intensity using the line-intercept method

intensity estimate, $\hat{\rho}$, can be obtained by finding $\hat{N}$, an estimate of N, and by writing $\hat{\rho} = \hat{N}/WL$. Suppose that the line transect intersects with n plants of widths $w_1, w_2, \ldots, w_n$. Before we can use these data we must find an expression connecting $\hat{N}$ to the widths.

The probability of a plant of width w_i being intersected by the transect is $p_i = w_i/W$. The expected number of plants that are intersected by the transect is simply

$$\mathrm{E}(k) = \sum_{i=1}^{N} p_i = \frac{1}{W}\sum_{i=1}^{N} w_i = \frac{N\bar{w}}{W}, \tag{2.66}$$

which leads to

$$N = W \cdot \mathrm{E}(k) \cdot \left(\frac{1}{\bar{w}}\right), \tag{2.67}$$

where $(1/\bar{w})$ is the reciprocal of the average width of the N plants. Evidently both $\mathrm{E}(k)$ and $(1/\bar{w})$ are unknown, but we do have immediate sample estimates, namely n and

$$\widehat{\left(\frac{1}{\bar{w}}\right)} = \frac{1}{n}\sum_{i=1}^{n}\left(\frac{1}{w_i}\right). \tag{2.68}$$

This gives us

$$\hat{N} = W \cdot n \cdot \frac{1}{n}\sum_{i=1}^{n}\left(\frac{1}{w_i}\right), \tag{2.69}$$

and hence

$$\hat{\rho} = \frac{1}{L}\sum_{i=1}^{n}\left(\frac{1}{w_i}\right). \tag{2.70}$$

For a general discussion of this method, and for an illustration of its use, see Eberhardt (1978).

Total plant area and boundary length

Consider the study area shown in Figure 2.17, in which the entire area has been subdivided into non-overlapping regions, some of which correspond to the plants previously illustrated in Figure 2.16. Our intention is to estimate the proportion of the study area that corresponds to the plants of interest.

Suppose first that we used quadrats of total area Q. Evidently the best estimate of the total plant area would be obtained by multiplying the observed plant area within these quadrats by WL/Q—a simple scaling. Suppose instead that our 'quadrats' take the form of a strip transect of width t and length L. Obviously the same procedure would work, and our multiplier would become W/t. Imagine now that t becomes so small that, whenever a plant is intersected, it occupies the full width of the transect. Suppose that the *length* of the transect intersected by the plant in this manner is L_p. Then the plant area observed is tL_p and hence the estimate of the total plant area is $tL_p \cdot (W/t) = WL_p$. If, instead, we estimate the *proportion* of the study area that is covered with the plant, then this is given by $WL_p/(WL) = L_p/L$.

Evidently this procedure would work when the study area was divided into many different plant phases—as suggested in Figure 2.17. Denoting by λ_j, the proportion of the total area that is assigned to phase j, we then have

$$\hat{\lambda}_j = L_j/L, \tag{2.71}$$

where L_j is the total length of the transect line(s) assigned to phase j, L is the (total)

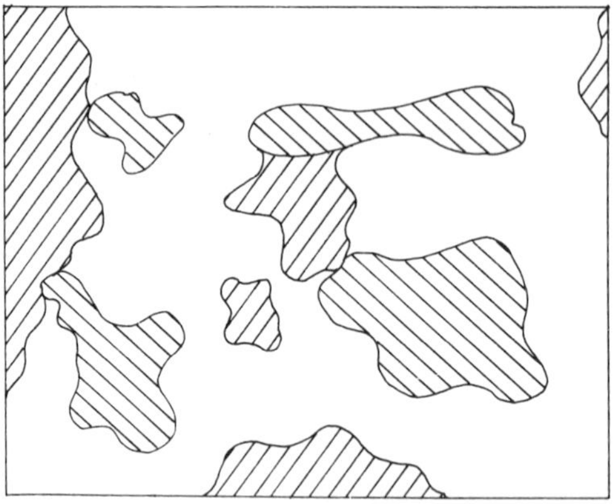

Figure 2.17 A complete partition of the study area

length of the transect line(s), and there is no longer any reason to confine attention to a rectangular study area.

Lackritz and Scheaffer (1983) consider the asymptotic properties of $\hat{\lambda}_j$. Suppose that the transect intersects n cells. Let the intersections be of lengths $X_1, X_2, \ldots, X_n$ and let the mean and variance of the distribution of the $\{X_i\}$ be μ and σ^2, respectively. Lackritz and Scheaffer show that sensible estimates of variances and covariances are given by

$$\text{var}(\hat{\lambda}_j) = \hat{\mu}\hat{\lambda}_j(1 - \hat{\lambda}_j)(1 + \hat{\sigma}^2/\hat{\mu}^2)/L \tag{2.72a}$$

and

$$\text{cov}(\hat{\lambda}_j, \hat{\lambda}_k) = -\hat{\mu}\hat{\lambda}_j\hat{\lambda}_k(1 + \hat{\sigma}^2/\hat{\mu}^2)/L, \tag{2.72b}$$

where $\hat{\mu}$ and $\hat{\sigma}^2$ are simply the sample mean and the sample variance, respectively.

The formulae (2.72a) and (2.72b) apply to the case illustrated in Figure 2.17, in which the entire study area has been subdivided into a mosaic of phases. Often, however, the information available will be at the level of 'plant' and 'not plant', as illustrated in Figure 2.16. In this case the observed intersections will be the aggregations of the unseen $\{X_i\}$, and different asymptotic formulae then result. Denoting the mean and variance of the aggregated lengths assigned to phase j by μ_j and σ_j^2, respectively, Lackritz and Scheaffer obtain an estimate of the variance of $\hat{\lambda}_j$ which can be expressed as

$$\text{var}(\hat{\lambda}_j) = \hat{\lambda}_j(1 - \hat{\lambda}_j)\left\{J - 2 + \sum_{i=1}^{J} (\hat{\sigma}^2/\hat{\mu}_i^2)\right\}\left\{\sum_{l=1}^{J} \hat{\lambda}_l(1 - \hat{\lambda}_l)\hat{\mu}_l\right\} \Big/ (J - 1)L, \tag{2.73}$$

where J is the number of identifiable phases. In the case where there are just two phases (present/absent) this formula simplifies and can be written as

$$\text{var}(\hat{\lambda}_j) = \hat{\lambda}_j^2(1 - \hat{\lambda}_j)^2(\hat{\sigma}_1^2/\hat{\mu}_1^2 + \hat{\sigma}_2^2/\hat{\mu}_2^2)\{(\hat{\mu}_1 + \hat{\mu}_2)/L\}. \tag{2.74}$$

With just two phases it is often the case that the number of intersections with phase 1 is equal to the number with phase 2, in which case the term in curly brackets in equation (2.74) further simplifies to become $1/n$, where n is the common number of intersections.

The general formula for the covariance of $\hat{\lambda}_j$ and $\hat{\lambda}_k$ takes the same form as that given in equation (2.73), although the first terms are $-\hat{\lambda}_j\hat{\lambda}_k$ in place of $\hat{\lambda}_j(1 - \hat{\lambda}_j)$.

***Example* 2.6**

Diggle (1981b) reports on some data concerning the presence of heather (*Calluna vulgaris*) at Jadraas in Sweden. A plot of dimensions 10 m × 20 m was divided into 20 000 10-cm square sub-plots and the heather was recorded as being present or absent for each sub-plot. Figure 2.18 presents a stylized version of part of these data, with the entire figure representing an area of heather 4 m

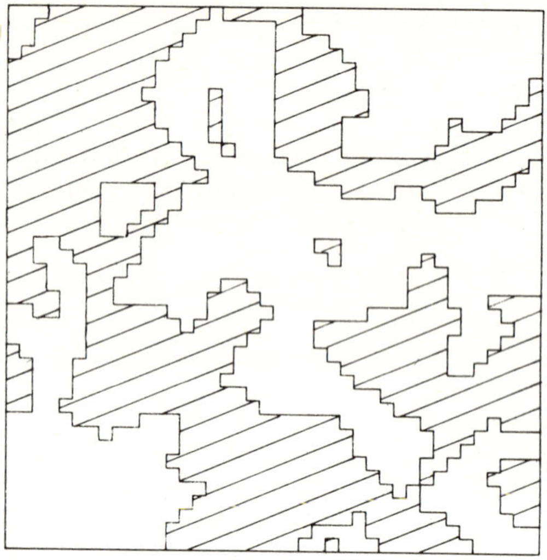

Figure 2.18 A stylised version of the incidence of *Calluna vulgaris*

square. Of the 1600 sub-plots included in the figure, 779 were recorded as containing heather and thus $\lambda = 779/1600 = 0.487$.

Four transect lines were placed at random across the stylized figure and the various intersection lengths were recorded. There were 21 intersections with heather regions and 21 with non-heather regions and hence we can use the simplified form of (2.74) with the final term replaced by 1/21. The observed estimate of λ for the heather phase was 0.509 and the observed values of $\hat{\sigma}_1^2/\hat{\mu}_1^2$ and $\hat{\sigma}_2^2/\hat{\mu}_2^2$ were 0.9843 and 1.0457, respectively, so that the estimated variance is 0.00604. Thus, an approximate 95% confidence interval would be given by $\hat{\lambda} \pm 2\{\mathrm{var}(\hat{\lambda})\}^{1/2}$ which, with these figures, is 0.509 ± 0.155. This, of course, comfortably includes the true value of 0.487.

CHAPTER 3

Spatial Autocorrelation

3.1 GENERAL THEMES

Spatial autocorrelation is a property that mapped data possesses whenever it exhibits an organized pattern or, as Cliff and Ord (1981a, p. 6) express it, whenever there is 'systematic spatial variation' in values across a map. Note that this definition stresses *values* and not position: one takes 'the locations of the observational units of the variable under consideration as given' (Cliff *et al.*, 1975, p. 145). This concern with patterns in the values recorded at locations, as opposed to patterns in the locations *per se*, is what distinguishes this chapter from much of what has already been discussed—just as an emphasis on the 'self-correlation' of a single variable forms a natural dividing line between this and later material more concerned with interrelations between variables. Needless to say, all of these approaches are closely intertwined so that, for example, in the present chapter we shall be reanalysing some of the same sets of data that have already been scrutinized.

While it may be easy to define the property of spatial autocorrelation in general terms, it is evident that there are many different ways in which mapped data can be organized, and it is therefore useful to have a precise definition of what is meant by a *lack* of spatial autocorrelation. The definition proposed by Cliff and Ord (1973, p. 2) amounted to the statement that if there were no connection between the variables (X_i, X_j) at any pair of regions (i,j) in the study area, then the data exhibited a lack of spatial autocorrelation. A more rigorous definition has been suggested by Haining (1980, p. 3). Haining demands complete independence; for quantitative variables this requirement may be written as

$$P(X_1 < x_1, \ldots, X_n < x_n) = \prod_{i=1}^{n} P(X_i < x_i), \tag{3.1}$$

where n is the number of regions in the study area. If the variable is categorical, then an appropriate definition of independence is

$$P(X_i = a, X_j = b) = P(X_i = a) \cdot P(X_j = b), \tag{3.2}$$

where a and b are two possible categories of the variable and $i \neq j$.

Haining (1980) also described three spatial stochastic processes, the first two of which yield variables that comply with the requirements of equations (3.1) and (3.2):

(i) continuous variables $\{X_j\}$ are assigned to each of n locations by drawing n times from a $N(0,\sigma^2)$ distribution;

(ii) discrete (dichotomous) values are assigned to each of n locations on the basis of n tosses of an unbiased coin;

(iii) the value Y_{ij} assigned to a location with coordinates (i,j) is to some extent influenced by values assigned to neighbouring locations; for example,

$$Y_{ij} = \theta(Y_{i-1,j} + Y_{i+1,j} + Y_{i,j-1} + Y_{i,j+1}) + e_{ij},$$

where $|\theta| < 1/4$, where $\{e_{ij}\}$ are uncorrelated error terms from a distribution with zero mean and constant variance, and where the locations are restricted, for notational and computational ease, to the intersection points in a lattice.

While processes (i) and (ii) will normally produce chaotic maps, process (iii) will create some semblance of order.

One difficulty with equations (3.1) and (3.2) is that the quantities $P(X_i < x_i)$ and $P(X_i = a)$ required estimation and that this in turn requires rather extensive knowledge of the form of the distribution of X. It is insufficient to simply assume that, for example, this distribution is $N(0,\sigma^2)$ on the basis of flimsy evidence—even though this may simplify the job of detecting the presence of spatial autocorrelation.

Randomization

An alternative to assuming the form of the underlying probability distributions for the mapped variables is to adopt the so-called randomization approach. As an example, consider the maps shown in Figures 3.1a and b: both maps contain the same n locations and the same x_i-values, the difference between them is simply a different matching of values to locations. In general $n!$ different matchings could be produced by random permutations of the x_i-values. As we shall see below, each map can be summarized by at least one coefficient, and thus there are potentially $n!$ values for each coefficient (though some values may in fact be repeated). Some of these values will turn out to be 'extreme' members of the reference set in the sense that their absolute values will only rarely be exceeded and we shall pay particular attention to such values.

The assignation of values to locations by means of a random permutation is essentially the negation of any master plan, and thus, if the actual arrangement of values that we observe leads to what would have been an extreme value of our coefficient if the values had been distributed at random, then we naturally interpret this fact as suggesting the some 'rule' has been invoked in the organization of the mapped distribution. Since randomization assigns a value to a location irrespective of the whereabouts of the other values, with the proviso that there is only one value per place and that each value is allocated just once,

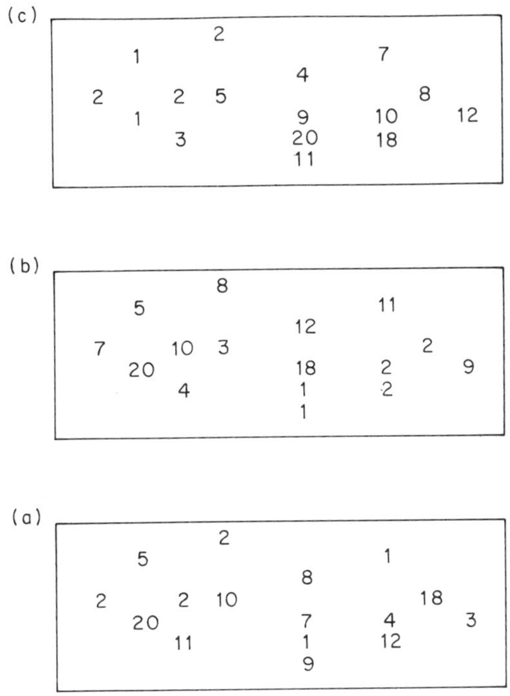

Figure 3.1 Alternative realizations of a randomization procedure

alternative allocation rules would require some consideration to be given to the spatial juxtaposition of values. Figure 3.1c shows a map that would be considered to be an unusual outcome of the randomization procedure, and it is more reasonable to assume that the rule governing this distribution of values placed some control over their spatial layout—in other words Figure 3.1c exhibits spatial autocorrelation.

The distributional outcome of randomization is valid even if one prefers to consider the observed data as a sample outcome of a stochastic process, for Ord (1980) has shown that (for the coefficient I) the results of the randomization procedure hold for stochastic processes with an unknown distribution function.

Randomization thus has considerable practical importance, a point emphasized by Sokal and Oden (1978a) who give it prime importance in their exposition, and by Hubert *et al.* (1981) who emphasize its ability to bypass 'the optimistic use of asymptotic distributional results of possibly unknown accuracy'. Thus, the procedure simultaneously allows one to avoid specifying stochastic processes and to deal with statistics that have awkward sampling distributions. In practice, if there are a large number of values available, then the results of invoking a stochastic process produce—as one would hope and expect—very similar conclusions to those based on the randomization analysis, though there are, of course, some differences of detail.

The general cross-product statistic

So far, coefficients have been mentioned without details being given of how a map can be distilled into a single statistic. Over the years a number of separate statistics have been developed, each with the ability to provide distinctive revelations about spatial distributions, but it is only recently that their close mutual affinity under the umbrella of the general cross-product statistic has been clearly spelt out in the literature (see Hubert *et al.*, 1981). We therefore commence our discussion in terms of this important unifying statistic before treating the better known special cases individually. The general cross-product statistic is given by equation (3.3):

$$r = \sum_i \sum_j W_{ij} Y_{ij}. \tag{3.3}$$

Here W_{ij} is a measure of the *spatial* proximity of locations i and j, while Y_{ij} is a measure of the proximity of i and j on some other dimension. The term W_{ij} is the element in the ith row and jth column of the summary matrix $\mathbf{W}$, while Y_{ij} occupies a corresponding position in a matrix $\mathbf{Y}$.

The statistic r was first proposed by Knox (1964) and generalized by Mantel (1967). A useful account is provided by Cliff and Ord (1981a, p. 26). If the observed values at locations i and j are x_i and x_j, then a common choice for the entries in the $\mathbf{Y}$ matrix is to use

$$Y_{ij} = (x_i - x_j)^2. \tag{3.4}$$

An alternative definition, used by Hubert *et al.* (1983) is $Y_{ij} = \cos(d_i - d_j)$, where d_i and d_j are angular directions at i and j. However, since this relates more to the analysis of directional data, which we describe in Volume 2, it does not feature in the subsequent discussion in this chapter.

Example 3.1

In Table 3.1a are shown the hypothetical x-values at nine contiguous quadrat locations designated as $a, b, \ldots, i$, as shown. The $\mathbf{W}$ matrix for these locations is given in Table 3.1b in which it will be seen that $W_{ij} = 1$ if the locations are contiguous, while $W_{ij} = 0$ otherwise. The entries in this matrix may be termed the *proximity values* of the locations. Using equation (3.4), the $\mathbf{Y}$ matrix is given in Table 3.1c. Using equation (3.3) it is readily seen that the value of r is 8. Now while the 1's in Table 3.1a ostensibly have been purposefully arranged to cluster in the south-east corner, we shall see below that the value $r = 8$ is a relatively likely outcome of a random arrangement of the five 0's and four 1's that constitute the values in the table.

Table 3.1 Examples of the W and Y matrices

(a) Hypothetical density diagram of x

Locations	x-values
a b c	0 0 0
d e f	0 1 1
g h i	0 1 1

(b) W matrix

	a	*b*	*c*	*d*	*e*	*f*	*g*	*h*	*i*
a	0	1	0	1	0	0	0	0	0
b	1	0	1	0	1	0	0	0	0
c	0	1	0	0	0	1	0	0	0
d	1	0	0	0	1	0	1	0	0
e	0	1	0	1	0	1	0	1	0
f	0	0	1	0	1	0	0	0	1
g	0	0	0	1	0	0	0	1	0
h	0	0	0	0	1	0	1	0	1
i	0	0	0	0	0	1	0	1	0

(c) Y matrix

	a	*b*	*c*	*d*	*e*	*f*	*g*	*h*	*i*
a	0	0	0	0	1	1	0	1	1
b	0	0	0	0	1	1	0	1	1
c	0	0	0	0	1	1	0	1	1
d	0	0	0	0	1	1	0	1	1
e	1	1	1	1	0	0	1	0	0
f	1	1	1	1	0	0	1	0	0
g	0	0	0	0	1	1	0	1	1
h	1	1	1	1	0	0	1	0	0
i	1	1	1	1	0	0	1	0	0

(d) Alternative hypothetical density diagram of x

0	1	0
1	0	1
0	1	0

The Monte Carlo approach

Ideally, the value of r should be checked against the complete reference set generated by simultaneously permuting the rows and columns of the **Y** matrix (Hubert *et al.*, 1983). Even with this very simple set of data, there are 9! such permutations, though there are only $9!/4!5! = 126$ different maps and there are even less discrete values in the r reference set (because of the simple binary structures of **W** and **Y**). Evidently the computational burden of complete enumeration will be prohibitive for only moderately large data sets. This problem led Barnard (1963) to suggest that a close approximation to the reference set distribution would be obtained by *sampling* from the 9! permutations—the Monte Carlo approach. Besag and Diggle (1977) strongly recommend Monte Carlo methods whenever full randomization tests 'appear desirable but computationally cumbersome', and details of the method are also given by Edgington (1969) and by Cliff and Ord (1973, 1981a). Hubert *et al.* (1983) mention other alternatives to complete enumeration of the reference set, one of which enables maximum significance levels to be calculated in certain circumstances. Another alternative, due to Siemiatycki (1978), enables the exact distribution to be approximated by a series of curves which are functions of the first four moments of the distribution. All these alternatives are, however, computationally cumbersome and we shall not use them in our analyses.

***Example* 3.1** (Cont.)

One of the many possible permutations of the rows and columns of Table 3.1c is the sequence 652947318 in place of the usual 1–9. This permutation results in, for example, Y_{12} becoming Z_{65}, where **Z** is the randomized version of **Y**. As it happens, this rearrangement also gives the value $r = 8$. This value, together with the outcomes of a further 98 randomly chosen permutations, is summarized in Table 3.2.

***Table* 3.2 *The distribution of r obtained from* 99 *random permutations of the Y matrix given in Table* 3.1c**

r	8	10	12	14	16	18	20	22	24	Total
f	8	11	25	25	11	7	10	0	2	99

Table 3.2 shows that, while the observed value $r = 8$ is an extreme value, it is nevertheless equalled on eight occasions and is therefore insufficiently extreme to be deemed significant. Far more extreme and much less likely is the result $r = 24$, which arises from the permutation of values shown in Table 3.1d.

The normal approximation

Cliff and Ord (1981a) investigate the question of appropriate sampling distributions in considerable detail. They find that if there are a large number of places under consideration, and particularly if the special versions of r, namely the join-count statistics c and I (to be discussed subsequently) are used, then normal approximations are usually accurate and are of practical value in testing the significance of departures from the null hypothesis.

It is the usual practice of statistics to reserve lower-case letters for observed values and to denote the corresponding random variables by upper-case letters. This is a practice to which we shall adhere in other chapters, but in the present chapter we shall use the same symbol in either context. There is actually very little chance of confusion of meaning.

Equations (3.5) and (3.6) give the mean and variance of r, and hence of its approximating normal distribution:

$$E(r) = S_0 T_0 / n(n-1) \tag{3.5}$$

$$\operatorname{var}(r) = \frac{S_1 T_1}{2n^{(2)}} + \frac{(S_2 - 2S_1)(T_2 - 2T_1)}{4n^{(3)}} + \frac{(S_0^2 + S_1 - S_2)(T_0^2 + T_1 - T_2)}{n^{(4)}} - \{E(r)\}^2 \tag{3.6}$$

Here,

$$S_0 = \sum_i \sum_j W_{ij} (i \neq j); \qquad S_1 = \tfrac{1}{2} \sum_i \sum_j (W_{ij} + W_{ji})^2 \qquad (i \neq j),$$

$$S_2 = \sum_i (W_{i0} + W_{0i})^2; \qquad W_{i0} = \sum_j W_{ij}; \qquad W_{0i} = \sum_j W_{ji}; \qquad n^{(2)} = n(n-1),$$

$$n^{(3)} = n(n-1)(n-2); \qquad n^{(4)} = n(n-1)(n-2)(n-3),$$

where n is the number of locations. The formulae for T_0, T_1, and T_2 are identical in form to those for S_0, S_1, and S_2, with the exception being that W is replaced by Y wherever it occurs in the formulae.

When the number of localities is large and there are no special features to the **W** and **Y** matrices, it is reasonable to use the normal approximation, but it is worth noting that the study by Mielke (1978) and the data analysis of Siemiatycki (1978) both show that the approximation 'cannot be taken for granted'.

***Example* 3.1** (Cont.)

As an example of the calculations, consider the data given in Table 3.1a (or, as it happens, Table 3.1d). For this data we have $S_0 = 24$, $S_1 = 48$, $S_2 = 272$, $T_0 = 40$, $T_1 = 80$, and $T_2 = 720$. Consequently, $E(r) = 13.33$ and $\operatorname{var}(r) = 9.52$, while the observed value was $r = 8$.

The distribution of r can only take even values (see Table 3.2) so that if we were to use a normal approximation, then the value of the test statistic would be

$$z = \{|r - E(r)| - 1\} / \{\operatorname{var}(r)\}^{1/2}, \tag{3.7}$$

where the 1 in the numerator is simply a continuity correction (equal to one-half of 2). For the present data $z = \{|8 - 13.33| - 1\}/(9.52)^{1/2} = 1.40$. The probability of a unit normal random variable exceeding 1.40 is (from standard tables) equal to about 8.1%. This is remarkably close to the empirical probability of 8/99 derived from the simulations summarized in Table 3.2— since the true distribution is limited to only a few discrete values (8–24) the apparent accuracy of the normal approximation has to be somewhat fortuitous, but is nevertheless reassuring.

3.2 JOIN-COUNT STATISTICS

The join-count statistics have a considerable literature effusing from the early work of Moran (1948) and Krishna Iyer (1949) and they bear a close relationship to r. The basis of these statistics is that the values $\{x_i\}$ observed at the various locations are coded as either 1 (black) or 0 (white), so that a map of values is represented as a mosaic of black and white areas.

A question of interest is whether neighbouring locations are more likely to display opposite colours or the same colour. This suggests classifying the joins between contiguous regions as being either *BB*, *WW*, or *BW* joins (where, for example, *BB* denotes a join between two black regions).

Following Example 3.1, we define the **W** matrix so that $W_{ij} = 1$ if two regions are contiguous and $W_{ij} = 0$ otherwise. If we use equation (3.4) to define the **Y** matrix, then the number of *BW* joins is simply $r/2$. Similarly, the number of *BB* joins is $r^*/2$, where r^* is the value of r obtained by using the following revised definition for Y_{ij}, namely

$$Y_{ij} = x_i x_j. \tag{3.8}$$

If the total number of joins in the system is J, then obviously the number of *WW* joins is just $J - BB - BW$.

There is little to choose between the three statistics as indicators of autocorrelation, but the results of Cliff and Ord (1981a, p. 176) suggest that *BW* is marginally more informative. Since the number of *BW* joins is given by $r/2$, the mean and variance of the corresponding random variable are respectively $\frac{1}{2}\,\mathrm{E}(r)$ and $\frac{1}{4}\,\mathrm{var}\,(\mathrm{r})$.

An easy method of evaluating *BW* is to count the number of distinct boundary edges (not corners) between contiguous places of different colours. The two maps shown in Figure 3.2 correspond to the data of Tables 3.1a and 3.1d and respectively generate values of 4 and 12 for *BW* (corresponding to the values of 8 and 24 for r). This definition of contiguity (the exclusion of 'corner touching') is referred to as the 'rook's' definition, by analogy with the game of chess. Alternative definitions of contiguity are considered below.

Because of the particularly simple form of the elements of the **Y** matrix, calculation of the T statistics is very easy for all three of these join-count statistics. Suppose that there are a total of n places, of which b are coloured black and $w =$

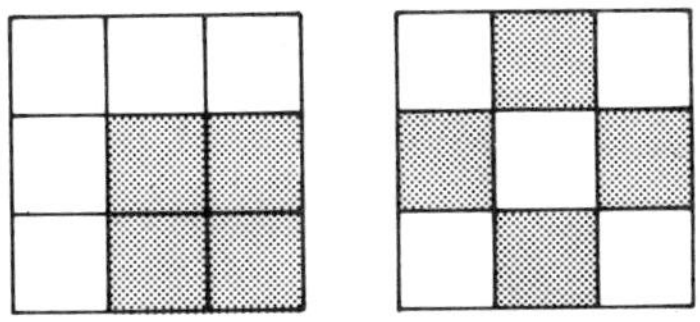

Figure 3.2 Two artificial black/white maps

$(n-b)$ are coloured white. It is easy to show that, for the BW case,

$$T_0 = 2bw, \qquad T_1 = 2T_0, \qquad T_2 = 4nbw, \tag{3.9}$$

and that, for the BB case,

$$T_0 = b(b-1), \qquad T_1 = 2T_0, \qquad T_2 = 4b(b-1)^2. \tag{3.10}$$

When the counts refer to locations on a rectangular (or square) lattice, the S statistics are also particularly easy to calculate. Suppose that the lattice has r rows and c columns. Then, for the rook's case,

$$S_0 = 2(2rc - r - c), \qquad S_1 = 2S_0, \qquad S_2 = 8(8rc - 7r - 7c + 4). \tag{3.11}$$

If the lattice is square, with $r = c$, these formulae simplify to give:

$$S_0 = 4c(c-1), \qquad S_1 = 2S_0, \qquad S_2 = 16(4c^2 - 7c + 2). \tag{3.12}$$

Thus, for a square $c \times c$ lattice, the expected values of the BW and BB statistics are given, for the rook's case, by

$$\mathrm{E}(BW) = 4bw/c(c+1), \tag{3.13}$$

$$\mathrm{E}(BB) = 2b(b-1)/c(c+1). \tag{3.14}$$

***Example* 3.2**

The data on the locations of the perennial shrub *Atriplex hymenelytra* in Death Valley, California, were examined above in Example 1.2. A simple preliminary analysis can be accomplished by coding the quadrats as black/white corresponding to the presence/absence of the shrub. The resulting diagram is Figure 3.3 which should be compared with Figure 1.14.

The value of BW for these data is 173. Since $b = 65$, $w = 191$, $n = 256$, and $r = c = 16$, equations (3.6), (3.9), (3.12), and (3.13) give $\mathrm{E}(BW) = 182.6$ and var $(BW) = 70.7$. The distribution of BW is discrete and the appropriate continuity correction is 0.5, so that we calculate

$$z = \{|173 - 182.6| - 0.5\}/(70.7)^{1/2} = 1.08.$$

Since n is large and neither b nor w are very small, we can legitimately use a

Figure 3.3 Black/white location map for *Atriplex hymenelytra*

normal approximation to assess the significance of this result. The two-sided 95% significance point of a normal distribution is 1.96, which is appreciably larger than our observed value of 1.08. We conclude that our value is by no means unusually extreme and consequently that, on the basis of the results of this test, we have no reason to reject the null hypothesis of a random distribution of plants in the valley. We have no significant evidence that the presence of a shrub in one quadrat has any effect on whether the shrub is present in an adjacent quadrat.

***Example* 3.3**

A similar exercise on the inner 32 rows and 32 columns of the savanna-burning data given in Table 1.2 leads to a rather different conclusion. For that data $BW = 817$, $E(BW) = 913.3$, $\text{var}(BW) = 421.1$, and $z = 4.7$. Since 4.7 is very much greater than 1.96, we conclude that the observed pattern would be a very unlikely one to have occurred as a result of a random arrangement of black and white quadrats. There is very significant evidence of positive spatial autocorrelation (because the BW value is much less than its expected value). A value of BW much greater than $E(BW)$ would be evidence of negative autocorrelation—this is the case for the data given in Figure 3.1d.

***Example* 3.4**

Analogies with chess-boards break down with irregularly shaped areas, though this in no way impedes the analysis. Figure 3.4 is taken from Sokal and

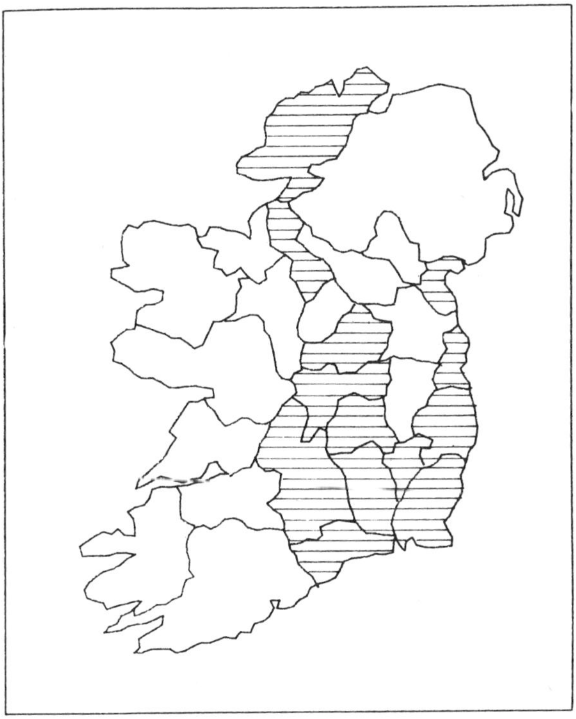

Key: ⊟ = Above median frequency

Figure 3.4 The distribution of the A allele gene

Oden (1978b) and pertains to human blood group distributions in the 26 counties of Eire. Counties with more than the median frequency of the A allele gene (the data refers to 1947) as reported by Mourant *et al.* (1958) are shaded. Sokal and Oden (1978b) state that 'the geographic patterns of gene frequencies ... are clearly non-random, being generally lower in the West and higher in the East of Ireland', and this certainly appears to be a reasonable comment.

To perform the analysis we suitably define contiguity by entering a 1 into the equivalent of Table 3.1b whenever a pair of counties possess a common boundary. The map yields the information that $BW = 24$, $E(BW) = 29.64$, and $\text{var}(BW) = 11.94$, giving $z = 1.49$. Assuming, for the present, that the normal approximation is reasonable, we argue that since $1.49 < 1.96$, we are unable to reject, at the 5% level, the hypothesis of a random distribution of black and white areas: in other words this test provides no significant evidence of spatial autocorrelation.

One-sided tests

In the examples so far, we have assumed that we are equally interested in detecting positive autocorrelation and negative autocorrelation. These situations correspond, respectively, to a clumping pattern and to a chequer-board pattern (i.e. to clusters and to regularity). It will generally be the case, however, that prior knowledge, or plain commonsense, will suggest that any deviation from randomness is certain to be in a particular direction.

Suppose, for example, that we suspect that there may be positive autocorrelation present, while negative autocorrelation is inconceivable. We are only interested in small values of BW. Our question is not 'Does the observed value of BW *differ* significantly from $E(BW)$?', but rather 'Is the observed value of BW significantly *less* than $E(BW)$?'. This is the standard scenario for a one-tailed test. We need compute the value of the test statistic, z, only if $BW < E(BW)$, and (assuming that the normal approximation is reasonable) we reject the null hypothesis of randomness only if $z > 1.645$, where 1.645 is the upper one-tailed 5% point of a unit normal distribution.

There is, of course, nothing sacrosanct about 5%. We could instead use the 4% point or any other significance point that takes our fancy. The value 5% has come to be regarded as a standard reference value, but it possesses no particular virtues. Of course, the implication of the 5% is that on 5% of the occasions that the distribution is random we shall falsely conclude that it is not random; in statistical jargon we have a Type I error probability of 5%.

***Example* 3.4** (cont.)

On commonsense grounds we could expect that, if there were regional variations in blood types, then these would not be delimited by artificial county boundaries, with the consequence that the people in neighbouring counties would not be dissimilar in blood-group distribution: in other words we expect positive autocorrelation and would find the idea of negative autocorrelation in the present context bizarre!

The observed z-value of 1.49 corresponds to a one-tailed normal probability of about 7%. This is not large, but it is not so small that we would feel obliged to reject the null hypothesis of randomness. This result is odd, in view of the strong visual impression of autocorrelation that the map suggests. Of course the fact that we cannot, on the basis of our test results, reject the randomness hypothesis does *not* mean that the hypothesis is true—simply that we have no strong evidence that it is false. An analysis invoking all 32 counties of the whole of Ireland could provide the missing evidence. Alternatively, a more sophisticated test (such as will be suggested below) might utilize the spatial information inherent in the data in a more intelligent fashion and hence be able to detect a departure from randomness. It is interesting to note that Sokal and Oden (1978b) were unable to detect significant evidence of a departure from randomness.

The Poisson approximation

The normal approximation is only reasonable when none of n, np, or $n(1-p)$ is very small, where p is the proportion of black localities out of a total of n localities. For small n, the normal approximation may be improved by the adjustment suggested by Cliff and Ord (1981a, p. 63), although it would be better to use the randomization distribution rather than to chance a doubtful approximation. However, if n is *not* small and either p or $(1-p)$ *is* small, then the upper tail of the BB distribution is closely approximated by that of the Poisson distribution having the same mean. In a similar way, on occasions when the form of the generalized cross-product statistic, r, resembles that of BB, then this too may be approximated by a Poisson distribution (Knox, 1964; Barton and David, 1966). More will be said below about this approximation when r reappears in the context of the space–time interaction tests.

***Example* 3.4** (cont.)

For the Irish counties example, $p = 0.5$ and $n = 26$, so that neither n nor np appear small enough to justify the Poisson approximation. On the other hand, their values are close to the lower limits suggested by Dacey (1965, 1968a) for the use of the normal approximation, and so, in Table 3.3, we present the outcomes of 99 random permutations of the rows and columns of the **Y** matrix for the Irish data.

Table 3.3 *The randomization distribution of BW for the Irish blood data*

BW	22	23	24	25	26	27	28	29	30	31	32	33	34	35	36	Total
f	2	3	4	2	8	4	6	10	19	8	12	8	7	4	2	99

By comparison with the symmetrical normal distribution the exact distribution of BW is evidently somewhat skewed and is rather similar to that obtained by Silk (1979, p. 119) for a 7×7 lattice. There are nine values equal to or smaller than the observed value, 24, and hence once again the observed value cannot be regarded as providing significant evidence of positive spatial autocorrelation.

***Example* 3.5**

As an example of a situation in which a Poisson approximation is likely to be relevant, consider the black/white map given in Figure 3.5. This is the inner 32×32 quadrat grid for the savanna-burning data of Table 1.2, with a quadrat coded as black only if there was in excess of 40 g of herb remains within that quadrat. This was not a frequent occurrence!

For this data $S_0 = 3968$, $S_1 = 7936$, and $S_2 = 61\,984$. For the BW statistic we have $BW = 82$, $T_0 = 48\,000$, $T_1 = 96\,000$, and $T_2 = 98\,304\,000$ so that $E(BW) =$

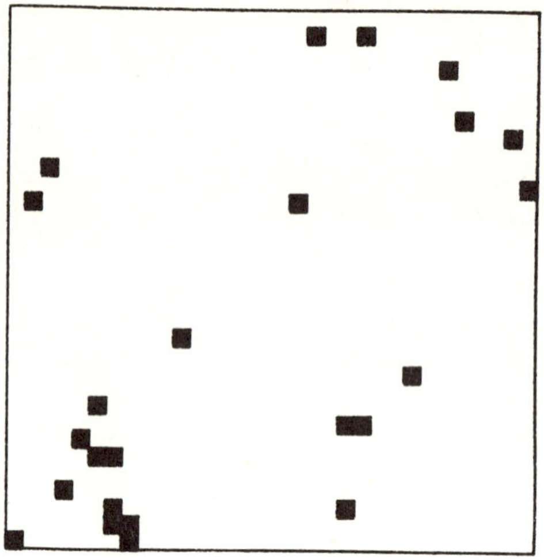

Figure 3.5 Areas of dense herb remains for the Nigerian savanna data

90.9 (a result consistent with 24 black quadrats each providing roughly 4 BW joins), var $(BW) = 6.5$, and $z = 3.3$. This value for z greatly exceeds 1.645 and we therefore conclude that there appears to be strong evidence of positive autocorrelation.

If there is positive autocorrelation, then we would expect an unusually large number of BB joints. It is easy to see that, for this data, $BB = 5$. Using the equations (3.10) we have $T_0 = 552$, $T_1 = 1104$, and $T_2 = 50\,784$ so that $\mathrm{E}(BB) = 1.045$ and var $(BB) = 0.998$. Evidently BB joins are rare events (because of the scarcity of black quadrats) and the suggestion of using a Poisson approximation is reinforced by noting that $\mathrm{E}(BB) \simeq \mathrm{var}\,(BB)$. Using 1.045 as an estimate of the Poisson parameter, λ, we find that

$$P(BB \geq 5) = 1 - \left(1 + \lambda + \frac{\lambda^2}{2!} + \frac{\lambda^3}{3!} + \frac{\lambda^4}{4!}\right) \cdot \mathrm{e}^{-\lambda} = 0.0044.$$

Evidently the observed value of 5 would be a *most* unusual value to have been obtained from a random arrangement of 24 black and 1000 white quadrats within a 32×32 lattice. There can be no reasonable doubt that this data displays a strong positive spatial autocorrelation.

Multicoloured maps

If a more probing analysis is required of a set of data, then this may be accomplished by allowing x to take more than just the two values 0 and 1. The corresponding pictorial representation is then 'multicoloured' rather than 'monochrome'! A review of this type of analysis is provided by Dacey (1968b). Sokal and Oden (1978b) provide an extreme example in which 17 colours are used, each colour corresponding to a different tree species.

The formulae (3.5) and (3.6) extend to this situation in a natural fashion. We need to distinguish between two types of joins and write S_{rr} as the number of joins between contiguous places coloured in the *same* way (with colour r), and D_{rs} as the number of joins between contiguous places coloured in *different* ways (with colours r and s). The results (Cliff and Ord, 1973, p. 14) are

$$\mathrm{E}(S_{rr}) = S_0 n_r^{(2)}/2n^{(2)}, \tag{3.15a}$$

$$\operatorname{var}(S_{rr}) = \frac{1}{4}\left[\frac{S_1 n_r^{(2)}}{n^{(2)}} + \frac{(S_2 - 2S_1)n_r^{(3)}}{n^{(3)}} + \frac{(S_0^2 + S_1 - S_2)n_r^{(4)}}{n^{(4)}} - 4\{\mathrm{E}(S_{rr})\}^2\right], \tag{3.15b}$$

$$\mathrm{E}(D_{rs}) = S_0 n_r n_s/n^{(2)}, \tag{3.15c}$$

$$\operatorname{var}(D_{rs}) = \frac{1}{4}\left[\frac{2S_1 n_r n_s}{n^{(2)}} + \frac{(S_2 - 2S_1)n_r n_s(n_r + n_s - 2)}{n^{(3)}} + \frac{4(S_0^2 + S_1 - S_2)n_r^{(2)} n_s^{(2)}}{n^{(4)}} - 4\{\mathrm{E}(D_{rs})\}^2\right], \tag{3.15d}$$

where n_r is the number of places coloured r. For a rectangular lattice the equations (3.11) may be used to calculate S_0, S_1, and S_2, as before. The expressions involving n_r and n_s that appear in the numerators of the terms in the equations (3.15) will be found to be extensions of the formulae for the T statistics given in equations (3.9) and (3.10). All the previous discussion concerning distributions and approximations carries over directly to the multicoloured situation, as Example 3.6 makes clear.

Example 3.6

Following our analyses of the savanna-burning data on an absence/presence basis in Example 3.3, and on an abundant/not abundant basis in Example 3.5, we now consider the multicoloured approach. We define x to take the value 4 for quadrat 'counts' of 4 or more, and otherwise to take the value of the quadrat count. The raw data was given in Table 1.2, and it is found that, with this coding, we have $n_0 = 657$, $n_1 = 215$, $n_2 = 98$, $n_3 = 30$, and $n_4 = 24$. The values of S_0, S_1, and S_2 are 3968, 7936, and 61 984, as before. The statistics given by the equations (3.15) are summarized in Table 3.4, along with the observed join-counts and, where appropriate, the corresponding continuity-corrected z-values. The z-values are not given for those cases where a Poisson approximation is appropriate.

Scrutiny of Table 3.4 confirms the previous analyses of Table 1.2 and adds more detail. The value of S_{00} is significantly high, reflecting the positive

Table 3.4 ***Statistics relating to the analysis of Table 1.2 as a multicoloured map***

x levels	Observed join frequency	Expected join frequency	Variance	z-value
(r,r)				
0,0	864	816.3	116.1	4.4
1,1	94	87.1	55.3	0.9
2,2	18	18.0	14.9	0.0
3,3	2	1.7	1.6	
4,4	5	1.1	1.0	
(r,s)				
0,1	503	535.1	227.8	2.1
0,2	213	243.9	97.2	3.1
0,3	61	74.7	28.6	2.5
0,4	40	59.7	22.8	4.0
1,2	99	79.8	59.0	2.4
1,3	28	24.4	19.0	0.7
1,4	23	19.6	15.3	0.7
2,3	15	11.1	9.8	1.1
2,4	14	8.9	7.9	1.6
3,4	5	2.7	2.6	

autocorrelation of the herbless areas which is manifest in large herbless plains. There is a commensurate reduction in the numbers of contiguous empty and occupied quadrats. Note that the degree of segregation (negative autocorrelation) between occupied and empty quadrats appears to become increasingly marked as the quantity of herb remains in the occupied quadrat increases. This simply reflects the fact—evident in the original data—that the areas of large herb remains are not isolated but are the centres of relatively unburnt areas.

The bishop's definition of contiguity

In the case of data organized as contiguous quadrats, the definition of contiguity employed up till now has been that quadrats abut only if they have touching *edges*. However, this is not the only possible definition of contiguity and there may be occasions when we are more interested in diagonal trends. In such cases a more relevant definition of contiguity would be that quadrats abut if they have touching *corners*—this is the bishop's case. Table 3.5a gives the bishop's case **W** matrix for the simple lattice diagram of Table 3.1a.

As in the rook's case, the values of the S statistics are easily calculated for rectangular $r \times c$ lattices. We find that

$$S_0 = 4(r-1)(c-1), \qquad S_1 = 2S_0, \qquad S_2 = 16(2r-3)(2c-3), \tag{3.16}$$

Table 3.5 W matrices for alternative contiguity rules applied to Table 3.1

(a) Bishop's case

0	0	0	0	1	0	0	0	0
0	0	0	1	0	1	0	0	0
0	0	0	0	1	0	0	0	0
0	1	0	0	0	0	0	1	0
1	0	1	0	0	0	1	0	1
0	1	0	0	0	0	0	1	0
0	0	0	0	1	0	0	0	0
0	0	0	1	0	1	0	0	0
0	0	0	0	1	0	0	0	0

(b) Queen's case

0	1	0	1	1	0	0	0	0
1	0	1	1	1	1	0	0	0
0	1	0	0	1	1	0	0	0
1	1	0	0	1	0	1	1	0
1	1	1	1	0	1	1	1	1
0	1	1	0	1	0	0	1	1
0	0	0	1	1	0	0	1	0
0	0	0	1	1	1	1	0	1
0	0	0	0	1	1	0	1	0

so that, for a square $c \times c$ lattice we have

$$S_0 = 4(c-1), \qquad S_1 = 2S_0, \qquad S_2 = 16(2c-3)^2, \tag{3.17}$$

$$\mathrm{E}(BW) = 4bw(c-1)/c^2(c+1), \tag{3.18}$$

$$\mathrm{E}(BB) = 2b(b-1)(c-1)/c^2(c+1). \tag{3.19}$$

***Example* 3.6** (cont.)

The previous multicoloured approach to the savanna-burning data adopted the rook's definition. If the bishop's definition of contiguity is employed, then an implicit assumption of diagonal trends is embodied in the alternative hypothesis. The results of the new analysis are given in Table 3.6, which should be compared with Table 3.4. It will be seen that, on the whole, the z-values in Table 3.6 are smaller than their counterparts in Table 3.4, suggesting that, for the most part, any apparent diagonal alignment is a secondary consequence of dominant horizontal or vertical clumping. An exception is the S_{11} value which is markedly larger than its rook's case counterpart, suggesting a significant diagonal alignment of the thinly populated $x=1$ quadrats. There is an excellent explanation for this which has been provided by Hopkins' original (1965) discussion: the prevailing wind in this part of the savanna blows from the south-west, which is represented by the top right-hand corner of the table, and the

Table* 3.6 *Statistics relating to the analysis of Table* 1.2 *as a multicoloured map, bishop's case

x levels	Observed join frequency	Expected join frequency	Variance	z-value
(r,r)				
0,0	823	790.8	144.4	2.6
1,1	101	84.4	56.0	2.2
2,2	19	17.4	14.6	0.3
3,3	3	1.6	1.5	
4,4	3	1.0	1.0	
(r,s)				
0,1	497	518.3	234.9	1.4
0,2	216	236.3	104.4	1.9
0,3	58	72.3	31.5	2.5
0,4	48	57.9	25.2	1.9
1,2	81	77.3	58.7	0.4
1,3	21	23.7	18.9	0.5
1,4	21	18.9	15.1	0.4
2,3	17	10.8	9.6	1.8
2,4	10	8.6	7.7	0.3
3,4	4	2.6	2.5	

physics of the burning mechanism is such that one would expect to find bands of vegetation at right-angles to the wind—in other words, in *diagonal* bands from the top left to the bottom right of the table.

The queen's definition of contiguity

A composite definition of contiguity is obtained if we acknowledge that two areas abut if they have either touching corners or touching edges: this is the queen's case. Because this definition has a more all-embracing alternative hypothesis it can fail to detect specialized types of pattern. For example, an analysis of the checkerboard pattern of Table 3.1d gives $E(BW) = 11.11$ compared with $BW = 12$. This represents a clear failure to detect the obvious organization that is present. The reason for this failure is that the queen's case amalgamates the proliferation of rook's case BW joins with the dearth of bishop's case BW joins to come up with an overall total that seems quite usual. Monte-Carlo studies by Cliff and Ord (1981a, p. 174) provide detailed evidence of the comparative power of the rook's and queen's definitions and they note that, for **W** matrices in general, increased coverage of possible patterns inevitably creates a less informative test.

For the queen's case the results for $r \times c$ and $c \times c$ lattices are as follows:

$$S_0 = 2(4rc - 3r - 3c + 2), \qquad S_1 = 2S_0, \qquad S_2 = 8(32rc - 39r - 39c + 46), \tag{3.20}$$

$$S_0 = 4(2c - 1)(c - 1), \qquad S_1 = 2S_0, \qquad S_2 = 16(16c - 23)(c - 1), \tag{3.21}$$

$$E(BW) = 4bw(2c - 1)/c^2(c + 1), \tag{3.22}$$
$$E(BB) = 2b(b - 1)(2c - 1)/c^2(c + 1). \tag{3.23}$$

Example **3.6** (cont.)

The results of the autocorrelation analysis of the savanna data, using the queen's definition of contiguity which employs the **W** matrix given in Table 3.5b, are given in Table 3.7 which can be compared with Tables 3.4 and 3.6. For the savanna data it is apparent that the contributions from the rook's and bishop's cases reinforce one another with the result that the composite z-values are roughly equal to the larger of the two separate z-values. We conclude that there are clumps of unburnt vegetation that have a diagonal orientation.

Multiple hypothesis tests

In examining tables such as Table 3.7 we have unintentionally become embroiled in the problem of the multiple testing of hypotheses. As we remarked above, if we perform a test at the 5% significance level, then, on 5% of the occasions on which the null hypothesis is true, we shall make a 'Type I Error' and will falsely conclude that it is false. In each of Tables 3.4, 3.6, and 3.7 we performed 12 tests of significance. If there was no autocorrelation present in the data we would nevertheless expect to have obtained a few 'significant' results. One simple ploy to alleviate this difficulty is to adhere to a Type I error rate rather lower than the

Table* 3.7 *Statistics relating to the analysis of Table* 1.2 *as a multicoloured map, queen's case

x levels	Observed join frequency	Expected join frequency	Variance	z-value
(r,r)				
0,0	1687	1607.0	303.5	4.6
1,1	195	171.6	114.2	2.1
2,2	37	35.4	29.7	0.2
3,3	5	3.2	3.1	
4,4	8	2.1	2.0	
(r,s)				
0,1	1000	1053.4	480.7	2.4
0,2	429	480.2	215.0	3.5
0,3	119	147.0	65.1	3.5
0,4	88	117.6	52.0	4.0
1,2	180	157.1	119.7	2.0
1,3	49	48.1	38.3	0.1
1,4	44	38.5	30.8	0.9
2,3	32	21.9	19.5	2.2
2,4	24	17.5	15.7	1.5
3,4	9	5.4	5.1	

usual 5% when searching through whole sets of statistics for significance. We should also look for supporting evidence elsewhere: the appearance of a single inexplicable 'significant' value amongst a stream of 'insignificant' values can be reasonably attributed to chance, whereas a mass of highly significant values—as with the savanna data—impress one by their sheer number.

3.3 MORAN'S *I*

The join-count statistics utilized in the previous section reduced the information in the data to its lowest level, i.e. present/absent. In this section we employ Moran's (1950) I statistic, given by equation (3.24), to seek out autocorrelation, using data which need not necessarily be at the lowest level of measurement. As before, we denote the observed value at location i by x_i, and we write $\bar{x}$ for the average of the $\{x_i\}$ over the n locations. Moran's statistic is defined by

$$I = \frac{n}{S_0} \frac{\sum_i \sum_j W_{ij}(x_i - \bar{x})(x_j - \bar{x})}{\sum_i (x_i - \bar{x})^2}, \tag{3.24}$$

while a closely related statistic suggested by Geary (1954) is

$$c = \frac{(n-1)}{2S_0} \frac{\sum_i \sum_j W_{ij}(x_i - x_j)^2}{\sum_i (x_i - \bar{x})^2}. \tag{3.25}$$

Hubert *et al.* (1981) show that, if the entries of the **Y** matrix of equation (3.3) representing the 'distance' between the x-values at i and j are calculated according to the equation

$$Y_{ij} = (x_i - \bar{x})(x_j - \bar{x}), \tag{3.26}$$

then we may obtain I by dividing the resulting cross-product statistic, r, by S_0 and by the sample variance of the observed x-values. In a similar fashion, the statistic c may be related to r using the definition:

$$Y_{ij} = (x_i - x_j)^2. \tag{3.27}$$

Note that equation (3.27) is the same as equation (3.4) which was used in the calculation of the BW statistic.

When there is no autocorrelation present the expectation of I is $-1/(n-1)$, while that of c is 1, whereas when there is a maximum positive autocorrelation present c will be near 0 and I will approach 1. The difference in the form of equations (3.24) and (3.25) implies that c is more sensitive to the absolute differences between pairs of values, while I is more sensitive to extreme x-values. Jumars *et al.* (1977) make a case for calculating both I and c because of this difference, and present an example in which the autocorrelation that is present is clearly detected by c and overlooked by I. However, this is a special case: in general the results of analyses using c and using I will provide broadly the same conclusions. We shall concentrate on I because Cliff and Ord (1975a, 1981a) have shown that tests based on I are consistently more powerful than those based on c.

For the sake of completeness we note that Haining (1977, 1978a) has developed a likelihood ratio statistic λ which is appropriate when the alternative hypothesis (to the hypothesis of no autocorrelation) specifies either a simultaneous moving average scheme or the simultaneous autoregressive scheme given as (iii) at the start of this chapter. These models and related models are discussed in Chapter 5. Cliff and Ord (1981a, pp. 167, 177) demonstrate that λ is often exactly equivalent to I, and their results suggest that their is little advantage to be gained from the considerable extra calculations involved.

The distribution of I under randomization

Cliff and Ord (1981a, p. 46) show that the mean and variance of the sampling distribution of I arising from the randomization process are given by

$$\mathrm{E}(I) = -1/(n-1), \tag{3.28}$$

$$\mathrm{var}(I) = \frac{n\{(n^2-3n+3)S_1 - nS_2 + 3S_0^2\} - k\{n(n-1)S_1 - 2nS_2 + 6S_0^2\}}{(n-1)^{(3)}S_0^2} - \frac{1}{(n-1)^2}, \tag{3.29}$$

where

$$k = m_4/m_2{}^2 \quad \text{and} \quad m_r = \frac{1}{n}\sum_i (x_i - \bar{x})^r. \tag{3.30}$$

As with the join-count statistics, it is possible to assume I to be approximately normally distributed under the null hypothesis unless fairly liberal conditions are violated (Cliff and Ord, 1971). Normality hinges on the number of places under consideration, and on the extent and manner in which they are interconnected in the **W** matrix: with this latter reservation we are probably justified in considering 20 places sufficient to assume normality. Cliff and Ord (1981a, p. 52) note in detail the circumstances under which the normal approximation may not hold, an example of which is outlined by Ripley (1981, p. 99) who describes a regional system that produces an uncooperative **W** matrix for which we must resort to a Monte Carlo approach.

A detailed treatment is provided by Sokal and Oden (1978a) who give a step-by-step breakdown of each stage of the calculation of I, E(I), and var (I), using, as an example, the spatial variations of the wing-length of the aphid *Pemphigus populitransversus* over a region of Kansas, USA. In their example they make use of the small-sample corrections suggested by Cliff and Ord (1971, 1973) as an alternative to the randomization distribution.

***Example* 3.7**

In a study of the drongo, an asiatic bird of the family *Dicruridae*, Mayr and Vaurie (1948) found apparent evidence of spatial gradients in characteristics such as wing lengths and the lengths of the birds' feathery crests. Such *clines*, as Huxley (1939) called them, are apparently a commonly occurring biological pattern, though this is not surprising since they are a rather loosely defined concept, with no implication, for example, that the gradient is constant. Because of this generality, the cline, like the patch, describes one important way in which nature is organized in a decidedly non-random pattern, though the 'discovery' of clines may occasionally not stand up to scientific analysis. It was with this in mind that Royaltey *et al.* (1975) set out to discover whether the patterns identified by Mayr and Vaurie (1948) were real or imaginary, and their analysis was based on the rank orders of the species' characteristics. Our analysis will employ I so that the specific location values can be used rather than their ranks, so that we may expect to obtain a more informative answer.

The lines in the maps of Figure 3.6 connect the localities for which measure-

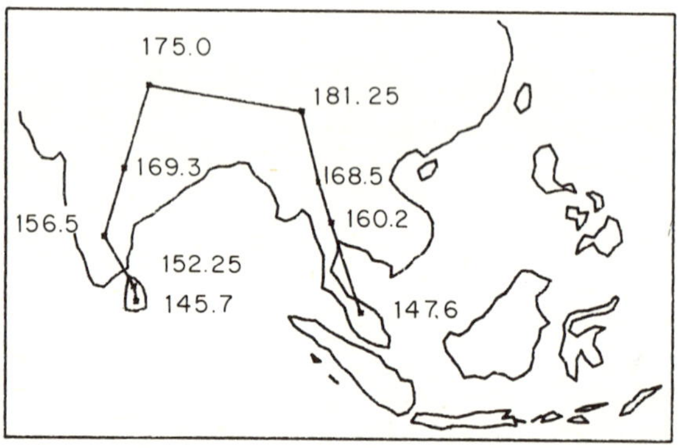

(a)

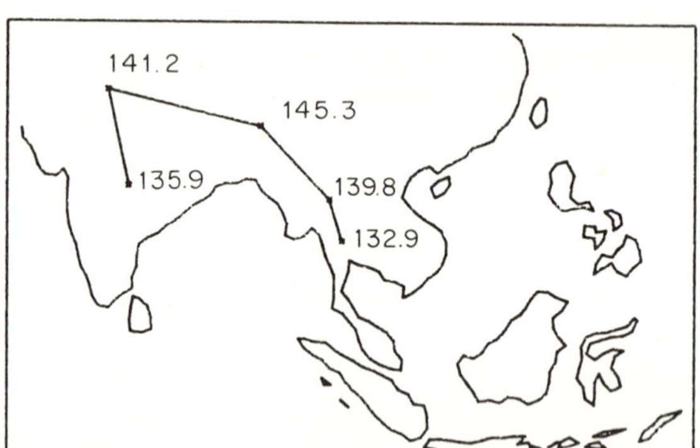

(b)

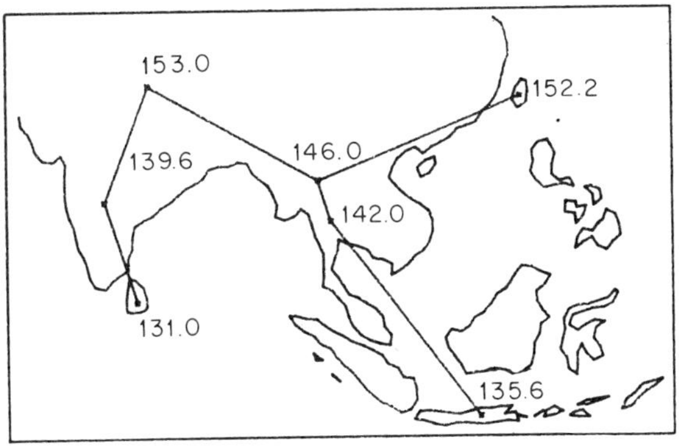

(c)

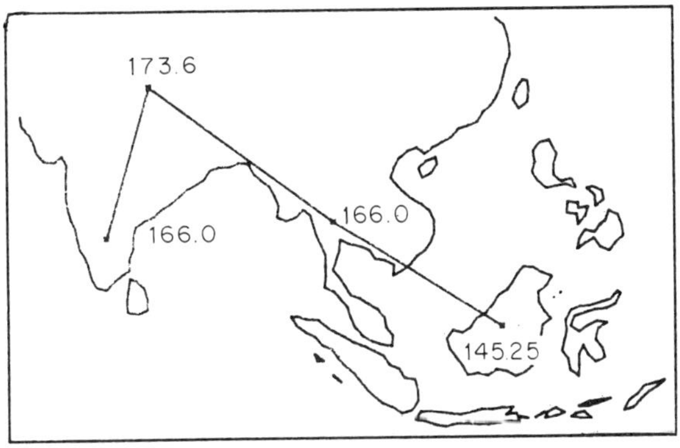

(d)

Figure 3.6 Clines in winglengths of species of the *Drongo*. (a) *paradiseus*. (b) *leucophaeus*. (c) *macrocerus*. (d) *hottentotus*. (Redrawn from Mayr and Vaurie, 1948)

ments are available and indicate the probable routes of bird-movement and gene-flow. These connecting links form the proximity values for the associated **W** matrices. Of course, these links represent just one theory of gene-flow and different alternative hypotheses would require different **W** matrices.

Table 3.8 displays the **W** matrix for the species *paradiseus* for which $S_0 = 16$, $S_1 = 32$, and $S_2 = 120$. Substituting these values together with the x_i-values indicated on the map for this species (Figure 3.6a) into equations (3.24), (3.28), and (3.29) we obtain $I = 0.603$, $E(I) = -0.125$, and $\text{var}(I) = 0.108$. If there had been more locations than just the nine that there actually are, we would use a normal approximation and calculate $z = \{I - E(I)\}/\sqrt{\text{var}(I)}$, which in this case is equal to 2.21, and exceeds the upper one-tailed 5% point (1.645) of the unit normal distribution. This would imply that there was indeed some significant

Table 3.8 A **W** *matrix pertaining to the species* **paradiseus**

Location	A	B	C	D	E	F	G	H	I
A	0	1	0	0	0	0	0	0	0
B	1	0	1	0	0	0	0	0	0
C	0	1	0	1	0	0	0	0	0
D	0	0	1	0	1	0	0	0	0
E	0	0	0	1	0	1	0	0	0
F	0	0	0	0	1	0	1	0	0
G	0	0	0	0	0	1	0	1	0
H	0	0	0	0	0	0	1	0	1
I	0	0	0	0	0	0	0	1	0

The locations are indicated in Figure 3.6a.

evidence of the existence of a wing-length cline. However, $n = 9$ and the **W** matrix has few non-zero entries so that the normal approximation is open to criticism and instead we should assess the significance of the observed value of I by comparison with its randomization distribution. In fact, in 99 random permutations of the **Y** matrix, on only one occasion is the observed value equalled or exceeded (see Table 3.9a), so that we do indeed conclude that there is significant evidence of a wing-length cline for the species *paradiseus*.

For the species *leucophaeus* we have information only for five simply connected localities; consequently there are only $5!/2$ distinct values for the permutation distribution of I. These values are summarized in Table 3.9b. The observed value of I for this data is 0.0953 and, from the randomization distribution, we discover that the chance of obtaining this or a larger positive value is $15/60 = 0.25$, which obviously is not an unusually small probability so that we conclude that the data is suggestive (in as much as it resembles that for *paradiseus*) but does not on its own provide significant evidence of autocorrelation.

A similar conclusion applies to the species *macrocerus*, for which the observed value of $I = 0.2059$ is exceeded by 14 of the 99 random permutations. Data for the wing-lengths of some other species are confined to so few localities that there is no possibility of obtaining a 'significant' result—for the species *hottentotus*, for example, there are only six possible values for I. Nevertheless, the similarity in the observed clines does provide cumulatively convincing evidence of their reality.

Example 3.8

Previous analyses of the burnt savanna data have established beyond doubt the presence of appreciable positive spatial autocorrelation; nevertheless, it is of interest to note the extra discriminatory power of the I statistic which is calculated from the quadrat counts directly and not from some coded substitutes.

Table 3.9 The cumulative distributions of* I *for the data sets given in Figure 3.6

(a) *Paradiseus* (99 random permutations)

Value of I	−1.0	−0.8	−0.6	−0.4	−0.2	0	0.2	0.4	0.6	0.8
Freq. $\geq I$	99	98	94	78	59	40	19	7	1	0

(b) *Leucophaeus* (complete enumeration)

Value of I	−1.2	−1.0	−0.8	−0.6	−0.4	−0.2	0	0.2	0.4	0.6
Freq. $\geq I$	60	59	54	47	37	27	17	10	5	0

(c) *Macrocerus* (99 random permutations)

Value of I	−1.2	−1.0	−0.8	−0.6	−0.4	−0.2	0	0.2	0.4	0.6
Freq. $\geq I$	99	98	95	89	69	52	33	15	5	0

(d) *Paradiseus* (99 random permutations—ranked data)

Value of I	−1.0	−0.8	−0.6	−0.4	−0.2	0	0.2	0.4	0.6
Freq. $\geq I$	99	97	86	68	45	28	15	6	0

We find that $I = 0.1876$, while $E(I) = -0.00098$ and $\text{var}(I) = 0.000487$ so that the z statistic (a normal approximation is certainly reasonable) takes the extremely high value of 8.55. We are left in no doubt that the I statistic has found striking evidence of autocorrelation!

Moran's *I* with ranked data

The application of I to ranked data is straightforward. In this case the 'value' for the localities consist of the ranks $1,2,\ldots,n$ and consequently the equation for I simplifies to become:

$$I = \frac{12}{S_0(n^2-1)} \sum_i \sum_j W_{ij}(x_i - \bar{x})(x_j - \bar{x}). \tag{3.31}$$

The expressions for $E(I)$ and $\text{var}(I)$ do not simplify, but the value of k may now be expressed as $3(3n^2 - 7)/5(n^2 - 1)$ which is approximately equal to 1.8 for values of n greater than 10.

***Example* 3.7** (cont.)

Obviously, if we have actual x-values available, then these should be used since they will lead to a more sensitive test. However, as an example, suppose that, for the *paradiseus* data we do not have the values 145.7, 152.25, 156.5, 169.3, 175.0, 181.25, 168.5, 160.2, and 147.6, but only their ranks which are 1, 3, 4, 7, 8, 9, 6, 5, and 2. Using the ranks we have $I = 0.5625$, $E(I) = -0.125$, and $\text{var}(I) = 0.1231$, so that $z = 1.96$. This is, indeed, slightly smaller than the z-value obtained previously. However, as before, we should resort to the randomization distribution in order to test significance, and this is summarized in Table 3.9d. As it happens our value of 0.5625 is not once exceeded by the values obtained from the 99 random permutations (which are not the same as those used previously), indicating strong evidence of spatial ordering.

3.4 GENERALIZED PROXIMITY VALUES

Up to this point our applications of the join-count and I statistics have presumed that the proximity values of the **W** matrix are confined to the integers 1 and 0. From the point of view of the general cross-product statistic, which it will be remembered unifies these disparate measures and comprises *two* distance matrices **W** and **Y**, this means that spatial closeness is a rather unsubtle measure compared with the apparent precision afforded to the entries in the **Y** matrix. There is, of course, no reason why, where necessary, spatial proximity cannot be measured on an equally discriminatory scale. It was one of the contributions of Cliff and Ord (1969) to make this realization and to work out the ramifications in the context of a generalized version of Moran's I and Geary's c.

There are, of course, any number of schemes for obtaining proximity values, and the rationale of any one of these is a good indication of the reasoning and logical basis of the alternative hypothesis with which it is synonymous. For example, a reason for replacing 0 and 1 as proximity values for contiguous counties, as in Example 3.4, by, say, the lengths of common boundary, is that the latter will more closely reflect the quantity of inter-county interaction. Another feasible alternative, applicable to both point data and to area-based data, is to allow proximity values to depend in some way on the geographical distance between the places. This will often mean that there are no off-diagonal zeroes in the **W** matrix. As long as some satisfactory and generally agreeable measure of 'distance' is used, this approach seems logically sound since it recognises the fact that, in the real world, interaction between places does not usually terminate sharply beyond places that are immediate neighbours. One satisfactory side-effect of using such non-binary proximity values is that the distribution of the test statistic is more closely approximated by a continuous function such as the normal distribution (Cliff and Ord, 1981a, p. 56).

***Example* 3.9**

Mayr (1942, 1963) provides evidence of 'speciation', the 'multiplication of species in space' (Romanes, 1897); among varieties of passerine birds in the Central Solomon islands and strongly suggests that the inter-island straits have presented such a barrier to communication that 'strikingly' different populations in several species have evolved. Cursory examination of the data in Figure 3.7 indicates that similar characteristics occur on adjacent islands, whereas elsewhere subspeciation and differentiation have taken place. If these conjectures are true, then significant positive autocorrelation should be detectable when the **W** matrix contains proximity values representing these geographical facts. To illustrate the use of generalized proximity values, it is apposite that the entries in the **W** matrix are calculated as the inverses of the minimum straight-line distance between each island pair. We assume, initially, that birds on even the most widely separated islands have had some opportunity to interbreed. This assumption runs contrary to the suggestion that some of the straits between the islands are *totally* impassable so far as passerine bird travel is concerned, and we shall examine this

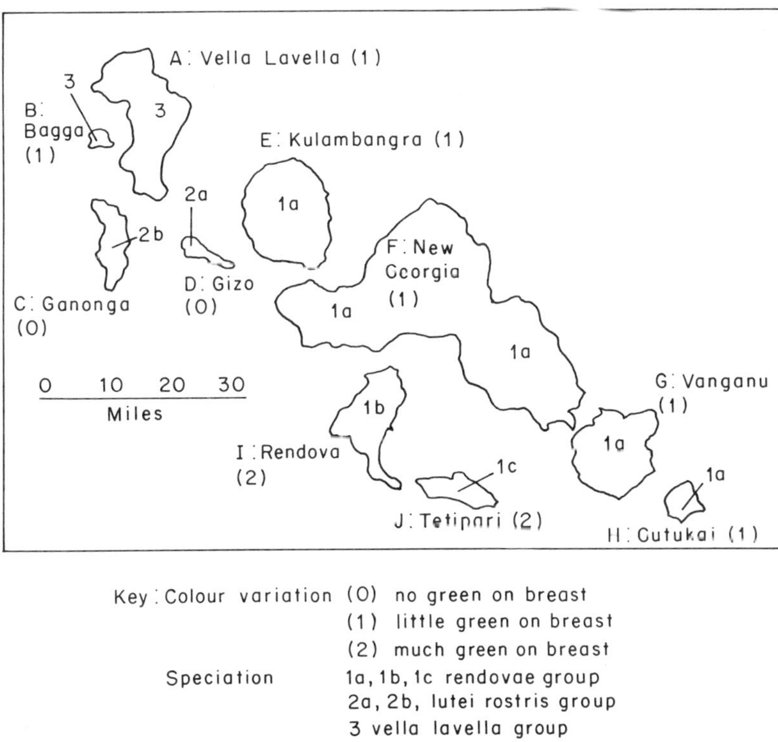

Figure 3.7 Colour variation and speciation among passerine birds in the Central Solomon Islands. (After Mayr; 1963. Reproduced by permission of Harvard University Press)

suggestion subsequently. For the present, therefore, we use the **W** matrix shown in Table 3.10, for which all the off-diagonal cells contain non-zero values.

We focus attention on two aspects of the *Zosterops rendovae* species, considering first the spatial variation of the varying amounts of greenness of their breasts and, later, the variation in an overall index of differentiation. There are ten islands in the group, and three intensities of colour have been recognized ranging from very green (which we code as $x = 2$), some green ($x = 1$), and green absent ($x = 0$). Adopting the definition of 'distance', $(x_i - x_j)^2$, given earlier as equation (3.4), the **Y** matrix quantifying inter-island breast-colour differentiation is as in Table 3.11.

The outcome of using Tables 3.10 and 3.11 is that we obtain $S_0 = 10.29$, $S_1 = 7.71$, $S_2 = 49.78$, $T_0 = 80$, $T_1 = 352$, and $T_2 = 3520$ so that $r = 4.73$, $E(R) = 9.15$, and $\text{var}(R) = 5.94$ which lead to $z = 1.81$. Since only positive spatial autocor-

Table* 3.10 *Generalized proximity values permitting long-range inter-breeding of* Zosterops *species in the Solomon Islands

Island	A	B	C	D	E	F	G	H	I	J
A	0.00	0.56	0.29	0.14	0.08	0.04	0.01	0.01	0.02	0.02
B	0.56	0.00	0.14	0.06	0.05	0.03	0.01	0.01	0.02	0.01
C	0.29	0.14	0.00	0.13	0.05	0.04	0.01	0.01	0.03	0.02
D	0.14	0.06	0.13	0.00	0.20	0.12	0.02	0.01	0.04	0.02
E	0.08	0.05	0.05	0.20	0.00	0.71	0.02	0.02	0.06	0.03
F	0.04	0.03	0.04	0.12	0.71	0.00	0.63	0.06	0.40	0.09
G	0.01	0.01	0.01	0.02	0.02	0.63	0.00	0.24	0.04	0.08
H	0.01	0.01	0.01	0.01	0.02	0.06	0.24	0.00	0.03	0.04
I	0.02	0.02	0.03	0.04	0.06	0.40	0.04	0.03	0.00	0.50
J	0.02	0.01	0.02	0.02	0.03	0.08	0.08	0.04	0.50	0.00

Table* 3.11 *A Y matrix for the green breasts of the Solomon Islands birds

Island	A	B	C	D	E	F	G	H	I	J
A	0	0	1	1	0	0	0	0	1	1
B	0	0	1	1	0	0	0	0	1	1
C	1	1	0	0	1	1	1	1	4	4
D	1	1	0	0	1	1	1	1	4	4
E	0	0	1	1	0	0	0	0	1	1
F	0	0	1	1	0	0	0	0	1	1
G	0	0	1	1	0	0	0	0	1	1
H	0	0	1	1	0	0	0	0	1	1
I	1	1	4	4	1	1	1	1	0	0
J	1	1	4	4	1	1	1	1	0	0

Table* 3.12 *Generalized proximity values confining interbreeding of the* Zosterops *species to a few neighbouring Islands

Island	A	B	C	D	E	F	G	H	I	J
A	0	0.56	0	0	0	0	0	0	0	0
B	0.56	0	0	0	0	0	0	0	0	0
C	0	0	0	0	0	0	0	0	0	0
D	0	0	0	0	0	0	0	0	0	0
E	0	0	0	0	0	0.71	0	0	0	0
F	0	0	0	0	0.71	0	0.63	0	0	0
G	0	0	0	0	0	0.63	0	0	0	0
H	0	0	0	0	0	0	0	0	0	0
I	0	0	0	0	0	0	0	0	0	0.50
J	0	0	0	0	0	0	0	0	0.50	0

relation is conceivable we carry out a one-sided significance test. It is reasonable to assume that the normal approximation is valid, so that, since 1.81 exceeds 1.645 (the upper 5% point of a unit normal distribution), we conclude that there is indeed significant evidence of spatial autocorrelation.

Suppose that we accept that there are indeed some straits which are impassable so far as these birds are concerned. In this case we need to use the revised **W** matrix given in Table 3.12 which allows communication only between adjacent islands separated by less than 2 km of water (Mayr, 1963). We now calculate r using Tables 3.11 and 3.12 and, since $S_0 = 4.8$, $S_1 = 5.86$, and $S_2 = 15.3$ we find $r = 0$, $E(R) = 4.27$, and $\mathrm{var}(R) = 5.47$. However, only the value of r concerns us, since the sparseness of the revised **W** matrix makes a normal approximation untenable and we have once more to resort to randomization with the results being those given in Table 3.13.

Since the observed value of 0 is exceeded by all 99 of the values obtained from randomization we once again obtain significant evidence that we should reject

***Table* 3.13 *The cumulative distribution of* r *based on* 99 *random permutations of Table* 3.11**

(a) Combined with Table 3.12

Value of r	1	2	3	4	5	6	7	8	9	10	11	12
Freq. $\le r$	99	86	64	45	32	18	14	8	4	2	2	0

(b) Combined with Table 3.10

Value of r	4	5	6	7	8	9	10	11	12	13	14	15	16	17
Freq. $\le r$	99	98	93	83	63	47	26	14	10	5	3	3	1	0

Table 3.14 *A* Y *matrix quantifying inter-island speciation*

Island	A	B	C	D	E	F	G	H	I	J
A	0	0	2	2	2	2	2	2	2	2
B	0	0	2	2	2	2	2	2	2	2
C	2	2	0	1	2	2	2	2	2	2
D	2	2	1	0	2	2	2	2	2	2
E	2	2	2	2	0	0	0	0	1	1
F	2	2	2	2	0	0	0	0	1	1
G	2	2	2	2	0	0	0	0	1	1
H	2	2	2	2	0	0	0	0	1	1
I	2	2	2	2	1	1	1	1	0	1
J	2	2	2	2	1	1	1	1	1	0

Table 3.15 *The cumulative distribution of* r *obtained from the combination of Tables* 3.14 *and* 3.12

Value of r	2	3	4	5	6	7	8	9	10
Freq. $\leq r$	99	97	92	83	68	54	33	13	0

the null hypothesis of no spatial pattern. Notice that the alternatives that we are accepting differ—one allows no cross-breeding for islands separated by more than 2 km, the other places no such limit on the flight ability of the birds. Our analyses have not been able to distinguish between these explanations: both are acceptable alternatives to the rejected hypothesis of there being no spatial arrangement.

Also shown in Figure 3.6 are the locations of various subspecies of *Zosterops rendovae*. These subspecies are denoted by 1, 2, etc. and are further divided into subgroups (1a, 1b, etc.) as classified by Mayr (1942, 1963). We use this information to construct the **Y** matrix given in Table 3.14. In this matrix a 0 denotes two islands supporting identical subspecies, a 1 denotes islands that are the homes of different subgroups of the same subspecies, and a 2 denotes a pair of islands occupied by different subspecies.

From Table 3.14 we obtain the statistics $T_0 = 132$, $T_1 = 488$, and $T_2 = 7312$. Combining these with the S statistics for the proximity matrix given in Table 3.10 we obtain $r = 8.83$ with $\mathrm{E}(R) = 15.09$, and $\mathrm{var}(R) = 2.94$ so that $z = 3.65$, which is much greater than the critical value (1.645). If we adopt the alternative proximity matrix given in Table 3.12 then we have $r = 1$ which proves to be an extreme value when compared to its randomization distribution (Table 3.15).

***Example* 3.10**

The previous analysis of *Atriplex hymenelytra* growing in Death Valley, California (Example 3.2) suggested that the presence or absence of the shrub in one quadrat was unconnected with its presence or absence in an adjoining quadrat. This analysis, however, was somewhat crude since it employed the simple join-count statistic and ignored the actual sizes of the quadrat counts.

In the present analysis we take the actual quadrat counts given in Figure 1.14 and use a proximity matrix which allows for the presence of long-range interactions between quadrats that are not immediate neighbours. We anticipate that the correlation between counts will diminish as the inter-quadrat distance increases and, to capture this hypothesized *distance–decay* effect, we estimate the proximity values using

$$W_{ij} = \{(i_r - j_r)^2 + (i_c - j_c)^2\}^{-b/2}. \tag{3.32}$$

Here, i_r and i_c are the row and column coordinates, respectively, of quadrat i, and can take any value between 1 and 16 (since the data consists of a 16×16 array of quadrat counts). Similarly, j_r and j_c give the location of quadrat j. The value of W_{ii} is, of course, taken as zero.

We set $b = 1.0$, although a higher value could be used if an exaggerated distance effect were required. It is impracticable to show the entire **W** matrix that results, but Table 3.16 illustrates the upper left-hand corner of the matrix. From the complete matrix we find $S_0 = 11\,222.96$, $S_1 = 6862.48$, and $S_2 = 2\,001\,103.25$. Also $\text{var}(x) = 0.29$, $n = 256$, $I = 0.01357$, $E(I) = -0.00392$, and $\text{var}(I) = 0.00004$ so that $z = 2.84$. Thus, using this more sensitive test procedure we are able to find the spatial autocorrelation which went undetected in our previous analysis.

The distinctive feature of the present analysis is its inclusion of a distance–decay effect. We can get some further insight into the distance effect by combining ideas from this chapter and from Chapter 1 in which quadrat counts were analysed by examining the variations in successively larger blocks. Table 3.17 shows the values of I and associated statistics for block sizes varying from 1 to 16. Calculations here are based on a binary **W** matrix with the rook's definition of contiguity. It appears that the autocorrelation present is most noticeable at about

Table* 3.16 *The top left corner of the* W *matrix of generalized proximity values for the* Atriplex hymenelytra *data

(r, c)	(1, 1)	(1, 2)	(1, 3)	(1, 4)
(1, 1)	0.000	1.000	0.500	0.333
(1, 2)	1.000	0.000	1.000	0.500
(1, 3)	0.500	1.000	0.000	1.000
(1, 4)	0.333	0.500	1.000	0.000

Table 3.17 Moran's **I** *and associated statistics for the combined counts in Figure* **1.15**

Block size	n	S_0	S_1	S_2	k	M_2	E(I)	var(I)	I	z
1	256	960	1920	14624	4.82	0.29	−0.0039	0.0021	0.0021	1.48
2	128	464	928	6880	3.68	0.58	−0.0079	0.0042	0.0042	2.56
4	64	224	448	3232	4.18	1.25	−0.0159	0.0085	0.0083	2.51
8	32	104	208	1408	2.44	2.73	−0.0323	0.0183	0.0173	1.81
16	16	48	96	608	3.56	8.19	−0.0667	0.0345	0.0306	0.88

block-size 3 and this gives some idea of the scale at which the interaction process is operating.

The discussion so far has focussed on distance–decay effects rather than discontinuous change in the form of patches, the search for which initially provoked the analysis of the *Atriplex hymenelytra* data by the methods of Greig-Smith. With this form of spatial organization in mind we observe that I will tend to be large and positive when patch diameter exceeds quadrat diameter and will dwindle to near zero as the quadrat diameter increases beyond the patch diameter.

Alternative models of proximity

The equation (3.32) that we have utilized in our example is that used by Jumars (1978) in a paper with the intriguing title of 'Spatial autocorrelation with remote underwater manipulator'. Jumars chose the value $b = 2$ and, using I, was able to detect autocorrelation in the density of the marine organism *Polyo phthalmus* at irregularly located underwater sample sites.

To date we have only considered using Euclidean 'bee-line' distance: it may be more sensible to use 'perceived' distance or distance by road or distance measured in time required to travel between the two points—furthermore, the distance between A and B need not be the same as that between B and A (uphill/downhill, upstream/downstream, etc.; cf. Sokal and Oden, 1978a). Sokal and Oden also mention a directional test akin to the use of the rook's and bishop's definitions encountered above that depends upon the proximity values being defined as the cosines of the angles that pairs of localities make with various hypothesized test directions.

Cliff and Ord (1981a, p. 229) provide proximity values for the Irish counties which are functions both of the distance between the county 'centres', d_{ij}, and of the proportion of the boundary of county i that is common to county j, b_{ij}. Notice that the value of b_{ji} will not be the same as that of b_{ij} so that this is a situation which involves an asymmetrical $\mathbf{W}$ matrix. In order to combine the distance and boundary effects Cliff and Ord suggest using $\mathbf{W}_{ij} = b_{ij}^a d_{ij}^{-b}$ and they take $a = b = 1$.

Searching for *W* matrices

Often strong reasons can be found for structuring the **W** matrix as, for example, by Sokal and Oden (1978a) in their study of the aphid *Pemphigus populitransversus* in which their proximity values take cognizance of the biological impossibility of trans-Appalachian gene flow. Similar biological barriers provided the rationale for the choice of **W** matrices in Example 3.9. Haining (1981c) strongly emphasizes the role of interaction data, such as migration flows, traffic, or telephone calls, as adjuncts to analysis (covariates) in a general modelling context. Gattrell (1979), on the other hand, uses multidimensional scaling to obtain inter-point 'distances' between Swedish towns in place of the more obvious straight-line or road distances.

Advances such as Cliff and Ord's (1969) generalization of the I statistic have given the analyst much more latitude in the choice of an alternative hypothesis, but this has as its concomitant that there is now an over-abundant choice with no clear directive as to how to proceed. If the alternative hypothesis is chosen in an arbitrary manner, then its acceptance or rejection can scarcely carry much weight. Ideally we should aim to provide a realistic alternative without reference to the data. An alternative solution suggested by Kooijman (1976) in the context of the correlogram is to attempt to find the values **W** which maximize I (or r). Another approach, which depends on the $\{x_{ij}\}$ having a normal distribution at the jth time period ($j = 1, 2, \ldots, J$), involves using the $\{x_{ij}\}$ to estimate the **W** matrix (Arora and Brown, 1977).

Aspatial *W* matrices

The proximity values of the **W** matrix have so far been considered as measures of *spatial* proximity. It is possible, however, to allow the **W** matrix exactly the same flexibility as the **Y** matrix and so to contain values reflecting proximity in *any* dimension. Haggett (1976) illustrates this implicitly in his analysis of the intensity of a measles epidemic in which he postulates as one possible alternative hypothesis that the disease spreads by aspatial diffusion through an urban hierarchy.

An example of the general procedure is illustrated in Table 3.18. There are nine localities, labelled a to i, and two density diagrams. The **Y** matrix is obtained from the density diagram of x by the relation $Y_{ij} = x_i x_j$, while the **W** matrix is obtained in exactly the same way from the density diagram for z. Since the two matrices are constructed in an identical fashion, it follows that the S statistics can be calculated using the formulae for the T statistics.

Table* 3.18 *Examples of general* W *and Y matrices

(a) locations	(b) x density diagram	(c) z density diagram
abc	110	100
def	110	110
ghi	111	111

(d) **W** matrix

	a	*b*	*c*	*d*	*e*	*f*	*g*	*h*	*i*
a	0	0	0	1	1	0	1	1	1
b	0	0	0	0	0	0	0	0	0
c	0	0	0	0	0	0	0	0	0
d	1	0	0	0	1	0	1	1	1
e	1	0	0	1	0	0	1	1	1
f	0	0	0	0	0	0	0	0	0
g	1	0	0	1	1	0	0	1	1
h	1	0	0	1	1	0	1	0	1
i	1	0	0	1	1	0	1	1	0

(e) **Y** matrix

	a	*b*	*c*	*d*	*e*	*f*	*g*	*h*	*i*
a	0	1	0	1	1	0	1	1	1
b	1	0	0	1	1	0	1	1	1
c	0	0	0	0	0	0	0	0	0
d	1	1	0	0	1	0	1	1	1
e	1	1	0	1	0	0	1	1	1
f	0	0	0	0	0	0	0	0	0
g	1	1	0	1	1	0	0	1	1
h	1	1	0	1	1	0	1	0	1
i	1	1	0	1	1	0	1	1	0

Note that, for the hypothetical data of Table 3.18, the value of $r(30)$ is most easily calculated using $r = m(m-1)$, where m is the number of localities that are valued as 1 in both density diagrams.

***Example* 3.11**

Figure 3.8 gives the joint distribution of oak, hickory, and maple trees in 576 quadrats in Lansing Woods, Clinton county, Michigan. The data is adapted from Diggle (1979b) and we are here solely concerned with the oak and hickory distributions with hickory corresponding to x and oak to z. The supposition is that the occurrence of hickory is related to the occurrence in other localities indexed by the presence of oak. A two-sided test is apposite in view of the possibility of either positive or negative interaction. Oak and hickory are both jointly present whenever a quadrat contains the numbers 1 or 5, which happens in 261 of the 576 quadrats and thus $r = 261 \times 260 = 67\,860$. Since 438 quadrats contain oak and 354 contain hickory, we have $S_0 = 438 \times 437 = 191\,406$, $S_1 = 382\,812$, $S_2 = 4 \times 438 \times 437^2 = 334\,577\,688$ and similarly, $T_0 = 124\,962$, $T_1 = 249\,924$, and $T_2 = 176\,446\,344$. Consequently, we find that $\mathrm{E}(r) = 72\,217.6$, $\mathrm{var}(r) = 7\,194\,814.7$, and $z = -1.62$. This result suggests that there is marginally less overlap between the oak and hickory distributions than would have been expected under the null hypothesis of two unrelated distributions. Note that the result of this analysis is very similar to that obtained by a conventional χ^2 analysis.

	1	2	3	4	5	6	7	8	9	10	11	12	13	14	15	16	17	18	19	20	21	22	23	24
1	5	5	5	5	7	7	5	1	5	7	1	5	7	8	5	1	5	5	1	7	1	5	5	6
2	6	6	6	5	5	5	1	6	7	5	7	5	3	3	3	5	5	1	5	7	5	5	5	5
3	5	5	6	5	5	6	6	5	7	7	7	1	4	7	3	5	4	6	7	7	5	7	5	5
4	5	5	6	5	6	2	8	5	5	5	5	5	3	7	6	5	7	7	1	7	5	5	6	5
5	5	8	5	5	6	5	5	6	5	5	6	1	7	2	3	1	8	4	5	6	5	5	5	5
6	5	5	5	5	5	6	5	5	6	5	5	7	3	5	5	8	5	7	7	5	5	6	5	7
7	6	1	5	1	5	7	7	6	2	1	6	3	1	4	1	5	1	2	1	1	7	5	6	6
8	5	6	5	5	5	5	1	4	2	3	4	8	4	7	3	4	2	5	1	5	7	5	5	5
9	7	5	7	5	7	7	4	6	1	4	1	7	3	4	3	3	3	1	1	5	1	2	1	2
10	5	5	5	6	1	7	2	1	5	4	4	7	7	7	5	3	1	5	5	5	3	2	6	2
11	5	5	6	6	5	5	3	4	1	3	1	1	7	4	6	6	6	8	5	6	6	3	1	4
12	5	7	5	5	5	6	6	3	1	3	5	2	4	1	6	2	5	1	1	2	4	5	1	6
13	5	5	5	5	7	2	3	1	1	3	1	5	1	1	1	1	5	6	5	5	5	6	8	5
14	7	5	6	6	7	6	3	5	5	5	5	3	2	5	7	3	5	5	5	5	5	5	6	6
15	5	7	7	5	3	5	8	7	5	5	6	5	1	3	2	3	2	2	5	3	1	5	5	7
16	5	5	7	1	7	7	7	7	8	5	6	4	3	2	7	5	7	3	3	7	5	6	5	7
17	5	5	1	1	1	1	7	7	7	5	6	1	4	3	3	3	3	3	3	3	5	6	5	5
18	7	5	3	3	3	1	3	7	7	1	1	3	2	7	3	4	3	4	2	4	5	1	5	5
19	5	1	3	3	3	7	7	3	3	7	3	7	3	3	3	2	1	2	7	1	1	4	1	5
20	7	5	1	3	3	7	7	3	7	3	5	1	3	8	2	1	5	1	5	5	7	1	3	6
21	5	1	5	1	3	5	7	7	5	3	3	8	5	5	5	3	6	1	3	5	3	2	4	4
22	6	1	1	3	3	3	2	1	8	7	3	2	3	1	1	7	8	3	3	3	3	3	1	6
23	5	6	1	3	1	5	2	1	1	3	3	4	3	7	3	7	4	3	1	6	3	7	2	8
24	7	6	5	6	6	3	5	5	3	2	1	3	7	3	3	7	1	7	5	2	7	3	4	4

Key:

	Oak	Hickory	Maple
1	p	p	p
2	a	p	p
3	p	a	p
4	a	a	p
5	p	p	a
6	a	p	a
7	p	a	a
8	a	a	a

p = present; a = absent

Figure 3.8 The joint distribution of oak, hickory, and maple trees in Lansing Woods, Michigan

3.5 CORRELOGRAMS

Correlograms are graphs (or tables) showing how autocorrelation changes with distance. Traditionally observations have been over time, and distance has meant time distance between a pair of observations. Spatial analysts following on from the pioneering work of Lebart (1966, 1969) have adapted this idea to spatial distance. In this context Sokal and Oden (1978a, 1978b) have analysed biological variables, Cliff *et al.* (1975) measles epidemics, and Ford (1976) the canopy of a Scots Pine forest.

Spatial correlograms

In order to more vividly establish the concept of the spatial correlogram, it is useful to entertain the idea of throwing a stick randomly around a field.

> Imagine throwing a stick of specific length over and over again at a certain study area, while noting the values of a variate at the points at which the ends of the stick land. Affirming spatial autocorrelation is tantamount to saying that the values at the ends of the stick are usually like each other, or that they are usually different. Suppose a study area is composed of squares arranged in a checkerboard fashion, with the colours of the squares alternating from black to white. Brief reflection will reveal that the stick length likely to yield the most negative autocorrelation is that which is roughly the length of one square, while a stick two squares long will yield positive autocorrelation (Sokal and Oden, 1978a).

In this case a graph showing the variations in autocorrelation with stick length would be one form of spatial correlogram.

In the case of the checker-board the correlogram would reveal a marked periodicity with period length equal to square length, but this artificial situation is far removed from reality. The typical behaviour of many spatial phenomena produces a correlogram that displays values that diminish rapidly towards zero as the inter-point distance increases—thus giving some empirical substantiation to the first 'law' of geography which states that 'everything is related to everything else, but near things are more related than distant things' (Tobler, 1970a). It is also possible to envisage correlograms with approximately zero autocorrelation among closely neighbouring localities but with a 'peak' or 'peaks' at greater distance. In such a case the conventional test based on contiguous localities will fail to detect autocorrelation. This reinforces the general warning that the acceptance of a null hypothesis does not imply that it is necessarily correct. It is obviously sensible to speculate about the likely shape of the correlogram *before* carrying out the data analysis.

In the checker-board and stick examples, Sokal and Oden (1978a) note that the expected correlogram 'alternates from positive to negative, with wavelength equal to average patch diameter'. Similarly, a map with a distinct trend or cline will produce a correlogram in which the positive short distance autocorrelation fails to fade away to zero at higher distances but continues to plummet downwards in a negative direction.

Pielou (1977) describes some theoretical correlograms deriving from random processes. She considers the one-dimensional Poisson process that results from a line transect that intersects a line tessellation (an L-mosaic) such as that illustrated in Figure 1.44. Here the Poisson 'events' are the crossing of boundaries. If these events occur with rate λ, then the autocorrelation function is simply $e^{-\lambda d}$, where d is the distance between events.

Sokal (1979) gives some detailed hints about probable patterns and processes inferable from observed correlograms. He notes that short distance positive autocorrelation, in which neighbouring localities assume usually similar values, could arise not only from clines but also from patches whose diameter exceeds the inter-point distance and is less than any inter-patch gaps. Short distance negative autocorrelation, involving usually different neighbours, suggests a mixture of heterogeneous patches smaller in average diameter than the inter-locality distance. Long distance positive autocorrelation will tend to be the outcome of a regular arrangement of patches so that values are approximately repeated—as on either side of a valley or a mountain. Long distance negative autocorrelation we have already dealt with in the context of the cline, but this could also arise from a situation in which there are sharp edged patches that display a trend.

The natural alliance of the spatial correlogram to time series analysis has meant that the work of Box and Jenkins (1970) has had some considerable influence, particularly in promoting the idea of comparing the observed correlogram with the correlograms to be expected under a variety of theoretical processes. Much of this literature is in the context of the spatio-temporal correlogram which is discussed subsequently.

Definition of distance classes

The construction of the correlogram requires the calculation of an appropriate measure of autocorrelation n times, once for each of the n sets of localities in n distance classes. Various ways of defining distance classes that are applicable to data organized on either a regular or an irregular basis are outlined by Tobler (1975) and by Cliff and Ord (1981a, p. 118). Allied to the sets of locality pairs occupying the same distance class is a matrix $\mathbf{W}_d, d = 1, \ldots, n$.

Consider an irregular scatter of point localities. Notions from graph theory may be employed to define the distance between any pair of localities as the number of intervening points *en route* from one to the other via the shortest route on the graph connecting nearby points. This definition can also apply to irregularly shaped areas connected by common boundaries, as in the analysis by Cliff *et al.* (1975, p. 150) of the measles epidemic. Alternatively, the distance class of a pair of point localities or a pair of centres of irregularly shaped areal localities may be the more obvious definition based on the straight-line physical distances. Obviously, the above definitions may also be applied to quadrats within a square or rectangular lattice, or to lattice intersection points.

When the location and shape of the localities are beyond the control of the analyst, the irregularity that often ensues need not deter investigation, for the above definitions are sufficiently general to handle irregular data. However, when there is some possibility of some control over the localities to which x-values are assigned, then purposeful alignment in a rectangular or square lattice does greatly simplify calculations.

Rook's definition S statistics for the correlogram on a regular lattice

Where the matrices $\mathbf{W}_d$ describe a regular lattice with 1, 0 proximity values, the S statistics for the rook's definition of contiguity are easily obtained using equations:

$$S_0 = 2\{c(r-d) + r(c-d)\}, \qquad S_1 = 2S_0, \qquad S_2 = 8(8rc - 7rd - 7cd + 4d^2). \tag{3.33}$$

In these equations d represents the number of quadrat steps in a horizontal or vertical direction between pairs of localities in the d distance class. Note that the equations are generalizations of equations (3.11) and only apply when both $c > 2d$ and $r > 2d$.

Example 3.12

Table 3.19a displays 25 localities with Table 3.19b presenting the $\mathbf{W}_d$ matrix for the case $d = 2$. In this matrix all entries are zero except for those denoted by the letter R (!). Applying the equations (3.33) we find $S_0 = 60$, $S_1 = 120$, and $S_2 = 608$.

Spatial correlograms for join-count statistics

Although most published correlograms involve Moran's I, there is no reason why the join-count statistics cannot be used to give a correlogram for colours on a lattice, or indeed for irregular localities. Application to the lattice requires the enumeration of the observed number of joins in distance classes $d = 1, \ldots, k$, where k is the smallest of $c/2 - 1$ and $r/2 - 1$. One then proceeds to calculate the expected number and variance and ultimately $z = (\text{obs} - \text{exp})/\sqrt{\text{var}}$, which is assumed, under the null hypothesis, to be a unit normal random variable. These statistics are obtained for each d by first obtaining the S statistics of the matrices $\mathbf{W}_d, d = 1, \ldots, k$.

Example 3.13

Previous analysis of the *Atriplex* shrubs on a presence/absence basis gave evidence of interaction between quadrats that were not immediate neighbours. This correlation should produce a peak (or peaks) at the appropriate distance on the correlogram. As in Example 3.2, $b = 65$, $w = 191$, and $n = 256$ so that, from equation (3.9), we have $T_0 = 24\,830$, $T_1 = 49\,660$, and $T_2 = 12\,712\,960$. Unlike the S statistics, these remain unchanged throughout the subsequent analysis.

***Table* 3.19 $\mathbf{W}_d$ *matrices, rooks and bishops definitions, for* d = 2**

(a) locations

a b c d e
f g h i j
k l m n o
p q r s t
u v w x y

(b)

	a	*b*	*c*	*d*	*e*	*f*	*g*	*h*	*i*	*j*	*k*	*l*	*m*	*n*	*o*	*p*	*q*	*r*	*s*	*t*	*u*	*v*	*w*	*x*	*y*
a			R								R		B												
b				R								R		B											
c	R				R						B		R		B										
d		R										B		R											
e			R										B		R										
f								R								R		B							
g									R								R		B						
h						R				R						B		R		B					
i							R										B		R						
j								R										B		R					
k	R		B										R								R		B		
l		R		B										R								R		B	
m	B		R		B						R				R						B		R		B
n		B		R								R										B		R	
o			B		R								R										B		R
p						R		B										R							
q							R		B										R						
r						B		R		B						R				R					
s							B		R								R								
t								B		R								R							
u											R		B										R		
v												R		B										R	
w											B		R		B						R				R
x												B		R								R			
y													B		R								R		

Table 3.20 gives the results of applying equations (3.33) to find the various sets of S statistics and then combining both S and T statistics in equations (3.5) and (3.6) to finally obtain $\mathrm{E}(BW)$, $\operatorname{var}(BW)$, and z.

Table 3.20 shows that the suspected peak exists at $d = 4$; at this distance there are significantly more BW joins than expected, pointing to negative autocorrelation and a patch edge.

Correlograms, the bishop's definition, and multicoloured maps

In a similar fashion it is possible to generalize the bishop's definition equations to obtain:

$$S_0 = 4(r - d)(c - d), \qquad S_1 = 2S_0, \qquad S_2 = 16(2r - 3d)(2c - 3d). \tag{3.34}$$

Table 3.20 *A correlogram and related statistics for* BW *rook's joins for the* Atriplex *shrubs data*

d	Observed BW joins	S_0	S_1	S_2	$E(BW)$	$var(BW)$	z
1	173	960	1920	14624	182.6	70.7	1.08
2	163	896	1792	12928	170.4	68.0	0.84
3	150	832	1664	11296	158.2	64.7	0.96
4	166	768	1536	9728	146.1	60.5	2.50
5	137	704	1408	8224	133.9	55.7	0.35

Clarification of what is meant by d steps apart, in the present context, is provided by Table 3.19b in which a letter B indicates the position of a 1 in the $\mathbf{W}_2$ matrix under this definition. With this definition we find that $S_0 = 36$, $S_1 = 72$, and $S_2 = 256$.

Example 3.14

The motivation for a 'diagonal' correlogram of the burnt savanna data is provided by the detection in Example 3.6 of diagonal alignment of the $x = 1$ quadrats. The present investigation is concerned with whether this similarity of burning intensity tails off beyond immediately diagonally contiguous localities. This example also illustrates how the multicoloured approach can be extended to correlogram analysis. As in Example 3.13, analysis for $d > 1$ demands new S statistics for each d, this time input into equation (3.15). Table 3.21 summarizes the various statistics and reveals a marked absence of diagonal alignment at any diagonal distance beyond the immediately adjacent cell.

Table 3.21 *A correlogram of diagonal alignment of the burnt savanna data*

d	S_{11}	S_0	S_1	S_2	$E(S_{11})$	$var(S_{11})$	z
1	101	3844	7688	59536	84.4	56.0	2.15
2	71	3600	7200	53824	79.1	55.2	1.02
3	74	3364	6728	48400	73.9	53.7	0.05
4	64	3136	6272	43264	68.9	51.8	0.61
5	64	2916	5832	38416	64.0	49.4	0.07

The S statistics for the correlogram using the queen's definition

The S statistics of the binary $\mathbf{W}$ matrix for lattice data and the queen's definition were shown to be rapidly calculated using equations (3.20) and these can be generalized to give

$$S_0 = 2(4rc - 3rd - 3cd + 2d^2), \qquad S_1 = 2S_0,$$
$$S_2 = 8(32rc - 39rd - 39cd + 46d^2). \tag{3.35}$$

The $\mathbf{W}_2$ matrix for the locations given in Table 3.19a is now that of Table 3.19b with both the B's and the R's replaced by ones and with zeros elsewhere.

Example 3.15

Table 3.22 gives the correlogram and associated statistics resulting from applying the composite queen's definition to the *Atriplex* data. The computations parallel those of Examples 3.13 and 3.14 except that the S statistics are obtained from equations (3.35). The main feature of the table is the preservation of the previously detected 'peak' at $d=4$, indicating significant negative autocorrelation among quadrats separated by this distance.

Table 3.22 *A correlogram and related statistics for* BW *Queen's joins for the* Atriplex *shrubs data*

d	Observed BW joins	S_0	S_1	S_2	E(BW)	var(BW)	z
1	343	1860	3720	55 920	353.7	151.6	0.8
2	323	1680	3360	47 040	319.5	151.8	0.2
3	284	1508	3016	38 896	286.8	144.9	0.2
4	289	1344	2688	31 488	255.6	132.5	2.9
5	244	1188	2376	24 816	225.9	115.9	1.6

Correlograms with Moran's I

A natural counterpart to the join-count analyses is the calculation of I_d, which is Moran's I applied to localities in distance class d. Although previous equations relating to I still hold, consideration must be given to the S statistics that vary with the $\mathbf{W}_d$ matrices and must therefore be substituted into equations (3.24) and (3.29) to give I_d and var(I_d) for each value of d. As with the join-count statistics, rapid calculation of the S statistics is facilitated by lattice data.

Example 3.16

Table 3.23 gives the results of calculating I_d using the actual number of *Atriplex* bushes in each quadrat. We again adopt the rook's definition of contiguity. As is to be expected, there are some differences of detail between this analysis and analyses of the simple presence/absence data. The previously discovered negative

Table 3.23 A correlogram for Moran's **I** *and the* **Atriplex** *data*

d	S_0	S_1	S_2	I_d	var (I_d)	z	r_d
1	960	1920	14624	0.064	0.002	1.5	0.113
2	896	1792	12928	0.100	0.002	2.2	0.191
3	832	1664	11296	0.113	0.002	2.4	0.195
4	768	1536	9728	−0.093	0.003	−1.7	−0.160
5	704	1408	8224	0.028	0.003	0.6	0.048
6	640	1280	6784	−0.055	0.003	−0.9	−0.082
7	576	1152	5408	−0.013	0.003	−0.2	−0.019

autocorrelation at $d = 4$ is now apparent for values of d from 1 to 4, with a notable decrease for larger values of d.

Example 3.17

In Example 3.8 significant positive autocorrelation was detected among contiguous quadrats of the burnt savanna using Moran's I. Retaining the rook's definition of contiguity and developing the analysis using the correlogram, there is every indication of significant autocorrelation at distances greater than $d = 1$.

The results of the analysis are summarized in Figures 3.9 and 3.10 which show

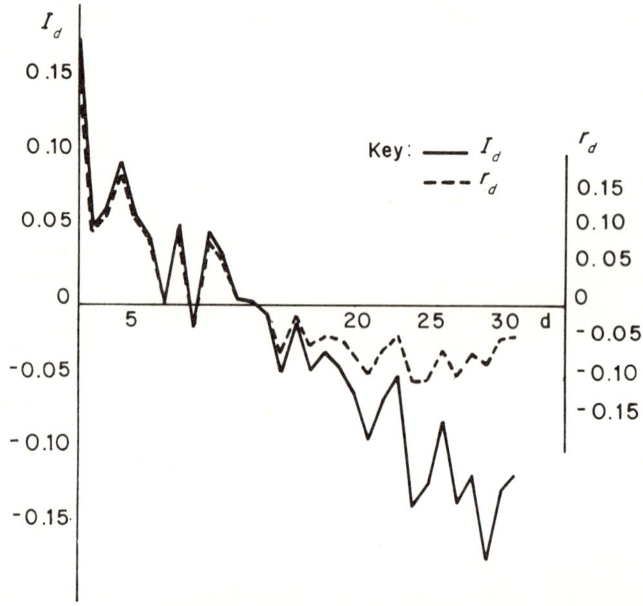

Figure 3.9 Correlogram of Moran's I, r; rook's definition. Burnt savanna data

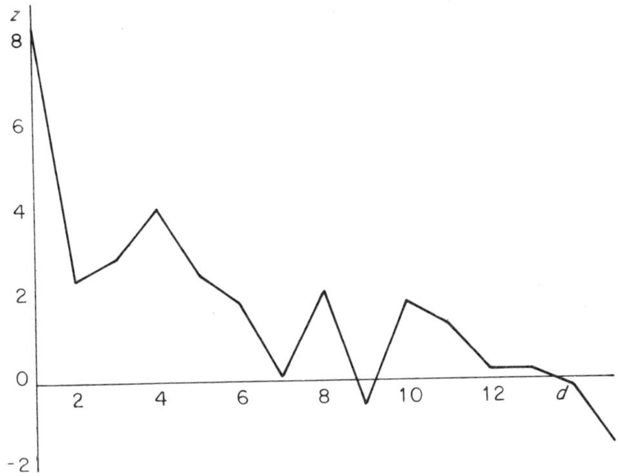

Figure 3.10 Correlogram of z based on Moran's I. Burnt savanna data

the variations in I_d and in z as d varies. The S statistics were calculated speedily using the equations (3.33). From the figures it is apparent that there is significant positive autocorrelation for values of d up to 5, with particularly significant peaks at $d=1$ and at $d=4$. In addition, there are near significant peaks at $d=8$ and at $d=10$, while the consistent stream of negative values beyond $d=13$ suggests a trend.

General distance classes

The adoption of particular definitions of distance classes on the regular lattice has facilitated the testing of the significance of some of the peaks (and troughs) of the correlogram. When the same procedures are applied to the case of irregularly scattered localities, then the corresponding $\mathbf{W}_d$ matrices do not so readily yield their S statistics and there will be far less information at any given value of d (though there will be *some* information at far *more* values of d than was previously the case) with the result that, while the overall profile of the correlogram will continue to be informative, information about individual d-values will be drowned in the general 'noise'. Examples of such correlograms are provided by Cliff *et al.* (1975) and by Jumars (1978).

Example 3.18

We continue the analysis of the burnt savanna data by allocating locality pairs (i, j) (which just happen to be lattice quadrats!) to distance classes on the basis of straight-line distances $d_{ij} = \{(x_i - x_j)^2 + (y_i - y_j)^2\}^{1/2}$, where x and y are locality

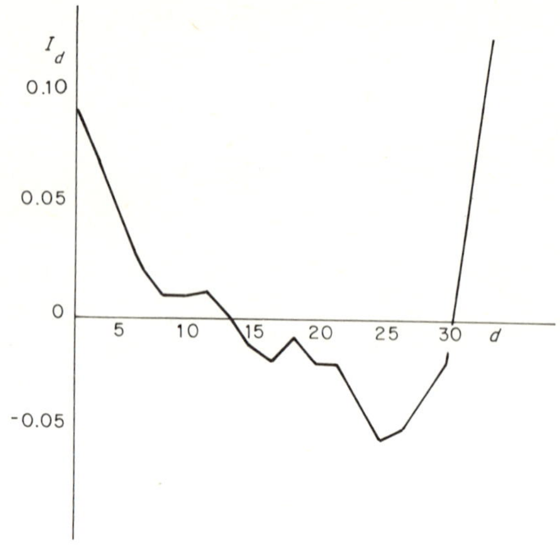

Figure 3.11 Correlogram of Moran's I using 20 distance classes for the burnt savanna data

coordinates. The resulting correlogram is shown in Figure 3.11. The 20 distance classes are each 1.65 units wide, where 1 unit is the side of a quadrat, and each class is labelled by its upper bound. Although the overall downward trend of Figure 3.9 is preserved in Figure 3.11, the sharpness of the direction-specific rook's definition correlogram has been replaced by a more equilibrated curve. There is the additional feature of a steep upward rise to high positive autocorrelation at about $d = 25$ or beyond, although there are relatively few observations at this range.

The modified autocorrelation coefficient

Assessment of the variations in I_d has been aided by significance tests using z. Various other adjuncts are available to supplement data interpretation and to produce a more rounded picture of the autocorrelation structure. The modified autocorrelation coefficient r_d is one such device that surmounts a difficulty in the interpretation of I_d. This difficulty is that the range of I_d is not ± 1 like the Pearson product-moment correlation coefficient, but rather, as noted by Cliff and Ord (1981a, p. 21) falls within the bounds

$$\pm \frac{n}{U_d}\left[\frac{\sum(x_i - \bar{x})_d^2}{\sum(x_i - \bar{x})^2}\right]^{1/2}, \tag{3.36}$$

where

$$U_d = \sum_i \sum_j W_{d_{ij}} \quad \text{and} \quad (x_i - \bar{x})_d = \sum_j W_{d_{ij}}(x_j - \bar{x}).$$

When I_d is scaled by its bounds we have the statistic r_d, which is restricted to the range ± 1 and which can provide a more informative correlogram or partial correlogram:

$$r_d = \frac{I_d U_d}{n}\left[\frac{\sum(x_i - \bar{x})^2}{\sum(x_i - \bar{x})_d^2}\right]^{1/2}. \tag{3.37}$$

Example **3.16** (cont.)

The final column of Table 3.23 gives r_d for the *Atriplex* bushes. Note that $\sum(x_i - \bar{x})_1^2 = 337.78$. In this case r_d runs more or less parallel to z and affirms the previous interpretation.

Example **3.17** (cont.)

It is evident from Figure 3.9 that is some deviation between the I_d and r_d correlograms for the burnt savanna data. The two statistics are in harmony for values of d up to 15, but thereafter the r_d correlogram is much more subdued, with the downward trend for larger values of d having been largely eliminated. Cliff and Ord (1981a, p. 21) note that the expected upper bound for I_d under the null hypothesis (of zero autocorrelation) is approximately $(n/S_0)^{1/2}$ for symmetric binary **W** matrices such as we are dealing with here. Thus, for example, when $d = 31$ we have $S_0 = 128$ and consequently the value of the bound is 2.83. Dividing I_d by this value does indeed give a value fairly close to that of r_d.

The partial correlogram

If there is a genuinely repeating pattern with period d, then this will be manifest in the correlogram not only at d but also at multiples of d. Thus, 'autocorrelation coefficients from a single correlogram are not independent since the structure at greater distances is in part a function of the structure at lower distances' (Sokal, 1979). We must, therefore, be wary of attributing significance to peaks in the correlogram corresponding to large values of d in cases where there are peaks for smaller values of d. This is tiresome since we may well wish to investigate autocorrelation at both large and small scales. The partial (spatial) autocorrelation coefficient eliminates the periodicity effect and gives a direct measure of the interaction at distance d. The methods involved are borrowed from time series and it is convenient to work with r_d rather than with I_d. The interdependence of the autocorrelations is summarized by the 'autoregressive process' defined in equation (3.38):

$$r_j = p_{d1} r_{j-1} + p_{d2} r_{j-2} + \ldots + p_{dd} r_{j-d} \qquad (j = 1, \ldots, d). \tag{3.38}$$

The complete set of these so-called Yule–Walker equations (Yule, 1927; Walker, 1931) can be summarized in matrix form and solved for **p** where p_{dd} is the partial autocorrelation at distance d. Here $\mathbf{p}$ is the vector $(p_{d1}, p_{d2}, \ldots, p_{dd})'$. If we write $\mathbf{q}$ as the vector $(r_1, r_2, \ldots, r_d)'$, and $\mathbf{R}$ as the matrix:

$$\mathbf{R} = \begin{bmatrix} 1 & r_1 & r_2 & \ldots r_{d-1} \\ r_1 & 1 & r_1 & \ldots r_{d-2} \\ \cdot & \cdot & \cdot & \cdots \\ \cdot & \cdot & \cdot & \cdots \\ \cdot & \cdot & \cdot & \cdots \\ r_{d-1} & r_{d-2} & r_{d-3} & \ldots 1 \end{bmatrix},$$

then the general matrix equation has the solution:

$$\mathbf{p} = \mathbf{R}^{-1}\mathbf{q}. \tag{3.39}$$

Equation (3.39) relates to a particular value of d: each separate value of d will therefore require separate calculations, with the entire process amounting to a considerable numerical effort. Durbin (1960) has suggested a method of avoiding inversion of the $\mathbf{R}$ matrix and details are given by Cliff and Ord (1981a, p. 135).

***Example* 3.19**

Our analysis of the *Atriplex* data has, by now, assumed marathon proportions. With each increase in sophistication we obtain further insight into the pattern displayed by the data. The results of the partial autocorrelation analysis using the Yule–Walker equations are summarized by the partial correlograms (for r and for I) shown in Figure 3.12. The two statistics follow very similar paths indicating positive peaks for d around 2, 3, and 9 a deep trough at $d = 4$. The latter confirms our earlier findings. The stringent conditions required for a formal test of significance of p_{dd} are detailed by Bennett (1979, p. 108).

***Example* 3.20**

Figure 3.13 shows the I and r partial correlograms for our other long-running saga, the burnt savanna data. Once again the I partial correlogram is affected by the non-unitary bounds for the statistic, and we therefore concentrate on the partial correlogram for r. It is useful to compare Figures 3.9 and 3.13. It will be seen that in the partial correlogram the peaks at $d = 4$ and $d = 8$ have been greatly diminished, while the peak at $d = 10$ is somewhat enhanced. In addition the valley at $d = 9$ is now accorded the magnitude of the adjacent peaks. Finally, the peaks at 25, 28, and 30 substantiate the suggestion of long-range interaction that was mentioned in Chapter 1.

The spatio-temporal correlogram

Defining the class to which a pair of observations belongs on the basis of their *spatial* distance is just one method of subdividing the data set. A natural extension

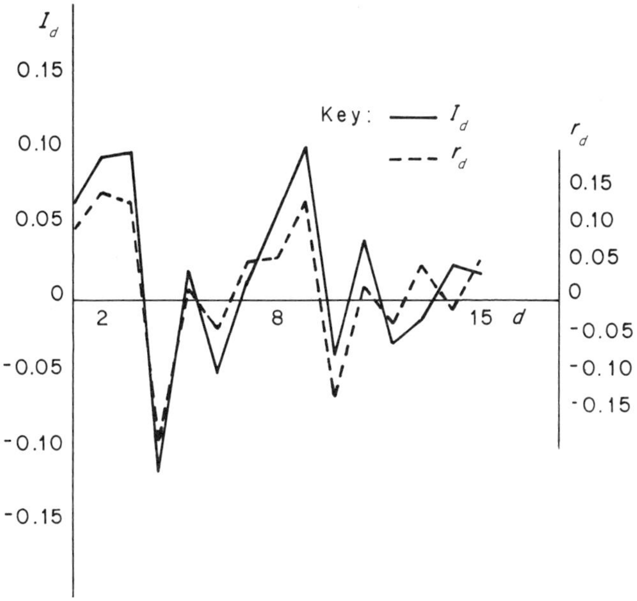

Figure 3.12 Partial correlogram for *Atriplex* bushes using I_d and r_d

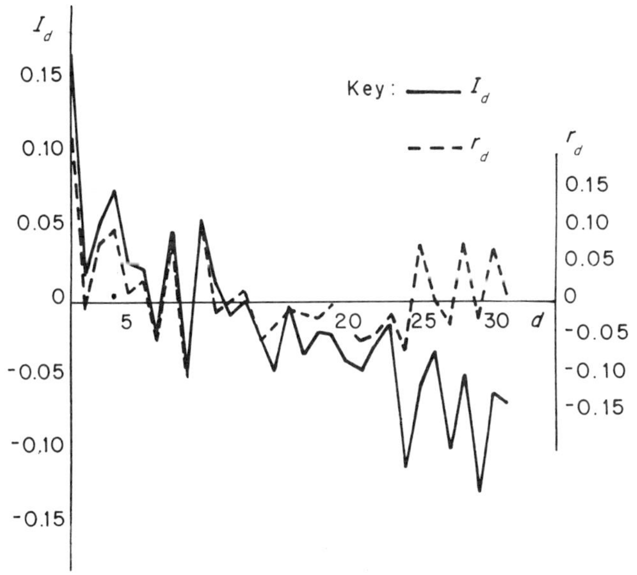

Figure 3.13 Partial correlogram for the burnt savanna data

***Table 3.24** The classification of some hypothetical lattice observations by space and time distance*

Observation	x coordinate	y coordinate	Time
1	1	1	2
2	1	2	2
3	2	1	2
4	1	1	2
5	1	1	1
6	1	2	1

Pair of observations	t	d	Class	
1,2	0	1	2	t = time distance
1,3	0	1	2	d = space distance
1,4	0	0	1	
1,5	1	0	3	
1,6	1	1	4	
2,4	0	1	2	
2,5	1	1	4	
2,6	1	0	3	
3,4	0	1	2	
3,5	1	1	4	
4,5	1	0	3	
4,6	1	1	4	
5,6	0	1	2	

when observations are made at the same localities on a number of different occasions is to segregate pairs of observations into separate classes on the basis of both their spatial and their temporal distance. As an example, Table 3.24 illustrates the situation in which there are four different classes indexed by their space and time distance using a rook's contiguity definition.

The spatio-temporal correlogram is essentially three dimensional since it demonstrates the variation in autocorrelation with varying space and time distance indices. Bennett (1975a), Martin and Oeppen (1975), Pfeifer and Deutsch (1980), and Hooper and Hewings (1981) construct various forms of spatio-temporal correlogram to describe the behaviour of real data, with the aim of comparing the empirical correlogram with that predicted by the various models that they consider.

One way to lay out spatio-temporal data for a spatial lattice is illustrated in Table 3.25, where each observation has its own space–time label. If the 36 labels given in Table 3.25 represented 36 different localities, then our previous spatial analysis would require us to form 36×36 **Y** and **W** matrices, with the aim of obtaining S and T statistics. In order to analyse this space–time data we start in

Table 3.25 Space–time labels on a 3 × 3 lattice

T_1	T_2	T_3	T_4
abc	*jkl*	*stu*	*b′c′d′*
def	*mno*	*vwx*	*e′f′g′*
ghi	*pqr*	*yza′*	*h′i′j′*

exactly the same fashion, using the same quick formulae (equations (3.33), (3.34), or (3.35) depending on contiguity definition) in order to calculate the S statistics. This is done for each spatial distance d, including $d = 0$ for which $S_0 = rc$, $S_1 = 2rc$, and $S_2 = 4rc$. Having obtained the S statistics for the spatial **W** matrix, we now calculate the required S^* statistics for the spatio-temporal **W*** matrix using equations (3.40):

$$
\begin{aligned}
&\text{if } t = 0, \quad && S_i^* = TS_i && (\mathrm{i} = 0, 1, 2); \\
&\text{if } t > 0, \quad && S_0^* = 2(T - t)S_0, && \\
& && S_1^* = 2S_0^*, && \\
& && S_2^* = (4T - 6t)S_2, && \text{for } T > 2t; \\
& && S_2^* = 2(T - t)S_2, && \text{for } T \leq 2t.
\end{aligned}
\tag{3.40}
$$

In these equations T is the number of separate points in time at which observations have been made (i.e. $T = 4$ for Table 3.25), while t is the time interval under consideration. There will be a different space–time **W*** matrix for each combination of d and t and we can usefully refer to a specific matrix by using

Table 3.26 A* W* *matrix for spatio-temporal lattice data

W* =

	T_1	T_2	T_3	T_4
T_1		**W**		
T_2	**W**		**W**	
T_3		**W**		**W**
T_4			**W**	

W =

	1	2	3	4	5	6	7	8	9
1		1		1					
2	1		1		1				
3		1				1			
4	1				1		1		
5		1		1		1		1	
6			1		1				1
7				1				1	
8					1		1		1
9						1		1	

subscripts. As an example, Table 3.26 gives the $\mathbf{W}_{dt}^*$ matrix corresponding to the situation illustrated in Table 3.25. For this matrix $S_0^* = 144$, $S_1^* = 288$, and $S_2^* = 2720$.

Example 3.21

Tinline (1971) analysed the spread of an epidemic of foot and mouth disease amongst livestock in the English Midlands in 1967–1968. Part of the data is reproduced in Table 3.27, which shows the number of outbreaks in 10 km × 10 km grid cells in a study area on two occasions separated by an interval of eight days. This time interval was chosen because it is equal to the mean incubation time of this virus. Tinline argues for a causal relation between distributions at eight-day intervals, although exact replication will be distorted by wind since 'the dominant mode of spread appears to have been windborne infection'. We should therefore expect significant autocorrelation among observations at zero time distance and small space distance because of the localized nature of the infection. We may also anticipate significant positive autocorrelation between observations eight days apart and some distance apart—the distance depending on the extent of the dissemination of the virus by the wind. Of course, if the original distribution is accurately replicated, this latter autocorrelation may simply reflect the autocorrelation in the original distribution rather than any causal connection over time at distances greater than zero.

Table 3.28 gives the spatio-temporal correlogram for the data of Table 3.27, and includes the z statistics (based on I). It shows that there is a significantly clustered pattern on both time maps, with positive autocorrelation extending to $d = 2$. There is also significant pattern transference from the first map to the second, with the value in a cell being closely related to the previous value in that cell (as one would expect). Furthermore, there is strong positive association between quadrats that are distant by 1 unit in both time and space.

Table* 3.27 *The number of outbreaks of foot and mouth disease at two different times

Time 1								Time 2							
0	0	0	0	0	0	0	0	0	0	0	0	0	1	0	0
0	0	0	0	1	0	0	0	0	0	2	1	0	0	0	0
0	0	1	0	0	0	0	0	0	0	1	0	0	0	0	0
0	0	0	0	1	0	0	0	1	1	2	1	0	0	0	0
0	2	2	0	2	0	0	0	0	1	5	2	3	0	0	0
1	0	0	0	0	0	0	0	1	2	1	2	0	0	0	0
5	0	0	1	0	0	0	0	3	2	0	0	0	0	0	0
0	0	0	0	0	0	0	0	1	1	0	0	0	0	0	0

Table 3.28 A spatio-temporal correlogram of the foot and mouth disease data

		d							
		0	1	2	3	4	5	6	7
$t=0$	r	1.00	0.46	0.21	0.09	−0.21	−0.18	−0.12	−0.03
	z	8.0	4.7	2.0	0.8				
$t=1$	r	0.50	0.26	0.09	0.01	−0.28	−0.16	−0.12	−0.03
	z	4.0	2.7	0.9	0.2				

Example 3.22

Hagerstrand (1953) studied the spread of improvements to pasture land owned by farmers in the Southern Swedish Uplands. The region studied by Hagerstrand was inconveniently irregular in shape and in order to simplify matters the data given in Table 3.29 relates to an interior rectangle of this region. The data refers to square quadrats of side 5 km with the number assigned to a quadrat being the

Table 3.29 The number of farmers improving pastures at different times in the Southern Swedish Uplands

1929											1930										
0	0	0	0	1	1	0	0	0	0	0	0	0	0	0	0	0	0	0	0	0	0
0	0	0	0	0	0	0	0	1	0	0	0	0	1	0	0	0	0	1	0	0	0
0	0	5	0	0	0	0	0	0	0	1	0	0	1	1	0	0	0	0	1	0	0
0	0	2	0	0	0	0	0	0	0	0	0	0	0	1	0	0	0	0	0	0	0
0	0	2	0	0	0	0	0	0	0	0	2	0	3	0	0	0	0	0	0	0	0
0	1	0	0	3	0	0	0	0	0	0	0	0	1	1	0	0	0	0	0	0	0
1	0	1	1	0	0	0	0	0	0	0	0	1	1	1	0	0	0	0	0	0	0
0	0	1	0	0	0	0	0	0	0	0	1	0	0	0	0	0	0	0	0	0	0
1931											1932										
0	0	0	0	0	1	0	0	0	0	0	4	0	0	1	1	1	0	2	0	9	1
1	3	1	0	0	1	0	0	0	0	0	4	5	2	1	0	0	0	0	0	16	3
0	0	5	3	0	0	0	3	0	1	0	0	3	3	0	1	0	1	1	0	1	3
0	1	1	0	0	0	0	2	0	0	0	0	0	0	0	0	0	1	1	0	0	0
2	1	2	0	0	0	0	0	0	0	0	5	4	2	3	3	0	1	0	0	0	0
4	4	5	6	2	0	1	0	0	0	0	2	4	4	3	2	2	1	1	0	0	0
1	8	7	3	1	0	1	1	0	0	0	2	15	5	1	2	0	0	2	0	0	0
0	3	7	6	0	0	0	0	0	0	0	1	7	6	1	0	0	0	0	0	0	0

number of farmers in that quadrat with less than 10 cultivated hectares who adopted improvements in that year. Tinline (1971) analyses this data from a different viewpoint.

The results of the analysis using the rook's definition of contiguity are summarized in Table 3.30. These results firmly establish that the number of improvers in a quadrat is strongly positively related to the number in quadrats that are close in time or space. Neither spatial or temporal closeness alone is sufficient. The counterpart of this pattern is consistent negative autocorrelation at higher distances. Almost irrespective of spatial distance, observations with $t \geq 2$ are negative, a response to the fact that there were more improvers in 1931 and 1932 than there were in 1929 or 1930.

Partial spatio-temporal autocorrelation

For reasons similar to those expressed in the discussion of partial correlograms, various authors have suggested the construction of partial spatio-temporal correlograms. The use of the spatio-temporal form of the Yule–Walker equations provides a natural parallel to our previous discussion and this approach has been suggested by, for example Bennett (1975a) and Pfeifer and Deutsch (1980). Unfortunately there are both practical and technical drawbacks to this approach (cf. Martin and Oeppen, 1975; Hooper and Hewings, 1981), and a preferable approach, suggested by Martin and Oeppen (1975), Bennett (1979) and Hooper and Hewings (1981), involves the calculation of partial autocorrelations. Regrettably the notation here is unavoidably clumsy and is potentially confusing.

Suppose, as an example, that we are interested in obtaining the partial autocorrelation for lags of 1 unit of time and 2 units of distance. We need not consider the effects of any correlations involving more extreme lags, but we do need to consider the (1, 0) and (1, 1) (time, distance) lags. Let X_1, X_2, X_3, and X_4 be the variables corresponding to lags of (0, 0), (1, 0), (1, 1), and (1, 2), respectively. We wish to calculate the (partial) correlation between X_1 and X_4 taking into account the values of the intervening variables, X_2 and X_3. This partial correlation is written as $\rho_{14.23}$ with the direct correlation between X_1 and X_4 being written as ρ_{14}.

To calculate $\rho_{14.23}$ we proceed in stages, calculating, in turn, $\rho_{14.3} = g(\rho_{14}, \rho_{13}, \rho_{43})$, $\rho_{12.3} = g(\rho_{12}, \rho_{13}, \rho_{23})$, $\rho_{24.3} = g(\rho_{24}, \rho_{23}, \rho_{43})$, and finally $\rho_{14.23} = g(\rho_{14.3}, \rho_{12.3}, \rho_{24.3})$, where

$$g(a, b, c) = \frac{a - bc}{\sqrt{(1 - b^2)(1 - c^2)}}. \tag{3.41}$$

The extension of this procedure to greater lags is both straightforward and tedious.

It should be noted that the form of the partial spatio-temporal correlogram (and, indeed, the form of the spatio-temporal correlogram itself) will be affected

Table 3.30 **A spatio-temporal correlogram of the Swedish farming data**

	d										
	0	1	2	3	4	5	6	7	8	9	10
$t = 0$											
r	1.00	0.66	0.30	0.19	0.13	0.15	0.00	−0.09	0.02	0.14	0.03
z	13.3	11.8	4.2	2.4							
$t = 1$											
r	0.43	0.43	0.28	0.14	0.07	0.08	−0.07	−0.16	−0.08	−0.08	−0.03
z	5.4	7.3	3.7	1.6							
$t = 2$											
r	0.01	−0.04	−0.04	−0.07	−0.09	0.12	−0.19	0.12	0.15	0.12	0.06
z	0.2	−0.6	−0.4	−0.9							
$t = 3$											
r	−0.05	−0.06	−0.08	−0.11	−0.09	−0.10	−0.17	−0.13	−0.10	−0.11	−0.08
z	−0.8	−1.1	−1.2	−1.6							

by the definition of contiguity and, crucially, by the choice of unit of time. If the unit of time chosen is too small, then there will be a massive inter-time correlation at a distance lag of zero. On the other hand, if the time lag is too large, then there may be no apparent correlation over time! An additional complication is that, on a small lattice, there will be very little information concerning large distance lags.

***Example* 3.21** (cont.)

As an example of the clarification obtained by studying the *partial* space–time correlogram, we reconsider the foot and mouth data. For the partial autocorrelation with (time, distance) lag (1, 1) we use equation (3.41) with a, b, and c referring to the (1, 1), (1,0), and (0, 1) autocorrelations that were given in Table 3.28. We therefore obtain the values $a = 0.26$, $b = 0.50$, and $c = 0.46$, from which we get a value for the partial autocorrelation of 0.04. Thus, contrary to the impression conveyed by Table 3.28, there is little evidence of a direct association between the value recorded for a plot P, at time t, and the values recorded at neighbouring plots at time $t-1$. The apparent association is simply a by-product of the association of both with the value recorded for P at time $t-1$.

3.6 RELATED APPROACHES

Space–time interaction tests

Although placed near the end of this chapter, this section could very easily have been located near the beginning, because it describes tests intimately associated with much of what has gone before. Indeed, many of the earliest and most interesting, if somewhat inconclusive, applications of the various versions of the cross-product statistic r were initially in the guise of space–time interaction tests, the earliest being that of Knox (1964), who searched for clusters in space and time (epidemicity) in the occurrence of childhood leukaemia in Northern England. Knox counted the number of pairs of affected children that were close together in both space (1 km) and time (60 days) and compared this with the number expected in the absence of epidemicity.

However, as Besag and Diggle (1977) and Green (1979) have observed, there will be clustering even in the absence of a leukaemia epidemic, since the majority of people huddle together in towns and since leukaemia is most frequent in summer. Knox's aim was therefore to detect any extra clustering over and above that already inherent in the data. Nevertheless, in the absence of contagion, the mechanisms which produce clustering in time and clustering in space might be expected to act independently. It is on this basis that we may consider the null hypothesis to represent the absence of the space–time interaction (Besag and Diggle, 1977).

An additional complication is that the presence of the space–time interaction is not, by itself, proof of contagion, for as Glick (1979) notes, it merely indicates that the distribution of a disease is influenced by movement of some sort, and this may not be, say, viral transmission. It is with this in mind that Glass *et al.* (1971) attributed significant space–time clustering of leukaemia cases in New Zealand to differential rapid population change rather than to a true process of contagion. This confusion between possible causes that have a comparable effect is yet another example of the virtual indistinguishability of heterogeneity and clustering, as was mentioned in Chapter 1.

Knox's (1964) analysis was based on a consideration of all $n(n-1)$ pairs that could be formed from the n cases observed. Suppose that S pairs were close in space and that T pairs were close in time. If closeness in space is independent of closeness in time, and if the $n(n-1)$ pairs are regarded as being independent of one another, then the mean number of pairs that will be simultaneously close in both time and space will be $ST/n(n-1)$. Under this hypothesis of independence the observed number of pairs, x, that are close in both time and space will have an approximate Poisson distribution. Thus, the hypothesis of independence would be rejected in favour of epidemicity if the observed value of x was unusally large. It is worth noting that we are here dealing with pairs and not cases and hence one pair may not be independent of another (since they may have a case in common). The $n(n-1)$ pairs are therefore not strictly independent and consequently Knox (1964) chose suitably stringent criteria of closeness. This also provides a partial explanation for the approximate nature of the Poisson distribution for sparsely distributed join-count statistics.

Many applications of space–time interaction tests have been either epidemiological, such as in studies of the incidence of leukaemia and cleft palate (see Glick, 1979, for references), or in the investigation of 'animal' behaviour. There have been some interesting ornithological applications. Besag and Diggle (1977), for example, tested whether kittiwakes nesting on window ledges of disused warehouses were stimulated to lay by the presence of immediate neighbours. Because of the exploratory nature of their study, they defined time adjacency in various ways ranging from laying dates within 2 days of each other to within 10 days of each other. Green (1979) examined the possibility of *interference*, the opposite of contagion, in the nesting behaviour of great tits and blue tits in Marley Wood, Oxford, England. He discovered intraspecific but not interspecific interference, the latter test being an extension to two variables of the Knox method.

Other developments of the space–time interaction test follow the same path as that already outlined for purely spatial autocorrelation tests, and involve the replacement of the zero–one proximity measures of the conceptually simple Knox statistic by more discriminatory measures (cf. Mantel, 1967). To avoid duplication, we simply note that in both the spatio-temporal and spatial versions of r similar remarks can be made about operational details and statistical distributions.

***Example* 3.23**

In addition to studying *Atriplex* bushes, Gulmon and Mooney (1977) also investigate the interrelated distribution of *Tidestromia oblongifolia*, which is another perennial shrub able to withstand conditions in Death Valley. Figure 3.14 is an amalgamation of three of their maps showing the number of new *Tidestromia* seedlings for a particular plot for the years 1973, 1974, and 1975.

A working definition of closeness in space is a straight line inter-seedling distance of less than 10 m. This is given a proximity value 1 in the **W** matrix, whilst more distant seedling pairs are given the value 0. Seedlings appearing in the same year are defined as being close in time and attain a proximity value 1 in the **Y** matrix, while other seedling pairs take the value 0. The outcome of these definitions is that, using equation (3.3), we obtain the value 46 for the general cross-product statistic, r. The adopted closeness criteria invalidate the Poisson as a relevant distribution under the null hypothesis and the adoption of the normal approximation can only be tentative, given that there are few plants. However, using equations (3.5) and (3.6), we find $E(r) = 39.48$ and $\text{var}(r) = 19.98$ so that, from equation (3.7), we have $z = \{|46 - 39.48| - 1\}/(19.98)^{1/2} = 1.23$. This z-value corresponds to an upper-tail probability of about 11%.

By way of checking the reasonableness of the normal approximation, we also

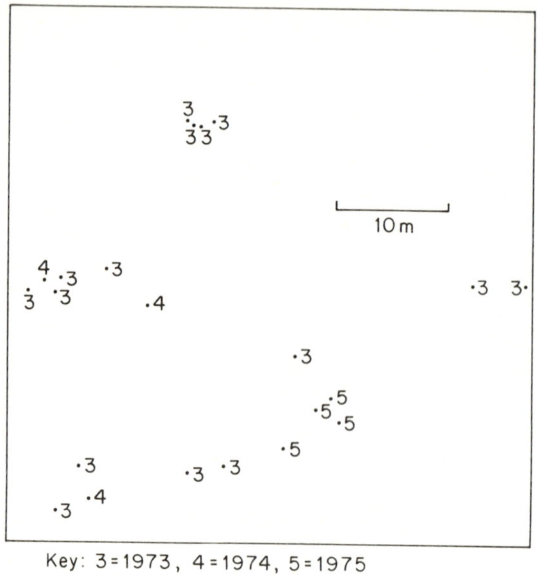

Figure 3.14 The three-year (1973–1975) distribution of new *Tidestromia* seedlings in a square plot in Death Valley, California. (From Gulmon and Mooney, 1977. Reproduced by permission of Blackwell Scientific Publications Ltd)

consider a Monte Carlo randomization: in 99 simulations the value 46, or a greater value, was obtained on 14 occasions. Our conclusion is therefore that the test has failed to detect significant evidence of space–time clustering. The visual impression one receives from Figure 3.14 is of strong spatio-temporal clustering—our non-significant result comes about partly because we are not measuring proximity in the test in the same way as we perceive it, and partially because there is not a great deal of information for the latter two years of the study.

Relative conditional intensity

Vere-Jones (1978) has described an approach to space–time autocorrelation which treats the observations as the possible outcome of a stochastic point process in m-dimensional space, where the dimensions are space, time, magnitude, and so on. This has proved useful in the study of microearthquakes and also (Kagan and Knopoff, 1976a, 1976b) the analysis of a world-wide survey of major earthquakes. To justify considering earthquakes as points, their time duration and spatial dimensions are neglected, and this seems a reasonable simplification given the time and space scales that are involved.

With this seismological data, the point at issue is whether the space–time locations are independent of one another (and can therefore be regarded as a realization of a Poisson process) or whether there is evidence of some autocorrelation function. Vere-Jones suggests that any autocorrelation may be detected by evaluating what he refers to as the relative conditional intensity, which is based on the number of pairs of observations valued i and j separated by a time distance t and a spatial distance d. We denote this number by n_{ijtd}, where occurrence i precedes occurrence j in time. Evidently the magnitude of n_{ijtd} will vary in proportion to n_i, the total number of occurrences valued i, and hence we consider the ratio n_{ijtd}/n_i. In a similar way, if wide time and distance classes are used to obtain the count n_{ijtd}, then n_{ijtd}/n_i will be correspondingly large. To take account of this feature, Vere-Jones suggested using the scaled statistic $n_{ijtd}/(n_i v_{td})$, where v_{td} represents the 'volume' in space and time within which the pairs are counted. This scaled statistic is therefore a conditional intensity.

It is appropriate to relate the conditional intensity to an estimate of unconditional local intensity, n_j/v, where n_j is the number of value j occurrences and v is the entire 'volume' of the space–time for the study period. In the absence of space–time autocorrelation, the conditional and unconditional intensities should be very similar, and hence the ratio $(n_{ijtd}v)/(n_i n_j v_{td})$, the *relative conditional intensity*, is an index of space–time clustering or dispersal with a value 1 for a Poisson process.

Vere-Jones (1978) adopts uneven classes in his study of Japanese microearthquakes, choosing successive time and distance intervals which increase at a constant ratio (doubling for time and increasing at the rate $\sqrt{2}$ for distance). The space–time volume is then calculated as the product td, where t and d are the time and distance intervals. Further refinements suggested by Vere-Jones include the disaggregation of the data according to the directional links between earthquake pairs and by earthquake depth. Vere-Jones also devises multiplicative adjustment factors to allow for edge effects, thus avoiding the downward bias due to 'nearest neighbours' that lie outside the study region.

Ordinal autocorrelation measures

Where there is an element of arbitrariness in the proximity values assigned to **W** and **Y** matrices, it is advisable, and possible, to adopt methods based solely upon the *ordinal* relationship between matrix cells. This avoids unrealistically attributing a high level of precision to proximity values and takes account of the fact that the outcome of a spatial autocorrelation test will be sensitive to its operational definitions. By concentrating on the ordinal properties of the proximity values one is equalizing a large number of different sets of values when there is no special reason to favour any one value above the rest. Thus, while it may be unrealistically precise to state that the spatial proximity of locality pair (i,j) is W_{ij} compared to W_{ik} for locality pair (i,k), it may be credible to make the simpler assertion that $W_{ij} > W_{ik}$.

Following, broadly but not in detail, the exposition and notation of Hubert *et al.* (1981), we write:

$$\begin{aligned} a(i,j,k) &= 1, \quad \text{if } W_{ij} > W_{ik}, \\ &= 0, \quad \text{otherwise;} \\ b(i,j,k) &= 1, \quad \text{if } Y_{ij} > Y_{ik}, \\ &= 0, \quad \text{otherwise;} \end{aligned} \tag{3.42}$$

$$\Omega = \sum_{i,j,k} a(i,j,k) \cdot b(i,j,k).$$

The value of Ω is therefore interpretable as the number of times that the order of proximity to locality i of localities j and k is the same for two different measures of proximity (which might be spatial proximity and temporal proximity). The summation in the definition of Ω is over all possible triples of locations, and thus, with n locations, the maximum possible value for Ω is $n(n-1)(n-2)$.

Note that since the values of the a and b functions are set solely by virtue of comparisons within a row of the proximity matrix, different criteria could be used to define proximity for each row (location) of the matrix. Usually, of course, the same criteria would be used throughout the matrix, in which case Hubert *et al.* suggest that it might be more informative to consider two pairs of locations (i,j), (k,l) and to calculate the statistic θ, given by equations (3.43):

$$\begin{aligned} r(i,j,k,l) &= 1, \quad \text{if } i \neq j, k \neq l, \text{ and } W_{ij} > W_{kl}, \\ &= 0, \quad \text{otherwise;} \end{aligned}$$

$$s(i,j,k,l) = 1, \quad \text{if } i \neq j, k \neq l, \textit{ and } Y_{ij} > Y_{kl}, \qquad (3.43)$$
$$= 0, \quad \text{otherwise;}$$
$$\theta = \sum_{i,j,k,l} r(i,j,k,l) \cdot s(i,j,k,l).$$

The significance of Ω or θ can be tested by applying the usual Monte Carlo methods to the **W** and **Y** matrices, so that reference distributions for Ω or θ under the null hypothesis can be obtained. Hubert (1978) gives more details of these methods, and Hubert *et al.* (1980) discuss a worked example.

Alternative sampling models

Throughout, the interpretation of observed spatial autocorrelation statistics has been founded upon sampling models generated, in theory or in practice, by permuting observations. This method of obtaining reference distributions under the null hypothesis embodies the proviso that no (aspatial) distribution of x_i other than that observed is realizable. Thus, the set $\{x_i\}$ is treated as a *de facto* population and no alternative values for the $\{x_i\}$ are considered. In contrast, alternative sampling models treat the set $\{x_i\}$ as being just one of the many alternative sets of values that might be envisaged as a potential outcome of some general process. With this alternative viewpoint the data is thought of as constituting a sample of values rather than as a population.

In the context of the (two-colour) join-count statistics, the usual constraint is that, for every permutation, the numbers of localities of each colour are equal to the numbers observed. With the alternative viewpoint, one assumes merely that there are fixed probabilities, p and $q = 1 - p$, that a locality will be coloured black or white. This is referred to as the *free-sampling* assumption. The consequence of this assumption is that all manner of maps, ranging from n entirely black localities (probability p^n) to n entirely white localities (probability q^n), are realizable, in contrast to the restricted fixed numbers situation previously considered. The analogue of free sampling in the case of ordinal or interval data, for which statistics such as Moran's I or Geary's c are appropriate, is generally taken to be that the set of n (interval) values $\{x_i\}$ is just one of the infinite sets of values that could be drawn from an appropriate normal distribution—this is the so-called *normality assumption.*

The equations (3.44a)–(3.44d) give the mean and variance of the free-sampling distribution of the join-count statistics generalized to k colours (Cliff and Ord, 1981a, p. 19). In these equations, p_r denotes the probability that a locality is coloured r, S_{rr} denotes the number of joins between contiguous localities both coloured with colour r, and D_{rs} is the number coloured in different ways. The equations may be compared with the 'non-free' equations(3.15):

$$E(S_{rr}) = \tfrac{1}{2} S_0 p_r^2, \qquad (3.44a)$$

$$\mathrm{var}\,(S_{rr}) = \tfrac{1}{4}\{S_1 p_r^2 + (S_2 - 2S_1)p_r^3 + (S_1 - S_2)p_r^4\}, \tag{3.44b}$$

$$\mathrm{E}(D_{rs}) = S_0 p_r p_s, \tag{3.44c}$$

$$\mathrm{var}\,(D_{rs}) = \tfrac{1}{4}\{2S_1 p_r p_s + (S_2 - 2S_1)p_r p_s(p_r + p_s) + 4(S_1 - S_2)p_r^2 p_s^2\}. \tag{3.44d}$$

In these equations the S statistics retain the definitions initially presented in equation (3.6).

The equations (3.45a) and (3.45b) give the mean and variance of I under normality—the value of the mean being unchanged:

$$\mathrm{E}(I) = -1/(n-1), \tag{3.45a}$$

$$\mathrm{var}(I) = \frac{n^2 S_1 - nS_2 + 3S_0^2}{S_0^2(n^2 - 1)}. \tag{3.45b}$$

Because of the comparative absence of restraint on the set of realizable maps, these approximately normal sampling distributions will usually possess a larger variance than their counterparts generated by randomization. As a consequence it could be said that the randomization approach is more sensitive to the presence of organization, although it would be unwise to make too much of this, since the inferences are conditional upon the observed data. In comparison, the wider inferences relating to some general process that may be derived from the alternative sampling model naturally demand harder evidence.

The true value of randomization lies in its simplicity and applicability to data from any source regardless of the distribution of any parent population that may exist. In contrast, the alternative sampling models, which are mentioned largely for the sake of completeness, demand the satisfaction of relatively narrowly defined conditions. For example, to evaluate the equations (3.45a) and (3.45b) requires reliable estimates of p_r from observation of previous outcomes of the generating process. In the case of I or c, there is the additional assumption of normality under the null hypothesis. These conditions may be satisfiable when the data describe, say, the incidence of a recurrent phenomenon such as a disease, but many data sets—for example election returns (Haggett *et al.*, 1977)—have no reliable precursor and are best treated as unique, so that randomization should be invoked. In fact it has consistently been the case that where both kinds of sampling models could sensibly be invoked, largely identical (though unreported) conclusions have been made about the data.

Difference maps

The analyses considered hitherto have largely been aimed at detecting and describing organization present in the distribution of a single variable represented by the values $\{x_i\}$ assigned to n localities. It may well be that organization is very evident in a mapped distribution and attention then focuses on the more pertinent question of the cause of this organized pattern. Thus, one invokes a model to account for the x_i distribution. Such a model, if successful, should generate a comparable set of expected values $\{y_i\}$ that approximates to that of the $\{x_i\}$, and the map of residuals $\{x_i - y_i\}$, the *difference map*, should itself display an absence of spatial autocorrelation.

An early attempt to analyse the differences between theoretical and observed maps is the work of Cliff (1970). There are various potential model structures that one might consider and the important case of the regression model will be considered in a later chapter. An alternative type of model is the stochastic spatial diffusion process initiated by Hagerstrand (1953); this is considered in detail in the context of difference maps by Cliff and Ord (1981a) who detail some of the special considerations that the use of this model involves.

The simplest way to analyse a difference map is, of course, to use the join-count statistics with colours being allocated to localities depending on their residual values. Comparable, but more intricate, analyses may be carried out using variants of Moran's *I*. A parallel application of the difference map idea is in the context where the $\{x_i\}$ and $\{y_i\}$ are values from two comparable variables. If one of the two variables can act as a surrogate for the other, then one would expect an absence of organization in the difference map.

The choice of a test

A variety of different but interrelated tests have been described in this chapter, each with the ability to convey distinctive information about a set of data as a consequence of utilizing an alternative perspective. In order to provide some advice as to which of the plethora of available tests is most suited to eliciting relevant information, it is necessary to come to some evaluation of the comparative worth of the various options. This involves, for each test, weighing some measure of its informativeness against the complexity, computational effort, and cost involved in extracting the information. For example, the *BW* statistic is preferred to the *BB* statistic since both are equally easy to calculate and the former is the more informative. Similarly, Moran's *I* is preferred to the likelihood ratio statistic λ, since the latter is considerably more complex and only marginally more informative. On the other hand, it is much more difficult to evaluate the relative merits of Moran's *I* and the *BW* statistic.

This theme is pursued further in Figure 3.15, in which the various tests discussed in this chapter (apart from the all pervasive r and Ω statistics) are located with respect to the axes 'informativeness' and 'simplicity'. Needless to say, the precise location of each test on the graph is fairly arbitrary, being a subjective evaluation from practical experience and not being based solely on the results of objective scientific tests of comparative power as reported by, say, Cliff and Ord (1975a, 1981a). The aim, in Figure 3.15, is to incorporate intuitive notions of the relevant simplicity (and hence cost) of the various tests together with the more readily quantified informativeness dimension.

The design of Figure 3.15 is such that statistics located relatively near to the

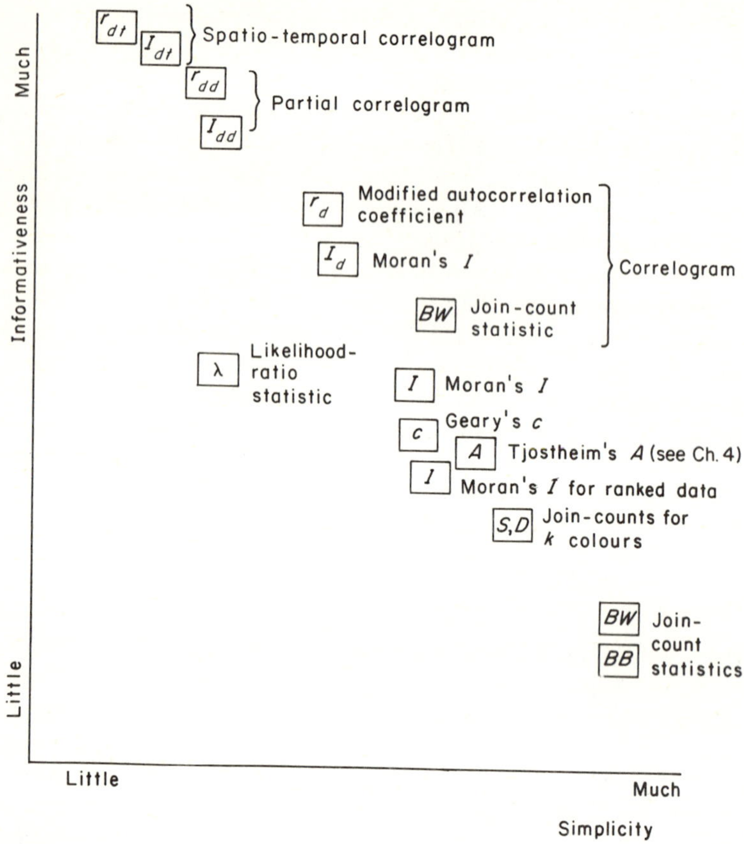

Figure 3.15 The approximate informativeness and simplicity of various autocorrelation measures

north-east corner are to be preferred on account of their combined virtues of simplicity and informativeness.

The nature of the available data is a major constraint influencing the test (or tests) ultimately chosen. For example, the absence of a time dimension to the data automatically excludes r_{dt} and associated statistics from the options available. On the other hand, observations may be made at different points in time and yet the time dimension may still be ignored, either by the selection of data pertaining to only one of the time points, or by pooling observations so that they appear to be simultaneous. If this is the case, then there has been a potentially important loss of information. Likewise, creating categorical data from what were originally interval or ratio level observations in order to facilitate join-count tests will also involve information loss.

In these situations it is up to the analyst to decide whether the consequent simplification compensates for the loss of information. There can be no objective answer to the question 'Which test?', for the benefits of simplification must be viewed in the context of the resources that the analyst can command, and these will vary.

A major cost to bear in mind is software implementation and development. A useful source of a wide variety of published programs are the journals *Environment and Planning, Series A*, *Computers and Geosciences*, *Computer Applications in the Natural and Social Sciences*, *Applied Statistics*, *Journal of the Association of Mathematical Geology*, *Communications of the Association of Computer Manufacturers*, *Computer Journal*, and *Geocom Bulletin*. There are also the *Computer Contributions* of the State Geological Survey, University of Kansas, selections of computer programs published by the geographers Baker (1971), Marble (1967), and Tobler (1970b), and a Geographic Program Exchange (GPE) scheme. Despite these sources, in 1984 no comprehensive package exists that handles all the approaches outlined in this chapter, and, for this reason, considerable attention has therefore been focused on simple, low-cost methods.

CHAPTER 4

Inter-type Relations

The five chapters of this volume are closely linked and the order in which they are presented reflects a slow change in emphasis from exploration to explanation. Thus, Chapter 1 was concerned with the location of points and the patterns displayed by these points. Each individual point (or plant) was indistinguishable from every other individual, except by reason of its position. Whilst some initial effort was made (in Section 1.1) to discuss the mechanisms leading to different types of pattern, the emphasis there lay with identifying and quantifying that pattern.

When we reached Chapter 3, the emphasis had changed. The analysis was still largely exploratory, but the interest now lay not in the pattern of the locations, but rather in the pattern of *values* at locations which, in themselves, were of no interest. Looking ahead to Chapter 5, we shall there be considering the exterior forces, such as rainfall, climate, or whatever, that lead to the observed variations in these values. These regression models could also be used to explain variations in plant intensity. The basis of Chapter 5, however, is that the explanations lie with immutable physical properties of the real world. In this chapter we straddle the gap between the first three chapters and the final chapter: here, we consider the effects that one point pattern has on another and the links between one arrangement of values and another arrangement of values.

One important limitation of the present chapter is that we shall confine our attention to static patterns. The specialized problems of the spread of rumours, species, or diseases, such as are represented by the spatial contact models of Mollison (1977), defy illustration in a simple fashion.

The layout of the present chapter is governed by the order in which topics have been introduced in the earlier chapter, and we therefore commence by considering bivariate point patterns.

4.1 BIVARIATE POINT PATTERNS

In this section we shall revert to the habit of referring to individuals as points rather than plants, for consistency with Section 1.1, although subsequently we shall make our usual distinction between sampling points and the individuals of interest (then termed plants).

We commence by recalling some arguments and definitions. First we recall that the phrase 'point intensity' refers to the number of points per unit area averaged over the entire region of interest. Next, we shall recall that in our introduction to the Poisson process we conceptualized the region of interest as being subdivided into an infinite number of infinitesimal sub-regions, some of which contained points whilst others did not. The particular characteristic of the Poisson process is that the point intensity associated with each of the sub-regions is constant.

If we permit the point intensity to vary from sub-region to sub-region, thereby implying that some sub-regions are more likely to contain points than others, then the resulting point distribution will take on a 'patchy' appearance. We termed this distribution the *compound Poisson* distribution. At the risk of confusing our readers, we would now like to remove this name and, following Diggle and Milne (1983), refer to this process of point generation as a *Cox process*. The nomenclature arises because the basic statistics of the occurrence of individuals under the situation of a positionally varying intensity were the subject of the classic paper by Cox (1955). It is useful to change the notation at this point because we shall want to consider various types of Cox process, to which we shall give names, and a name with three adjectives would be a considerable mouthful!

The two main classes of point distribution were those for *clusters* of points (Cox processes, centre–satellite processes) and those that defined *regularity* (inhibition processes, the Central Place model). Consider now the possibilities open to us when we consider *bivariate* processes. Essentially there are four elements of the point pattern with which we shall be concerned. These are:

(i) the pattern of type A points viewed on its own;
(ii) the pattern of type B points viewed on its own;
(iii) the combined pattern of the intermingled type A and type B points; and
(iv) the interrelation between the type A pattern and the type B pattern.

Since each of (i), (ii), and (iii) refers to a single point pattern, the methods of analysis that were discussed in Chapter 1 can be used to draw broad conclusions about each pattern. Evidently, each individual pattern might be classified as clustered, random, or regular. We know that such classifications reflect our level of measurement, since what appears clustered at one scale may seem regular at another. This should, perhaps, warn us that, for example, intermingling two random patterns can produce a pattern which shows unmistakable evidence of either clustering or regularity. Since there are three possible classifications for each individual pattern, and since the combined pattern may also be classified as any of 'clustered', 'random', or 'regular', it follows that there are some 27 possible combinations of patterns possible at any scale of measurement, though we shall only consider a few of these!

The appearance of the combined pattern reflects the fourth element in the list above, namely the interrelation between the two patterns. It is this interrelation that welds the first three elements together. We can conceive of three possibilities.

(1) *Attraction* between the two point processes.
(2) *Independence.*
(3) *Repulsion* between the two point processes.

Suppose we consider the relation between the (possibly varying) point intensities that have given rise to the two point patterns. If it is the case that as one point intensity rises then the other point intensity also rises, then the processes will exhibit attraction—where there are many type A points there will also be many type B points. We could anticipate that two processes would display attraction if, for example, one process generated the locations of parents and the other process generated the locations of offspring. Alternatively, the two patterns might reflect different facets of a single underlying process: thus, the process of habitation gives rise both to public libraries and to doctors' surgeries. At the scale of a 1:100 000 Ordnance Survey map, these two patterns will probably appear closely linked but of differing intensity.

On the other hand, if the two point processes refer to *competing* species, then the combined pattern is likely to exhibit regularity due to the repulsion between species. This was the case the trees in Lansing Woods that we examined earlier in Examples 2.3 and 3.11. Lee (1979) suggests that similar 'spatial avoidance' is to be found amongst rival chains of food stores.

It is clear that with bivariate point patterns much of the interest—and virtually all the new material in this chapter—centres on the investigation of the relation *between* the two patterns. In the remainder of this section, however, we consider some possible mathematical descriptions of bivariate point processes.

Processes displaying attraction

With the simple Poisson process, the intensity is everywhere the same. In order to express this mathematically we need some notation. Following Diggle and Milne (1983), we write $\Lambda(l, m)$ to mean the intensity of points within the tiny region centred on the position whose coordinates are (l, m). Thus, the Poisson process itself is defined by

$$\Lambda(l, m) = \lambda, \quad \text{for all } (l, m). \tag{4.1}$$

The general Cox (variable intensity) process is then defined by

$$\Lambda(l, m) = \lambda(l, m), \tag{4.2}$$

which simply states that the intensity at (l, m) is not a constant, but is a function of the location (l, m).

With a bivariate point process we have two point intensities, $\Lambda_1(l, m)$ and $\Lambda_2(l, m)$, to consider. If we had $\Lambda_1(l, m) = \lambda$ and $\Lambda_2(l, m) = \mu$, then this would imply that both the individual patterns were random and hence that the overall pattern was random with intensity $(\lambda + \mu)$.

Consider now the case where both point patterns are generated by linked

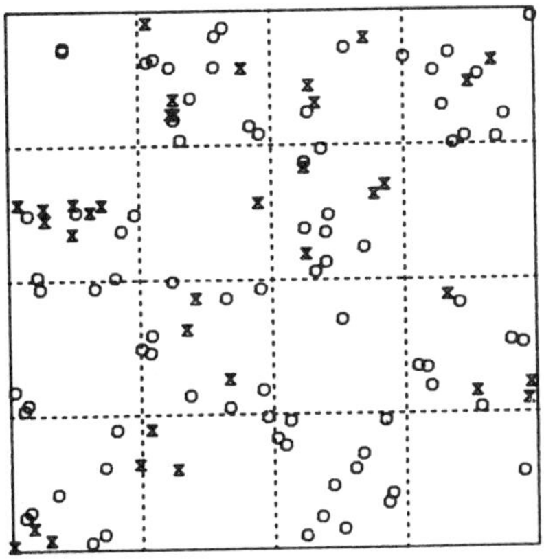

Figure 4.1 A chess-board pattern underlying a linked Cox process

attractions (such as might be the case with variable soil fertility and two species of plant). In this case we might have

$$\Lambda_1(l, m) = k\Lambda_2(l, m). \tag{4.3}$$

Diggle and Milne (1983) refer to this as a *linked Cox process.* Since the constant k is everywhere the same, it is easily and efficiently estimated by computing the ratio N_1/N_2, where N_1 and N_2 are the total numbers of the two types of points that occur in the study region.

A somewhat degenerate example of a linked Cox process occurs when $\Lambda_2(l,m) = \lambda$, since this then implies that all the patterns are random. A more interesting example is provided by imagining a 4×4 chess-board in which $\Lambda_2(l,m) = 1$ on 'white' squares and 4 on 'black' squares, with $k = 2$. This is illustrated by Figure 4.1, in which the underlying lattice is illustrated. It is fascinating to redraw this diagram with the dotted lines omitted (judicious use of photocopier and whitener would suffice); it is then surprisingly difficult to detect the underlying grid. Clearly, with an irregular underlying tessellation the pattern mechanism would be still more difficult to detect.

An example of a univariate clustered process was the centre–satellite model, which is a natural model for mother/daughter patterns. Diggle and Cox (1983) introduce a process of this type which they refer to as a *linked pairs* process. Here $\Lambda_1(l,m) = \lambda$, while, for each mother point that is realized there is a daughter point that is located in the vicinity of the mother. Diggle and Cox (1983) model this process by assuming that the mother–daughter distances are observations from a normal distribution, with the orientation of the daughter from the mother being equally likely to be in any direction. We have adopted the same mother–daughter distance distribution to derive Figure 4.2.

Figure 4.2 A realization of a linked pairs process

Notice that, since $\Lambda_1(l, m) = \lambda$ and since the position of each daughter point is close to that of a mother point, with one daughter to each mother, it follows that $\Lambda_2(l, m) = \lambda$. However, the combined pattern is clearly *not* Poisson, so that we can deduce that

$$\Lambda_+(l, m) \neq \Lambda_1(l, m) + \Lambda_2(l, m), \tag{4.4}$$

unless $\Lambda_1(l, m)$ and $\Lambda_2(l, m)$ are independent of one another. Here, of course, $\Lambda_+(l, m)$ denotes the intensity of the combined process.

An example of a real data set, that appears superficially remarkably similar to the data generated by the linked pairs process, is illustrated in Figure 4.3. This is a 3 m × 3 m section of the 4.5 m × 9 m section of bramble patch illustrated in full by Diggle and Milne (1983). Further details and the background to the data are given by Hutchings (1979). Figure 4.3 illustrates the positions of 30 newly emergent canes and 36 one-year-old canes: it is plain that the positions of the new canes are governed by the positions of the older canes.

Various authors (Brown, 1979; Stoyan, 1979; Diggle and Milne, 1983) have discussed so-called *thinned* point processes. The idea here is that, with the passage of time, individuals die or are otherwise removed from the pattern. As an example of this idea, Figure 4.4a displays the pattern of mothers given previously in Figure 4.2, while Figure 4.4b displays the pattern of mothers for the following year (or generation). The pattern given in Figure 4.4b is simply the linked pairs pattern of Figure 4.2, with about 75% of the mothers removed and about 25% of the children removed.

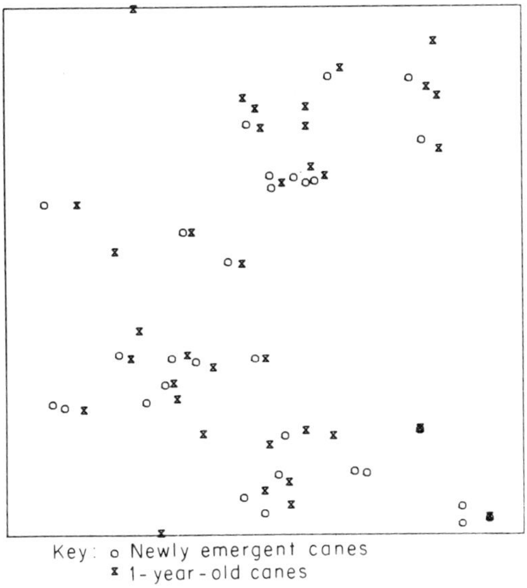

Figure 4.3 Bramble canes (*Rubus fruticosus*) in a 3 m × 3 m plot. (Redrawn from Diggle and Milne, 1983. Reproduced by permission of the Editors of the *Journal of the Royal Statistical Society, Series B*)

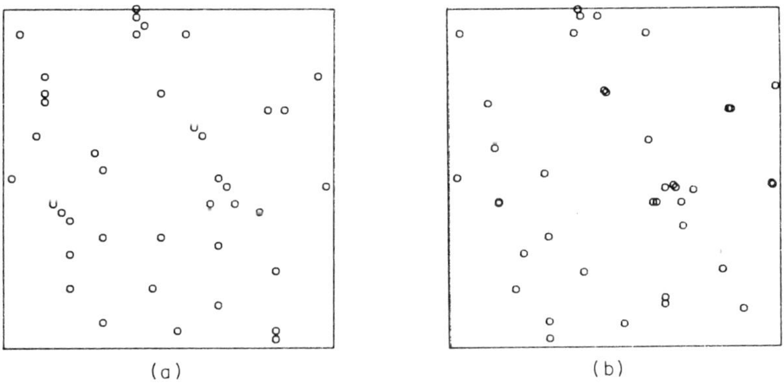

Figure 4.4 Disproportionately thinned mother–daughter process. (a) Mothers only. (b) Thinned mothers + daughters

Processes displaying repulsion

One of the first 'regular' patterns that we considered was Diggle's variant on Matern's dynamic process, and this was illustrated in Figure 1.8c. In Diggle's variant, potential points were generated sequentially from a Poisson process, but were only retained providing that they fell a distance of at least δ (say) from all previously retained points. Diggle and Cox (1983) have extended this idea in a straightforward fashion to a bivariate process. Points are generated alternately from two separate Poisson processes. A provisional point is only retained providing that it falls a distance of at least δ from all previously retained points of the *opposite* type. If this is not the case, then another point is generated from the same process, and this continues until an acceptable point is found.

Clearly, with this procedure, since both sets of points were generated from Poisson processes, inhibition regions for one process against the other will be randomly placed and hence the *composite* pattern will appear random, whereas each separate pattern will appear clustered.

As described, with exact alternation of the two processes, each process will have the same effective mean intensity, but this could easily be varied. If, say, we generated two points from one process for every one point from the other process, then the inhibition regions would contain points at different mean intensities and the compound pattern would then appear inhomogeneous.

We could also refer to the process we have described as 'selectively hard core'. A 'softer' variant was also proposed by Diggle and Cox (1983). In this variant there was a non-zero probability of retaining a point from one process that lay within a distance d ($< \delta$) of a point from the other process. This probability would vary between 1 at $d = \delta$ and 0 at $d = 0$.

While these bivariate *sequential inhibition* models of Diggle and Cox lead to easily generated patterns, they do not seem particularly plausible as an explanation of any real-life bivariate situation. An example of the pattern produced using the equal-intensity hard-core variant is provided by Figure 4.5.

Another somewhat artificial model—although again a model that leads to a simply generated pattern—was proposed by Hanisch and Stoyan (1979). They envisaged a two-type composite Poisson pattern which was subdivided into its component types according to the realized remoteness of the generated points. Specifically, one generates a realization of a Poisson process and then examines each point in turn. If a point is found to be at a distance δ or more from all the surrounding points, then this point is deemed to belong to type 1, and is otherwise deemed to belong to type 2. An example of this kind of *isolated Poisson* pattern is given in Figure 4.6. It is evident that the type 1 points must display regularity of the hard-core type, while the type 2 points will display clustering.

We complete this subsection by considering two suggestions based on the idea of compensating Cox processes. Brown *et al.* (1981) envisaged that $\Lambda_1(l, m)$ and $\Lambda_2(l, m)$ would be linked as follows. Consider a sub-region of area A which is such

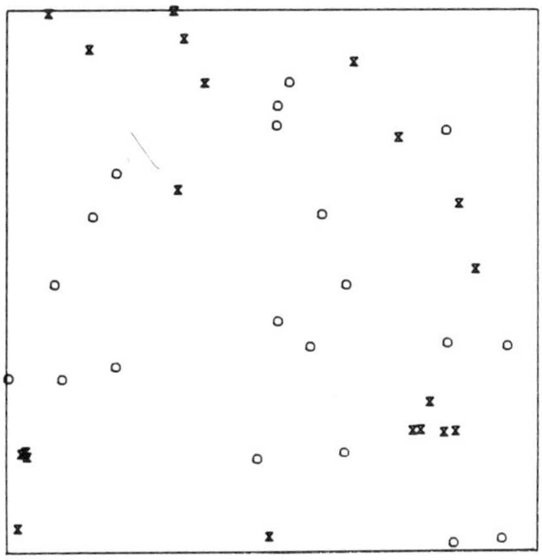

Figure 4.5 Diggle and Cox's bivariate sequential process

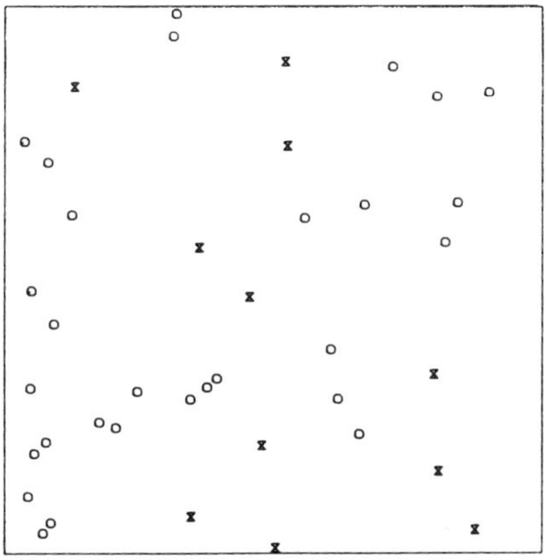

Figure 4.6 Hanisch and Stoyan's isolated Poisson process

that throughout this area $\Lambda_1(l,m) = \lambda_1$. Type 1 events are distributed over this sub-region according to a Poisson process. Let N_1 be the number of type 1 events actually observed in this sub-region. Then N_1 is an observation from a Poisson distribution with parameter $\lambda_1 A$. Suppose that $P[X \leq N_1 | X \sim \text{Poisson}(\lambda_1 A)] = p$. We turn now to type 2 events and suppose that over the sub-region these events are also distributed according to a Poisson process with $\Lambda_2(l,m) = \lambda_2$. However, the number of type 2 events is linked to the value of N_1, since it is required that $P[Y \leq N_2 | Y \sim \text{Poisson}(\lambda_2 A)] = 1 - p$.

Suppose now that the declared intensity functions, $\Lambda_1(l,m)$ and $\Lambda_2(l,m)$, are

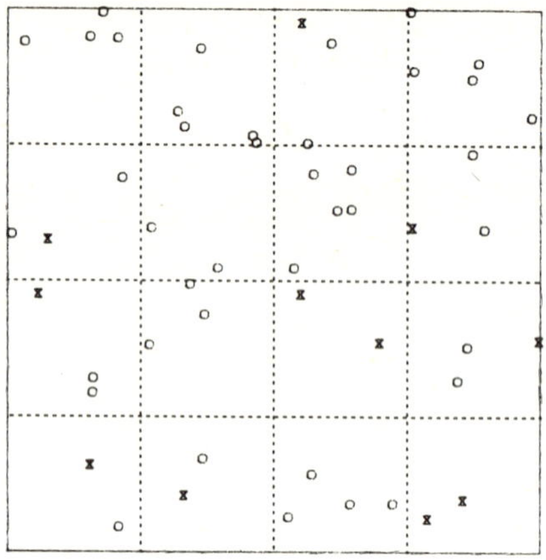

Figure 4.7 Compensating Cox processes on a 4 × 4 chess grid

both constant (and equal to λ_1 and λ_2, respectively) not only for the specified sub-region, but for the entire region of interest. The two separate patterns can then each be regarded as a realization of a Poisson process. However, suppose that the link that connects the intensities of the specified sub-region holds also for all such minute areas throughout the entire region of interest. Thus, a tiny region that fails to contain a type 1 point will have an increased chance of containing a type 2 point—and vice versa. The net result is that the composite pattern will therefore appear more homogeneous (i.e. regular) than would be expected if the points had all been generated from a single random process. The authors of this process remark that this 'demonstrates that the Poisson process is a plausible model for a single-type point process...even if there is a second, possibly unobserved, Poisson process negatively correlated with the first'. An example of these compensating Cox processes is given in Figure 4.7, in which we consider separate realizations of the two point processes for a 4 × 4 chess grid (as in Figure 4.1), with $\lambda_1 = 2\lambda_2$.

The final process that we shall illustrate was termed a *balanced* Cox process by Diggle and Milne (1983). They envisage a composite Poisson process formed by the relation $\Lambda_1(l, m) + \Lambda_2(l, m) = \lambda$. Thus, if for some sub-region type 1 points are distributed at random according to a Poisson process with parameter $\mu(<\lambda)$, then the type 2 points are distributed at random in that sub-region according to a Poisson process with parameter $\lambda - \mu$. Since μ will not be the same for all sub-regions, the two separate point processes will be Cox processes, and only the composite process will appear random. This process is illustrated for the 4 × 4 chess grid, with the ratio μ/λ alternating between 1/4 and 3/4, in Figure 4.8.

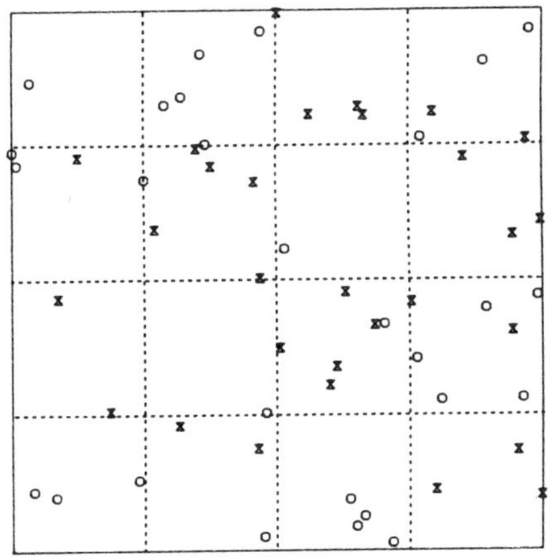

Figure 4.8 The balanced Cox process on a 4 × 4 chess grid

Summary

We have seen that the two parts of a bivariate pattern may be quite dissimilar in character, both from each other and from their composite pattern. Separate random patterns may display mutual attraction or repulsion. Only if the two random patterns are independent of one another will the combined pattern be itself random. Table 4.1 summarizes the properties of the various schemes that have been discussed in this section.

4.2 ANALYSIS USING QUADRATS

In Chapter 1 we saw that the study of pattern using quadrats, either as a square or in a transect, led to a comparatively dull appreciation of the pattern that was

Table* 4.1 *Some alternative bivariate point patterns

Name	Figure	Separate processes	Inter-correlation	Combined processes
Double Poisson		Random + random	Independence	Random
Linked Cox	4.1	Clustered + clustered	Attraction	Clustered
Linked pairs	4.2	Random + random	Attraction	Clustered
Thinned	4.4	Any + any	Any	Any
Sequential inhibition	4.5	Clustered + clustered	Repulsion	Random
Isolated Poisson	4.6	Regular + clustered	Repulsion	Random
Compensating Cox	4.7	Random + random	Repulsion	Regular
Balanced Cox	4.8	Clustered + clustered	Repulsion	Random

present. The study of inter-plant and point–plant distance gave much more penetrating insights into the nature of the pattern. Naturally, this also holds for bivariate patterns, but, nevertheless, we spend some time on quadrat-based measures because handling quadrat counts is comparatively simple, whereas locating individual points accurately and obtaining precise distances can cause considerable practical difficulties.

Although we write in terms of quadrats, the sample localities in question may be extremely small quadrats which could be regarded as points! Furthermore, it will sometimes be convenient to regard the quadrat value as being assigned to the central point of that quadrat. In some of what follows, therefore, we shall allude to both quadrats and points.

Presence/absence cross-classifications

A very simple procedure for detecting possible attraction or repulsion involves noting, for each quadrat, whether a species is present in that quadrat or is absent from that quadrat. The resulting categorical data can then be summarized in a 2^k contingency table, where k is the number of species being considered. Methods for handling categorical data are treated in detail in Volume 2 of this book, but it would be perverse not to consider the case $k = 2$ at this point.

Some notation is useful, and this is presented in Table 4.2. From this table we see that, of a total of N quadrats that were considered, m contained species 1, r contained species 2, and a contained both species 1 and species 2.

Evidently, the estimated chance of a quadrat containing species 1 is $\hat{p}_1 = m/N$, while the chance of a quadrat containing species 2 is $\hat{p}_2 = r/N$. If the locations of the two species were independent of one another, then the proportion of quadrats that would be found to contain *both* species would be $p_1 p_2$, which is estimated by $\hat{p}_1 \hat{p}_2$. Since we are examining N quadrats, it follows that the expected number of these N quadrats that would be found to contain both species would be estimated by $N\hat{p}_1\hat{p}_2 = mr/N$. In a similar fashion we can see that the estimates of the expected frequencies, under the assumption of independence of the locations of the two species, would be ms/N, nr/N, and ns/N.

***Table 4.2** Notation for a two-species presence/absence cross-classification*

		Species 2 Present	Species 2 Absent	Total
Species 1	Present	a	b	m
	Absent	c	d	n
	Total	r	s	N

In order to compare the observed frequencies (a, b, c, d) with the corresponding expected frequencies $(mr/N,\ ms/N,\ nr/N,\ ns/N)$ we can use the Pearson goodness-of-fit test, for which the statistic X^2 was defined by equation (1.21). As it happens, because of the interrelations between the frequencies $(a + b = m,$ $m + n = N$, etc.) the formula for the goodness-of-fit test greatly simplifies in the present situation, and it is easiest to use

$$X^2 = N(ad - bc)^2/mnrs. \tag{4.4}$$

Under the hypothesis of independence, X^2, given by equation (4.4), has a complicated discrete distribution which is closely approximated by the continuous χ_1^2 distribution. This approximation may be improved by making a slight continuity correction to the formula for X^2. The nature of the correction depends upon the context in which we are performing the test: we need to distinguish between two situations, as follows.

Situation 1. The entire region has been subdivided into N quadrats, and these are found to contain m quadrats in which species 1 is present and r quadrats in which species 2 is present.

Situation 2. The N quadrats (or points) refer to a sampled portion of the region of interest. As it happens, this sample contains m species 1 quadrats and r species 2 quadrats. We want to draw inferences from this sample concerning the region as a whole.

Loosely speaking the two situations might be summarized as 'scrutinizing the whole population' and 'scrutinizing a random or systematic sample'. The distinction between the corrections for these two situations arises because, in situation 1, all of m, n, r and s are effectively fixed quantities, whereas in situation 2, only N is fixed.

There have been frequent outbreaks of discussion amongst statisticians concerning the optimal method for the treatment of this 'simple' 2×2 table, and a recent summary is provided by Upton (1982). From Upton's results it appears that improved tests are provided by

$$X_p^2 = (N - 1)\{|ad - bc| - N/2\}^2/mnrs, \tag{4.5}$$

in the 'population' case described as situation 1, and by

$$X_s^2 = (N - 1)\{|ad - bc| - 1/2\}^2/mnrs, \tag{4.6}$$

in the 'sample' case described as situation 2. The expression $|ad - bc|$ means, of course, the absolute value of $(ad - bc)$. The correction factor of $N/2$ in (4.5) is due to Yates (1934) and is familiarly referred to as Yates' correction. The correction factor of 1/2 in (4.6) is due to Pirie and Hamdan (1972). Yates (1984) clearly feels that the distinction between the two situations is specious, and that X_p^2 should *always* be used.

For very small values of N, even the adjustments shown on the right-hand side of equation (4.5) do not suffice to make the χ_1^2 approximation viable. However, with small N, it is possible, in this 'population' situation, to enumerate the distribution exactly. Equivalently, we can examine the distribution of any one (A, say) of the individual quadrat counts. For the population case, under the hypothesis of independence of the row and column classifications it can be shown that

$$\Pr(A = a) = m!n!r!s!/a!b!c!d!N!, \tag{4.7}$$

which, for small N, can be evaluated with reasonable ease. The hypothesis of independence is then rejected if either of $\Pr(A \leq a)$ or $\Pr(A \geq a)$ is found to be unusually small. The use of equation (4.7) was first suggested by Sir Ronald Fisher and the resulting test has come to be known by the (somewhat misleading) title of *Fisher's Exact Test.*

A possible alternative to the χ_1^2 approximation, in the case of intermediate values of N, for which equation (4.7) would be difficult to evaluate, is to use, as an approximation to $x!$ (where x is any of a, b, c, d, m, n, r, s, or N), the result known as *Stirling's formula*:

$$\log(x!) \simeq \log\{(2\pi)^{1/2}\} + (N + 0.5)\log N - N, \tag{4.8}$$

where all the logarithms are natural logarithms.

The distinction that was made above between the treatment of the two situations applies to 2×2 tables in general. There is, however, another consideration which arises specifically in the present context because of the spatial nature of the data. If we consider an entire population, then we must use a closely knit grid of adjacent quadrats which we may expect to be spatially autocorrelated. Alternatively, if the sample quadrats are close together, then they too will tend to be spatially autocorrelated. In other words, the information obtained from two adjacent or nearby quadrats will not amount to two separate observations. The most usual circumstance is when neighbouring quadrats are positively autocorrelated, i.e. are similar. If there are many adjacent sample locations, then the effect of this autocorrelation will be that X_s^2 takes on an inflated value which may suggest association when it does not actually exist. Fingleton (1983a, 1983b) has shown that for a systematic sample of localities forming a grid layout, the observed value of (what amounts to) X_s^2 should be deflated by division by $\{1 + 2D(D+1)\}$, where D is a measure of the distance over which it is believed that significant positive autocorrelation exists (and beyond which observations are presumed to be independent).

With a grid of sampling locations, a natural measure of distance is the *city block* metric, familiar to the urban inhabitants of North America. The mathematical form for D is that $D = |x_1 - x_2| + |y_1 - y_2|$, where the coordinates of the two points being considered are respectively (x_1, y_1) and (x_2, y_2). If we believe that correlation extends only as far as neighbouring points on the sampling grid (i.e. neighbouring quadrats in a block of quadrats), then $D = 1$ and the appropriate divisor is 5. If we extend the correlation as far as $D = 2$, then the divisor rises to 13, and so on.

Note that in deflating X_s^2 in this way, we are assuming that autocorrelation exists to the most intense degree. In fact, the data may have contracted only a mild case of autocorrelation, so that we may actually be over-deflating. However, if X_s^2 retains its significance in such a situation, we can rest assured that the species really are associated.

We now turn to the question of a suitable modification for X_p^2. To see that the presence of underlying autocorrelation renders the exact test invalid, consider the following example. Suppose that a, b, c, and d are all equal to 1, so that m, n, r, and s are all equal to 2 and N is equal to 4. Imagine that the population that we are considering is a 2×2 lattice so that each cell will have properties very similar to its neighbour. Since species 1 appears in just two of the quadrats, then in the presence of strong spatial autocorrelation these are much more likely to have a common edge ($D = 1$) than to abut cornerwise ($D = 2$). It follows therefore that the possible configurations of presence and absence for the two species are not all equally likely. Given that $m = 2$ we know that two of the four cells contain species 1. These two cells may be chosen from the four in $\binom{4}{2} = 6$ ways, of which two include pairs which are at a distance $D = 2$. Exactly the same numbers apply to species 2 so that there are a total of 36 combinations which, under the hypothesis of independence, are all equally likely. Examining these configurations we find that in 24 cases we have $(a, b, c, d) = (1, 1, 1, 1)$, as observed. Thus, the probability that $A = 1$ is $24/36 = 2/3$, which is the value given by equation (4.7). However, suppose that the spatial autocorrelation is so strong that we can ignore the chance of the two cells in which a species appears being at a distance $D = 2$. Thus, the number of arrangements for each species is reduced from six to four and so the total number of configurations is reduced from 36 to $4 \times 4 = 16$. Examination of these 16 configurations shows that only eight correspond to the case $(a, b, c, d) = (1, 1, 1, 1)$ and hence the chance of this occurrence is now $8/16 = 1/2$.

Since we chose $\{1 + 2D(D + 1)\}$ as a suitable deflating factor for a grid of sampling locations, and since one can usually think of a statistical population as the limit of a sample as the sample size increases, we conclude that for a population laid out as a lattice a suitable deflater will again be $\{1 + 2D(D + 1)\}$, although we repeat our caution that this may give rise to a conservative statistic.

Irrespective of the problems introduced by our attempts to eliminate the innate spatial autocorrelation, there are other limitations on this presence/absence analysis. Most obviously no account is taken of the actual quantities in which the species are present. Also, the results of the test are likely to be influenced by the quadrat size (or by the spacing of the grid of sampling points). However, the most potentially damaging problem is concerned with the number of (absent, absent) cells. If the two species are only present over a small proportion of the study area (or if quadrats are erroneously placed outside the reasonable region of interest) then the dominant cell will be the (absent, absent) cell and the consequent large value of d will inevitably lead to the conclusion that the species attract one another, unless the (present, present) cell, a, is zero. Some authors (e.g. Bray, 1956) have sought to get around this problem by ignoring the value of d and computing some measure of association from the remaining three cells of the table. However, this does not really solve the problem, which appears insoluble for the case of just two species: one must simply take care that one's quadrats are of an appropriate size and are positioned at relevant places.

For more than two species the (all absent) cell poses much less of a problem: with k species there will be 2^k cells in the cross-classification and we can easily afford the luxury of ignoring the (all absent) cell. However, consideration of the methods required for the solution of this problem must await Volume 2.

***Example* 4.1**

In Example 2.3 we examined the obviously clustered pattern of maples in Lansing Woods, Michigan, and we remarked that it was clear from a scrutiny of the figures given by Diggle (1979b) that, for the three species oaks, hickories, and maples, there was (in our present terminology) evidence of repulsion between the maples and the hickories. The map of the locations of the maples was reproduced as Figure 2.6.

Subsequently, in Example 3.11 we presented the joint presence/absence distribution of the three species in Figure 3.8. In that example we concluded that there was marginally less overlap between the oak and hickory distributions than would have been expected under independence. We shall now use the methods of the present section to examine each pair of tree species in turn, though being conscious that, by so doing, we shall overlook any potential *combined* influence of two of the three species on the locations of the third species.

The data given in Figure 3.8 lead to the three summary 2×2 tables that are presented in Table 4.3. In these tables the bracketed figures correspond to the estimated expected frequencies under the hypothesis of independence, calculated separately for each table.

It is obvious from these tables that any repulsion that exists between the oaks and either of the other trees is very mild, whereas the repulsion apparent between the maples and the hickories is very obvious. Using equation (4.5) we obtain values for X_p^2 of 2.37 (oaks *v.* hickories), 0.87 (oaks *v.* maples), and 25.68 (maples *v.* hickories). Since the upper 5% point of a χ_1^2 distribution is 3.84, it is plain that neither of the first two 2×2 presence/absence tables displays any significant departure from independence.

On the other hand, if we assume that the extent of the autocorrelation is limited to adjacent quadrats ($D = 1$), then there is still significant evidence (revised X_p^2

***Table 4.3** Separate presence/absence tables for the hickories, oaks, and maples of Lansing Woods, Michigan*

		Hickories	
		Present	Absent
Oaks	Present	261 (269)	177 (169)
	Absent	93 (85)	45 (53)

		Maples	
		Present	Absent
Oaks	Present	175 (180)	263 (258)
	Absent	62 57	76 (81)

		Maples	
		Present	Absent
Hickories	Present	116 (146)	238 (208)
	Absent	121 (91)	101 (131)

= 5.14) of repulsion between the hickories and the maples at the 5% level. However, if we were to take $D - 2$, then the evidence would no longer be significant (revised $X_p^2 = 1.98$).

If we suspect that the results from neighbouring locations are correlated, then as an alternative to recording the results from every quadrat we could take a systematic sample of quadrats in which we ensure that no two quadrats are touching ($D > 0$). The results of one such systematic sample are given in Table 4.4.

The results in Table 4.4 refer to a systematic one-in-four sample from the entire grid. Since we are now sampling from a larger population of interest, we must calculate the value of X_s^2 given by equation (4.6), and we obtain the values 2.61, 2.29, and 7.90 for the three sub-tables. As with the earlier analysis it appears that the oak locations are essentially independent of the locations of the maples and hickories, with the latter two species tending to avoid one another. These results are in agreement with the multi-way analysis conducted by Fingleton (1983b) (see also Volume 2 of the present work).

Table* 4.4 *The results of a systematic* 1-*in*-4 *survey of the Lansing Woods quadrats

		Hickories Present	Hickories Absent
Oaks	Present	62	45
	Absent	27	10

		Maples Present	Maples Absent
Oaks	Present	37	70
	Absent	18	19

		Maples Present	Maples Absent
Hickories	Present	26	63
	Absent	29	26

***Example* 4.2**

Figure 4.9, which has been redrawn from Hanisch and Stoyan (1979), shows the locations of sinkholes in an area of the south-west of the German Democratic Republic. The sinkholes have been classified as either large or small. Since we may expect large sinkholes to display some regularity, the Isolated Poisson process comes to mind as a possible description of the joint pattern of large and small sinkholes.

Subdividing the area shown in Figure 4.9 into a 4×6 array of equi-sized quadrats, we obtain the presence/absence data summarized in Table 4.5. If we treat the data of Figure 4.9 as being the entire population of interest, then we need to calculate X_p^2, which turns out to have the value 0.0965 which is clearly much smaller than 3.84 and certainly does not indicate any departure from independence.

Recalling that the square root of a χ_1^2 random variable is a unit normal random variable, it is interesting to compare the square root of the calculated value, 0.0965, with tables of the unit normal distribution function. Since $(0.0965)^{1/2} = 0.311$, we find that the equivalent tail probability is 37.8%, which is obviously not significantly small.

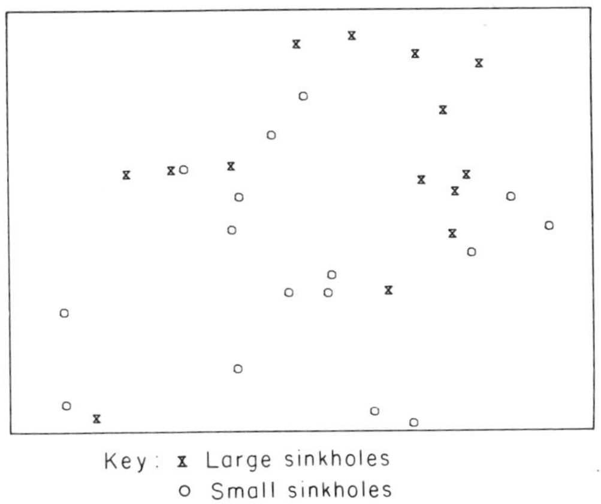

Figure 4.9 Midpoints of sinkholes (Erdfälle) in a region of south-west Germany. (Redrawn from Hanisch and Stoyan, 1979. Reproduced by permission of Akademie-Verlag, Berlin.)

Table 4.5 Presence/absence table for the sinkholes pattern

		Large sinkholes	
		Present	Absent
Small sinkholes	Present	5	6
	Absent	4	9

Since the number of quadrats, 24, is not very large we might be worried that the χ_1^2 approximation was not valid. It is certainly a feasible proposition to carry out the exact calculations based on equation (4.7). The four more extreme configurations (given the observed marginal totals) are shown in Table 4.6. Associated with these configurations, we find $\Pr(A=9)=0.00004$, $\Pr(A=8)=0.0016$, $\Pr(A=7)=0.0197$, and $\Pr(A=6)=0.1011$. For the observed table we have $\Pr(A=5)=0.2526$ and hence $\Pr(A \geq a)=0.3750$, while $\Pr(A \leq a)=0.7361$. Clearly, neither is unusually small, and we can therefore accept the hypothesis of independence.

Notice that the exact $\Pr(A \geq a)=0.3750$ is roughly equal to the tail probability, 0.378, obtained from the χ_1^2 approximation. This indicates that the approximation is excellent for N as small as 24. Indeed, it can reasonably be used for much smaller N (Upton, 1982).

***Table* 4.6 *More extreme configurations than the data of Table* 4.5**

9	2		8	3		7	4		6	5
0	13		1	12		2	11		3	10

Inter-quadrat correlations

The reduction of quadrat counts to the simple presence/absence categorization trades simplicity for loss of precision. We may well obtain a more informative analysis by considering the actual quadrat counts ($\{x_k\}$ for species 1 and $\{y_k\}$ for species 2, with $k = 1, 2, \ldots, n$, where n is the number of quadrats). In the event of repulsion we would expect negative correlation between X and Y, with positive correlation for attraction. The ordinary (product-moment) correlation coefficient is given by

$$\mathrm{r} = \frac{\sum xy - (\sum x)(\sum y)/n}{\left[\left\{\sum x^2 - \frac{1}{n}(\sum x)^2\right\}\left\{\sum y^2 - \frac{1}{n}(\sum y)^2\right\}\right]^{1/2}}. \tag{4.9}$$

We assume here that X and Y are not autocorrelated. The effects of the presence of autocorrelation in one or both variables are described by Bivand (1980) who shows that r will tend to be inflated when both X and Y are highly autocorrelated. When one of the variables depends on the other, we have a regression situation and can deal with autocorrelation using the methods described subsequently in Chapter 5.

So long as we are sampling within the region of interest, this formula for the correlation is likely to be valid. However, we cannot always be sure of the locations of the boundaries of the region. For example, a naturally occurring forest will not have clearly delineated boundaries—it simply becomes progressively less dense. A potential problem here is that we may obtain a large number of completely empty quadrats simply because we are not sampling at an appropriate place (this is the (absent, absent) problem once again).

An appealing solution is simply to ignore these (absent, absent) quadrats and to calculate r as before, but with a reduced value of n. However, that this is not valid is easily seen by arranging the data as a two-way table of counts $\{f_{ij}\}$, where f_{ij} is the number of quadrats containing i members of species 1 and j members of species 2. If we treat f_{00}, the observed (absent, absent) frequency as though it were 0, then we shall obviously induce an off-diagonal trend (i.e. a negative correlation) into the data.

It is possible to ignore the f_{00} cell and to fit the so-called *quasi-independence* model to the remaining data. We shall consider this possibility in Volume 2. For the meantime we mention the (less desirable) alternative procedure which involves the calculation of

$$r' = \frac{\sum' xy - (\sum' x)(\sum' y)/n'}{\left[\left\{\sum' x^2 - \frac{1}{n'}(\sum' x)^2\right\}\left\{\sum' y^2 - \frac{1}{n'}(\sum' y)^2\right\}\right]^{1/2}}, \tag{4.10}$$

where n' is the number of (present, present) quadrats, and the summations refer to these cells alone.

If we are in doubt as to whether the f_{00} count has been inflated by the inclusion of irrelevant quadrats, then a partial check is obtained by calculating $\hat{f}_{00} = n_1 n_2/n$, where n_i is the number of quadrats that contained no members of species $i (i = 1,2)$, n is the total number of quadrats, and $\hat{f}_{00}$ is the estimate of f_{00} under the hypothesis of independence. If $(\hat{f}_{00} - f_{00})^2/\hat{f}_{00}$ is much in excess of 1 (say 4, or more), then there is evidence of an inflated number of (absent, absent) cells. Note that this inflation *might* be due to attraction between the two species, so we must beware of blind application of magic formulae!

To test the significance of r we can calculate

$$z = r\{(n-2)/(1-r^2)\}^{1/2}, \tag{4.11}$$

which, under the hypothesis of independence and with the joint distribution of X and Y being bivariate normal, has a t distribution with $(n-2)$ degrees of freedom. A corresponding statistic z', is associated with n'.

Two final remarks: the correlation that we examine here is, of course, a *linear* correlation between the counts of the two species. If a non-linear correlation is suspected, then some transformation would be required. Finally, the correlation we obtain is conditional on the quadrats used. This is graphically illustrated by Openshaw and Taylor (1979) who partition their study region in many different ways to obtain literally thousands of different r-values ranging from -0.97 to 0.99—all for the *same* population. This phenomenon has been termed the *modifiable areal unit problem*, although it need not necessarily be a problem and can be used with good effect to elicit useful information from the population. This remark also relates, of course, to the earlier uses of quadrats, but it is noted here because of the discussion in the literature (e.g. Yule and Kendall, 1965), which has traditionally been in the context of the correlation coefficient, with some spin-off to regression (e.g. Williams, 1976; Openshaw, 1977).

Example 4.3

Figure 4.10 gives the observed counts for the hickories and maples data from Lansing Woods. The 16 × 16 grid of counts refers to a 'block sample' from the south-west corner of the 24 × 24 grid of presence/absence data given in Figure 3.8. The restriction to a side of 16 quadrats allows us to examine, in a simple

(a) Hickories

0	4	3	3	0	0	2	1	1	1	0	0	0	0	0	0
2	2	3	2	2	0	0	1	1	0	0	0	0	0	0	0
2	5	4	4	1	0	3	0	1	0	0	1	0	0	2	1
1	0	4	2	1	5	1	1	0	0	2	3	1	0	3	2
1	4	1	1	1	2	1	0	2	0	0	3	1	2	1	3
1	1	3	1	0	2	0	2	2	1	4	0	2	2	0	0
0	1	1	1	0	3	1	0	1	3	3	1	1	0	2	0
4	0	0	2	0	0	0	0	1	0	1	0	0	1	1	1
3	0	1	1	1	1	0	0	0	3	1	2	0	0	0	0
0	1	0	0	1	1	1	0	0	1	1	0	0	0	0	0
0	1	0	0	0	0	0	0	1	0	0	0	1	0	0	0
1	0	1	0	0	0	0	0	0	1	0	1	0	0	2	0
1	1	1	0	0	1	0	0	1	2	2	0	0	0	1	1
1	1	0	1	0	1	4	0	0	0	1	2	0	2	2	1
2	3	0	0	1	1	1	2	0	0	0	0	0	0	0	0
0	1	2	1	1	1	3	1	0	1	0	0	0	0	0	0

(b) Maples

0	0	0	0	0	0	1	0	4	3	4	0	3	0	2	4
0	0	0	0	0	0	1	0	2	2	4	0	1	2	0	3
0	0	0	0	1	0	3	2	2	3	5	0	0	1	0	1
0	0	0	0	0	0	0	1	3	2	1	2	2	2	0	0
0	0	0	0	0	0	2	1	0	2	1	0	1	2	5	2
0	0	0	0	1	0	1	0	1	0	2	2	2	1	1	1
0	0	0	0	0	0	1	0	0	0	0	1	2	1	0	2
0	0	0	0	1	0	0	0	0	0	0	1	2	2	0	1
0	0	0	1	0	0	1	0	0	0	0	1	1	1	1	3
0	0	3	2	4	2	1	0	1	0	1	1	2	1	2	0
0	2	1	1	1	0	1	0	0	2	0	2	2	0	3	2
0	0	5	3	4	0	0	3	1	2	2	1	2	2	3	0
1	0	2	2	3	0	0	2	1	1	0	4	1	1	3	3
0	1	0	1	3	2	0	0	0	2	2	1	1	1	2	1
0	1	2	2	2	0	2	3	2	2	0	1	2	1	0	2
0	0	1	1	2	1	0	0	2	3	1	5	0	1	3	1

Figure 4.10 Observed counts of hickories and maples in a 16×16 grid

fashion, the effect of changing the quadrat size. We commence by considering the joint frequency distribution for the 256 counts, which is given as Table 4.7.

Initially we focus attention on the (0, 0) frequency. If the hickory and maple distributions were independent of one another, then we would have $\hat{f}_{00} = 115 \times 123/256 \simeq 55$. Since the observed value is appreciably *less* than 55, this suggests an unexpectedly (under the independence model) high number of cells of the (present, absent), and (absent, present) types—in other words we immediately begin to suspect that the data will display repulsion. Using all 256 quadrats we find that $\sum x = 227$ (hickories) and $\sum y = 267$ (maples), with $\sum x^2 = 519$, $\sum y^2 = 647$, and $\sum xy = 154$. Using equation (4.9) we then have $r = -0.242$ and hence, using equation (4.11), we have $z = -4.09$. Since the upper and lower 0.1% points of a t_{254} distribution are equal to about ± 3.12, we plainly have highly significant evidence of a negative association (repulsion) between the two point distributions.

Table* 4.7 *Joint frequency distribution of quadrat counts for hickories and for maples

		Hickories 0	1	2	3	4	5	Total
	0	43	35	16	12	7	2	115
	1	32	18	9	1	0	0	60
Maples	2	29	16	3	2	1	0	51
	3	12	4	2	1	0	0	19
	4	5	2	0	0	0	0	7
	5	2	2	0	0	0	0	4
Total		123	77	30	16	8	2	256

Table 4.8 summarizes the effect of changing the quadrat size upon our interpretation of the data. Evidently the disjoint 'patches' of trees are at least of a size equal to 8 × 8 quadrats. If we were able to choose a quadrat size much larger than the 'patch' size, then the repulsion between the trees within a quadrat would be obscured and the value of z would fall off. Notice that this is not the case with the present data and notice also that, although the negative correlation becomes progressively more marked with increasing quadrat aggregation, its significance stays at very much the same level, because there is a concurrent reduction in sample size.

Example 4.4

Diggle and Cox (1983) present the bivariate pattern of broad-leaved and coniferous trees in a plot in Vastmanland, Central Sweden. The plot forms part of a large stand which had been allowed to regenerate following a clearance some 17

Table* 4.8 *The effect of changing the quadrat size on the apparent correlation

	Quadrat size	n	$\sum x^2$	$\sum y^2$	$\sum xy$	r	z
Hickories/maples data	1	256	519	647	154	− 0.242	− 4.09
	4	64	1 327	1 735	669	− 0.488	− 5.05
	16	16	4 467	5 953	2 880	− 0.665	− 4.46
	64	4	15 209	21 587	12 726	− 0.853	− 4.44
Types:and data	1	256	970	342	172	− 0.071	− 1.14
	4	64	2 132	706	804	0.020	0.16
	16	16	6 378	2214	3 029	− 0.267	− 1.08
	64	4	23 394	8010	12 681	− 0.049	− 0.07

(a) Broad leaved

0	0	0	4	2	1	0	0	0	1	2	2	2	1	4	0
0	0	3	4	2	0	0	0	4	0	0	1	1	3	0	0
0	1	7	1	2	1	0	0	5	0	0	0	0	1	0	2
1	2	2	4	3	2	1	4	0	1	2	3	0	1	0	0
5	2	2	4	6	4	1	0	1	0	1	5	1	0	0	2
0	3	2	0	0	0	2	0	1	1	1	1	3	1	0	6
1	0	0	2	7	1	4	0	4	1	0	0	2	3	2	3
0	0	0	0	6	0	2	0	0	0	0	0	2	1	2	0
0	0	0	0	1	0	0	3	2	0	2	1	1	1	0	0
0	0	1	0	1	0	4	0	1	0	0	0	0	1	0	0
0	0	0	0	0	0	0	3	2	2	0	2	0	1	0	1
4	0	0	0	1	0	2	1	2	1	0	0	0	1	0	1
0	0	0	4	0	0	3	3	1	0	1	1	3	3	1	2
0	3	0	0	0	0	0	0	0	0	0	8	1	0	3	1
0	0	0	0	1	6	0	0	3	1	1	1	1	0	4	3
1	1	0	0	1	3	0	1	0	0	0	4	1	1	0	0

(a) Coniferous

0	0	2	0	0	0	0	0	1	0	1	1	0	0	0	1
0	0	1	0	0	2	3	0	1	0	0	0	0	2	0	2
1	1	0	1	1	2	1	0	0	0	1	1	2	0	2	1
1	0	1	0	0	0	2	0	1	1	0	2	0	0	0	1
0	1	0	1	3	1	0	0	0	0	1	1	0	0	0	1
0	0	1	1	1	0	0	0	0	1	0	4	0	2	1	0
0	0	0	0	0	4	0	0	0	0	1	1	0	0	0	1
0	1	1	1	2	0	0	0	2	0	0	0	0	0	2	0
0	0	1	0	1	0	0	1	4	0	1	1	1	1	2	1
0	0	0	0	0	1	0	1	0	0	0	0	1	4	3	1
0	0	0	2	3	1	0	0	0	0	1	1	1	2	1	0
0	0	2	0	0	0	1	0	0	2	1	1	2	4	2	1
1	0	0	0	1	1	0	0	0	0	1	2	3	1	0	0
0	0	2	0	0	0	1	0	0	4	1	0	1	4	0	2
0	2	2	1	1	1	0	3	1	0	1	0	1	0	0	0
0	1	1	0	0	0	0	0	0	0	0	0	1	1	1	0

Figure 4.11 Observed counts of broad-leaved and coniferous trees

years previously—further details are to be found in Gustavsson (1974). Figure 4.11 gives counts in a 16 × 16 grid of quadrats based on Diggle and Cox's diagram. The length of a quadrat side is approximately 1.4 m.

As with the previous data set we commence by examining the joint frequency distribution (given in Table 4.9) and, in particular, the (absent, absent) frequency, f_{00}. In this case the observed value of 68 is almost identical to the independence estimate $\hat{f}_{00} = 124 \times 142/256 \simeq 69$. On the basis of this single calculation we can anticipate that the two types of tree may display independence. The calculations summarized above in Table 4.8 confirm this first impression: the z-values fluctuate around zero, but never remotely approach significance.

There were 296 broad-leaved and 172 coniferous trees in the area surveyed — thus, in the notation of Table 4.8, $\sum x = 296$ and $\sum y = 172$. We can therefore calculate that, for the smallest quadrats, the sample mean quadrat counts were 1.156 and 0.672 trees per quadrat, while the variances were estimated as 2.462 and 0.888, respectively. Thus, for both types of tree, the ratio $s^2/\bar{x}$ is noticeably greater

Table 4.9 Joint frequency distribution of quadrat counts for the typestand data

		Broad-leaved trees									
		0	1	2	3	4	5	6	7	8	Total
Coniferous trees	0	68	28	17	11	12	2	1	2	1	142
	1	36	19	11	5	3	1	1	0	0	76
	2	14	7	1	2	0	0	1	0	0	25
	3	4	0	0	1	0	0	1	0	0	6
	4	2	4	1	0	0	0	0	0	0	7
Total		124	58	30	19	15	3	4	2	1	256

than the value 1 which is to be expected with a random pattern. Using the Index of Dispersion test (equation (1.4)) we obtain ID-values of 543 and 337, which are, of course, greatly in excess of the upper 5% point of a χ^2_{255} distribution.

The pattern described by this typestand data is therefore not one of those specified in Table 4.1 (except, perhaps, the Thinned Process (!)), since it appears to consist of two independent clustered processes.

4.3 ANALYSIS USING TRANSECTS

In Chapter 1 we used the term 'transect' to refer to a row of contiguous quadrats. In the present section we shall vary this definition somewhat by occasionally ignoring the quadrat boundaries and regarding the transect as a thin belt along which the observer passes. This is akin to the line transects discussed in Chapter 2. We shall consider three rather different situations:

(i) a transect of quadrats in which each quadrat is simply classified by its dominant species;

(ii) a belt transect which we traverse noting the sequence of species that we encounter; and

(iii) a belt transect which we traverse noting the sequence of diseased and healthy trees.

A special case of (i) occurs when we consider a single species and classify each quadrat according to the presence (1) or absence (0) of that species. In this section we shall confine attention to cases involving no more than two species, because methods for handling larger numbers of species are more appropriately handled in Volume 2. With this restriction, all three situations listed above can be analysed in the same way, though there are special features of interest in the third situation.

The runs test

Remembering that we are restricting attention to dichotomous situations, let us denote the two alternatives by A and B. The information supplied by the transect is therefore coded as a sequence such as

AAA B A BB AAAA B AA.

Two extreme examples of the type of sequence that we might encounter are provided by:

(i) AAAAAAAAAA BBBB, and
(ii) AA B AA B AA B AA B AA.

Suppose that we are dealing with the special case in which we are considering a single species that may be absent (A) or present (B) in a quadrat. The sequence (i) is clearly to be interpreted as providing strong evidence that the species occurs in tight clusters, while (ii) indicates a regular pattern. These interpretations are, of course, only relevant to our particular choice of quadrat size, since sequence (i) may be part of a larger regular pattern and sequence (ii) may conceal smaller scale clustering.

Alternatively, if A and B refer to two alternative species, then the sequence (i) provides an extreme example of inter-species repulsion, whereas sequence (ii) provides an example of attraction, with both species having (at this scale of measurement) very regular patterns.

Suppose that, in the entire population of interest, the proportion of A's is p and the proportion of B's is $1 - p$. In other words, if we choose an item at a random point in the sequence of A's and B's, there is a probability p that the item will be an A and a probability of $1 - p$ that it will be a B. If the distributions of A's and of B's are independent of one another, then the probability of a selected item being an A will be p, *independently of all other items in the sequence.* We can therefore calculate the probability that the first k items in the sequence are all A's, with the $(k+1)$th item being a B, as $p \cdot p \cdot \ldots \cdot p \cdot (1-p) = p^k(1-p)$. The probability would be exactly the same wherever the run started and so we can make the general statement that

$$\Pr(\text{a run of A's has a length of exactly } k) = p^k(1-p). \qquad (4.12)$$

The probability distribution defined by (4.12) is called the *geometric distribution.* The mean of this distribution is given by $\sum kp^k(1-p) = 1/p$, while the variance is $1/p^2$.

However, p is unkown. The only information that we have available is the sequence of A's and B's. Since the population mean run length for the A's is $1/p$, it seems reasonable to expect that the observed sample mean run length for the A's will approximate $1/p$. Suppose that the sequence contains a total of N_A A's and N_B B's, and suppose that the numbers of runs of A's and B's are r_A and r_B, respectively. Then an estimate of $1/p$ is provided by N_A/r_A, and an estimate of $1/q$, where $q = (1-p)$, is provided by N_B/r_B. Considering the reciprocals we therefore arrive at the (maximum likelihood) estimates of p and q, namely $\hat{p} = r_A/N_A$ and

$\hat{q} = r_B/N_B$, respectively. Strictly speaking, the above discussion should refer to *completed* runs, so that the first and last sequences in any transect would have to be neglected and the values of both the $\{N_i\}$ and the $\{r_i\}$ would be correspondingly reduced.

Since $p + q = 1$, we clearly anticipate that $\hat{p} + \hat{q}$ will be close to 1! However, this is not the case, since these estimators are governed by the nature of the underlying bivariate pattern. Consider, as an example, the two sequences given above. Disregarding the need to eliminate the first and last sequences, we have $N_A = 10$ and $N_B = 4$, in both cases. However, in the first case, $r_A = r_B = 1$, so that $\hat{p} + \hat{q} = 0.35$, while in the second example, $r_A = 5$ and $r_B = 4$, so that $\hat{p} + \hat{q} = 1.5$. Thus, the observed value of $\hat{p} + \hat{q}$ could act as an index of the association between the two component patterns.

The most convenient test of association is provided by $r(= r_A + r_B)$, the total number of runs in the sequence. We have $N(= N_A + N_B)$ items in our sequence which, under the hypothesis of independence, might have appeared in any order. By a straightforward consideration of the possible combinations (see Pielou, 1977, for the detailed argument) we find that the probabilities of r taking the values $2k$ or $2k + 1$ are given by

$$p_{2k} = 2\binom{N_A - 1}{k - 1}\binom{N_B - 1}{k - 1}\bigg/\binom{N}{N_A}, \tag{4.13a}$$

$$p_{2k+1} = \left\{\binom{N_A - 1}{k}\binom{N_B - 1}{k - 1} + \binom{N_A - 1}{k - 1}\binom{N_B - 1}{k}\right\}\bigg/\binom{N}{N_A} \tag{4.13b}$$

The bounds on r are 2 and $r_{\max} = 1 + 2\min(N_A, N_B)$. To decide whether an observed value of r is atypical we compute the sum of the $\{p_j\}$ either from 2 to r, or from r to $r_{\max}$, as appropriate. If this cumulative probability proves unreasonably small (less than 2.5%, for instance) then we could conclude that there was significant evidence of a departure from randomness.

Note that, in equation (4.13b), when $r = r_{\max}$, k will be equal to the smaller of N_A and N_B, and consequently one of

$$\binom{N_A - 1}{k} \quad \text{or} \quad \binom{N_B - 1}{k}$$

will be nonsensical. Whichever of these occurs should be assigned the value zero.

For cases where both N_A and N_B are less than 21, tables of the $\{p_j\}$ have been given by Swed and Eisenhardt (1943). However, with modern scientific calculators, these tables are scarcely needed. Moreover, even for quite small values of N_A and N_B, an excellent normal approximation exists. The mean and variance of r are given by $(1 + 2N_AN_B/N)$ and $2N_AN_B(2N_AN_B - N)/N\ (N - 1)$, respectively.

Using a continuity correction, we therefore have the test statistic

$$z=\left\{|N(r-1)-2N_AN_B|-\frac{N}{2}\right\}\Big/\{2N_AN_B(2N_AN_B-N)/(N-1)\}^{1/2}. \quad (4.14)$$

The corresponding cumulative tail probability is given by $P=1-\Phi(z)$, where $\Phi(\cdot)$ is the unit normal distribution function.

If we wish to combine the information from several transects, then this should be done by calculating the separate cumulative probabilities, $\{P_i\}$, for each transect. In each case the P-value must refer to the *same* tail of the distribution. With m transects, the information can then be combined using the Fisher statistic

$$V=-2\sum_{i=1}^{m}\log(P_i), \quad (4.15)$$

where, as usual, the logarithm is to the base e. Under the null hypothesis of independence, V will be an observation from a χ^2 distribution with $2m$ degrees of freedom.

***Example* 4.5**

In Figure 4.3 we illustrated part of a bramble patch analysed by Diggle and Milne (1983). The bivariate pattern appears similar to the pattern to be expected with the linked pairs process, and we can anticipate that a transect would give rise to a large number of runs. Denoting the newly emerged canes by A's and the 1-year-old canes by B's, the outcome of a traverse along a randomly placed belt transect was the sequence:

AA B AAA BB A B A B AA B A B A BBB A B A.

We see that $r_A=9$, $r_B=8$, $N_A=13$, and $N_B=11$, so that $r=17$, $r_{\max}=23$, and $N=24$. In order to calculate $\hat{p}+\hat{q}$ we must eliminate the two end runs so that we get $\hat{p}=(9-2)/(13-3)$, while $\hat{q}=8/11$, so that $\hat{p}+\hat{q}=1.43$. Since this is considerably in excess of 1 we can deduce that there is evidence of some attraction between A and B.

To quantify the significance of this result we calculate P using equations (4.13a) and (4.13b). The values of $\binom{24}{13}p_j$, for $j=17, 18, \ldots, 23$, are respectively 95 040, 44 550, 14 850, 4 400, 880, 132, and 12. Thus, the cumulative P-value is 6.4%, which is not unreasonably small. Notice, however, that with one further run the P-value would have dropped to 2.6%.

Clearly, the single transect provides rather little information and we would normally expect to have rather more information available. Three further transects gave the sequences:

A B A B A B AAA BB A BBBBB,
B A B A B A B A B AAA B A,
AAAAA BB A.

Carrying out the equivalent calculations to those made for the first sequence we find that the upper-tail probabilities are, respectively, 0.451, 0.016, and 0.929. The third sequence clearly shows signs of repulsion rather than attraction. Combining the information from these four transects, using equation (4.15), we find that $V = 15.47$, which just fails to achieve significance at the 5% level [$\chi^2_{8,0.05} = 15.51$]. Notice that, because the P's are all *one-tail* probabilities, we are *only* interested in the upper tail of the χ^2 distribution.

As an illustration of the excellence of the normal approximation, the value of z for the first sequence is $86/(74932/23)^{1/2} = 1.5067$, which gives $P = 6.6\%$—very close to the true value. The approximate probabilities for the other three sequences are similarly close, being 0.446, 0.019, and 0.947, respectively.

Runs of healthy and diseased plants

Pielou (1963) examined the problem of a forest containing infected trees. Here there are two quantities of particular interest. The first is the proportion of trees which are infected: this is easy to estimate. The second is the proportion of the total area of the forest which is occupied by infected sub-areas. This is more difficult to assess because of the possibility that these sub-areas contain trees which do not yet show signs of infection. Pielou suggests that the trees can be divided into three classes: diseased (D); apparently healthy, but in an infected sub-area (I); and healthy (H). Using a belt transect we could anticipate a sequence of trees such as

HHHH II D I DD I HH I DDD I H,

although, because of our inability to distinguish between I's and H's, this sequence would actually be recorded as

HHHHHH D H DD HHHH DDD HH.

The basis of Pielou's approach is that the sequence forms a Markov chain. That is to say, the chance of the next tree being an H (or an I or a D) is dependent upon the type of the present tree, but not on the type of any previous trees. Since this dependence must apply equally to a traverse of the transect from left to right or from right to left the chain must be *reversible* and this suggested to Pielou that an appropriate transition matrix would be

$$\begin{array}{c} \\ D \\ I \\ H \end{array} \begin{array}{c} \begin{array}{ccc} D & I & H \end{array} \\ \begin{pmatrix} av & a(1-v) & (1-a) \\ av & a(1-v) & (1-a) \\ yv & y(1-v) & (1-y) \end{pmatrix} \end{array}.$$

where a, y, and v are probabilities to be defined below. Thus, the probability of a D being followed by another D is av, and so on.

The three probabilities a, y, and v are defined as follows. The probability that a healthy tree is followed by another healthy tree is $(1 - y)$, the probability that a tree in an infected region is followed by another such is a, and the probability that a tree in an infected region is diseased is v. As we might expect from the rather awkward transition matrix, estimation of a, y, and v is not an easy business and the relevant formulae are provided by Vithayasai (1971).

4.4 SOPHISTICATED DISTANCE METHODS

We saw in Chapter 1 that methods involving inter-plant distances provided us with the most comprehensive and penetrating insights into the pattern displayed by an array of points, and we can expect that the same will be true for bivariate patterns. The price that we pay for this information is the need to make a considerable quantity of accurate measurements.

However, before considering the more precise distance measures, we consider a very simple distance-based test that had no analogue in Chapter 1.

The nearest-neighbour table

Consider two point patterns that are separately consistent with the hypothesis of a random arrangement. We have seen that the combined bivariate point pattern may appear random, regular (Compensating Cox process), or clustered (Linked Pairs process). The test that follows is designed to distinguish between these three situations, but it relies on the individual species having near-random patterns. The test *cannot be used with a species that is distributed in an obviously non-random fashion.*

Suppose that the spatial intensity of species i is $\lambda_i (i = 1, 2)$. The chance that a randomly chosen plant belongs to species 1 is therefore $\lambda_1/(\lambda_1 + \lambda_2)$, while the corresponding probability of a species 2 plant being selected is $\lambda_2/(\lambda_1 + \lambda_2)$. Suppose that we choose a plant at random and enquire about the nature of its nearest neighbour. Plainly, since the plants of each species are apparently distributed at random according to a Poisson process, if the two plant distributions are independent of one another, then the chance of the nearest neighbour (to any plant) being of species 1 is again $\lambda_1/(\lambda_1 + \lambda_2)$. If we consider every plant in turn then we can summarize our results in a table such as Table 4.10, in which the notation follows that of the superficially similar Table 4.2.

Table 4.10 refers to a total of m species 1 plants and n species 2 plants. These $m + n = N$ plants comprise either

(i) all the plants of interest, or

(ii) all the plants in the study area for which the nearest neighbour can be established with certainty, or

Table* 4.10 *The nearest-neighbour table

		Nearest neighbour Species 1	Species 2	
Base plant	Species 1	a	b	m
	Species 2	c	d	n
		r	s	N

(iii) all the plants within certain sampling areas taken from the study area (using, for example, semi-systematic sampling).

The distinction between cases (i) and (ii) is simply as to whether there are boundary effects to be considered. Case (iii) extends the test to situations in which it is not feasible to consider the entire population.

A fundamental statistical difference distinguishes the figures in Table 4.10 from those in Table 4.2. In the present case m and n are undoubtedly fixed quantities, while r and s are not fixed (though under the hypothesis of independence we can expect r to be close to m and s to be close to n). Thus, Table 4.10 has only the one set of fixed marginal totals, whereas Table 4.2 had either two sets fixed (population) or just N fixed (sample).

If the null hypothesis (of independence between the two separate point patterns) is correct, then $a/m \simeq c/n \simeq \lambda_1/(\lambda_1 + \lambda_2)$. An appropriate test statistic is therefore $a/m - c/n$. After some algebraic manipulation and the introduction of an appropriate continuity correction (Upton, 1982, gives details) we arrive at the test statistic

$$X_{NN}^2 = (N-1)\{|ad - bc| - h/2\}^2/mnrs, \tag{4.16}$$

which has an approximate χ_1^2 distribution. In this expression, h is the highest common factor of m and n. Thus, if $m = n$, then $h = n$, while if m and n are co-prime (i.e. have no factors in common, other than 1), then $h = 1$. Note that Yates (1984) argues that X_P^2, given by equation (4.5) is again appropriate.

If the two species exhibit attraction, then the values of b and/or c will appear larger than expected, whereas if the two species exhibit repulsion, then it is the values of a and/or d that will appear inflated. Clearly, there are many possible indices of association that might be cited for a 2 × 2 table. One such is Pielou's (1961) *Index of Segregation*, defined by

$$S = 1 - \frac{\text{observed number of mixed pairs}}{\text{expected number of mixed pairs}},$$

where a 'mixed pair' denotes an individual of one species having an individual of the opposite species as its nearest neighbour. The expected number of mixed pairs

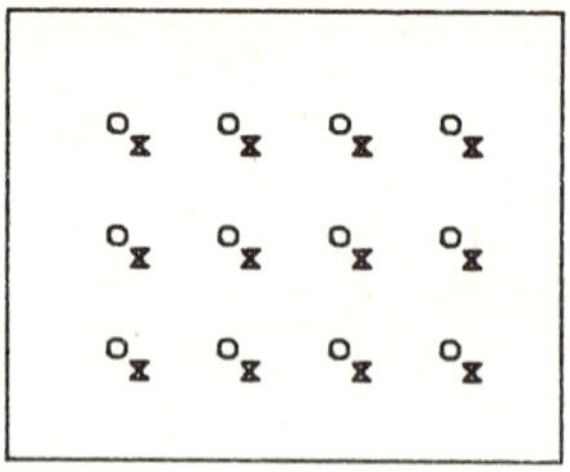

Figure 4.12 Independently located overlapping lattices

is easily seen to be $ms/N + nr/N$, so that the definition becomes

$$S = 1 - N(b + c)/(ms + nr) = 2(ad - bc)/(ms + nr). \qquad (4.17)$$

The usefulness of S and of X^2_{NN} as a means of detecting any lack of independence between point patterns is confined to patterns that are consistent with the Poisson process. if we use these statistics in conjunction with clustered or regular patterns, then this clustering or regularity is liable to influence the value of the statistics. As a simple example, suppose that the two species exhibit regularity, each pattern taking the form of a rectangular lattice (an orchard of trees, for example) with the distance between neighbouring lattice points being d for every pair of neighbouring points in each lattice. The two patterns are now superimposed, with the location of the second pattern being quite independent of the location of the first. The situation is illustrated in Figure 4.12, from which it will be seen immediately that the nearest neighbour of each plant of species i is a plant of species $(3 - i)$ $(i = 1,2)$. In other words cells a and d of Table 4.10 will be zero and Table 4.10 will give rise to an unusually large value of X^2_{NN} and a very large (negative) value of S. However, the patterns were positioned independently of one another: the apparently significant association is an artifact induced by the non-random nature of the component patterns.

***Example* 4.6**

Figure 4.13 shows the distributions of two types of Bronze Age ornament, as recorded by Hansel (1968) and reported by Hodder and Orton (1976), who summarize the data using a nearest-neighbour table. If we regard each ornament pattern separately in the context of the study area, then it is evident that the ornaments are not randomly distributed over the study area. Both sets of ornaments are strongly clustered, and the clusters are apparently non-overlapping. If we were to use the nearest-neighbour table approach, then we should find that $a = 9$, $b = 2$, $c = 2$, and $d = 3$, so that $X^2_{NN} = 2.51$, which is not a significantly large value. This lack of significance reflects both the lack of sufficient data and the spatial clumsiness of the test, which is clearly not appropriate for this situation.

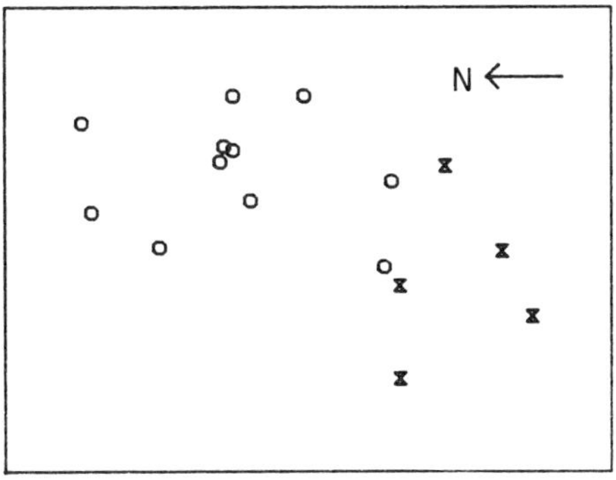

Figure 4.13 Distributions of two Bronze Age ornaments in the Carpathian Basin (Hansel, 1968). (Redrawn from Hodder and Orton, 1976. Reproduced by permission of Cambridge University Press)

***Example* 4.7**

Table 4.11 presents the nearest-neighbour results for the sinkhole data illustrated earlier in Figure 4.9. Five large sinkholes and six small sinkholes were nearer the edge of the study area than they were to any other sinkhole, and therefore these have been excluded from the table. Since $m = 9$ and $n = 10$ have no factors in common, $h = 1$ and hence $X^2_{NN} = (18 \times 22.5)/90.88 = 1.15$, which is not significant. Assuming that the individual distributions represent realizations of a Poisson process, we can accept the hypothesis that they are distributed independently of one another.

Notice that for none of the other data sets discussed in this section (hickories/maples, typestand, and brambles) is the nearest-neighbour test appropriate, since each displays clustering to a greater or lesser extent.

Table* 4.11 *Nearest-neighbour table for the sinkhole data

		Nearest neighbour Large	Nearest neighbour Small	
Base sinkhole	Large	5	4	9
	Small	3	7	10
		8	11	19

Point–plant and plant–plant distributions

Probably the most widely used distaince-based test of randomness for a single plant pattern is the Clark–Evans test described in Chapter 1 (cf. equations (1.39)–(1.43)). Suppose that we have a bivariate plant pattern in which the separate plant patterns are independent with intensities ρ_1 and ρ_2. Let W_{12} be the distance between a randomly chosen plant and its nearest neighbour of the opposite type. Lee (1979) shows that

$$\mathrm{E}(W_{12}) = p_1/2\sqrt{\rho_2} + p_2/2\sqrt{\rho_1}, \tag{4.18}$$

where $p_i = \rho_i/(\rho_1 + \rho_2)$, for $i = 1,2$, and that

$$\mathrm{E}(W_{12}^2) = p_1/\rho_2\pi + p_2/\rho_1\pi. \tag{4.19}$$

These results are only exact for the infinite plane, and with a finite study region we can anticipate the need for edge corrections such as those embodied earlier in equations (1.42) and (1.43).

Lee proposes that a comparison of the observed mean inter-plant distance with the theoretical value given in equation (4.18) would provide a test of independence of the two separate plant patterns. However, such a test would crucially depend on the underlying assumptions of univariate Poisson processes, and it seems unlikely that the test would be well-behaved when used with markedly clustered or markedly regular distributions.

In the univariate case more refined statistics were obtained by considering $\widehat{G(w)}$ and $\widehat{F(x)}$, the observed distribution functions of the sample nearest-neighbour distances and the sample point–plant distances, respectively. With two species of plant present we need to extent our notation somewhat. We shall write $F_i(x)$ to be the distribution function of the distances from randomly chosen points to the nearest plants of species $i(i = 1,2)$, and we shall write $G_{ij}(w)$ to be the distribution function of the distances from randomly chosen plants of species i to the nearest neighbour of species $j(i,j = 1,2)$.

Note that if the two species occur at random, then, from equation (1.48), $F_i(x) = 1 - \exp(-\rho_i\pi x^2)$, while if species j is distributed at random with respect to species i, then $G_{ij}(w) = 1 - \exp(-\rho_j\pi w^2)$. Note that, in general, $G_{ij}(w) \neq G_{ji}(w)$.

Goodall (1965) observed that if the two species are distributed independently of one another, then, *whatever the type of distribution of species j* (random, regular, or clustered), it should appear just the same when viewed from random sampling points as when viewed from randomly chosen plants of the other species. in other words, if the two plant patterns are independent of one another, then we will have

$$F_1(x) = G_{21}(w), \tag{4.20a}$$

$$F_2(x) = G_{12}(w). \tag{4.20b}$$

The procedure for calculating estimates of these distribution functions, taking account of edge effects, was set out in Chapter 1 (cf. equation (1.52) and Table 1.13). An obvious first step is then to plot the estimates $\widehat{F}_1(v)$ and $\widehat{G}_{21}(v)$ (or the alternative pair of functions) on a single diagram with v varying between 0 and the greatest observed point–plant or plant–plant distance. We can then examine the two graphs for signs of any significant deviation from one another.

The most direct way of quantifying the deviations between $F_i(x)$ and $G_{ji}(w)$ is to calculate the value of the Kolmogorov–Smirnov test statistic:

$$D = \max_v |F_i(v) - G_{ji}(v)| \qquad (i = 1,2; j = 3 - i). \tag{4.21}$$

Suppose that this maximum value occurs at $v = v^*$. The effective sample sizes for $F_i(v^*)$ and $G_{ji}(v^*)$ are simply the values of the denominator of equation (1.52) applied to each in turn. The denominator is the total number of observations in the (point–plant or plant–plant) sample reduced by the number of points (or plants) that were within a distance v^* of a boundary. Suppose that the resulting effective sample sizes are m^* and n^*. An approximate test is provided by calculating the value of

$$K = D\{m^*n^*/(m^* + n^*)\}^{1/2}. \tag{4.22}$$

If K exceeds 1.36, then there is significant evidence of a difference between the two distribution functions at the 5% level, while for significance at the 1% level, the critical value becomes 1.63. This approximate test is intended for use when m^* and n^* are both greater than 40 and is somewhat conservative for smaller values of m^* and n^*.

Judging from the results of Diggle and Cox (1983), it would seem that a more powerful test than that based on D is the well-known Mann–Whitney two-sample rank sum test (sometimes known as the Wilcoxon test). Suppose that we have n^* point–plant distances and m^* plant–plant distances (where observations in which the base point or base plant is nearer to the boundary than it is to another plant have been discarded). These $(n^* + m^*)$ values are then arranged in order and assigned the ranks $1, 2, \ldots, (n^* + m^*)$. We then sum the ranks of, say, the n^* retained point–plant distances. Denoting this sum by R, the Mann–Whitney test statistic is

$$M = \{R - n^*(m^* + n^* + 1)/2\}/\{m^*n^*(m^* + n^* + 1)/12\}^{1/2}, \tag{4.23}$$

which has an approximate unit normal distribution provided m^* and n^* are both greater than 10.

With completely mapped bivariate populations we can use the complete set of data to estimate $G_{21}(w)$ and $G_{12}(w)$, as described above. However, if the population is not mapped, then an entirely satisfactory procedure for the estimation of $G_{ij}(w)$ is as follows. From a randomly chosen point we locate the nearest type i plant (which gives us an observation on X for this type of plant), and we then obtain the distance from that plant to the nearest plant of type j. Because

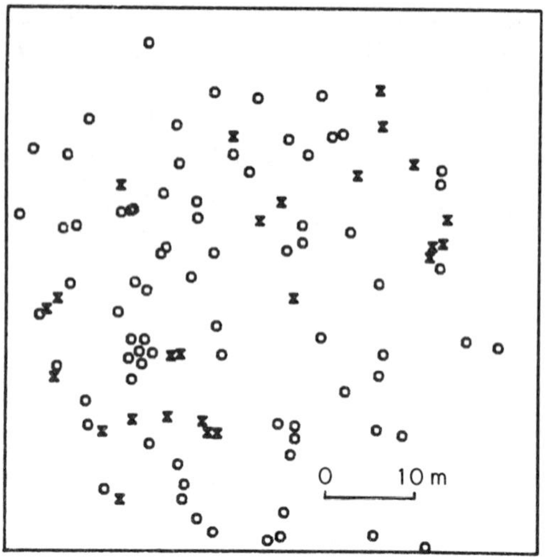

Figure 4.14 Joint distribution of *Atriplex* and *Tidestromia*. (Reproduced by permission of Blackwell Scientific Publications Ltd)

we are then looking for a plant of the *opposite* type, the area swept out in the search for the plant of the first type is irrelevant and hence the problems that were illustrated by Figure 1.24 do not arise.

Example 4.8

In Chapter 1, and again in Chapter 3, we considered various aspects of the distribution of the *Atriplex* bushes in Death Valley, California. In Example 3.23 we considered the distribution of the *Tidestromia* bushes at the same location. We now consider the *joint* distribution of these two shrubs, as illustrated in Figure 4.14, which is redrawn from Figure 1 of Gulmon and Mooney (1977) and refers to the distribution of plants within their site 2 for the year 1971. There were 77 *Atriplex* plants and 25 *Tidestromia* plants within this study area. Note that while Figure 4.14 suggests a square study area, the true study area was a somewhat larger circular area. All references to boundaries in the subsequent discussion refer to the restricted square area shown. The effect of the reduction of the area will amount to a reduction in the usable sample sizes.

Figure 4.15a shows the estimated distribution functions for the distances from a random point to the nearest *Atriplex* bush, and from a *Tidestromia* bush to the nearest *Atriplex* bush. The calculations follow those set out in Table 1.13 and are omitted. It is apparent that the maximum deviation between the two estimated distribution functions occurs at a distance of about 4 m. Examining the calculations we find that the separate estimates (from the equivalent of equation

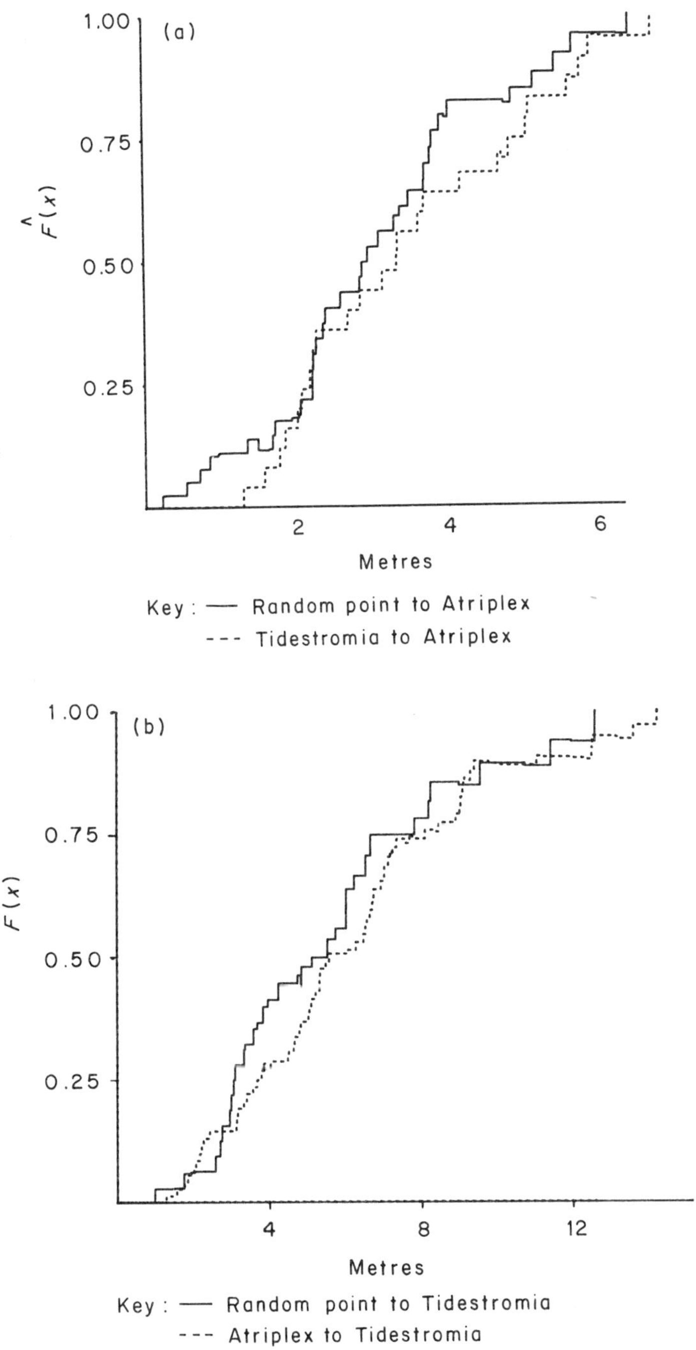

Figure 4.15 Cumulative distributions. (a) Distances to *Atriplex* bushes. (b) Distances to *Tidestromia* bushes.

(1.52)) were 24/29 and 17/25, so that m^* and n^* are 25 and 29, respectively, and so that the value of D is $(24/29 - 17/25)\ (29 \times 25/54)^{1/2} = 0.54$. Since this is very much less than the approximate 5% critical value (1.36), we can safely conclude that there is no evidence of any significant discrepancy between the two functions.

Figure 4.15b displays the corresponding results with respect to the *Tidestromia* bushes nearest to random points and to the *Atriplex* bushes. In this case the two functions appear quite similar and calculations reveal a value of $D = 0.56$ which is, reassuringly, in excellent agreement with the previous estimate.

The more powerful Mann–Whitney statistic, M, hints at some displacement of the two distribution functions, as is evidenced by Figure 4.16, which summarizes the observed distances (from the left-hand end of the line) between random points and *Atriplex* bushes and between each *Tidestromia* bush and the nearest *Atriplex* bush. It will be seen that the shorter distances are dominated (seven out of the shortest nine) by distances from points, whereas the larger distances are dominated (nine out of the largest twelve) by distances from *Tidestromia* bushes. Of course, the same effects are visible in Figure 4.15a, but that figure is less easily interpreted.

A total of 40 randomly located points were considered, but, of these, 11 were discarded because they proved too close to the neighbouring boundary. Thus $n^* = 29$, while $m^* = 25$ and the value of R (taking the shortest distance as having rank 1) is found to be $1 + 2 + 3 + \ldots + 50 + 53 = 718$, which, taken with the mean value of 797.5 and the variance of 3322.92 leads to the value of 1.38 for the test statistic, M. Consultation of tables of the unit normal distribution discloses that this corresponds to a two-tail probability of about 17%.

Repeating the procedure for distances to *Tidestromia* bushes, using 40 random points and 40 randomly chosen *Atriplex* bushes, leads to the values $n^* = 20$, $m^* = 34$, $R = 464$, and thus $M = -1.54$, corresponding to a two-tail probability of about 12%. Neither test provides more than a suggestion of a difference in the point–plant and plant–plant distributions, although the inference to be drawn from Figure 4.16 is that there may be some very slight repulsion between the two species. This is, after all, what common sense would lead us to expect.

Paired point–plant distances

An alternative procedure that, like the Mann–Whitney test, does not involve the estimation of $F(\cdot)$ and $G(\cdot)$ distribution functions, has been proposed by Diggle and Cox (1983). They suggest that we choose points at random within the study region and calculate $\{(x_{1i}, x_{2i})\}$, the distances from each point to the nearest plants of both species, discarding any point which lies too close to the boundary of the study region. If the two species are located in an independent fashion, then there should be no correlation between X_1 and X_2. If the two species are attracted to one another, then we can expect x_{1i} to be about the same distance as

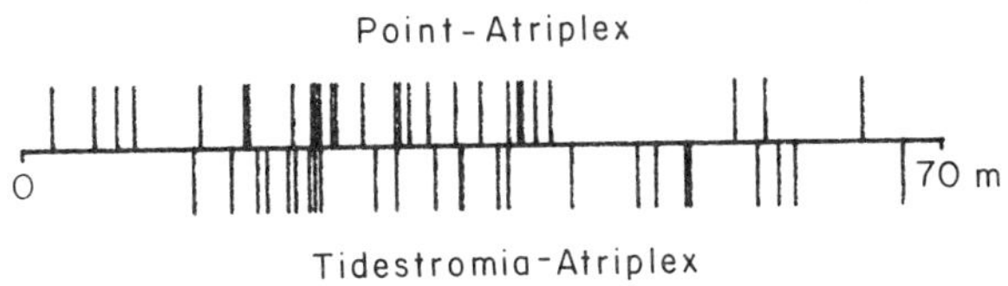

Figure 4.16 Jointly ordered point–*Atriplex* and *Tidestromia–Atriplex* distances

x_{2i}, so that the pairs of values will be positively correlated. On the other hand, if two species exhibit repulsion, then if a member of species 1 is close to our sampling point, members of species 2 will be distant (and vice versa), so that the pairs of observations will display negative correlation.

The individual $\{x_{ji}\}$ cannot be considered to be observations from a normal distribution and the product-moment correlation coefficient is therefore not appropriate. Diggle and Cox suggest using Kendall's coefficient of rank correlation, but, because it is a little easier to calculate and to describe, we shall use the alternative Spearman coefficient.

To calculate Spearman's rank correlation coefficient it is necessary to calculate the separate rankings of the $\{x_{1i}\}$ and of the $\{x_{2i}\}$. Suppose that, after taking account of the boundaries, we have retained n pairs of distances $\{(x_{1i}, x_{2i})\}$. We now assign ranks (r_{1i}, r_{2i}) to these distances, with $1 \leq r_{ji} \leq n$, and then calculate the rank correlation coefficient as

$$r_s = 1 - \frac{6 \sum_{i=1}^{n} (r_{1i} - r_{2i})^2}{n(n^2 - 1)}. \tag{4.24}$$

Under the null hypothesis of independence this statistic has mean 0 and variance $1/(n-1)$, so that, for reasonably large n, we can test the significance of r_s by calculating

$$z = r_s(n-1)^{1/2}, \tag{4.25}$$

and comparing the value of z with tables of the unit normal distribution function. Further details concerning this statistic, the product-moment correlation coefficient, and Kendall's statistic are to be found in Siegel (1956).

Diggle and Cox's results suggest that a test based on r_s will be rather more powerful than that based on the Mann–Whitney procedure, although, as usual, it is sensible to examine the data in as many different ways as are available, since each test is tuned to subtly different nuances in the data. The extra power of the present test is almost certainly attributable to the *pairing* of the distances to the two types of plants. Diggle and Cox mention some other tests that also exploit this pairing, although the use of rank correlation appears to be the most successful.

Example 4.9

Table 4.12 gives the details of the paired procedure applied to the Death Valley data. Initially there were 40 sample points, but 21 of these had to be discarded because the distance of the point from the nearest boundary was less than the distance to the nearest specimen of one (or both) of the shrub types. The remaining 19 distances were ranked for each shrub and we immediately notice that the shortest point–*Atriplex* distance has been paired with the longest point–*Tidestromia* distance, suggesting that we shall now be able to detect some repulsion between the species.

The value of $\sum(r_{1i}-r_{2i})^2$ is 1804, with $n=19$, so that the value of Spearman's rank correlation coefficient, r_s, is -0.5825, and the value of the test statistic, z, is -2.47. This value is decidedly extreme, corresponding to a two-tail probability of about 1.5%. As anticipated, the paired distance approach has provided a more powerful test than its predecessors and we can now confirm that there is significant evidence of some (obviously fairly slight) repulsion between the two species.

Table 4.12 *The retained paired point–plant distances for the Death Valley data*

Point to *Atriplex* distance in metres (x_{1i})	Rank of x_{1i} (r_{1i})	Difference in ranks	Rank of x_{2i} (r_{2i})	Point to *Tidestromia* distance in metres (x_{2i})
3.50	16	14	2	1.77
0.55	2	9	11	3.86
2.06	6	2	8	3.11
2.60	11	7	18	2.21
3.10	14	11	3	2.60
1.69	4	6	10	3.61
2.27	8	1	9	3.36
3.71	17	10	7	3.08
6.40	19	15	4	2.78
1.35	3	11	14	5.55
1.72	5	10	15	6.03
2.37	9	4	5	2.98
2.39	10	7	17	6.56
2.22	7	5	12	4.26
2.85	12	4	16	6.04
3.81	18	12	6	3.01
0.23	1	18	19	12.56
3.31	15	14	1	1.00
2.96	13	0	13	4.87

Toroidal shifts

In our discussion of the use of the nearest-neighbour table, we emphasized that the test procedure depended upon the separate point patterns being reasonably consistent with underlying random processes. If the separate processes are not random, then the distribution of the test statistic X^2_{NN} (equation (4.16)) is no longer certain. However, X^2_{NN} (or its simpler relative, $|ad - bc|$) remains a sensible summary statistic, so that we are led to consider methods for obtaining its distribution with non-random plant distributions.

Following from the work in previous chapters, it is natural to consider a Monte Carlo approach. In the present case we must preserve the separate plant distributions in their observed form, since that form represents the entirety of our certain knowledge concerning their pattern. It is possible that the separate patterns are realizations of one of the processes described in Chapter 1, but we have no means of knowing which process is most appropriate.

The restriction on preserving the separate observed patterns led Lotwick and Silverman (1982) to consider the use of toroidal shifts. Suppose that the study area is a rectangular or square region. We now imagine that this region has been obtained by unfolding a torus—with the result that the top of the region should be regarded as abutting the bottom of the region, while the left of the region abuts the right of the region. The net effect was illustrated in Figure 1.32. We now hold one of the plant distributions in its observed position and move the other plant distribution by a random amount on the torus. This is easily effected by adding Δx to the x coordinate and Δy to the y coordinate of each and every member of that plant distribution, so that the whole collection of these plants is shifted by the same amount. If, as a result, any of the revised plant positions appear to lie outside the original study region, then this new position will be interpreted as lying in an opposite corner of the study region.

An an example, suppose that, within a rectangle of dimensions 50×100, a plant is located at (43, 96). The result of the choice $\Delta x = 15$, $\Delta y = 14$ would be that the new 'position' of the plant would be (8, 10), whereas the choice $\Delta x = 4$, $\Delta y = 10$ would 'move' the plant to (47, 6). An example of a shifted pattern is given in Figure 4.17, which refers to the effect of the shift $\Delta x = 0.05$, $\Delta y = 0.10$ on the plants lying within a unit square.

An illustration of the application of toroidal shifts in conjunction with the study of nearest neighbours is provided by Harkness and Isham (1984), who study the interaction between competing species of ants (*Cataglyphis* and *Messor*). They choose as test statistics the number of *Cataglyphis* nests that have a *Cataglyphis* nest as nearest neighbour and also the average inter-species nearest-neighbour distance. They also use the $K_{12}(t)$ statistic, which we discuss in the next subsection.

If the study region is neither rectangular nor square, then it cannot be regarded as a representation of a toroid. However, the idea of exploring the effect of shifting

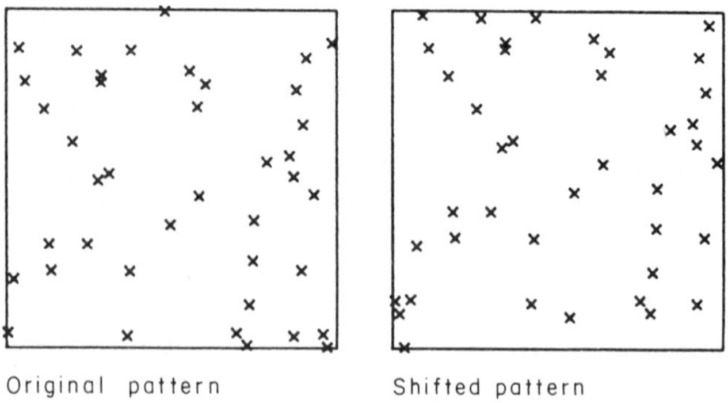

Figure 4.17 An example of a toroidal shift. (a) Original pattern. (b) Shifted pattern. ($\Delta x = 0.05$, $\Delta y = 0.10$ in unit square)

one plant pattern with respect to the other remains appealing. A possible approach is to shift the boundary of the study region at the same time (and by the same amount) as we shift one of the two plant distributions. After this shift we then confine attention to that subsection of the original study region that is also contained within the shifted study region.

The effect of shifting the region as well as the plants is illustrated in Figure 4.18. It is clear that, for this method to be practicable, the values of Δx and Δy must be carefully chosen so as not to be so large that there is little or no overlap between the original and shifted positions of the study region. On the other hand, too small a set of shifts will result in values of the test statistics which are virtually identical to those actually observed. Some sort of intermediate course is required, although it is not clear whether the resulting 'randomization' distributions would necessarily have desirable properties.

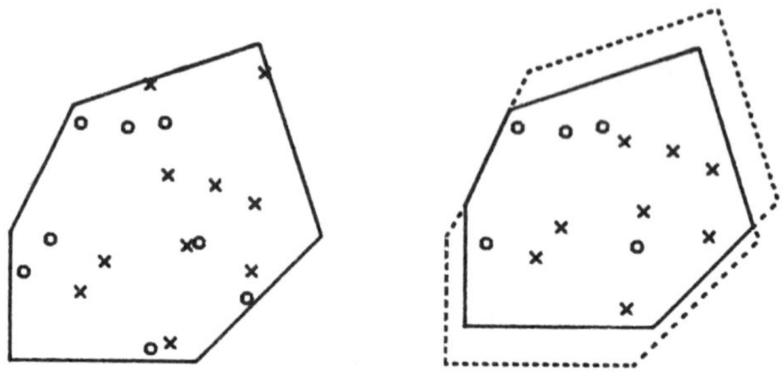

Figure 4.18 Shifting patterns in irregular regions. (a) Original pattern. (b) Shifted pattern and restricted region

Combined count and distance analysis

Probably the most subtle and revealing of all the methods considered in Chapter 1 was that based on the function $K(t)$, which was defined by

$$\rho K(t) = \text{E(number of plants within a distance } t \text{ of a randomly chosen plant)},$$

with ρ being the intensity of the plant distribution. Changes in the value of $K(t)$ as t increases give insights into the clustering or regularity of patterns at different inter-plant distances, and make it possible to describe a pattern as having aspects of clustering at one value of t and aspects of regularity at another value of t (as evidenced in the stylized patterns presented in Figures 1.3 and 1.4).

A natural extension of the function $K(t)$ to the bivariate (or multivariate) point process is obtained by writing

$$\rho K_{ij}(t) = \text{E(number of plants of type } j \text{ within a distance } t \text{ of a randomly chosen plant of type } i).$$

Thus, for a bivariate point process, the individual point processes are characterized by $K_{11}(t)$ and $K_{22}(t)$.

In Chapter 1 we noted that, for a random pattern, we would have $K(t) = \pi t^2$, which suggested that a statistic of interest would be $L(t) = t - \{K(t)/\pi\}^{1/2}$. The argument that led to this result for a random pattern (equation (1.53)), can be as easily applied to the placing of individuals of one species of plants with respect to the positions of the individuals of a second species of plants. Thus, if species j plants are placed (with average intensity ρ_j) at random with respect to species i plants, then the expected number of species j plants lying within a distance t of a plant of species i is $\rho_j \pi t^2$, and thus, if the two species are independent, we will have

$$K_{ij}(t) = \pi t^2, \tag{4.26}$$

with

$$L_{ij}(t) = t - \{K_{ij}(t)/\pi\}^{1/2} = 0. \tag{4.27}$$

Studying the definition of $L_{ij}(t)$, we see that a large positive value implies repulsion between the two processes at that value of t, whereas a large negative value implies attraction between the two processes at that value of t.

Diggle and Milne (1983) develop expressions for $K_{ij}(t)$ for various specialized bivariate point processes. In particular, for the *linked Cox process* (equation (4.3)) they show that $K_{11}(t) = K_{22}(t) = K_{12}(t)$. Thus, if a linked Cox process is suspected—which might be the case if it was felt that the observed 'hand-in-hand' variations in the intensity of the two plant processes were dependent upon some 'unseen' variable such as soil fertility, rainfall, or human population intensity—then a suitable diagnostic step would be to superimpose the graphs of the three K functions and examine the extent of their correspondence.

In a similar way, for a pattern exhibiting repulsion, a guide as to whether a suitable model could be a *balanced Cox process* (for which the *joint* pattern is Poisson, while the separate patterns have variable intensities), is provided by examining whether, for any t, there is any noticeable departure of $M(t)$ from zero, where $M(t)$ is defined by

$$M(t) = t - [\{\rho_1^2 K_{11}(t) + 2\rho_1\rho_2 K_{12}(t) + \rho_2^2 K_{22}(t)\}/\pi]^{1/2}/(\rho_1 + \rho_2). \tag{4.28}$$

Here, the term in the square brackets is just the square root of a weighted sum of $K_{11}(t)$, $K_{22}(t)$, $K_{12}(t)$ divided by π. With a balanced Cox process, the unusually large $K_{12}(t)$-values would be offset by the unusually small $K_{11}(t)$ and $K_{22}(t)$ values.

Notice that, throughout the above, we have written $K_{12}(t)$, but could equally well have written $K_{21}(t)$, since the two must be identical for any realization of a process in an infinite plane. In practice, however, unless we regard our study region as toroidal (!) we shall be working within a finite bounded region wherein exact equality of the estimates of $\hat{K_{12}}(t)$ and $K_{21}(t)$ is not feasible. While the separate estimates are unlikely to differ very greatly, we clearly need a single combined estimate for insertion in equations (4.26)–(4.28).

Lotwick and Silverman (1982) show that a minimum variance estimator of $K_{12}(t)$ is provided by choosing $\hat{K_{12}}(t) = \alpha(t)\tilde{K_{12}}(t) + \{1 - \alpha(t)\}\tilde{K_{21}}(t)$, where $\tilde{K_{12}}(t)$ and $\tilde{K_{21}}(t)$ are the separate sample estimates whose forms are given later in equations (4.30a) and (4.30b). Since $\alpha(t)$ is a function of t, it is clear that that the minimum variance estimator will be difficult to obtain (especially since $\alpha(t)$ will itself require estimation). However, in the case of a bivariate Poisson process, with n_i plants of species $i (i = 1, 2)$ lying within the region of interest, Lotwick and Silverman show that $\alpha(t) = n_2/(n_1 + n_2)$, for all t. It therefore seems that a reasonable compromise as an estimator of $K_{12}(t)$ will be

$$\hat{K_{12}}(t) = \{n_2 \tilde{K_{12}}(t) + n_1 \tilde{K_{21}}(t)\}/(n_1 + n_2). \tag{4.29}$$

The definitions of $\tilde{K_{12}}(t)$ and $\tilde{K_{21}}(t)$ follow that of $K(t)$ given above by equation (1.56). Let the function $k(x_r, y_s)$ be defined through the inverse function $1/k(x_r, y_s)$ which is the proportion of the perimeter of the circle centered on the plant x_r (of species 1) and passing through the plant y_s (of species 2) which lies within the region of interest. These quantities were illustrated for the single species case in Figure 1.33. With this definition of $k(x_r, y_s)$, we have

$$\tilde{K_{12}}(t) = A \sum_{r=1}^{n_1} \sum_{s=1}^{n_2} k(x_r, y_s)/n_1 n_2, \tag{4.30a}$$

$$\tilde{K_{21}}(t) = A \sum_{r=1}^{n_1} \sum_{s=1}^{n_2} k(y_s, x_r)/n_1 n_2, \tag{4.30b}$$

where A is the area of the study region.

Example 4.10

We now apply the combined count and distance analysis to the Death Valley data illustrated in Figure 4.14. We have already detected some evidence of repulsion between the *Atriplex* and *Tidestromia* bushes, and we can expect that this will now be revealed with greater detail. Our earlier analyses of the *Atriplex* data suggested that there was some short-range repulsion at a distance of about 4 m, and this too we can expect to see here more clearly.

Figure 4.19 shows the results of plotting $\hat{L_{11}}(t)$, $\hat{L_{22}}(t)$, and $\hat{L_{12}}(t)$ against t. The graphs of $\hat{L_{11}}(t)$ and $\hat{L_{22}}(t)$ may be compared with the (strikingly different) corresponding graphs given in Figure 1.38a (Bodmin Moor tors) and Figure 1.38b (Southampton junior schools). Both types of desert bush display regularity, which is particularly marked in the case of the *Tidestromia* bushes, for which it seems likely that one of the sequential hard-core processes (cf. Figures 1.8b and c) would be appropriate. The core radius appears to be about 1.5 m, which agrees well with Gulmon and Mooney's (1977) recorded average nearest-neighbour distance of 4.1 m for these bushes. For the *Atriplex* bushes the graph of $L(t)$ displays a more gentle rise from zero, suggesting that, for these bushes, a soft-core process would be preferable, with an average core diameter of around 1 m. In plain English it appears that *Tidestromia* bushes occupy territories that are both larger and more regular in size than those of the *Atriplex* bushes.

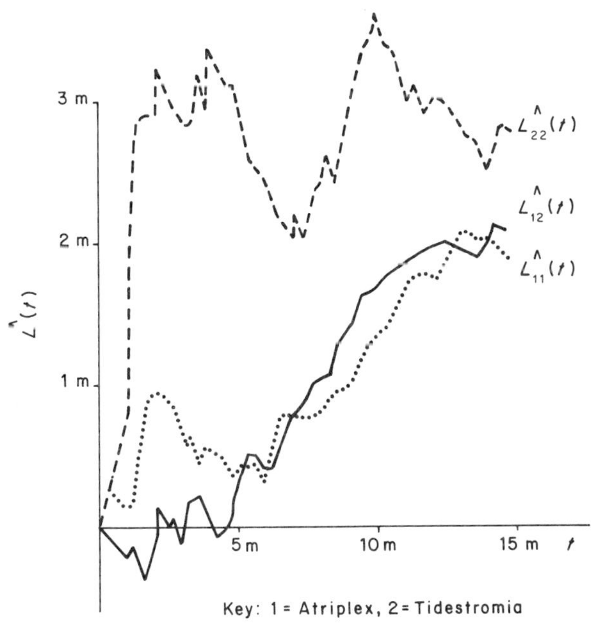

Figure 4.19 Plot of the $\hat{L(t)}$ functions against t for the Death Valley data

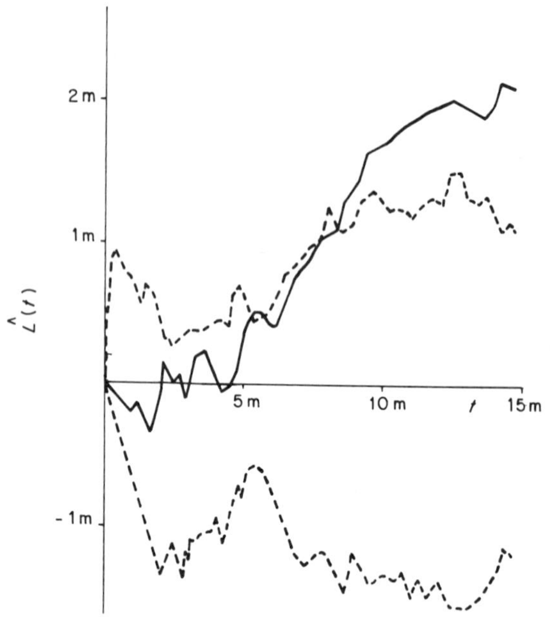

Figure 4.20 Comparison of $\hat{L_{12}}(t)$ with the envelope formed using 19 toroidal shifts

The graph of $\hat{L_{12}}(t)$ against t suggests that, at short ranges, there is little evidence of any repulsion between the species, whereas at distances of more than 5 m there is increasingly strong evidence of repulsion. Figure 4.20, which reproduces $\hat{L_{12}}(t)$, together with the envelope formed from 19 simulations derived using toroidal shifts, confirms the significance of this repulsion for distances of more than 10 m. This suggests that the underlying fertility of the study region is not uniform, but forms patches of diameter 20 m and above.

Since the joint process exhibits a combination of regularity and repulsion, so that all the $K(t)$ functions are positive, there is no point in calculating the 'average' statistic $M(t)$, which will take on the same general form as the $K(t)$ functions. Of the processes listed in Table 4.1, only a 'thinned' process could be appropriate. Some evidence for this process is provided by the figures given in Gulmon and Mooney (1977) which illustrate the positions of live plants in successive years. One aspect of this data was considered briefly in our earlier discussion of space–time interaction tests and a composite map of *Tidestromia* locations was given as Figure 3.14.

4.5 SPATIAL RANK CORRELATION

We turn now to the rather different type of data that we considered in Chapter 3, where our concern is not with the locations of points *per se*, but with the values of some characteristic at certain points (which may have been deliberately chosen, or may be haphazard). In this chapter we are concerned with two types of spatially distributed 'object', and this implies that some of the recorded values refer to one variable (variable 1!) and some to another (variable 2).

In cases where we have information at each point location concerning each of the two variables, we would normally consider using the product-moment correlation coefficient r, given by equation (4.9). However, we may well have misgivings concerning the applicability of the assumptions that underpin this statistic or we may have only a rather hazy idea of the precise locality values. In either case a sensible solution would be to use a rank correlation coefficient such as that due to Spearman and given earlier by equation (4.24). In the present instance, of course, the $\{r_{ji}\}$ would refer to the ranks of the observations on variable j.

If our information consists of (paired) quadrat counts, then it is quite likely that there will not be a unique ordering of the observed counts, but rather that there will be many tied counts. There are a number of possible strategies for dealing with tied counts, of which the simplest (but not necessarily the most appropriate) is to give each tied observation the average of the ranks that are tied. The exact distribution of r_s is altered by this procedure, and for further details the reader should consult Siegel (1956).

The above discussion has related to the case where the observations on the two variables are paired, but it takes no explicit account of the actual positions of those locations, and is therefore not exploiting the spatial properties of the data. Tjostheim (1978), however, has developed an index, labelled A, that is designed to measure correlation in a manner which specifically takes account of the physical positions of the reported data. The index would be equally appropriate for two sets of identically located observations on the same variable at two distinct points in time, and, in this sense, A relates directly to the univariate spatial and spatio-temporal analyses previously encountered in Chapter 3. Tjostheim's index may therefore be considered as yet another variant on the general cross-product statistic, here denoted as Ω, which pervaded that chapter.

Tjostheim's index is a function of the ranks of the (ordinal) observations for the two variables and of the (Euclidean) distance between locations. Suppose that the locations of the ith ranked values for variables 1 and 2 are, respectively, (x_{1i}, y_{1i}) and (x_{2i}, y_{2i}), and suppose that each set of x- and y-values is standardized so as to have mean 0 and equal variance. The index A is then defined by

$$A = \frac{\sum_i (x_{1j}x_{2i} + y_{1i}y_{2i})}{\sum_i (x_{1i}^2 + y_{1i}^2)} \tag{4.31}$$

Notice that, since the two variables are measured at the same locations, the denominator could equally well be written as $\sum(x_{2i}^2 + y_{2i}^2)$. If we define $r_{x_1x_2}$ as the (Pearson) correlation between the x coordinates of identically ranked observations, with $r_{y_1y_2}$ being similarly defined for the y coordinates, then equation (4.31) may be rewritten as

$$A = \tfrac{1}{2}(r_{x_1x_2} + r_{y_1y_2}). \tag{4.32}$$

We now revert to the use of proximity matrices, as in Chapter 3. Consider two proximity matrices **Y** and **W** with rows and columns equivalent to the ordered ranks at times 1 and 2. To relate to Tjostheim's A, 1's are allotted to the main diagonal of **Y**, with 0's elsewhere. This defines non-spatial proximity very simply as a variable with two levels, 'identical rank' and 'not identical rank'. The **W** matrix contains (Euclidean) distances between pairs of ranks, that is

$$W_{ij} = \{(x_{1i} - x_{2j})^2 + (y_{1i} - y_{2j})^2\}. \tag{4.33}$$

Glick (1982) makes it apparent that the familiar quantity $\Omega = \sum\sum Y_{ij}W_{ij}$, the general cross-product statistic, constitutes the kernel of Tjostheim's index; it is simply the sum of the spatial distances between localities with identical ranks. Thus, for localities with row and column coordinates (5, 7), (6, 8), (7, 5), (8, 8), and ranks 1, 2, 3, 4 at time 1, and 2, 4, 3, 1 at time 2, we get

$$\Omega = \{(5-8)^2 + (7-8)^2\}^{1/2} + \ldots + \{(8-6)^2 + (8-8)^2\}^{1/2} = 6.58$$

Similarly, the Spearman rank correlation coefficient is related to the general cross-product statistic by setting $Y_{ij} = (i-j)^2$, with $W_{ij} = 1$ for ranks with identical spatial locations and $W_{ij} = 0$ otherwise. Thus, this coefficient may be viewed as a function of the sum of the squared differences between ranks at each locality.

The shortcomings of employing traditional (non-spatial) rank correlation, initially outlined by Glick (1982), are highlighted by the perspective offered by the general cross-product statistic, since the latter demonstrates that rank correlation effectively ignores the broader spatial positioning of the observations by reducing spatial proximity to a crude zero–one measure indexing identical and non-identical locations in the **W** matrix. On the other hand, it does utilize a relatively sophisticated measure of non-spatial proximity in the **Y** matrix. By contrast, Tjostheim's index has a sensitive measure of spatial proximity whilst the non-spatial proximities of the **Y** matrix are comparatively crude.

Experience with other variants of the general cross-product statistic suggests that a hybrid measure that treats both types of proximity with equivalent exactitude is possible. Hubert and Golledge (1982a, 1982b) detail many extensions of Tjostheim's A and consider it in the context of rectangular matrix comparison methods. Included amongst their interesting discussion is the notion of k-dimensional distances being input into the **W** matrix. In this case, one would average over k correlations, rather than just the two of equation (4.32).

As formulated, Tjostheim's index is confined in its application to ordinal data. However, Hubert and Golledge (1982a) demonstrate that this restriction can relatively easily be circumvented if it is felt desirable to use the raw data rather than its ranked form. Glick (1982) suggests that the emphasis on ranks implies that the index is particularly appropriate for ordinal data.

Given the normalization involved in equation (4.31) and the Euclidean distance definition of equation (4.33), Tjostheim (1978) observes that, under randomization,

$$\mathrm{E}(A) = 0, \qquad \mathrm{var}(A) = (1 + r_{xy}^2)/\{2(n-1)\}, \tag{4.34}$$

and that, under fairly liberal conditions, the reference distribution is asymptotically normal. Hubert and Golledge (1982b) observe that Tjostheim's (1978) proof is a special case of an earlier and more general theorem due to Puri and Sen (1971). Note that the value of var(A) depends upon the point locations. The variance is least $(1/\{2(n-1)\})$ when $r_{xy} = 0$, which occurs when the locations form a rectangular grid, and is greatest $(1/(n-1))$ when $r_{xy} = \pm 1$, which occurs when the locations are co-linear.

The extension of this index to the case where the observations on variable 2 are measured at *different* locations to those at which variable 1 has been measured is an obvious future development. We can anticipate that, provided that there is a substantial overlap between the sampling regions for the two variables, then an appropriate test statistic would be the sum of the distances between equally ranked locations, with randomization of the ranks within the variable locations providing the sampling distribution to be used for assessing the significance of the observed result.

Example 4.11

As a further illustration of the properties of the index A for the case where each sampling point provides a value for both variables, consider the artificial data given in Table 4.13. The table displays the ranked values at 16 locations on a square (x, y) grid which is centered on $(0, 0)$, for two points in time. The data in Table 4.13a were generated by using a random permutation of the numbers 1 to 16. A comparison of Table 4.13b with Table 4.13a will indicate that the data are

***Table* 4.13 *Artificial lattice data used to calculate Tjostheim's* A**

a)	Time 1				b)	Time 2			
y					y				
3	4	3	1	8	3	5	4	3	1
1	5	11	7	12	1	11	7	12	13
−1	13	14	10	6	−1	14	10	6	2
−3	2	9	15	16	−3	9	15	16	8
x	−3	−1	1	3	x	−3	−1	1	3

Entries are locality ranks

broadly similar, but with slight shifts which are, for the most part, sideways. Thus, row coordinates are usually retained while column coordinates are subject to slight shifts. Since the ranks are randomly located in Table 4.13a, there is no reason why the rank at locality i should bear any resemblance to its replacement rank at the later time period and therefore we would not expect the value of Spearman's coefficient to be significantly large. On the other hand, we would expect that the value of Tjostheim's A would be very large indeed.

Table 4.14 summarizes the 16 ranked locations for the two time periods together with the calculations that determine the value of $\sum(x_{1i}x_{2i} + y_{1i}y_{2i})$ as $20 + 56 = 76$. Since $\sum x_{1i}^2 = \sum x_{2i}^2 = 80$, we see that the value of A is 0.475. From equation (4.32) it is clear that this is the average of a row correlation of 0.7 and a column correlation of 0.25 and thus faithfully mirrors the artificial process used in constructing Table 4.13b.

The two hypotheses of interest are the null hypothesis of independence and the alternative of positive correlation. Assuming approximate normality, we therefore perform a one-tailed test. From the nature of the calculations in Table 4.14 it is clear that, for a rectangular lattice, $\sum(x_{1i}x_{2i} + y_{1i}y_{2i})$ is an observation from a discrete distribution having relatively few possible values. It is therefore sensible to improve the fit of the normal approximation by introducing a continuity correction. An appropriate continuity correction is $h/2$, where h is the highest common factor of $(\delta x)^2$ and $(\delta y)^2$, and δx, δy are the sizes of the lattice steps in the

***Table* 4.14 *The calculation of Tjostheim's index for the data of Table* 4.13**

Rank at time 1	Coordinates at two time points x	y	x	y	Cross-products $x\ x$	$y\ y$	
1	1	3	3	3	3	9	
2	−3	−3	3	−1	−9	3	
3	−1	3	1	3	−1	9	
4	−3	3	−1	3	3	9	
5	−3	1	−3	3	9	3	
6	3	−1	1	−1	3	1	
7	1	1	−1	1	−1	1	
8	3	3	3	−3	9	−9	Thus $A = 76/160$
9	−1	−3	−3	−3	3	9	
10	1	−1	−1	−1	−1	1	
11	−1	1	−3	1	3	1	
12	3	1	1	1	3	1	
13	−3	−1	3	1	−9	−1	
14	−1	−1	−3	−1	3	1	
15	1	−3	−1	−3	−1	9	
16	3	−3	1	−3	3	9	
				Totals	20	56	

x and y directions (an analogous result was used in deriving X^2_{NN} given by equation (4.16)).

For the lattice data $r_{xy} = 0$ and hence, from equation (4.34), $\text{var}(A) = 1/30$. Hence, the appropriate normal value is $(30)^{1/2}(76 - 4/2)/160 = 2.53$, which corresponds to a tail probability of about 0.5% and therefore gives highly significant evidence of spatial correlation between the ranks at the two time points. By contrast, Spearman's rank correlation coefficient has the (larger) value of 0.512, but is much less significant.

Hubert and Golledge (1982b) give a contrasting example that pertains to the preferences (in rank order) of males and females for 15 localities. They obtain a significantly high value for r_S, indicating that the sexes give identical localities similar ranks. On the other hand, the value of A does not differ significantly from zero, indicating that it is not possible to predict the position of the locality ranked i by the females from knowledge of the position of the locality ranked i by the males (or vice versa). This apparently paradoxical result is due to the similarly ranked locations not being spatially clustered.

CHAPTER 5

Regression and Autoregression

5.1 REGRESSION, STRUCTURAL, AND SAMPLING MODELS

Regression models are models of processes involving usually one dependent response variable (Y) and one or more independent variables ($X_1, X_2, \ldots$). The assumption made by a regression model is that the value observed for Y can be simply explained by a function (often a simple linear combination) of the X variables. Such models have widespread application in the social, biological, and behavioural sciences, and there are numerous books devoted to both theoretical (e.g. Plackett, 1960; Seber, 1977) and applied (e.g. Draper and Smith, 1981; Chatterjee and Price, 1977) aspects of this vast subject. Regression analysis must surely be one of the most widely used statistical techniques and it is clearly impossible for us to cover in any detail the entire spectrum of regression theory and its applications within a single chapter of a book on spatial data analysis. Within this chapter we therefore focus attention specifically on certain aspects of regression analysis that come to the fore with spatial data. We shall, however, include a somewhat sketchy outline of the basic principles of regression analysis during our discussion, in the hope that this will suffice as an *aide memoire* to those readers for whom regression analysis is not a daily ritual!

We shall refer to the two components of any regression model as the structural and sampling models. The structural component is the specification of the functional link between the Y and X variables. This link is normally linear (e.g. $Y = aX_1 + bX_2 + c$), but is sometimes non-linear (e.g. $Y = aX^2 + b$ or $Y = cX^d$). A commonly used non-linear link is one that specifies a linear link between the logarithms of the X and Y variables (e.g. $\log Y = \log c + d \log X$). Since simple explanations are the easiest to understand and interpret, regression analysis is dominated by these very simple structural models.

The sampling component of the model is the theoretical probability distribution assumed for the Y variable (or, equivalently, for the error structure). This is usually, though not necessarily, a normal distribution. The advent of sophisticated modern software packages such as GLIM and GENSTAT has made practical the use of alternative sampling distributions (cf. Nelder and Wedderburn, 1972). The option of a non-normal distribution should only be invoked for cases where there are theoretical objections to the use of the normal distribution. We shall return to this point later in the chapter.

5.2 ORDINARY LEAST SQUARES

Consider a model that describes the dependence of the variables Y on the $(k-1)$ independent X variables by invoking the simple linear structural relationship.

$$\mathrm{E}(Y_i) = b_0 + b_1 X_{1i} + b_2 X_{2i} + \ldots + b_{k-1} X_{k-1,i} \quad (i = 1, 2, \ldots, n). \tag{5.1a}$$

Here $\mathrm{E}(Y_i)$ denotes the expected value of Y_i, the $\{b_j\}$ are a set of unknown constant coefficients whose values are to be determined, and the $\{X_{ji}\}$ are the set of values of the X variables. In the spatial context we may describe Y_i as the value of the quantity of interest at locality i.

An entirely equivalent formulation of the model (5.1a) employs vector notation:

$$\mathrm{E}(\mathbf{Y}) = \mathbf{Xb}. \tag{5.1b}$$

Here $\mathbf{Y}$ is the $n \times 1$ column vector $(Y_1, Y_2, \ldots, Y_n)'$, $\mathbf{b}$ is the $k \times 1$ column vector $(b_0, b_1, \ldots, b_{k-1})'$, and $\mathbf{X}$ is an $n \times k$ matrix, the first column of which contains 1's and the remaining columns the (fixed) values of the $(k-1)$ variables in each of the n localities. Note that these may be simply (1, 0: presence/absence) variables in which case they may be referred to as *dummy* variables.

When all the X variables take zero values, the expected value of Y_i is simply b_0, for all i. An interpretation of $b_j (j > 0)$ is that it represents the change in Y that would result from a unit change in X_j, given that all the remaining X variables were held constant. A key issue is whether any of the $\{b_j, j > 0\}$ can be set equal to zero. If this is possible, then it implies that the corresponding X variable may be eliminated from the model, since that variable has no bearing on Y, the variable of interest.

Instead of formulating the relation between Y and the X variables in terms of the expected value of Y, we could equivalently work with the observed value of Y together with the deviation of the observed value from that expected. Thus, an alternative rendering of (5.1), in matrix form, is:

$$\mathbf{Y} = \mathbf{Xb} + \mathbf{e}. \tag{5.2}$$

Here $\mathbf{e}$ is an $n \times 1$ column vector of unobserved errors.

If our model is a realistic representation of the relation between Y and the X variables, then the values of Y at the various localities will be very close to those predicted by the model. In other words the individual values of $Y_i - \mathrm{E}(Y_i) = e_i$ will be very small. A natural method of summarizing the lack of fit of the model is to consider the size of $L = \sum e_i^2$. The smaller the value of L, the better the fit of the model. The ordinary least squares (OLS) procedure amounts to the decision to choose that set of values $\{\hat{b}_j\}$ which minimizes L. The OLS procedure may therefore be regarded as being no more than the application of common sense—but, as we shall see, it does have some optimal properties which make it preferable, under certain conditions, to any alternative procedure.

In matrix terminology we seek to find the vector $\hat{\mathbf{b}}$ which minimizes

$$L = \mathbf{e}'\mathbf{e} = (\mathbf{Y} - \mathbf{Xb})'(\mathbf{Y} - \mathbf{Xb}). \tag{5.3}$$

Expanding the right-hand side of equation (5.3), and bearing in mind that each of the terms is in fact a scalar (though written as a product of vectors), we have

$$L = \mathbf{Y}'\mathbf{Y} = 2\mathbf{b}'\mathbf{X}'\mathbf{Y} + \mathbf{b}'\mathbf{X}'\mathbf{Xb},$$

so that

$$\partial L/\partial \mathbf{b} = -2\mathbf{X}'\mathbf{Y} + 2\mathbf{X}'\mathbf{Xb}.$$

The minimum occurs when this derivative is zero, that is, when

$$(\mathbf{X}'\mathbf{X})\mathbf{b} = \mathbf{X}'\mathbf{Y}.$$

Consequently, provided the matrix $(\mathbf{X}'\mathbf{X})$ has an inverse, the vector of ordinary least squares estimates is given by

$$\hat{\mathbf{b}} = (\mathbf{X}'\mathbf{X})^{-1}\mathbf{X}'\mathbf{Y}. \tag{5.4}$$

The Gauss–Markov theorem

We remarked that the OLS procedure can have optimal properties. These properties and the associated restrictions are encapsulated within the Gauss–Markov theorem. Before stating the theorem we must first introduce some further terminology. In general we will not expect a given combination of values of the X variables to always lead to precisely the same observed Y-value (though they would, of course, lead to the same value of $\mathrm{E}(Y)$). The variation that we observe in these Y-values reflects the variation in the distribution of the error terms $\{e_i\}$. Corresponding to Y_i there will be an error e_i which may be regarded as an observation from a distribution with variance $\mathrm{var}(e_i)$. In addition, since the $\{e_i\}$ need not be uncorrelated, we need to take into account the $n(n-1)$ covariances denoted by $\mathrm{cov}(e_i, e_k)$. In the matrix formulation these variances and covariances are summarized within a so-called *variance–covariance* or *dispersion* matrix, $\mathbf{V} = \sigma^2\mathbf{D}$.

The conditions of the Gauss–Markov theorem fall into two groups, one concerned with the dispersion matrix, and the other with the X variables. So far as the dispersion matrix is concerned the conditions specified are that $\mathbf{V} = \sigma^2\mathbf{I}$. In other words, all the Y measurements are made with equal accuracy and each is independent of the rest. The assumptions concerning the X variables are that there is no correlation between the X variables and the errors, that the X variables are measured without error, that there are no missing covariates, and that the correct functional link between Y and the X variables has been adopted.

Given the fulfilment of these various conditions the Gauss–Markov theorem states that the OLS procedure provides those estimators which have smallest variance (i.e. are most precise) amongst the class of all possible linear unbiased estimators. The term 'unbiased' implies that $\mathrm{E}(\hat{\mathbf{b}}) = \mathbf{b}$, which is an obviously desirable property. The term 'linear' implies that the estimators can be expressed as a linear combination of the observed Y-values, which is again a pleasing property in view of its simplicity. Lovers of acronyms will be enchanted to learn that estimators having these properties are termed BLUE (best linear unbiased estimators)!

The conditions of the Gauss–Markov theorem are rather tough ones, but, fortunately, it appears that the OLS procedure is very robust to mild departures from the conditions. This reflects the commonsense basis for the procedure. One condition that, while not spelt out by the Gauss–Markov conditions, is nevertheless very desirable, is that there should not be linear dependencies between the X variables. Linear dependencies within $\mathbf{X}$ mean that the rank of the $k \times k$ matrix is less than k and hence that $(\mathbf{X}'\mathbf{X})$ does not have an inverse. Estimation using the so-called generalized inverse is still possible, but is outside the scope of this chapter. A test of whether $(\mathbf{X}'\mathbf{X})$ has an inverse is provided by computing its determinant: if the determinant is zero, then no inverse exists. If the determinant is very close to zero, then the matrix is said to be *ill-conditioned*. This means that the model is not very clearcut and this has the effect that a small change in an X-value is likely to result in major changes to the $\{\hat{b}_j\}$-values.

***Example* 5.1**

This is one of several examples meant ultimately to raise the issues involved in the regression analysis of spatial data. However, the preliminary analysis described below follows the traditional lines. For this set of data, the Y variable under consideration is the percentage of the adult population of 26 Irish counties possessing blood group A. The source of the data is a large survey of some 55 000 people amounting to approximately 1 in 37 of the adult population of Eire, the results of which were published by Hackett and Dawson (1958). Example 3.4 concerns a closely related data set that refers to 1947 rather than to 1958. Sokal and Oden (1978b) discuss the present data from the perspective of univariate spatial autocorrelation analysis and they discover a very definitely organized pattern. The question under consideration here is whether or not the pattern is a simple autocorrelation phenomenon or whether the variation in Y reflects the change in value(s) of some, as yet unspecified, X variable(s).

There are allusions in the literature (e.g. Hooper, 1947; Sokal and Oden, 1978b) as to possible reasons for the highly organized nature of the blood group A distribution. Thus, Hackett and Dawson (1958) observed that ‘the regions of the Republic richest in group A correspond very well with the areas of most intense Anglo-Norman settlement’. A likely X variable that quantifies the intensity of Anglo-Norman settlement is provided by the number of place names per unit area that use the English suffix ‘town’. The choice of this variable is based on the fact that an identifiable ‘town’ zone exists and represents ‘that part of Ireland which experienced the most durable impact of Anglo-Norman colonisation from 1166 onward’ (Jones Hughes, 1970). The source of the data for this variable is the

Table 5.1 ***The blood group A distribution* (1958) *and other variables for* 26 *Irish counties***

County	Y	X_1	X_2	X_3	n	Weight
1 Carlow	34.22	0.1214	8	1	1 087	244.68
2 Cavan	29.68	0.0110	22	0	2 133	445.18
3 Clare	26.54	0.0122	21	0	535	104.31
4 Cork	23.92	0.0344	16	0	1 476	268.61
5 Donegal	27.91	0.0268	18	0	989	198.99
6 Dublin	32.79	0.6134	1	1	18 105	3990.01
7 Galway	26.53	0.0201	19	0	2 254	439.34
8 Kerry	25.14	0.0066	25	0	740	139.27
9 Kildare	30.69	0.3853	4	1	1 437	305.67
10 Kilkenny	31.41	0.1771	7	1	1 885	406.11
11 Laoghis	28.38	0.0392	15	0	378	76.83
12 Leitrim	28.84	0.0085	24	0	1 300	266.79
13 Limerick	28.24	0.0839	11	0	740	149.96
14 Longford	28.57	0.0496	14	1	868	177.14
15 Louth	31.02	0.4637	3	1	3 466	741.64
16 Mayo	27.01	0.0106	23	0	1 640	323.32
17 Meath	30.68	0.5858	2	1	1 307	277.96
18 Monaghan	28.24	0.0040	26	0	2 450	496.49
19 Offaly	30.18	0.0571	13	0	1 047	220.62
20 Roscommon	27.96	0.0326	17	0	1 023	206.06
21 Sligo	27.32	0.0188	20	0	743	147.53
22 Tipperary	30.24	0.0864	10	1	1 250	263.69
23 Waterford	31.04	0.0831	12	1	2 081	445.46
24 Westmeath	29.82	0.3627	5	1	1 405	294.03
25 Wexford	35.86	0.2753	6	1	2 108	484.85
26 Wicklow	35.47	0.0908	9	1	1 776	406.50

Key: $Y = \%$ of sample with blood group A; $X_1 =$ 'towns' per unit area;
$X_2 =$ county rank based on X_1; $X_3 =$ within Pale (1) or beyond (0);
$n =$ sample size; weight is used in the subsequent logit analysis.

N.B. County denotes county of birth.

map given by Jones Hughes (1970). Also given in Table 5.1 are the values of X_2, which is simply the ranked version of variable X_1, with the actual 'town' intensities replaced by their ranked values. The variable X_3 is a dummy variable which labels counties as either being within (1) or beyond (0) the 'Pale', which was the area of *de facto* political and administrative control by the colonists. The labelling here is that of Freeman (1969). The values of n that form the final column of Table 5.1 are merely the sample sizes upon which the Y-values were based. The variation in the sizes of the n-values reflects the corresponding variation in the county populations.

Figures 5.1a and b suggest that a linear model may be a more apt description of the link between X_2 and Y than between X_1 and Y, while it is undeniable that X_1 is much more liable to measurement error. It therefore seems preferable to restrict attention to the variables Y, X_2, and X_3, noting that X_2 and X_3 are strongly

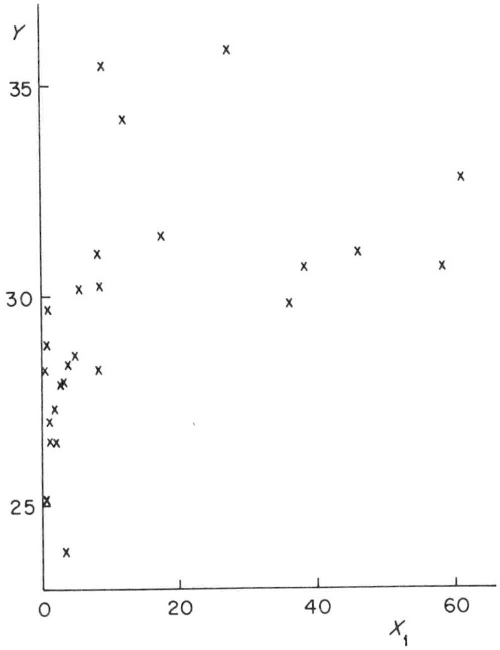

Figure 5.1a The relation between the percent of blood group A (Y) and the intensity of 'towns' (X_1)

correlated ($r = -0.833$), since they measure closely related aspects of the early English influence.

Equations (5.7a) and (5.7b), given below, are (simple linear) regression models linking Y to each in turn of the variables X_2 and X_3 of Table 5.1. These equations are simple versions of the general equation (5.1a). Equation (5.7a) ignores the influence of variable X_3 (i.e. sets b_3 equal to 0), while equation (5.7b) similarly ignores the influence of variable X_2. The numbers given in parentheses beneath each parameter is the estimated standard error of that estimate.

$$\hat{Y} = \underset{(0.8805)}{32.9977} - \underset{(0.0570)}{0.2571}\, X_2, \tag{5.7a}$$

$$\hat{Y} = \underset{(0.5285)}{27.5636} + \underset{(0.7779)}{4.2539}\, X_3. \tag{5.7b}$$

From these equations (ignoring any effect of the other X variable) we can state that a move from any one county to the next most 'anglicized' increases the expected percent of blood group A by about 0.25%. Similarly, if a randomly chosen person comes from within the 'Pale' rather than from beyond it, then the

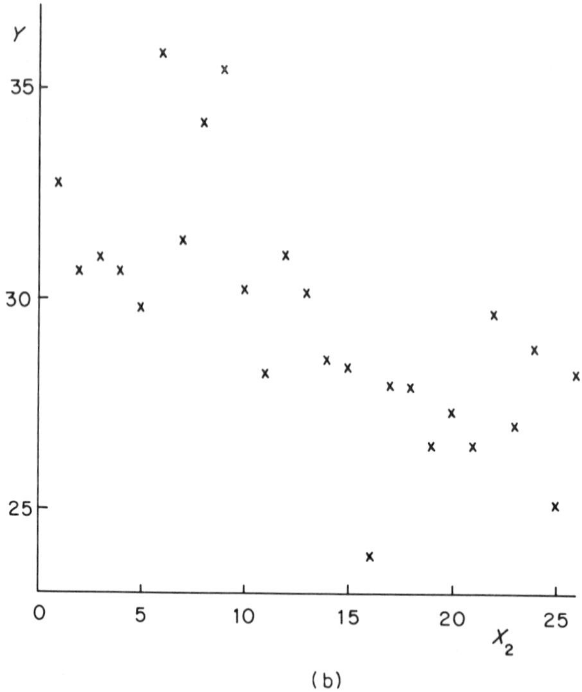

Figure 5.1b The relation between the percent of blood group A (Y) and the ranked intensity of 'towns' (X_2)

probability of that person having blood group A is increased by about 4.25%. The calculation of the standard errors of the b parameters is discussed in the next subsection.

Inference under Gauss–Markov conditions

Assuming, for the present, non-violation of the Gauss–Markov conditions, the OLS $\hat{\mathbf{b}}$ estimates possess optimal properties. The significance (or lack of significance) of the departure from zero of these estimates may be assessed in a straightforward fashion once the sampling distribution of the $\{\hat{b}_j\}$ has been determined. Now the matrix equation (5.4) demonstrates that each of the $\{\hat{b}_j\}$ is no more than a linear combination of the Y-values (as was indicated by the Gauss–Markov theorem) and consequently the distribution of the estimates $\{\hat{b}_j\}$ is dependent upon the distribution of the $\{Y_i\}$ or, equivalently, of the $\{e_i\}$.

The usual assumption that is made is that the $\{e_i\}$ are independent observations from a normal distribution with mean 0 and variance σ^2. The normality assumption derives support in many instances from the central limit theorem, which (very loosely) specifies that the normal distribution is the natural outcome of an infinite number of minor independent influences—which is precisely what the errors in our regression model are intended to represent.

A sensible precaution is always to test one's assumptions, if at all possible. In the present instance, having obtained the parameter estimates $\{\hat{b}_j\}$, we can calculate the Y-values, $\hat{\mathbf{Y}}$, that our model has predicted. The difference $\hat{\mathbf{e}} = \mathbf{Y} - \hat{\mathbf{Y}}$ is then an estimate of the unknown $\mathbf{e}$. If we expect that the $\{e_i\}$ have a joint normal distribution, then we should expect that the estimates $\{\hat{e}_i\}$ will also have an approximate normal distribution. Any marked departure from such a distribution will certainly point to some error in our original assumptions.

Straightforward matrix manipulation reveals that, when the conditions for OLS are satisfied, the variance–covariance matrix of the $\{\hat{b}_j\}$ is given by $\mathbf{V}(\hat{\mathbf{b}}) = (\mathbf{X}'\mathbf{X})^{-1}\sigma^2$. The variance of a particular parameter estimate, $\mathrm{var}(\hat{b}_j)$, will be the entry in cell (j,j) of this matrix, while the covariance of two parameter estimates $\hat{b}_j$ and $\hat{b}_k$ will be the entry in cell (j,k) of this matrix.

The variance, σ^2, of the $\{e_j\}$, and hence of the $\{Y_j\}$, is unknown, but an unbiased estimate is available in the form of the sum of the squared estimates of experimental error divided by the number of degrees of freedom available. Thus, $\hat{\sigma}^2 = \hat{\mathbf{e}}'\hat{\mathbf{e}}/(n-k)$. To test whether a particular estimate, $\hat{b}_j$, differs significantly from 0, it is therefore necessary to compute the value of

$$t = (\hat{b}_j - 0)/(\text{estimated } \mathrm{var}(\hat{b}_j))^{1/2}. \tag{5.8}$$

The observed value of t is then compared with the appropriate two-sided percentage point of a Student's t distribution having $(n-k)$ degrees of freedom.

For the particularly simple case in which there is a single independent variable, X, the estimated variances of $\hat{b}_0$ and $\hat{b}_1$ are given by

$$\text{estimated } \mathrm{var}(\hat{b}_0) = \hat{\sigma}^2 \cdot \sum x_i^2/nm(x^2), \tag{5.9a}$$

$$\text{estimated } \mathrm{var}(\hat{b}_1) = \hat{\sigma}^2/m(x^2), \tag{5.9b}$$

where

$$m(x^2) = \sum x_i^2 - (\sum x_i)^2/n. \tag{5.10}$$

Variance ratio tests

The sum of the squared estimated residuals, $D = \hat{\mathbf{e}}'\hat{\mathbf{e}}$, used to estimate the error variance, σ^2, and hence $\mathrm{var}(\hat{b}_j)$, is an important quantity in assessing the overall performance of a model and is sometimes referred to as the *deviance* of a model. Evidently the values of Y at the n locations will not usually all be equal: we can describe the extent of their variability by calculating the *total variation*, $T = \sum(Y_i - \bar{Y})^2$, where $\bar{Y}$ is the sample average value. Now, if T is large compared with D, then this implies that $(T-D)$ is large compared with D and hence that the major part of the variation in the values of Y has been explained by the regression model. The ratio $(T-D)/T$ is often denoted by r^2, and is sometimes called the *coefficient of determination.* The closer that r^2 is to 1, the better is the fit of the model.

A useful test statistic for assessing the importance of a group of parameters consists of calculating the average amount of variation in Y that is explained per parameter in this group, and comparing it with the ratio of the total amount of unexplained variation to its degrees of freedom $(n-k)$. Thus, in the present context, to test whether the model involving $(k-1)$ X variables accounts for a significantly large proportion of the total variation in Y, we compare the ratio of $(T-D)/(k-1)$ with $D/(n-k)$, i.e. we calculate the value of $(n-k)(T-D)/(k-1)D$. If it is the case that the regression model is of no assistance in explaining the variation in Y, then the amount of variation that appears to be explained by the model will be due to chance alone. Under these circumstances the value of the ratio (a so-called *variance ratio*) will be close to 1. On the other hand, if it is the case that the model genuinely provides an explanation of an appreciable part of the variation in Y, then the ratio will be noticeably greater than 1.

When we make the assumption of normality for the error variance, since all of D, $(T-D)$, and T are expressible as sums of squares, it follows immediately that the distribution of each can be a simple multiple of a chi-squared distribution under some circumstances. In particular, D/σ^2 can be shown to always have a χ^2_{n-k} distribution, while $(T-D)/\sigma^2$ has (independently) a χ^2_{k-1} distribution when Y is unrelated to the X variables (i.e. when the regression model is completely inappropriate) (cf. Bartholomew 1981). Since the suitably scaled ratio of two independent quantities having chi-squared distributions has a Fisher's F distribution, it follows that the ratio $(n-k)(T-D)/(k-1)D$ may be regarded as a potential observation from an $F_{k-1,n-k}$ distribution. Thus, a test of the importance, or otherwise, of the model as an explanation of the observed variation in Y is obtained by comparing the observed value of the ratio with the upper-tail significance points of the F distribution.

It should be noted that, if t has a Student's t distribution with, say, f degrees of freedom, then t^2 has a Fisher's $F_{1,f}$ distribution. Consequently, the t-test specified by equation (5.8) is equivalent to the comparison of the ratio $\hat{b}_j^2/\text{var}(\hat{b}_j)$ with an $F_{1,n-k}$ distribution.

***Example* 5.1** (cont.)

Applying the t-test specified by equation (5.8) to each of the simple linear regression relations given by equations (5.7a) and (5.7b), we obtain t-values of 4.51 (for $\hat{b}_2$) and 5.45 (for $\hat{b}_3$), both of which exceed 3.75, the two-sided 0.1% point of a t_{24} distribution. There is clear evidence of a significant linear relation between Y and either of X_2 or X_3.

In either case the value of T is, of course, the same. In fact $T = 210.8$. For the regression involving X_2 alone, the error sum of squares, D, is equal to 114.1, and so the value of the F statistic is $24 \times 96.7/1 \times 114.1 = 20.35$. Of course, since there is but one X variable in the model, the value 20.35 is the square of the t-value, 4.51, and is therefore equally as significant. The coefficient of determination, r^2, is equal to $96.7/210.8 = 0.459$. For the regression on X_3 alone, the error sum of squares, D, is equal to 93.8. Thus, in this case $r^2 = 0.555$.

***Example* 5.2**

In this example a multiple linear regression model is fitted to the data given in Table 5.2. These data relate to islands and to mainland localities along the Californian coast (see Figure 5.2.). Johnson *et al.* (1968) and Johnson and Raven (1970) provide an initial justification for this assemblage of variables, and they also include a further X variable (X_4, distance from coast) and two extra and much more extensive localities. One of these localities was the entire Californian coast (subsuming the other mainland areas) and the other was part of Lower

Table* 5.2 *The number of plant species and other characteristics of* 26 *Californian locations

Locality	Y	X_1	X_2	X_3
1 Cedros Island	205	134	3950	28.2
2 Guadalupe Island	163	98	4600	29.0
3 Santa Cruz Island	420	96	2470	34.0
4 Santa Rosa Island	340	84	1560	34.0
5 Santa Catalina Island	392	75	2125	33.3
6 San Clemente Island	235	56	1965	32.9
7 San Nicolas Island	120	22	910	33.2
8 San Miguel Island	190	14	830	34.0
9 Natividad Island	42	2.8	490	27.9
10 Santa Barbara Island	40	1.0	635	33.4
11 San Martin Island	62	0.9	470	30.5
12 San Geronimo Island	4	0.2	130	29.8
13 South Farallon Island	12	0.1	360	37.7
14 Ano Nuevo	40	0.02	60	37.1
15 San Benito Island	39	2.5	660	28.3
16 Anacapa	70	1.1	930	34.0
17 Los Coronadas	83	1.0	670	32.4
18 Todos Santos Island	72	0.5	315	31.8
19 San Diego County	1450	4260	6535	33.0
20 Monterey County	1400	3324	5860	36.2
21 Marin County	1060	529	2610	38.1
22 Santa Cruz Mountains	1200	1386	3810	37.3
23 Santa Monica Mountains	640	320	3110	34.1
24 Santa Barbara Area	680	110	3985	34.4
25 San Francisco County	640	45	930	37.8
26 Tiburon Peninsula	370	5.9	750	37.9

Key: Y = number of plant species; X_1 = area (square miles);
X_2 = maximum elevation (feet); X_3 = latitude (degrees North).

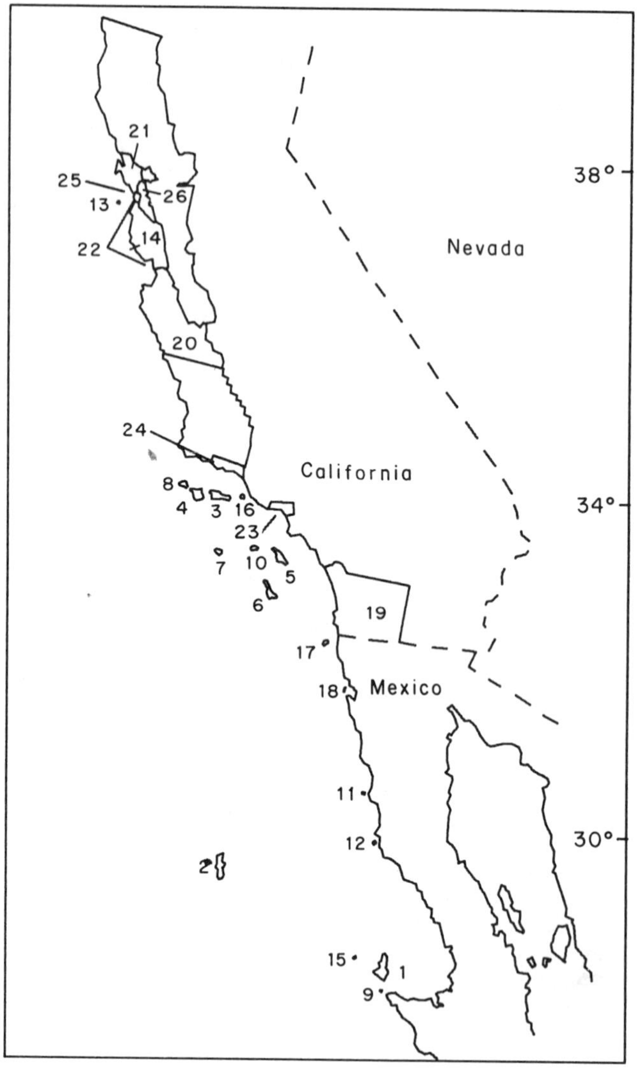

Figure 5.2 Californian coastal localities. (After Johnson and Raven, 1970)

California. Since there is reason to believe that the area–species link is non-linear once analysis is extended to areas of vast proportions, it is sensible to omit these extreme observations in the hope of preserving some semblance of linearity. Also, the allotting of location values to these areas cannot be done with any real confidence, as is required for later modelling attempts (e.g. see Example 5.5).

The OLS procedure, fitting a model of the form specified by equation (5.1), gives rise to the estimated relation given as equation (5.11):

$$\hat{Y} = -\underset{(370.0)}{1668.2} + \underset{(0.0502)}{0.1642}X_1 + \underset{(0.0281)}{0.1165}X_2 + \underset{(10.8296)}{52.5103}X_3. \tag{5.11}$$

The t-ratios (equation (5.8)) for these b parameters are 4.51, 3.28, 3.87, and 4.85, each of which is considerably in excess of 2.82, the two-tailed 1% point of a t distribution having $(26-4)=22$ degrees of freedom. Evidently, increased area, altitude, and latitude each increases the number of plant species present at a locality. The total variation in Y is given by $T = 4.870 \times 10^6$, whilst the error variation after fitting this model is $D = 0.606 \times 10^6$, so that the coefficient of determination for this model is $r^2 = 0.876$. The variance ratio for this model is $22 \times 4.264/3 \times 0.606 = 51.63$. This is obviously much greater than the value of 1 that would be expected if the model provided no explanation of the variation in Y, and indeed 51.63 is far in excess of the upper 0.1% point (7.80) of the relevant $F_{3,22}$ distribution.

The above inferences are all based on the proviso that the Gauss–Markov conditions hold good. As in Example 5.1, this may well not be the case and therefore caution should be exercised. Note that, in this example, we have chosen to include all the X variables in our model at the same time, whereas, in Example 5.1, each variable was treated in a separate model. Both approaches provide useful insights and both could be invoked for any given data set. For the present example it would also be possible to fit models including just two of the three X variables: we address this possibility in the next subsection.

Stepwise procedures

In most of the situations in which we will be attempting to explain the observed variation in Y by means of a linear model, we shall be unsure as to whether

(i) all the relevant X variables have been included in the model, and

(ii) all the X variables that have been included in the model are actually relevant!

In this subsection we consider the second of these two problems.

Suppose that we have a linear model involving just two X variables, X_1 and X_2. A natural extension of our previous notation is to write D_1 as the variation in Y that is left unexplained by the model that links Y to X_1 alone, with D_2 as the corresponding value for the model that links Y to X_2 alone, and D_{12} as the value obtained when both X variables are fitted. Since the model that includes both X variables must fit as least as well as the model that fits just one of the X variables, it follows that $D_{12} \leq D_1$ and $D_{12} \leq D_2$.

In the previous subsection we noted that, when Y is unrelated to the X variables, $(T-D)/\sigma^2$ has a χ^2_{k-1} distribution. We can now extend this result and state that both $(T-D_1)/\sigma^2$ and $(T-D_2)/\sigma^2$ have χ^2_1 distributions, while

$(T - D_{12})/\sigma^2$ has a χ_2^2 distribution. Furthermore, by subtraction, $(D_1 - D_{12})/\sigma^2$ and $(D_2 - D_{12})/\sigma^2$ also have χ_1^2 distributions under these conditions. These last results enable us to assess the significance of the extra explanation afforded by including further X variables in a model. To be precise, the value of the variance ratio $(D_1 - D_{12})/\{D_{12}/(n-k)\}$ should be compared with tables of the $F_{1,n-k}$ distribution, in order to obtain a direct test of whether there is a significant improvement in the explanation of the variation in Y that is provided by the model including both X variables, as opposed to the model that includes X_1 alone. Thus, the difference in the deviances of a pair of models provides a direct indication of the importance of the variable(s) that appear in one model and not in the other.

If a test of the conditional importance of an X variable suggests that the variable provides little extra information about the variation in Y, then clearly we can omit the variable from the model. Sometimes it will be the case that two X variables appear to be alternatives to one another.

There will often be many possible X variables that could be included in a model. The model that included all these variables might provide an excellent quantitative explanation of the variation in Y, whilst being virtually impossible to interpret. The analyst would wish to find some subset of these X variables which provided nearly as good a numerical explanation whilst having a relatively simple interpretation. To select such a model from the myriad of possible models is no easy task, but various computational algorithms have been produced. Many algorithms rely on stepwise selection in which X variables are added to the existing model, one at a time, until there are no variables remaining which make any significant improvement to the fit. This is known as *forward selection*. An alternative is to repeatedly delete X variables from a complex model whilst retaining an adequate explanation of the variation in Y. This is called *backward elimination*. Many hybrid schemes exist.

***Example* 5.1** (cont.)

Our earlier analyses of the Irish blood data provided alternative models that partially explained the observed variation in the proportions having blood group A. We envisaged that some readers would prefer the model that involved the quantitative variable, X_2, whereas others would prefer the descriptive sociological 'dummy' variable, X_3. It would, however, be perfectly feasible to include both X_2 and X_3 in our model, and we now explore that possibility.

Using our revised notation for the model deviances, our earlier results are summarized as $T = 210.8$, $D_2 = 114.1$, and $D_3 = 93.8$. When we fit the model of the joint regression of Y on both X_2 and X_3, we obtain $D_{23} = 91.6$. The extra variation in Y that is explained by X_2 (after X_3 has been included in the model) is the difference $D_3 - D_{23} = 93.8 - 91.6 = 2.2$. To test the significance of this extra contribution, we calculate the value of the variance ratio $23 \times 2.2/1 \times 91.6 = 0.55$.

This value is quite typical of the values to be expected from an $F_{1,23}$ distribution, and we therefore conclude that there is little point in including X_2 in the model in addition to X_3. It would appear from this initial analysis that, once it is known whether a county is within or beyond the 'Pale', there is little to be gained from the information about the intensity of 'towns' per unit area. We should perhaps add that, to the best of our knowledge, the 0/1 coding utilized by Freeman (1969) was not specifically derived from our own variable X_2!

This is not the final analysis of this data set, since the error structure for the observations is not really in accord with that demanded by the Gauss–Markov conditions. However, anticipating the subsequent results, we can confirm at this point that in all the later analyses X_3 continues to play the dominant role. We shall nevertheless continue to report both separate regressions for the reasons outlined above.

***Example* 5.2** (cont.)

Using our revised deviance notation, our earlier results amounted to $T = 4.870 \times 10^6$ and $D_{123} = 0.606 \times 10^6$. To these results we now add the deviance values for the six possible sub-models: $D_1 = 1.579$, $D_2 = 1.807$, $D_3 = 3.848$, $D_{12} = 1.253$, $D_{13} = 1.078$, and $D_{23} = 0.900$ (all $\times 10^6$). As an example of a conditional test for this rather complex situation, consider the assessment of the value of including X_1 in the model which already includes both X_2 and X_3. The appropriate variance ratio is $22 \times (0.900 - 0.606)/1 \times 0.606 = 10.66$, which greatly exceeds 7.95, the upper 1% point of an $F_{1,22}$ distribution. Thus, X_1 provides a significant extra contribution to the explanation of the variation in Y beyond that provided by X_2 and X_3.

In a similar fashion, we can assess the conditional importance of all the variables. The results suggest that, while X_1 and X_2 are to some extent alternatives to one another, the optimal model is that which includes all three variables. The elimination of any one of the variables from the model produces a significant worsening in the fit of the model. As with the previous example, this is not our final word, since a more precise analysis is available.

5.3 GENERALIZED LEAST SQUARES

An explicit assumption of the OLS procedure was that the variance–covariance matrix $\mathbf{V}$ was of the form $\sigma^2 \mathbf{I}$, where $\mathbf{I}$ is the identity matrix. We consider now the more general situation in which $\mathbf{V} = \sigma^2 \mathbf{D}$, where $\mathbf{D} \neq \mathbf{I}$. The remainder of this short section sets out the matrix manipulation that is required in order to introduce a transformation to new variables $\mathbf{Y}^*$ and $\mathbf{X}^*$ so that $\mathbf{Y}^* = \mathbf{X}^*\mathbf{b} + \mathbf{e}$ and $\mathbf{V}(\mathbf{Y}^*) = \sigma^2\mathbf{I}$—in other words a transformation that brings us once again to the straightforward OLS situation. The non-mathematical reader may wish to omit the next few lines of mathematics.

The transformation rests upon the fact that $\mathbf{D}$ must be a symmetric matrix with a positive determinant. A consequence is that there exists a matrix $\mathbf{P}$ (of eigenvectors) such that $\mathbf{P}'\mathbf{P} = \mathbf{D}$. We now make the transformation $\mathbf{X}^* = \mathbf{P}^{-1}\mathbf{X}$ and $\mathbf{Y}^* = \mathbf{P}^{-1}\mathbf{Y}$. Thus, $E(\mathbf{Y}^*) = \mathbf{X}^*\mathbf{b}$ and $\mathbf{e}^* = \mathbf{P}^{-1}\mathbf{e}$. Now $E(\mathbf{e}^*\mathbf{e}^{*\prime}) = E(\mathbf{P}^{-1}\mathbf{e}\mathbf{e}'(\mathbf{P}^{-1})') = \sigma^2\mathbf{P}^{-1}\mathbf{D}(\mathbf{P}^{-1})' = \sigma^2\mathbf{I}$. Hence, the GLS estimator of $\mathbf{b}$ is given by

$$\hat{\mathbf{b}} = (\mathbf{X}^{*\prime}\mathbf{X}^*)^{-1}\mathbf{X}^{*\prime}\mathbf{Y}^*. \tag{5.12}$$

If we substitute for $\mathbf{X}^*$ and $\mathbf{Y}^*$ into equation (5.12) we get the equivalent expression

$$\hat{\mathbf{b}} = (\mathbf{X}'\mathbf{D}^{-1}\mathbf{X})^{-1}\mathbf{X}'\mathbf{D}^{-1}\mathbf{Y}, \tag{5.13}$$

while the variance–covariance matrix for these estimators is

$$\mathbf{V}(\hat{\mathbf{b}}) = (\mathbf{X}^{*\prime}\mathbf{X}^*)^{-1}\sigma^2 = (\mathbf{X}'\mathbf{D}^{-1}\mathbf{X})^{-1}\sigma^2, \tag{5.14}$$

where σ^2 is estimated by

$$\hat{\sigma}^2 = \hat{\mathbf{e}}^{*\prime}\hat{\mathbf{e}}^*/(n-k) = \hat{\mathbf{e}}'\mathbf{D}^{-1}\hat{\mathbf{e}}/(n-k). \tag{5.15}$$

Weighted least squares

A common special case of generalized least squares occurs when the Y_i-values are independent, but are measured with unequal accuracy. Thus, var $(Y_i) = \sigma_i^2$, with the $\{\sigma_i\}$ not all being equal. Since the $\{Y_i\}$ are independent, the covariances are all 0 and consequently the variance–covariance matrix, $\mathbf{V}$, is easily invertible and the inverse has on-diagonal elements $\{1/\sigma_i^2\}$. Thus, those observations having a large error variance carry little weight in the estimation process and this rather sensible arrangement gives rise to the term weighted least squares (WLS).

There are many potential reasons for variations in the accuracy of the Y-values, but there is one particularly tractable and common situation. This is when Y is either a *proportion* or a multiple of a proportion and hence is an observation from a binomial distribution rather than a normal distribution. Suppose that Y has the restricted range $(0, n)$—in other words, Y/n is the proportion having some characteristic of interest—then, in this case, the variance of Y will be simply estimated as $Y(\mathrm{n} - Y)/n$.

***Example* 5.1** (cont.)

We now re-examine the blood data of Table 5.1 with binomial errors in mind. The value of Y_i given in the table was obtained by examining a sample of n_i people and classifying each person as being either of type A (n_{1i}) or of some other blood type ($n_{2i} = n_i - n_{1i}$). The value given for Y_i is equal to $100n_{1i}/n_i$ and therefore has an estimated variance of $10\,000n_{1i}n_{2i}/n_i = n_iY_i(100 - Y_i)$. With large values of n, such as are present in this example, it is reasonable to invoke the normal approximation to the binomial and so to utilize the standard inference procedures of OLS. However, rather than pursue this here, we look first at a somewhat different method of handling proportions.

Logit regression

A common transformation of a proportion p_i is to form the response variable $Z_i = \log(p_i/q_i)$, where $q_i = 1 - p_i$. One advantage of this so-called *logit* transformation is that the range $(-\infty, \infty)$ for Z_i corresponds to the range $(0, 1)$ for p_i and hence it is impossible to predict values of p_i outside the feasible range (for a full discussion, see Theil, 1969, 1970). An additional advantage is that, as n_i becomes large, the distribution of Z_i tends to the normal with variance $1/n_i p_i q_i$ (Cox, 1970). Reciprocals of variance estimates may be entered as weights in the main diagonal of the matrix **D** and the parameter estimates may be calculated using equations (5.13) and (5.14). Note that if p (or q) is close to 0 (or 1), as is (hopefully) often the case in mortality studies, then, writing $p_i = d_i/n_i$, the variance is given approximately by $1/d_i$ (cf. Pocock *et al.*, 1981).

Of course this transformation could be used with logarithms to the base 10 or any other base, but it is customary to use the so-called "natural" logarithms which use the base e. Thus, if $Z = \log\{p/(1-p)\}$, then exponentiating both sides and cross-multiplying, we have that $(1-p)\exp(Z) = p$. Consequently, p is given as $\exp(Z)/\{1 + \exp(Z)\}$. It is not, therefore, too difficult to transorm a logit model back into one which models the original Z variable that was of interest.

Example 5.1 (cont.)

We now use the logit transformation for the blood group example. The value of the variable Y was simply 100 times the proportion of the sample size, n, and therefore the logit transformation amounts to writing $z = \log\{Y/(100 - Y)\}$. The appropriate weights, which were given as the final column of Table 5.1, are the reciprocals of the variance estimates $10\,000/n_i Y_i(100 - Y_i)$. In this particular instance, since all the proportions are very similar in value, the weights will be roughly proportional to the sample sizes. Thus, by far the greater attention is paid to the Dublin figure since this is much the largest sample, while rather little attention will be paid to the Laoghis figure in view of its sparse population.

Equations (5.16a) and (5.16b) are the estimated logit regression lines using WLS and OLS, respectively. Notice that while the parameter estimates are more or less identical for the two lines (and, indeed, both the procedures give unbiased estimates), the WLS estimates are appreciably more precise as is evident from their smaller standard errors:

$$\hat{z}_i = \underset{(0.026)}{-0.706} - \underset{(0.0021)}{0.011}X_{2i}, \tag{5.16a}$$

$$\hat{z}_i = -0.707 - 0.012X_{2i}. \qquad (5.16b)$$
$$\phantom{\hat{z}_i = } (0.042) \quad (0.0027)$$

The corresponding WLS and OLS equations for the logit regression on X_3 are given by equations (5.16c) and (5.16d) and these broadly confirm the earlier remarks:

$$\hat{z}_i = -0.962 + 0.225X_{3i}, \qquad (5.16c)$$
$$\phantom{\hat{z}_i = } (0.028) \quad (0.034)$$

$$\hat{z}_i = -0.968 + 0.204X_{3i}. \qquad (5.16d)$$
$$\phantom{\hat{z}_i = } (0.025) \quad (0.037)$$

The interpretation of these logit models is, at first sight, somewhat complicated. As an example, consider equation (5.16c). The figure 0.225 indicates that the logarithm of the odds on an individual having blood group A (as opposed to some other group) is increased by 0.225 if the individual comes from within the 'Pale' as opposed to beyond the 'Pale'. If we take exponentials of both sides then, since $\exp(0.225) = 1.25$, we can state that the odds on an individual having blood group A are increased by 25% if that individual comes from within the 'Pale' as opposed to beyond the 'Pale'.

Since $\exp(-0.962) = 0.382$, using the result $p = \exp(Z)/\{1 + \exp(Z)\}$ we see that the model given as equation (5.16c) predicts that the percentage of blood group A for each of the counties beyond the 'Pale' will be $100 \times 0.382/1.382 = 27.65\%$, while, since $\exp(-0.962 + 0.225) = 0.479$, the corresponding estimated proportion for those counties within the 'Pale' is 32.37%. The model evidently provides a very useful summary of the observed data.

Intermediate weighted least squares

In the context of the spatial variation of death-rates, weighted least squares has been found to over-weight the larger towns. This is because a considerable proportion of the variation in Y not explained by the model is due to the effects of variables that have been omitted from the analysis. The studies by Gardner (1973) and Pocock *et al.* (1981) found that only between 15 and 20% of the residual variation in Y could be attributed directly to the binomial sampling.

Pocock *et al.* (1981) divide the variation in Y into three parts:

(i) error variation,
(ii) explained variation, and
(iii) unexplained variation.

The error variation is simply the binomial sampling variation, the explained variation is the variation in Y induced by the X variables considered in the model, and the unexplained variation is the variation in Y due to X variables not included in the model. Thus, the distinction between the error and unexplained variations is that the former is concerned simply with the repeatability of the Y-

values under repeat sampling with fixed X variables. Suppose that the true proportion (death-rate) at location i is p_i, and that the observed proportion is Y_i. Then $Y_i = p_i + e_i$, where e_i is an observation from a distribution (asymptotically normal) with variance $p_i(1 - p_i)/n_i$. The unconsidered X variables and those considered both play a part in causing p_i to vary from location to location. Using a linear model, we can write, in matrix form, that $\mathbf{p} = \mathbf{Xb} + \mathbf{r}$, where $\mathbf{r}$ is the unexplained portion of the variation in $\mathbf{p}$ and $\mathbf{Xb}$ is that portion due to the observed $\mathbf{X}$ variables.

Pocock *et al.* (1981) take the unexplained variations $\{r_i\}$ to be independent observations from a normally distributed random variable with mean zero and variance σ^2. Assuming e_i and r_i to be independent, and combining all three sources of variation into a single equation, we have

$$\mathbf{Y} = \mathbf{Xb} + \mathbf{u}, \tag{5.17}$$

where u_i is an observation from a normal distribution with mean zero and variance $\sigma^2 + p_i(1 - p_i)/n_i$. Provided that p_i is small, $(1 - p_i) \simeq 1$, so that the binomial variance is approximated by Y_i/n_i. The estimation of the σ^2 component is more problematical and Pocock *et al.* (1981) suggest a maximum likelihood procedure.

Generalized least squares with spatially autocorrelated errors

Apart from the variances along the main diagonal of the error covariance matrix $\mathbf{V}$ failing to be equal, another deviation from the Gauss–Markov conditions that rises to prominence with the analysis of spatial data is the presence of autocorrelated errors. This means that the off-diagonal cells of $\mathbf{V}$ contain non-zero values and, of course, this violates the conditions for the OLS procedure. However, provided the structure of the autocorrelation is known, analysis via generalized least squares provides no problem.

We assume, as usual, the regression relationship

$$\mathbf{Y} = \mathbf{Xb} + \mathbf{e}, \tag{5.18}$$

with the elements of $\mathbf{e}$ being correlated. One formulation of this correlation is to write

$$\mathbf{e} = \rho\mathbf{We} + \mathbf{u}, \tag{5.19}$$

which may be rewritten as

$$\mathbf{u} = (\mathbf{I} - \rho\mathbf{W})\mathbf{e} = \mathbf{Ae}. \tag{5.20}$$

We now have two 'error' vectors $\mathbf{e}$ and $\mathbf{u}$, although only the latter satisfies the Gauss–Markov conditions.

Equation (5.19) specifies the links between the errors e_i and e_k and hence the elements of the covariance matrix $\mathbf{V}$. If $\mathbf{e} \sim N(\mathbf{0}, \sigma^2\mathbf{D})$ then, from (5.20), we have

$\mathbf{D}^{-1} = (\mathbf{I} - \rho\mathbf{W}')(\mathbf{I} - \rho\mathbf{W}) = \mathbf{A}'\mathbf{A}$ and $\mathbf{V} = \sigma^2((\mathbf{I} - \rho\mathbf{W}')(\mathbf{I} - \rho\mathbf{W}))^{-1}$, which can be rewritten as either $\sigma^2(\mathbf{I} - \rho\mathbf{W})^{-1}(\mathbf{I} - \rho\mathbf{W}')^{-1}$ or $\sigma^2(\mathbf{I} - \rho\mathbf{W})^{-1}((\mathbf{I} - \rho\mathbf{W})')^{-1}$. This is the so-called simultaneous scheme for error autocorrelation (Ripley, 1981; Cliff and Ord, 1981a). An alternative to this is the conditional scheme in which $\mathbf{D}^{-1} = (\mathbf{I} - \rho\mathbf{W}) = \mathbf{A}$ so that $\mathbf{e} \sim N(\mathbf{0}, \sigma^2(\mathbf{I} - \rho\mathbf{W})^{-1})$ and $\mathbf{V} = \sigma^2(\mathbf{I} - \rho\mathbf{W})^{-1}$. Other specifications are also possible (Hepple, 1976) but the simultaneous scheme provides a useful introduction.

The matrix **W** which has appeared from equation (5.19) onward is an $n \times n$ proximity matrix indexing the (error) interactions between the n localities. Similar matrices play an important role in Chapter 3. The **W** matrix is operated on by ρ, the magnitude and sign of which determines the level and direction of error autocorrelation. Unlike the previous **W** matrices of Chapter 3, this **W** matrix is usually designed with row totals equal to 1. This has the effect of making **W** asymmetrical and usefully restrains ρ within the range -1 to 1 when it is estimated. Note, however, that for the purposes of generalized least squares the value of ρ is assumed to be known. If it is necessary to estimate ρ, then this can be accomplished via the method of maximum likelihood as described subsequently.

Spatial differencing

A way to avoid these relatively complex numerical estimation techniques is to assign the maximum possible value (i.e. 1) to ρ. Martin (1974), Ord (1975) (and Lebart, 1969, in a factor analysis context) give details of this ploy and references to parent procedures in the econometrics literature. This method requires one to believe that the errors are indeed related, as described by equation (5.19), and that the consequences of fixing $\rho = 1$ and then working with generalized least squares are that we obtain more desirable estimates than those obtained by ignoring any autocorrelation and fitting the regression equation using OLS.

Spatial differencing is essentially the transformation given between equations (5.12) and (5.13) applied with $\mathbf{P}^{-1} = (\mathbf{I} - \rho\mathbf{W})$ and $\rho = 1$. Hence, the transformation reduces to premultiplying **Y**, **X**, and **e** by $(\mathbf{I} - \mathbf{W})$ so that $\mathbf{Y}^* = \mathbf{Y} - \mathbf{WY}$ and $\mathbf{X}^* = \mathbf{X} - \mathbf{WX}$. This is tantamount to subtracting the weighted averages in surrounding localities from the value at each locality and is analogous to the differencing method frequently used to mitigate the influence of error autocorrelation in time series.

Martin (1974) carried out a Monte Carlo investigation of the comparative effects of spatial differencing and OLS in the presence of positive spatial autocorrelation. Not surprisingly, he found that spatial differencing provided the more accurate estimates of $\mathbf{V}(\mathbf{b})$. As the level of autocorrelation (ρ) approaches 1, the need for differencing becomes stronger, although the impact of error autocorrelation on the bias in $\mathbf{V}(\mathbf{b})$ is also conditional upon the extent of X variable autocorrelation. With $\rho > 0$, the bias produced by using the OLS estimator of $\mathbf{V}(\hat{\mathbf{b}})$ and σ^2 is downward, since this assumes $\mathbf{D}^{-1} = \mathbf{I}$ rather than $(\mathbf{I} - \rho\mathbf{W}')(\mathbf{I} - \rho\mathbf{W})$. Both these biases tend to over-emphasize the significance of departures of $\mathbf{b}$ from zero. For example, b_i divided by its deflated standard error inflates t, and the numerator of the F-ratio likewise becomes inflated. Furthermore, the conventional t and F tests are invalidated by the dependence among the errors, which are supposed to be independent normal variables, and t and F fail to adhere to the theoretical distributions under the null hypothesis.

This is not to claim, however, that spatial differencing totally eliminates bias—after all, setting $\rho = 1$ is sensible, but only arbitrary. Using differencing, the estimator of $\mathbf{V}(\hat{\mathbf{b}})$ is only actually unbiased if the true value of ρ is indeed equal to 1, in just the same way that the OLS estimator of $\mathbf{V}(\hat{\mathbf{b}})$ is only unbiased if the true value of ρ is 0. The estimator of $\mathbf{b}$ remains unbiased under either regime. As Martin (1974) observes, 'the problem, then, is to determine just how far ρ can vary from unity before the extent of the bias becomes intolerable and greater than that which would result from applying ordinary least squares'. The study by Martin suggests that differencing is to be preferred if $0 < \rho < 1$, when $\mathbf{V}(\hat{\mathbf{b}})$ is biased downwards. This deflates the apparent significance of any t tests concerning a b parameter, whereas the OLS procedure would lead to an inflated type I error rate. Haggett *et al.* (1977) note that $\rho < 0$ reverses the situation and that in this case OLS is certainly to be preferred. Of course, whilst both methods have the virtue of relative simplicity, neither actually produces estimates of ρ, and we now consider the possibility of obtaining such an estimate.

5.4 MAXIMUM LIKELIHOOD

This method involves the calculation of the joint likelihood of the observed $\{Y_i\}$-values given values assigned to certain parameters. The maximum likelihood estimates (MLEs) are those values that maximize L. This approach is an alternative to the least squares methodology and is equally appealing to one's common sense—given that the data has occurred it must surely be sensible to presume that the parameters take values that make that data likely. An essential prerequisite of the maximum likelihood procedure is the assumption of some stated family of probability distribution for the $\{Y_i\}$ or, equivalently, the $\{e_i\}$. As we shall see, the assumption of independence and a normal error distribution lead to parameter estimates identical to those produced by OLS.

The probability density function of the normal distribution is written generally as

$$f(x) = \frac{1}{6\sqrt{2\pi}} \exp\left\{\frac{-(x-\mu)^2}{2\sigma^2}\right\}, \tag{5.21}$$

where μ and σ are the mean and standard deviation of the distribution of X. Written in terms of the normally distributed error at locality i, e_i, we have

$$f(e_i) = \frac{1}{\sigma\sqrt{2\pi}} \exp\left(\frac{-e_i^2}{2\sigma^2}\right). \tag{5.22}$$

Considering the entire set of n localities, and assuming independent and identically distributed errors, the joint likelihood function is the product of the individual density functions:

$$L = \prod_{i=1}^{n} \frac{1}{\sigma\sqrt{2\pi}} \exp\left(\frac{-e_i^2}{2\sigma^2}\right), \tag{5.23}$$

or, in terms of matrix algebra,

$$L = \frac{1}{\sigma^n (2\pi)^{n/2}} \exp\left(\frac{-\mathbf{e}'\mathbf{e}}{2\sigma^2}\right) \tag{5.24}$$

It is usually convenient to work with the log of the likelihood function (the two maxima coincide), hence

$$\log L = -\frac{n}{2}\log(2\pi\sigma^2) - \frac{1}{2\sigma^2}(\mathbf{e}'\mathbf{e}), \tag{5.25}$$

and

$$\log L = \text{constant} - \frac{n}{2}\log\sigma^2 - \frac{1}{2\sigma^2}(\mathbf{e}'\mathbf{e}) \tag{5.26}$$

$$\log L = \text{constant} - \frac{n}{2}\log\sigma^2 - \frac{1}{2\sigma^2}(\mathbf{Y}'\mathbf{Y} - 2\mathbf{b}'\mathbf{X}'\mathbf{Y} + \mathbf{b}'\mathbf{X}'\mathbf{X}\mathbf{b}), \tag{5.27}$$

from the expansion of $\mathbf{e}'\mathbf{e}$ in equation (5.3). To maximize $\log L$, we need to differentiate the right-hand side of equation (5.27) and it will readily be seen that this is precisely equivalent to the algebra leading to equation (5.4).

Maximum likelihood estimators have desirable properties. They are very often unbiased and have a comparatively small variance. In large samples the distribution of the estimates is approximately normal and proofs abound in the literature of these properties (cf. Norden, 1972). Zellner (1983), Hepple (1976), and others have further thoughts on the appropriateness of the maximum likelihood method and reservations about the small sample properties of the estimators.

Maximum likelihood with spatially autocorrelated errors

A regression model with spatially autocorrelated errors was introduced earlier in equations (5.18)–(5.20) and was dealt with via generalized least squares. The maximum likelihood approach to fitting this model calls for a likelihood function equivalent to equation (5.24), although the problem now is that the errors $\{e_i\}$ are not independent, but are interrelated. It is assumed, from equation (5.20), that $\mathbf{u} = (\mathbf{I} - \rho\mathbf{W})\mathbf{e} = \mathbf{A}\mathbf{e}$, where $\mathbf{u} \sim N(\mathbf{O}, \sigma^2\mathbf{I})$, and so a standard result for the likelihood function written in terms of $\mathbf{e}$ is

$$L = |\mathbf{A}| \frac{1}{\sigma^n (2\pi)^{n/2}} \exp\left\{\frac{-(\mathbf{A}\mathbf{e})'\mathbf{A}\mathbf{e}}{2\sigma^2}\right\}, \tag{5.28}$$

as demonstrated by Hepple (1976), Doreian (1980), Cliff and Ord (1981a), and Ripley (1981).

Compared with equation (5.24) there is an extra term in equation (5.28), namely the determinant of $\mathbf{A}$, $|\mathbf{A}|$, which is sometimes referred to as the Jacobian of the transformation. This is introduced because of the transformation from $\mathbf{u}$ to $\mathbf{e}$ in equation (5.28). In fact $\mathbf{A}$ is a matrix of partial derivatives $\{\partial u_j/\partial e_i\}$ (Hoel, 1971, p. 368). On taking (natural) logarithms we obtain:

$$\log L = \text{constant} - \frac{n}{2}\log\sigma^2 - \frac{1}{2\sigma^2}(\mathbf{e}'\mathbf{A}'\mathbf{A}\mathbf{e}) + \log|\mathbf{A}|, \tag{5.29}$$

or, equivalently, on substituting for $\mathbf{e}$, we get:

$$\log L = \text{constant} - \frac{n}{2}\log\sigma^2 - \frac{1}{2\sigma^2}(\mathbf{Y}'\mathbf{A}'\mathbf{A}\mathbf{Y} - 2\mathbf{b}'\mathbf{X}'\mathbf{A}'\mathbf{A}\mathbf{Y} + \mathbf{b}'\mathbf{X}'\mathbf{A}'\mathbf{A}\mathbf{X}\mathbf{b}) + \log|\mathbf{A}|, \tag{5.30}$$

where the term in parentheses is the expansion of $\mathbf{u}'\mathbf{u}$. Note that there are three types of parameter embodied in equation (5.30): the error variance σ^2, the regression coefficients $\mathbf{b}$, and the autocorrelation coefficient ρ, although the latter is implicit.

Assuming ρ to be known, the MLEs of $\mathbf{b}$ and σ^2 could be obtained from setting to zero the partial derivatives of log L with respect to $\mathbf{b}$ and σ^2. From equation (5.30) these partial derivatives are:

$$\frac{\partial(\log L)}{\partial\mathbf{b}} = -\frac{1}{2\sigma^2}(-2\mathbf{X}'\mathbf{A}'\mathbf{A}\mathbf{Y} + 2\mathbf{X}'\mathbf{A}'\mathbf{A}\mathbf{X}\mathbf{b}), \tag{5.31}$$

$$\frac{\partial(\log L)}{\partial(\sigma^2)} = -\frac{n}{2\sigma^2} + \frac{1}{2\sigma^4}(\mathbf{u}'\mathbf{u}). \tag{5.32}$$

The estimating equations are therefore

$$\hat{\mathbf{b}} = (\mathbf{X}'\mathbf{A}'\mathbf{A}\mathbf{X})^{-1}\mathbf{X}'\mathbf{A}'\mathbf{A}\mathbf{Y}, \tag{5.33}$$

and, on expanding $\mathbf{u}'\mathbf{u}$ or $\mathbf{e}'\mathbf{A}'\mathbf{A}\mathbf{e}$,

$$\hat{\sigma}^2 = \frac{1}{n}(\mathbf{Y}'\mathbf{A}'\mathbf{A}\mathbf{Y} - 2\mathbf{b}'\mathbf{X}'\mathbf{A}'\mathbf{A}\mathbf{Y} + \mathbf{b}'\mathbf{X}'\mathbf{A}'\mathbf{A}\mathbf{X}\mathbf{b}). \tag{5.34}$$

Note that equation (5.33) is equivalent to equation (5.13) with $\mathbf{D}^{-1} = \mathbf{A}'\mathbf{A}$. Ripley (1981, p. 90) suggests a more general specification of $\mathbf{A}$, but we shall continue with the definition $\mathbf{A} = (\mathbf{I} - \rho\mathbf{W})$. To evaluate equations (5.33) and (5.34) we must therefore first obtain an estimate of ρ.

The estimation of ρ can be obtained as part of the general maximum likelihood process. The log-likelihood given by equation (5.30) involves ρ as part of the definition of the matrix **A**. Thus, an estimate of ρ is obtained by simultaneously maximizing log L with respect to ρ as well as with respect to **b** and σ^2. One possible scheme for achieving this end involves the substitution of equation (5.34) into equation (5.30). This reduces the latter to the simple equation

$$\log L = \text{constant} - \frac{n}{2}\log\hat{\sigma}^2 + \log|\mathbf{A}|. \tag{5.35}$$

We now maximize log L to obtain the value of $\hat{\rho}$ for given $\hat{\sigma}^2$ and $\hat{\mathbf{b}}$. Of course, an equivalent procedure is to minimize

$$M = \log\hat{\sigma}^2 - \frac{2}{n}\log|\mathbf{A}|. \tag{5.36}$$

Now $\hat{\sigma}^2 = (\mathbf{Y}'\mathbf{A}'\mathbf{PAY})/n$, where the matrix **P** is given by

$$\mathbf{P} = \mathbf{I} - (\mathbf{AX})\{(\mathbf{AX})'\mathbf{AX}\}^{-1}(\mathbf{AX})', \tag{5.37}$$

and thus $\hat{\rho}$ is the value that minimizes

$$M^* = \log(\mathbf{Y}'\mathbf{A}'\mathbf{PAY}) - \frac{2}{n}\log|\mathbf{A}|. \tag{5.38}$$

The treatment given above follows that set out by Doreian (1980). Further details are given by Cliff and Ord (1981a), Ord (1975), and Ripley (1981).

Computational problem

The most straightforward method of minimizing M^* is simply trial and error: M^* is evaluated for an intelligent choice of values of ρ within the bounds ± 1 which are imposed by the design of **W** (Doreian, 1980, 1981; Cliff and Ord, 1981a). As Hepple (1976) observes, the quantity $-2\{\log|\mathbf{A}|\}/n$ acts as a penalty or weighting function (in Whittle's, 1954, terms) so that M^* increases sharply as ρ takes values near to its bounds. Consequently, this second term cannot be ignored and it is not sufficient simply to minimize the first term of (5.36) using least squares (cf. Griffith, 1980, 1981; Ord, 1981).

We have found the GENSTAT package useful for performing the above minimization. One major computational difficulty is the repeated evaluation of $|\mathbf{A}|$, and this is one reason to prefer the simple error dependence structure of equations (5.17) and (5.18) giving $\mathbf{A} = \mathbf{I} - \rho\mathbf{W}$, an observation that has been made repeatedly (Ord, 1975; Hepple, 1976; Ripley, 1981). With small n the direct evaluation of $|\mathbf{A}|$ poses no problem for GENSTAT, although we retain this error structure for convenience with subsequent calculations. Ord (1975), Cliff and Ord (1973), and Hepple (1976) mention alternative iterative methods, but the method described is one that has been found to work well. Local minima were never encountered. A more radical alternative is to consider the entire likelihood function rather than simply its maximum. Ripley (1981) remarks that this is close in spirit to Hepple's (1979a) Bayesian approach.

Example 5.3

In this example the autocorrelated errors regression model (equations (5.18)–(5.20)) is fitted to the blood data of Table 5.1. The estimates are based on the **W** matrix published by Ord (1975) and Cliff and Ord (1973, 1981a) which was originally applied to another Irish county data set. Table 5.3 reproduces their matrix of proximity values **W** with typical cell frequency $w_{ij} = w_{ij}^*/\sum_j w_{ij}^*$, where $w_{ij}^* = d_{ij}^{-1} b_{ij}$, b_{ij} is the proportion of county i's total boundary length with

Table* 5.3 *Proximity values applied to the Irish blood data

County (i)	W_{ij} (j in parentheses)				
1	(9) 0.0874;	(10) 0.3207;	(11) 0.0699;	(25) 0.2540;	(26) 0.2680
2	(12) 0.2690;	(14) 0.1658;	(17) 0.2426;	(18) 0.3073;	(24) 0.0153
3	(7) 0.4808;	(8) 0.0617;	(13) 0.3590;	(22) 0.0985	
4	(8) 0.5056;	(13) 0.3411;	(22) 0.0327;	(23) 0.1206	
5	(12) 1.0000				
6	(9) 0.2226;	(17) 0.5442;	(26) 0.2332		
7	(3) 0.3392;	(16) 0.2639;	(19) 0.0394;	(20) 0.3076;	(22) 0.0499
8	(3) 0.0866;	(4) 0.7218;	(13) 0.1916		
9	(1) 0.0820;	(6) 0.1123;	(11) 0.1839;	(17) 0.1760;	(19) 0.1761;
	(26) 0.2697				
10	(1) 0.2919;	(11) 0.2312;	(22) 0.2259;	(23) 0.1345;	(25) 0.1166
11	(1) 0.0610;	(9) 0.1808;	(10) 0.2217;	(19) 0.4558;	(22) 0.0807
12	(2) 0.3057;	(5) 0.0229;	(14) 0.1604;	(20) 0.2277;	(21) 0.2833
13	(3) 0.2827;	(4) 0.2732;	(8) 0.1075;	(22) 0.3365	
14	(2) 0.1628;	(12) 0.1387;	(20) 0.2169;	(24) 0.4816	
15	(17) 0.7080;	(18) 0.2920			
16	(7) 0.3881;	(20) 0.2000;	(21) 0.4119		
17	(2) 0.2109;	(6) 0.1604;	(9) 0.1465;	(15) 0.2239;	(18) 0.0167;
	(19) 0.0161;	(24) 0.2254			
18	(2) 0.7075;	(15) 0.2416;	(17) 0.0508		
19	(7) 0.0348;	(9) 0.1245;	(11) 0.3742;	(17) 0.0177;	(20) 0.0289;
	(22) 0.1748;	(24) 0.2451			
20	(7) 0.2871;	(12) 0.1586;	(14) 0.1748;	(16) 0.1269;	(19) 0.0306;
	(21) 0.1749;	(24) 0.0471			
21	(12) 0.3114;	(16) 0.4126;	(20) 0.2760		
22	(3) 0.0602;	(4) 0.0203;	(7) 0.0432;	(10) 0.1743;	(11) 0.0649;
	(13) 0.2611;	(19) 0.1713;	(23) 0.2045		
23	(4) 0.1797;	(10) 0.2488;	(22) 0.4901;	(25) 0.0814	
24	(2) 0.0134;	(14) 0.4288;	(17) 0.2196;	(19) 0.2861;	(20) 0.0521
25	(1) 0.4254;	(10) 0.2146;	(23) 0.0810;	(26) 0.2791	
26	(1) 0.3036;	(6) 0.1985;	(9) 0.3099;	(25) 0.1880	

Table 5.4 Estimates as a result of fitting the autocorrelated errors model to the Irish blood data

Fit using $Y = \%$ blood group A:		
$\hat{Y} = 30.6480 - 0.0750X_2$	$\hat{\rho} = 0.85$	$\hat{\sigma}^2 = 1.8706$
(1.9493) (0.0548)	(0.0831)	(0.5493)
$\hat{Y} = 28.9380 + 1.3903X_3$	$\hat{\rho} = 0.83$	$\hat{\sigma}^2 = 1.8810$
(1.6323) (0.8609)	(0.0917)	(0.5523)
Fit using $Z = \log\{Y/(100 - Y)\}$:		
$\hat{Z} = -0.8194 - 0.0038X_2$	$\hat{\rho} = 0.84$	$\hat{\sigma}^2 = 0.0044$
(0.0894) (0.0027)	(0.0877)	(0.0013)
$\hat{Z} = -0.9061 + 0.0722X_3$	$\hat{\rho} = 0.81$	$\hat{\sigma}^2 = 0.0045$
(0.0714) (0.0415)	(0.0995)	(0.0013)

contiguous counties that is shared with county j, and d is the distance between the centres of counties i and j.

Various versions of the autocorrelated errors model have been fitted to the data, using as the dependent variable either Y, or the logit, $Z = \log\{Y/(100 - Y)\}$. In fitting these models we are implicitly assuming that the maximum likelihood approach is robust to such minor departures from normality as might be experienced with large samples. Following our previous reasoning, we relate Y(or Z) separately to each of X_2 (the 'town' ranking) and X_3 (the 0/1 'Pale' dummy). The results are summarized in Table 5.4.

To illustrate the estimation procedure, values are given in Table 5.5 of the quantities $|\mathbf{A}|$, σ^2, and M^* deriving from equations (5.20), (5.37), and (5.38) on incrementing ρ from -0.9 to 0.9 in steps of 0.2, taking the model relating Z to X_2 as an example.

Table 5.5 indicates that the minimum value of M^* occurs in the vicinity of

Table 5.5 Error autocorrelation and related quantities for the logit model applied to the blood group data using X_2 as the sole explanatory variable

ρ	$\|A\|$	σ^2	M^*
-0.9	0.0373	0.0208	-0.3603
-0.7	0.1588	0.0178	-0.6276
-0.5	0.4068	0.0152	-0.8595
-0.3	0.7275	0.0129	-1.0689
-0.1	0.9652	0.0109	-1.2601
0.1	0.9643	0.0092	-1.4329
0.3	0.7100	0.0077	-1.5878
0.5	0.3577	0.0063	-1.7255
0.7	0.1017	0.0051	-1.8379
0.9	0.0080	0.0041	-1.8611

$\hat{\rho}=0.9$. The more precise estimate $\hat{\rho}=0.84$ is the result of incrementing first by 0.1 and subsequently by 0.01 in the region of the minimum. Figure 5.3 shows the results.

Figure 5.3 suggests that the standard error of $\hat{\rho}$, which would be needed for a complete interpretation of the results, is relatively small. There is therefore every reason to believe that there is significant positive error autocorrelation and that the autocorrelated errors model is therefore appropriate. We shall consider the sampling distribution and standard error of $\hat{\rho}$ later in the chapter.

Diggle (1982) mentions that it is best if some process causing error autocorrelation can be identified, and cites as an example Freeman's (1953) study of contagious disease on a lattice of hop plants. In fact, after some mild matrix manipulation, it is possible to write out the autocorrelated errors model as a single equation, and this approach will be considered below (equation (5.50)).

Of course, only residuals and not true errors are observed. These residuals comprise both independent random error components and a component

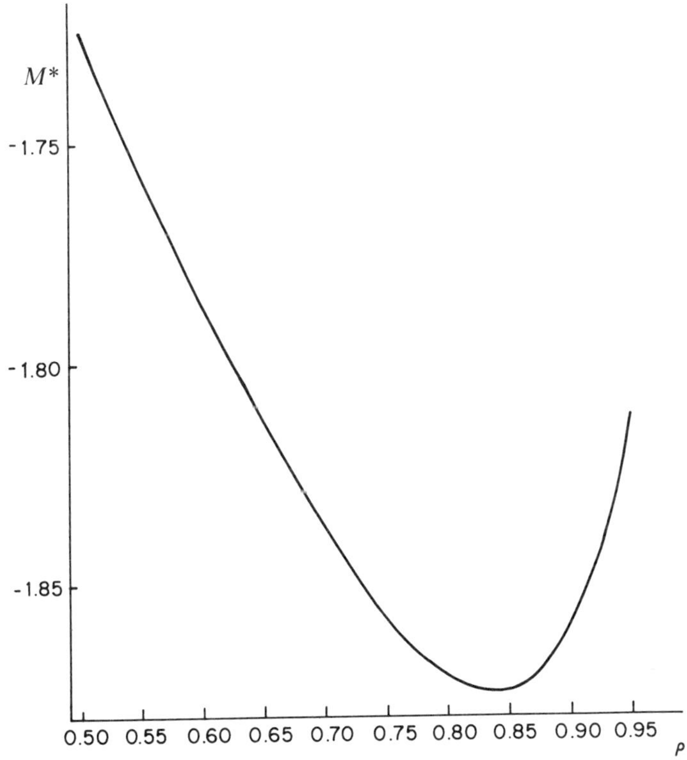

Figure 5.3 The relation between M^* and ρ for the logit model (using X_2 only) applied to the blood group data

attributable to the (spatially autocorrelated) latent X variables. Another contribution to the observed residuals emanates from a misspecification of the functional link connecting Y to the X variables—but this is mathematically equivalent to the effect of latent variables. In this case the autocorrelated errors model will continue to handle the genuinely autocorrelated part of the residuals in a reasonable fashion, but cannot, of course, compensate for the incorrect model specification.

***Example* 5.4**

This example is a reanalysis, using the autocorrelated errors model, of the Californian plant species data of Table 5.2. As with the previous example, normal errors are presumed in concert with the original least squares analysis. The motivation for fitting the model comes from the spatial component noted in earlier analyses (Johnson and Raven 1970).

Before we can commence the present analysis we must have a **W** matrix, which should be chosen to reflect our preconceived ideas of the spatial linkages present in the data. As ever, the choice of **W** is essentially arbitrary and we now consider two alternative **W** matrices, which differ in their cognizance of island locations.

The first **W** matrix that we consider might be termed a 'two-part distance-based' weights matrix:

$$\begin{aligned} w_{ij}^* &= d_{ij}^{-2}, \quad && d_{ij} \leq k, \\ w_{ij}^* &= 0, && d_{ij} > k, \end{aligned} \qquad w_{ij} = w_{ij}^* / \sum_j w_{ij}^*.$$

This assumes that the weight that the error value at i carries with respect to the value at j decreases non-linearly with increasing interlocality (mid-point) distance d_{ij} up to a cut-off distance k (taken to be 187.5 miles), beyond which the autocorrelation mechanism is taken to be inoperative. The weights that result are given in part (a) of Table 5.6.

The second **W** matrix, which might be termed an 'island isolation' matrix, is given by:

$$\begin{aligned} w_{ij}^* &= d, \quad && d_{ij} \leq k \\ w_{ij}^* &= 0, && d_{ij} > k, \text{ or if either locality } i \text{ or} \\ & && \text{locality } j \text{ is an island.} \end{aligned} \qquad w_{ij} = w_{ij}^* / \sum_j w_{ij}^*$$

This matrix is based straightforwardly upon the presumption that the physical isolation of the islands renders inoperative the autocorrelation mechanism that operates among interconnected mainland localities (see Figure 5.2).

The results of fitting these autocorrelated errors models are summarized in Table 5.7, while Figures 5.4a and b demonstrate the effect of changes in ρ upon

Table* 5.6 *Proximity values assigned to Californian localities

(a) Two-part distance based weights

Location (i)	W_{ij} (j in parentheses)
1	(2) 0.0067; (9) 0.6663; (11) 0.0088; (12) 0.0185; (15) 0.2998
2	(1) 0.1928; (11) 0.2672; (12) 0.2740; (15) 0.2660
3	(4) 0.2200; (5) 0.0217; (6) 0.0122; (7) 0.0435; (8) 0.0880; (10) 0.0396; (16) 0.3911; (17) 0.0051; (19) 0.0041; (20) 0.0056; (23) 0.0338; (24) 0.1354
4	(3) 0.2406; (5) 0.0154; (6) 0.0104; (7) 0.0397; (8) 0.4813; (10) 0.0266; (16) 0.0786; (17) 0.0046; (20) 0.0066; (23) 0.0193; (24) 0.0770;
5	(3) 0.0400; (4) 0.0259; (6) 0.1753; (7) 0.0801; (8) 0.0187; (10) 0.3814; (16) 0.0554; (17) 0.0336; (18) 0.0124; (19) 0.0224; (23) 0.1297; (24) 0.0249
6	(3) 0.0379; (4) 0.0297; (5) 0.2960; (7) 0.1095; (8) 0.0226; (10) 0.1889; (16) 0.0438; (17) 0.1095; (18) 0.0322; (19) 0.0421; (23) 0.0633; (24) 0.0244
7	(3) 0.1046; (4) 0.0873; (5) 0.1046; (6) 0.0847; (8) 0.0540; (10) 0.3258; (16) 0.0941; (17) 0.0212; (18) 0.0106; (19) 0.0132; (23) 0.0569; (24) 0.0430
8	(3) 0.1246; (4) 0.6232; (5) 0.0144; (6) 0.0103; (7) 0.0318; (10) 0.0226; (16) 0.0587; (20) 0.0103; (23) 0.0183; (24) 0.0860;
9	(1) 0.7997; (11) 0.0085; (12) 0.0163; (15) 0.1755
10	(3) 0.0621; (4) 0.0381; (5) 0.3252; (6) 0.0953; (7) 0.2126; (8) 0.0250; (16) 0.0813; (17) 0.0191; (18) 0.0083; (19) 0.0125; (23) 0.0906; (24) 0.0299
11	(1) 0.0522; (2) 0.0550; (9) 0.0420; (12) 0.5254; (15) 0.0619; (17) 0.0618; (18) 0.1527; (19) 0.0489
12	(1) 0.1131; (2) 0.0580; (9) 0.0832; (11) 0.5406; (15) 0.1387; (18) 0.0664
13	(14) 0.0790; (20) 0.0104; (21) 0.1579; (22) 0.1117; (25) 0.4579; (26) 0.1832
14	(13) 0.0676; (20) 0.0218; (21) 0.0270; (22) 0.7836; (25) 0.0612 (26) 0.0388
15	(1) 0.6140; (2) 0.0188; (9) 0.2995; (11) 0.0213; (12) 0.0463
16	(3) 0.4640; (4) 0.0852; (5) 0.0357; (6) 0.0167; (7) 0.0464; (8) 0.0491; (10) 0.0614; (17) 0.0068; (19) 0.0057; (20) 0.0062, (23) 0.0788; (24) 0.1440
17	(3) 0.0257; (4) 0.0213; (5) 0.0929; (6) 0.1794; (7) 0.0448; (10) 0.0619; (11) 0.0310; (16) 0.0294; (18) 0.2242; (19) 0.2242 (23) 0.0452; (24) 0.0199
18	(5) 0.0570; (6) 0.0874; (7) 0.0372; (10) 0.0446; (11) 0.1270; (12) 0.0536; (17) 0.3715; (19) 0.1858; (23) 0.0359
19	(3) 0.0320; (5) 0.0957; (6) 0.1064; (7) 0.0432; (10) 0.0626; (11) 0.0379; (16) 0.0379; (17) 0.3458; (18) 0.1729; (23) 0.0657
20	(3) 0.0563; (4) 0.0601; (8) 0.0725; (13) 0.0795; (14) 0.1953; (16) 0.0520; (21) 0.0563; (22) 0.1591; (23) 0.0364; (24) 0.0827; (25) 0.0806; (26) 0.0691

Table 5.6 (cont.)

(a) Two-part distance based weights

Location (i)	W_{ij} (j in parentheses)
21	(13) 0.1459; (14) 0.0292; (20) 0.0068; (22) 0.0407; (25) 0.2488; (26) 0.5287
22	(13) 0.0867; (14) 0.7113; (20) 0.0161; (21) 0.0342; (25) 0.0961; (26) 0.0556
23	(3) 0.0953; (4) 0.0496; (5) 0.1983; (6) 0.0573; (7) 0.0665; (8) 0.0364; (10) 0.1624; (16) 0.1870; (17) 0.0250; (18) 0.0120; (19) 0.0235; (20) 0.0103; (24) 0.0763
24	(3) 0.2789; (4) 0.1450; (5) 0.0279; (6) 0.0162; (7) 0.0368; (8) 0.1250; (10) 0.0392; (16) 0.2501; (17) 0.0080; (20) 0.0171; (23) 0.0558
25	(13) 0.2477; (14) 0.0387; (20) 0.0057; (21) 0.1457; (22) 0.0669; (26) 0.4953
26	(13) 0.1019; (14) 0.0252; (20) 0.0050; (21) 0.3185; (22) 0.0398; (25) 0.5096

(b) Island isolation weights

Location (i)	W_{ij} (j in parentheses)
19	(23) 1.0000
20	(21) 0.1162; (22) 0.3285; (23) 0.0752; (24) 0.1708; (25) 0.1665; (26) 0.1427
21	(20) 0.0082; (22) 0.0493; (25) 0.3016; (26) 0.6409
22	(20) 0.0797; (21) 0.1693; (25) 0.4759; (26) 0.2751
23	(19) 0.2139; (20) 0.0933; (24) 0.6928
24	(20) 0.2342; (23) 0.7658
25	(20) 0.0080; (21) 0.2041; (22) 0.0938; (26) 0.6941
26	(20) 0.0057; (21) 0.3649; (22) 0.0456; (25) 0.5838

Table* 5.7 *Parameter estimates and estimated standard errors of the autocorrelated errors model fitted to the Californian data

(a) Using two-part distance-based **W** matrix of Table 5.6a

Parameter	Estimate	Standard error	Estimate/Standard error
b_0	−1618.9	173.5	
b_1	0.1628	0.0361	4.51
b_2	0.1040	0.0224	4.64
b_3	51.8470	4.8765	10.63
ρ	−0.72	0.2200	−3.27
σ^2	0.0153	0.0045	

(b) Using island-isolation **W** matrix of Table 5.6b

Parameter	Estimate	Standard error	Estimate/Standard error
b_0	−860.9	382.4	
b_1	0.1479	0.0374	3.95
b_2	0.1017	0.0247	4.12
b_3	27.0610	11.4761	2.36
ρ	0.75	0.1342	5.59
σ^2	0.0182	0.0052	

the value of the likelihood. It is immediately apparent from the two figures that, while ρ is estimated as having approximately the same magnitude for each model, the effect of changing the distance based **W** matrix to the island isolation **W** matrix is to reverse the sign of $\hat{\rho}$.

Obviously large positive and large negative values of $\hat{\rho}$ would appear to have entirely different implications for the autocorrelation structure. It is a salutory experience to encounter two almost equally good 'explanations' of a set of data with such apparently diametrically opposed properties. In fact, the explained proportion of the variation in Y, r^{*2} (see equation (5.39)), for the distance-based model is 0.918, whilst that for the island isolation model is 0.903.

In the present context a genuine negative value for ρ seems scarcely credible, since it would imply that areas close together were particularly likely to have unequal numbers of species. For mainland areas this could only be true if we went out of our way to select extraordinarily different types of neighbouring ecosystems. On the other hand, it could well be true for the adjacent, but separated, islands.

We conclude therefore that the two models are really both in agreement! Island locations are liable to be very different in their characteristics, even when being physically close together, while even non-contiguous but land-connected mainland areas would display considerable similarity. Thus, the negative value of $\hat{\rho}$ for the first **W** matrix may be regarded as a signal that the islands should be treated in a distinctive fashion.

We shall discuss the significance of the various parameter estimates later in this chapter.

Example 5.5

The data in Table 5.8, which relates to the flora of the Galapagos Islands, was initially analysed, using OLS, by Johnson and Raven (1973), and an equivalent analysis of similar data has been conducted by Hamilton *et al.* (1963). We now carry out an analysis which incorporates the autocorrelated errors. There are five X variables, of which two, namely island area (X_1) and elevation (X_2), are direct measures of each island's environmental homogeneity and hence of the variety of

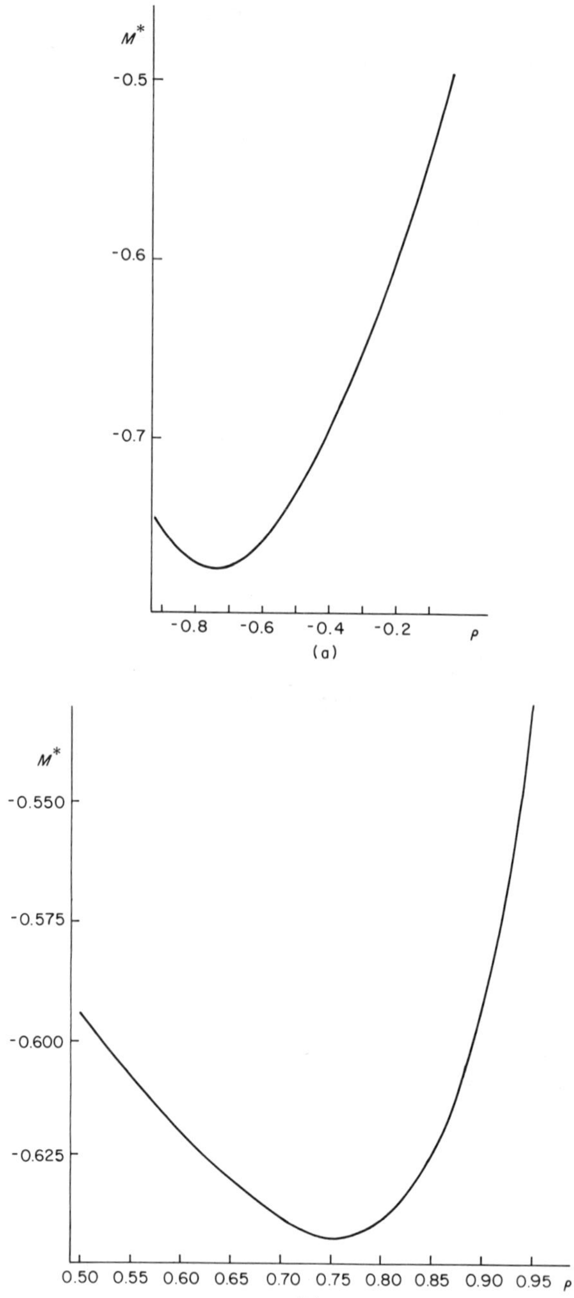

Figure 5.4 The relation between M^* and ρ for the Californian data. (a) Using the distance weights of Table 5.6a. (b) Using the isolated island weights of Table 5.6b

Table 5.8 Data relating to the variation in the numbers of plant species on various of the Galapagos Islands

Island	Y	X_1	X_2	X_3	X_4	X_5
1 Bartolome	31	1.24	109	0.6	26.3	572.33
2 Caldwell	3	0.21	114	2.8	58.7	0.78
3 Champion	25	0.10	46	1.9	47.4	0.18
4 Darwin	10	2.33	168	34.1	290.2	2.85
5 Enderby	2	0.18	112	2.6	50.2	0.10
6 Espanola	97	58.27	198	1.1	88.3	0.51
7 Fernandina	93	634.49	1494	4.3	95.3	4669.32
8 Gardner (1)	58	0.57	49	1.1	93.1	58.27
9 Gardner (2)	5	0.78	227	4.6	62.2	0.21
10 Genovesa	40	17.35	76	47.4	92.2	129.49
11 Isabela	347	4669.32	1707	0.7	28.1	634.49
12 Marchena	51	129.49	343	29.1	85.9	59.56
13 Onslow	2	0.01	25	3.3	45.9	0.10
14 Pinta	104	59.56	777	29.1	119.6	129.49
15 Pinzon	108	17.95	458	10.7	10.7	0.03
16 Rabida	70	4.89	367	4.4	24.4	572.33
17 San Cristobal	280	551.62	716	45.2	66.6	0.57
18 San Salvador	237	572.33	906	0.2	19.8	4.89
19 Santa Cruz	444	903.82	864	0.6	0.0	0.52
20 Santa Fe	62	24.08	259	16.5	16.5	0.52
21 Santa Maria	285	170.92	640	2.6	49.2	0.10
22 Tortuga	16	1.24	186	6.8	50.9	17.95
23 Wolf	21	2.85	253	34.1	254.7	2.33

Key: Y = number of plant species; X_1 = island area (km^2);
X_2 = elevation (metres); X_3 = distance from nearest island (km);
X_4 = distance from Santa Cruz; X_5 = area of adjacent island (km^2).

Notes: 1. Information on flora was abstracted from Wiggins and Porter (1971).
2. Johnson and Raven (1972) give incomplete data for a further six islands.
3. Gardner (1) is near Espanola, Gardner (2) is near Santa Maria.

ecological niches available for occupation by different plant species. The three remaining variables, distance to the nearest island (X_3), distance from the centre of the archipelago (X_4), and area of the adjacent island (X_5), indicate in various ways the relative position of each island in the island group. The locations of the islands are shown in Figure 5.5.

Of course, with data of this type, any variables that we select must constitute a somewhat arbitrary selection from those that appear relevant. Some observers might prefer to replace X_3 by distance to the nearest *major* island, or to replace X_5 by, for example, the proportion of the neighbouring region, within some specified range, which is covered by land.

The present selection of variables does, however, provide a basis for future model construction. Our present ambition is limited to an assessment of the

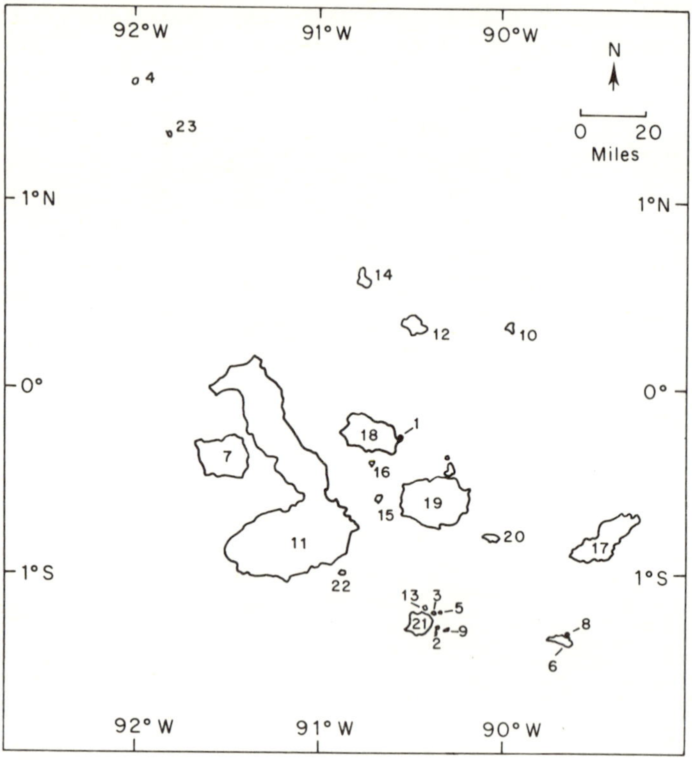

Figure 5.5 The Galapagos Islands. (After Johnson and Raven, 1973)

relative informativeness of the five variables chosen. We give, in Table 5.9a, the parameter estimates and standard errors that result from fitting the OLS multiple (linear) regression model and the corresponding values for the autocorrelated errors model. The latter values are based on a **W** matrix with elements $w_{ij} = w^*_{ij}/\sum_j w^*_{ij}$, where $w^*_{ij} = d_{ij}^{-2}$ and d_{ij} is the inter-island distance.

The results summarized in Table 5.9a show that, because the value of ρ is estimated as being close to 0, the two methods give very similar parameter estimates. It might be thought that the near zero value for $\hat{\rho}$ was due to the presence of so many locational X variables. To answer this point, these variables have been omitted from the simpler model reported in Table 5.9b. This model also omits the variable X_1, which, from Table 5.9a, appears to be of little significance. We see that, while the estimate of b_2 is considerably altered in this new model, the estimated value of ρ remains near zero. This is suggestive of independent inter-island errors similar to those implied by the island isolation **W** matrix of the previous example. Note that latitude, the inclusion of which would have made the models of Table 5.9b and Table 5.7 very similar, does not appear

Table 5.9 Parameter estimates of models fitted to the Galapagos data

(a) All five X variables

	Multiple regression (OLS)			Autocorrelated errors (ML)		
Parameter	Estimate	S.E.	Est/S.E.	Estimate	S.E.	Est/S.E.
b_0	17.98	29.90		18.01	26.13	
b_1	−0.0270	0.0263	−1.03	−0.0270	0.0224	−1.21
b_2	0.3221	0.0677	4.76	0.3223	0.0583	5.53
b_3	−0.0938	1.2110	−0.08	−0.0768	1.0462	−0.07
b_4	−0.2983	0.2662	−1.12	−0.3031	0.2322	−1.31
b_5	−0.0782	0.0216	−3.62	−0.0780	0.0173	−4.51
ρ				0.03	0.2893	0.10
σ^2	0.4784			0.3534	0.1044	

(b) Only X_2

	Multiple regression (OLS)			Autocorrelated errors (ML)		
Parameter	Estimate	S.E.	Est/S.E.	Estimate	S.E.	Est/S.E.
b_0	18.74	26.32		18.44	25.50	
b_2	0.1942	0.0420	4.62	0.1947	0.0400	4.87
ρ				0.02	0.2902	0.07
σ^2	0.8117			0.7410	0.2184	

in our present model simply because it is essentially constant over the area occupied by the Galapagos Islands.

We discuss the relative importance of the X variables subsequently, first justifying the standard error values that have been quoted in the tables. Another issue, which we address subsequently, is the question of whether we can improve on our assumption of a normal distribution for the errors. We may also mention at this point that, of the two models reported in Table 5.9, the simpler explains only about 50% of the variation in Y, whilst the more complex explains 76% of the variation. We shall consider the problem of selecting a 'best' model for these data later in the chapter.

Inference and maximum likelihood

To supplement point estimates obtained by maximum likelihood, it is desirable to be able to assess the significance of departures from zero of individual estimates, to make inferences about sets of parameters, and to have available some overall measure of goodness-of-fit. When using least squares analysis these functions are performed by t, F, and r^2, respectively.

From equations (5.18) and (5.19) it is evident that the effect of autocorrelation is that $\hat{e}_i$ is linked via $\hat{\rho}$ (which is usually positive) and $\mathbf{W}$ to $\hat{e}_j$ $(i \neq j)$ so that the residual sum of squares is a downwardly biased estimate of the error sum of squares. If we were to ignore this bias and use the error mean square, $D/(n-k)$, to estimate the error variance, then we would obtain inflated values of the t, F, and r^2 statistics.

An unbiased estimate of the error variance is $\hat{\sigma}^2$, given by equation (5.34) which, it is noted, is an expansion of the *independent* error term $\mathbf{u}'\mathbf{u}$. Thus, although the residuals $\hat{\mathbf{e}} = \mathbf{Y} - \mathbf{X}\hat{\mathbf{b}}$ and the fitted values $\hat{\mathbf{Y}}$ are autocorrelated, an unbiased estimate is nevertheless available. If the model proposed is a good fit to the data, then $\hat{\sigma}^2$ will be comparatively small when compared to the total variability of Y. A useful summarizing quantity is therefore

$$r^{*2} = \left(\sum(Y_i - \bar{Y})^2/n - \hat{\sigma}^2\right)\Big/\left(\sum(Y_i - \bar{Y})^2/n\right). \tag{5.39}$$

When $\hat{\sigma}^2$ is estimated via OLS, then this amounts to the definition of r^2 that was given above. The quantity r^{*2} is simply the proportion of the variation in Y that is explained by the model. Doreian (1980, 1981) derives r^{*2} as the square of the observed correlation between $\mathbf{Y}$ and $\hat{\mathbf{Y}}$.

Although the t test is, strictly speaking, invalidated by dependence, an equivalent test is available that relies on the asymptotic normality and independence of the maximum likelihood estimators. This involves comparing the value of the ratio of an estimate to its estimated standard error with the appropriate critical value from an $N(0,1)$ distribution. The standard errors are contained within the covariance matrix $\mathbf{V}$, which, it is well known (Kendall and Stuart, 1979), is provided by the second partial derivatives of the log-likelihood function (equation (5.30)). Thus,

$$\mathbf{V} = \left[\frac{-\mathrm{E}(\partial^2 \log L)}{\partial\theta_r \partial\theta_s}\right]^{-1}, \tag{5.40}$$

where θ_r and θ_s are typical parameters. The first partial derivatives with respect to $\mathbf{b}$ and σ^2 are given in equations (5.31) and (5.32), while the first partial derivative with respect to ρ is given by equation (5.41):

$$\partial(\log L)/\partial\rho = -\sum \lambda_i/(1 - \rho\lambda_i) + 1/\sigma^2 \mathbf{V}'\mathbf{W}\mathbf{e}, \tag{5.41}$$

where λ_i is the ith eigenvalue of the matrix $\mathbf{W}$. Since $\mathbf{A} = (\mathbf{I} - \rho\mathbf{W})$, this means that the eigenvalues of $\mathbf{A}$, and therefore their product to give $|\mathbf{A}|$, can be expressed in terms of λ_i as:

$$|\mathbf{A}| = \prod_{i=1}^{n} (1 - \rho\lambda_i), \tag{5.42}$$

or, taking logs, as

$$\log|\mathbf{A}| = \sum_{i=1}^{n} \log(1 - \rho\lambda_i). \tag{5.43}$$

Substituting this result into equation (5.30), and differentiating, yields the first part of equation (5.41) (cf. Doreian, 1980, for further details).

The expected values of the second partial derivatives obtained from equations (5.31), (5.32), and (5.41) are

$$E(\partial^2(\log L)/\partial(\sigma^2)^2) = -n/2\sigma^4, \tag{5.44}$$

$$E(\partial^2(\log L)/\partial\rho^2) = \alpha - \mathrm{tr}(\mathbf{B}'\,\mathbf{B}), \tag{5.45}$$

where $\mathbf{B} = \mathbf{W}\mathbf{A}^{-1}$, tr(·) denotes the trace of a matrix, and $\alpha = -\sum_{i=1}^{n} \lambda_i^2/(1-\rho\lambda_i)^2$,

$$E(\partial^2(\log L)/\partial\mathbf{b}^2) = -\mathbf{X}'\mathbf{A}'\mathbf{A}\mathbf{X}/\sigma^2, \tag{5.46}$$

$$E(\partial^2(\log L)/\partial(\sigma^2)\partial\rho) = -\mathrm{tr}(\mathbf{B})/\sigma^2. \tag{5.47}$$

The remaining expectations, such as $E(\partial^2(\log L)/\partial\rho\partial\mathbf{b})$, can be shown to be vectors of zeros. Ord (1975) and Cliff and Ord (1981a) give results equivalent to these but without the intervening algebra provided by Doreian (1980).

The quantities given by equations (5.44)–(5.47) are computed from sample data and entered into equation (5.40) to provide an approximate covariance matrix. Thus, following Ord (1975) and Doreian (1980),

$$\mathbf{V}(\hat{\sigma}^2, \hat{\rho}, \hat{\mathbf{b}}) = \sigma^4 \begin{bmatrix} n/2 & \sigma^2\mathrm{tr}(\mathbf{B}) & 0 \\ \sigma^2\mathrm{tr}(\mathbf{B}) & \sigma^4\mathrm{tr}(\mathbf{B}'\,\mathbf{B}) - \sigma^4\alpha & 0 \\ 0 & 0 & \sigma^2\mathbf{X}'\mathbf{A}'\mathbf{A}\mathbf{X} \end{bmatrix}^{-1}. \tag{5.48}$$

The square roots of the main diagonal elements of equation (5.48) are the asymptotic standard errors that were given in Tables 5.4, 5.7, and 5.9.

Although calculating the elements of the variance–covariance matrix given by equation (5.48) may appear to be a formidable task, we have found that the GENSTAT package renders it less than onerous. Table 5.10 gives some of the necessary intermediate values for the island-isolation weights matrix (Table 5.6b) applied to the California data and also the final estimated variance–covariance matrix (5.48) relating to Tables 5.2 and 5.7b. The eigenvalues $\{\lambda_i\}$ were obtained

***Table* 5.10 *Intermediate values leading to equation* (5.48)**

$\lambda_1 = 1$; $\lambda_2 = 0.90$; $\lambda_3 = -0.87$; $\lambda_4 = 0.14$; $\lambda_5 = -0.01$; $\lambda_6 = -0.27$
$\lambda_7 = -0.19$; $\lambda_8 = -0.70$; λ_9 to λ_{26} all equal to 0
$\hat{\sigma}^2 = 0.0182$ $\alpha = -24.26$
$\mathrm{Tr}(\mathbf{B}) = 5.5071$ $\mathrm{Tr}(\mathbf{B}'\mathbf{B}) = 33.3850$

$$\mathbf{V}(\hat{\sigma}^2, \hat{\rho}, \hat{\mathbf{b}}) = \begin{bmatrix} 0.000027 & -0.000140 & 0 & 0 & 0 & 0 \\ -0.000140 & 0.018080 & 0 & 0 & 0 & 0 \\ 0 & 0 & 0.146165 & 0.003228 & -0.003120 & -4.358443 \\ 0 & 0 & 0.003228 & 0.001433 & -0.000675 & -0.075538 \\ 0 & 0 & -0.003120 & -0.000675 & 0.000608 & 0.072259 \\ 0 & 0 & -4.358443 & -0.075538 & 0.072259 & 131.702438 \end{bmatrix}$$

using the subroutine F02AFF of the NAG library and implanted into the GENSTAT program.

For readers who wish to check their results, Table 5.11 gives the eigenvalues of the **W** matrices given in Tables 5.3 and 5.6a and of the **W** matrix used for the Galapagos Islands data.

One very simple way around the eigenvalue extraction problem is to calculate the conditional variance–covariance matrix of the $\hat{\mathbf{b}}$, although this does not give any clues as to the likely value of the standard error of ρ. This can be calculated using equation (5.12), with $\hat{\sigma}^2 = \hat{\mathbf{e}}'\mathbf{D}^{-1}\hat{\mathbf{e}}/n$ and with **D** dependent upon the estimate of ρ. Actually, since this is a conditional estimate, it is wise to estimate $\mathbf{V}(\hat{\mathbf{b}})$ at each iteration, in which case **D** changes according to the trial ρ-value. The amount of variation in $\mathbf{V}(\hat{\mathbf{b}})$ illustrates the degree of dependence of the conditional estimates on ρ (cf. Cook and Pocock, 1983).

When the number of localities, n becomes large, the numerical extraction of eigenvalues may be difficult. Some progress has been made by Ord (1975) in deriving analytical results that hold in certain conditions with large regular **W** matrices. The most applicable results apply when **W** is of dimension $pq \times pq$, where p and q are the numbers of rows and columns of a rectangular lattice. Then, if one assumes that the elements of the **W** matrix are consistent with, say,

$$\mathrm{E}(e_{rs}) = 0.25\rho(e_{r-1,s} + e_{r+1,s} + e_{r,s-1} + e_{r,s+1})$$

where e_{rs} is the error at r, s on the lattice and ρ is the autocorrelation, then the pq eigenvalues are

$$\lambda_{rs} = 0.5[\cos\{r\pi/(p+1)\} + \cos\{s\pi/(q+1)\}], \tag{5.49}$$

where $r = 1, \ldots, p$ and $s = 1, \ldots, q$. Haining (1981a,b) makes use of equation (5.49), assuming that the approximation entailed by corner and boundary localities with less than four neighbours is negligible and is comparable to the machine-induced errors of numerical matrix manipulation.

Attention has been focused here on testing the significance of individual parameters. This can also be accomplished by the likelihood-ratio test that are described below. These tests have the capacity to simultaneously test sets of parameters if this seems desirable.

***Example* 5.3** (cont.)

From Table 5.4 it is evident that $\hat{\rho}$ deviates very significantly from its expectation (zero under the null hypothesis). In each of the four models, dividing $\hat{\rho}$ by its standard error and referring to the $N(0, 1)$ distribution produces a value that lies outside the approximate (95%) two-sided critical values of ± 2. The fact that the autocorrelated errors model incorporates a location effect into its error structure has a profound effect upon the apparent significance of the X variables. Consider, for example, the dummy variable X_3. Using OLS the ratio of $\hat{b}_3$ to its

Table 5.11 *Eigenvalues of W matrices*

Irish counties												
$\{\lambda_i\}$:	1.00	0.89	0.86	0.73	0.57	0.49	0.45	0.34	0.25	0.20	0.06	−0.01
	−0.06	−0.11	−0.18	−0.21	−0.30	−0.34	−0.39	−0.44	−0.49	−0.51	−0.61	−0.65
	−0.74	−0.80										
Californian coast												
$\{\lambda_i\}$:	1.00	0.99	0.98	0.69	0.66	0.57	0.38	0.28	0.00	−0.03	−0.04	−0.06
	−0.07	−0.14	−0.18	−0.19	−0.21	−0.24	−0.34	−0.41	−0.43	−0.54	−0.57	−0.60
	−0.75	−0.76										
Galapagos Islands												
$\{\lambda_i\}$:	1.00	0.89	0.84	0.72	0.39	0.37	0.20	0.11	0.02	−0.05	−0.07	−0.08
	−0.10	−0.14	−0.18	−0.27	−0.28	−0.32	−0.43	−0.48	−0.52	−0.73	−0.87	

estimated standard error was (equation (5.7b)) $4.2539/0.7779 = 5.47$, using OLS on the logit version of the model the ratio (equation (5.16c)) was 6.62, and using WLS on the logit version (equation (5.16d)) we had 5.51. However, using the autocorrelated errors model (Table 5.4) we have a paltry 1.61 or (logit version) 1.74. Since we have reasonable *a priori* knowledge of the direction that the effect of X_3 (or, indeed, of X_2) will have on Y, it is reasonable to consider one-sided tests of significance. The one-sided 5% point of the unit normal distribution is 1.645, while the one-sided 10% point is 1.28. Using the logit model (preferable because of its handling of the restricted range for Y) we would therefore continue to conclude that there was significant evidence linking the incidence of blood group A to membership of the 'Pale', although the relation between blood group A and 'town' intensity is not likely to be considered significant.

The revised interpretation of the X variable effects is therefore that the odds on a randomly chosen individual having blood group A, as opposed to some other blood group, change from 2 to 5 for counties outside the 'Pale' ($\exp(-0.9061) = 0.4041$), to about 4 to 9 for counties inside the 'Pale' ($\exp(-0.9061 + 0.0722) = 0.4344$). The final model of Table 5.4 effectively estimates that the probability of a randomly chosen individual having blood group A is $0.4041/1.4041 = 28.8\%$ for individuals from outside the 'Pale' and $0.4344/1.4344 = 30.3\%$, for individuals from within the 'Pale'.

The results of fitting these models are more plausible and less dramatic $(30.3 - 28.8) \ll (32.4 - 27.7)$ than those obtained earlier. Not only is something consciously being done about error autocorrelation in the interests of more valid estimates, but also the autocorrelated errors models assign a small proportion of the variation in Y to what is supposedly random error. Table 5.12 gives the details.

In summary, therefore, the interpretation offered by these models is that, while blood group A percentages are a response to a county's 'colonial history', the impact of Anglo-Norman colonization is only barely detectable. This is somewhat

Table* 5.12 *Proportion of variance explained by various models of Irish blood distribution

Method	Dependent variable	Independent variable	% of variance explained
OLS	Y	X_2	45.9
		X_3	55.5
	Z	X_2	46.1
		X_3	55.6
WLS	Z	X_2	53.9
		X_3	64.3
ML	Y	X_2	76.9
		X_3	76.9
	Z	X_2	76.3
		X_3	75.7

contrary to the purely visual evidence, which tends to emphasize the coincidence of the two distributions. Statistically, there is also a highly significant effect whereby neighbouring counties possess very similar deviations from the predictions provided by the 'colonial hypothesis'. There appears to be a considerable element of inter-county interaction involved in the overall process governing the blood group A distribution, although we have already mentioned that autocorrelated residuals could be due to a number of causes.

***Example* 5.4** (cont.)

When we last considered the data concerned with the incidence of plant species in Californian localities, we noted that the two alternative weights matrices that we had proposed gave rise to markedly different (and clearly non-zero) values for ρ (see, though, the analysis of the residuals in Section 5.6 and the discussion in Brandsma and Ketellapper (1979a)). We argued that the more realistic weights matrix was that which implied island isolation. We now consider the significance of the parameter estimates, $\hat{\mathbf{b}}$. The values of these estimates, their standard errors calculated via equations (5.44)–(5.47), and the ratios of estimate to standard error have already been given in Table 5.7.

With either weights matrix, it is noticeable that all three variables appear clearly significant, since all have estimate to standard error ratios which are markedly greater than 2 in magnitude. An important difference that results from the introduction of the island isolation weighting is the remarkable diminution in the apparent importance of X_3, the latitude of the locality. This is presumably an artificial effect caused originally by the fact that almost all the islands were to the south of almost all the mainland areas, so that latitude and insularity were accidentally confounded.

In the present context it is worth recalling that the Galapagos data provided further evidence ($\hat{\rho} \simeq 0$) in favour of the thesis that islands should indeed be regarded as being insular!

Likelihood-ratio tests

The significance of either individual parameter estimates or sets of parameter estimates can be assessed by the likelihood-ratio test. Essentially one compares a pair of nested models, the second of which is identical to the first apart from the k (≥ 1) restrictions that have been placed on the parameters of the first model. The pair of models can therefore be termed restricted and unrestricted. The test effectively places on a scale from zero to one the ratio of the likelihoods of the two models. The numerator is the likelihood of the model restricted under the null hypothesis, and is therefore at most equal to the denominator which is the likelihood of the unrestricted model. When both numerator and denominator are functions of independent χ^2 distributions, as is the case under the null hypothesis in linear regression with independent normal errors, then the likelihood-ratio test is equivalent to the F test.

In practice one works with $\log L_R$ and $\log L_U$, the (natural) logarithms of the likelihoods of the restricted and unrestricted models. The null hypothesis is that the restrictions are true, in which case $-2\log(L_R/L_U) = 2\log(L_U/L_R) = 2\{\log(L_U) - \log(L_R)\}$ is distributed as χ_k^2. [Note that all the logarithms here are *natural* logarithms rather than logarithms to base 10.] As Bartholomew (1981) notes, this is an asymptotic result relying on large sample sizes. With smaller sample sizes the distribution of the likelihood ratio may deviate from χ^2: this probably accounts for the poor performance of Brandsma and Ketellapper's (1979a) application of the likelihood-ratio test to the autocorrelated errors model mentioned by Ripley (1981).

The unrestricted autocorrelated errors model

We have noted that it is possible to rewrite the autocorrelated errors model given by equations (5.18) and (5.19) as a single equation

$$\mathbf{Y} = \rho\mathbf{WY} + \mathbf{Xb} - \mathbf{WX}\gamma + \mathbf{u}, \tag{5.50}$$

and we now regard the constraint $\gamma = \rho\mathbf{b}$ as an optional restriction. Equation (5.50), with this constraint relaxed, is referred to as the unrestricted autocorrelated errors model. Ord (1975) estimates both restricted and unrestricted models using least squares and Burridge (1981) compares them by various means including the likelihood-ratio test. Under the null hypothesis, twice the logarithm of the likelihood ratio has a χ_{k-1}^2 distribution.

When γ is set equal to $\rho\mathbf{b}$, $\mathbf{Y}$ may be seen as a response to several influences. The value at locality i, Y_i, depends on $\{Y_j\}$ $(j \neq i)$, the 'surrounding' Y-values, via the term $\rho\mathbf{WY}$. It also depends on X_i via $\mathbf{Xb}$ and it depends on the 'surrounding' $\{X_j\}$ $(j \neq i)$ via $\rho\mathbf{WXb}$.

More complex autocorrelated errors models

It has already been hinted that more complex error structures are possible, although this intensifies the numerical estimation problem. As an example, Brandsma and Ketellapper (1979b) consider the situation where Y may be regarded as a migration stream under two separate influences, namely the origins and destinations of the migrants. This calls for the use of two $\mathbf{W}$ matrices and two concomitant ρ parameters and results in the replacement of the usual equation (5.19) by the more complex

$$\mathbf{e} = \rho_1\mathbf{W}_1\mathbf{e} + \rho_2\mathbf{W}_2\mathbf{e} + \mathbf{u}. \tag{5.51}$$

They give a log-likelihood function akin to equation (5.48), therefore requiring second partial derivatives $\partial^2 L/\partial\rho_1^2$ and $\partial^2 L/\partial\rho_2^2$ plus others to build up the covariance matrix. The maximization problem now becomes biparametric and numerical methods described by, say, Hepple (1976), are advocated as a means of obtaining the maximum likelihood estimates $\hat{\rho}_1$ and $\hat{\rho}_2$. In order to test the significance of departures of $\hat{\rho}_1$ and $\hat{\rho}_2$ from zero, they prefer the likelihood-ratio test on account of its simultaneity. The restriction under the null hypothesis is that $\hat{\rho}_1 = \hat{\rho}_2 = 0$, so that $\log L_R$ is derivable from the least squares (and hence ML) estimates $\hat{\mathbf{b}}$ and $\hat{\sigma}^2$. The quantity $2 \log (L_U/L_R)$ is distributed approximately as χ_2^2 under the null hypothesis.

A further suggestion is to estimate parameters pertaining to **W**. One example would be to resolve the proximity values so that $w_{ij} = d_{ij}^{\hat{f}}$, where $\hat{f}$ is a maximum likelihood estimate. This would call for the partial derivatives $\partial L/\partial f$ and $\partial^2 L/\partial f^2$. A broader approach, following the same line of reasoning but doing away entirely with *a priori* **W** matrices, is to be found in the analysis of geographical variation in cardiovascular mortality by Pocock *et al.* (1982). They suggest an autocorrelated errors model in which

$$\operatorname{var}(e_i) = \sigma^2; \qquad \operatorname{cov}(e_i, e_j) = \sigma^2 \lambda \exp(-\mu d_{ij}) \qquad (i \neq j), \tag{5.52}$$

where d_{ij} is the distance between locations i and j.

Note that the error autocorrelation structure is controlled by parameters σ^2, λ and μ. By invoking the usual normal errors assumption, they obtain maximum likelihood estimates $\hat{\lambda}$, $\hat{\mu}$, $\hat{\sigma}^2$, and $\hat{\mathbf{b}}$. Since they specify that autocorrelation falls off everywhere as a negative exponential function, it is constant for different locality pairs in the same relative position. As Marshall (1982) observes, this stationarity assumption is not a prerequisite of the **W** matrix method, which permits differential weighting of pairs at identical distances. This inherent flexibility was demonstrated by the Irish blood **W** matrix.

Cook and Pocock (1983) add their weighting scheme (described earlier in the subsection dealing with intermediate weighted least squares) to the autocorrelated errors model, so that $\operatorname{var}(e_i) = \sigma^2 + 1/d_i$, where d_i is the observed death rate at locality i. However, this refinement makes little difference to their results because the sampling error variance is small.

Poisson responses

Up to this point in the chapter we have assumed that the errors **e** or **u** associated with our models could be regarded as observations from a normal distribution. These assumptions lead directly to tests based on the t and F distributions and governed our choice of likelihood function. Usually, because of the Central Limit theorem, the normality assumption is a reasonable approximation, although sometimes there are conceptual difficulties because of the potential for obtaining negative estimates for some essentially non-negative variable. This is particularly likely to arise when Y represents a count (as with the species examples examined earlier). Of course, when the observed counts are large, any reasonable model is unlikely to come up with a negative Y-value, but for low Y-values this will not be the case. Moreover, when the locality counts are low, it is difficult to gloss over the essentially discontinuous nature of Y's variation, and some more realistic distributional assumption is required.

An immediately obvious alternative in such circumstances is the Poisson distribution which has already featured widely in this book, commencing with the discussion surrounding equation (1.1). The outcome of a Poisson process at any locality was shown there to depend on the parameter λ, the overall intensity of points per unit area, and the set of outcomes, the counts in n localities, comprised a random distribution. Clearly, as it stands, this is an inadequate prescription of Y's variation. Non-negative counts are desirable, but a spatially random response is of limited interest to the spatial analyst! Thus we modify equation (1.1) to formulate the more appropriate model

$$P(Y_i = r) = \lambda_i \exp(-\lambda_i)/r! \qquad (i = 1, \ldots, n), \tag{5.53}$$

$$E(\mathbf{Y}) = \boldsymbol{\lambda} = \mathbf{Xb}, \tag{5.54}$$

$$\operatorname{var}(\mathbf{Y}) = \boldsymbol{\lambda}, \tag{5.55}$$

where $\mathbf{X}$, $\mathbf{b}$, and $\mathbf{Y}$ take on the meanings given above for equation (5.1).

In comparison with equation (1.1), $\boldsymbol{\lambda}$ is no longer constant but consists of a vector of values determined as a weighted combination of designated X variables. Estimation of $\mathbf{b}$, and hence of $\boldsymbol{\lambda}$, is accomplished by maximizing the log-likelihood. Since, from equation (5.53), the probability of the outcome Y_i is $\lambda_i^{Y_i} \exp(-\lambda_i)/Y_i!$, the joint probability (the likelihood) of the set of outcomes $\{Y_i\}$ over n localities is

$$L = \prod_{i=1}^{n} \lambda_i^{Y_i} e^{-\lambda_i}/Y_i! \tag{5.56}$$

and the log-likelihood is

$$\log L = \sum_{i=1}^{n} \{-\lambda_i + Y_i \log \lambda_i - \log(Y_i!)\}. \tag{5.57}$$

In the case of the unrestricted model, it is evident that the best estimate of λ_i will be the observed frequency at that location, namely Y_i. Thus, for the unrestricted model, substituting Y_i for λ_i in equation (5.57), we obtain:

$$\log L_U = \sum_i \{Y_i(\log Y_i - 1) - \log(Y_i!)\}. \tag{5.58}$$

In the case of the restricted model (equation (5.54)) the estimates of the $\mathbf{b}$ parameters are easily obtained numerically by the application of any standard function maximization computer program.

Table 5.13 Quadrat counts displaying a possible north–south trend

0	0	0	1	0
3	3	5	2	0
2	3	6	2	5
1	5	4	6	7
6	2	4	3	4

Example 5.6

Consider the hypothetical quadrat data given in Table 5.13. The arrangement of quadrat counts indicates a possible departure from simple randomness (and hence the probable presence of a modified Poisson process) on account of the evident North–South and possible East–West trends in the data. To incorporate these effects, equation (5.54) is adopted with X_1 and X_2 specified as quadrat coordinates ranging from (1, 1) in the north-west corner to (5, 5) in the south-east corner.

The maximum likelihood estimates $\mathbf{\hat{b}}$ in Table 5.14a provide evidence of a highly significant North–South trend (X_1), but there does not appear to be any significant evidence of an East–West trend. Supplementary tests of these hypotheses, and a measure of the overall goodness-of-fit (or deviance, Nelder and Wedderburn, 1972) of various models can be obtained using the likelihood ratio. The unrestricted model in this case has 25 parameters, one per cell, while the null hypothesis replaces these 25 parameters by the three parameters b_0, b_1, and b_2. It is convenient to denote the likelihood for this model by $L_{R(12)}$. Then $2\log(L_U/L_{R(12)})$ is distributed approximately as χ^2_{22} under the assumption that the restrictions hold true. Since $2\log(L_U/L_{R(12)}) = 28.33 < \chi^2_{22,0.05} = 33.92$, the

Table 5.14 Maximum likelihood estimates of Poisson response models fitted to the data of Table 5.13

(a)

Parameter	Estimate	Standard error	Ratio
b_0	−1.2791	0.4794	
b_1	1.2882	0.1953	6.60
b_2	0.1248	0.1494	0.84

(b)

Parameter	Estimate	Standard error	Ratio
b_0	−0.8539	0.4098	
b_1	1.2713	0.1989	6.39

gap between the log-likelihoods is insufficiently large to cause us to reject the null hypothesis.

The ratios of estimates to standard errors given in Table 5.14 suggest that X_2 (the East–West trend) is irrelevant. When X_2 is omitted from the model, we obtain the revised parameter estimates given in Table 5.14b. The likelihood for this model may be denoted by $L_{R(1)}$. Since this simpler model cannot fit better than its more complicated predecessor, it follows that $L_{R(1)} \leq L_{R(12)}$. The likelihood-ratio statistic for testing the fit of this model is $2\log(L_U/L_{R(1)})$ which has an asymptotic χ^2_{23} distribution when the model is true. Since $2\log(L_U/L_{R(1)}) = 28.66 < \chi^2_{23,0.05} = 35.17$, there is no reason to believe that the model of a single North–South trend is invalid.

In both cases the model specified by the null hypothesis is accepted as a result of the likelihood-ratio test. To decide between the models we need to examine the marginal significance of the East–West trend, given the acceptance of the North–South trend. This is the direct analogue of the stepwise procedures for multiple linear regression that we studied above. In the present case, the quantities to be compared are $\log(L_{R(1)})$ and $\log(L_{R(12)})$. The difference, $\log(L_{R(12)}/L_{R(1)}) = 0.33$, is associated with the single extra restriction on the λ parameters that is embodied in the East–West trend. Since $0.33 \ll \chi^2_{1,0.05} = 3.84$, we conclude that there is no necessity to retain this parameter in the model.

In a similar manner, we can establish the significance of the North–South trend by comparing $\log(L_{R(1)})$ with $\log(L_{R(0)})$, where the latter is the log-likelihood corresponding to the model in which λ is presumed to be constant over the 25 quadrats. The obvious estimate of λ is the observed quadrat mean, $\bar{Y} = 74/25 = 2.96$. The value of $2\log(L_U/L_{R(0)})$ is 50.15, which is considerably greater than $\chi^2_{24,0.05} = 36.42$, indicating that there is a significant lack of fit to this model. Also, $2\log(L_{R(1)}/L_{R(0)}) = 50.15 - 28.66 = 21.49 \gg \chi^2_{1,0.001} = 10.83$, which leaves little room to doubt the reality of the North–South trend.

Finally, we note that there is available an analogous statistic to r^2. If we regard the model of constant λ as the null model—that is, as being the simplest model that we could conceivably consider—then 50.15 could be regarded as being equivalent to T, the total deviance, and the proportion to the total variation explained by the model would be given as $21.49/50.15 = 43\%$.

Poisson responses with logarithmic link

An alternative to equation (5.54) is to specify a logarithmic link between the X and Y variables, so that

$$\mathrm{E}(\mathbf{Y}) = \exp(\mathbf{Xb}). \tag{5.59}$$

This has the effect of replacing an additive model by a similarly structured multiplicative model, and might make more sense in some applications.

***Example* 5.6** (cont.)

In non-matrix terms, the fitted values derived from equation (5.59) are given by $e^{\hat{b}_0}(e^{\hat{b}_1})^{X_{1i}}(e^{\hat{b}_2})^{X_{2i}} = \hat{Y}_i$. Following a maximization of the restricted likelihood, we obtain the estimates $\hat{b}_0 = -0.1788$, $\hat{b}_1 = 0.3248$, and $\hat{b}_2 = 0.0608$, for the data of Table 5.13. Thus, the estimated Poisson parameter λ for, say, the quadrat in the second row and third column is $0.8363 \times 1.3838^2 \times 1.0628^3 = 1.92$. We see that, in this model, a movement down r rows of the table implies an increase in λ by the multiplicative factor 1.3838^r, instead of the additive increase of $1.2882r$ implied by the previous model. There is, of course, a corresponding pair of results for changes across columns.

The deviance for this multiplicative (log) link model is $34.99 > \chi^2_{22,0.05} = 33.92$, the proportion of deviance explained is 0.30, and the restriction $b_2 = 0$ increases the deviance by 1.62. Both additive models appear superior in as much as they explain a greater proportion of the observed variation in Y.

The source of Table 5.13 can now be revealed. The data were generated by drawing at random from Poisson distributions having $\lambda = i$, where i is the row of the table. It therefore is only to be expected that the best model that we found to explain the data was, in fact, of exactly the form used to generate the data.

***Example* 5.7**

The Galapagos Islands data of Table 5.8 were analysed earlier in Example 5.5 on the assumption of normal errors, although since the Y variable, the number of different plant species, is a count, Poisson errors may be more appropriate. Table 5.15 gives the values of the deviance, the proportion of deviance explained, and other statistics, for various models with each type of error structure. The subsets chosen roughly reflect the apparent importance of the variables that was suggested by the results in the final column of Table 5.9a.

Table 5.15a is concerned with the additive models that assume normal errors. The values of r^2 suggest that the most appropriate model of those listed in this part of the table is that involving X_2 (elevation) and X_5 (area of adjacent island) alone. Using the F tests described in Section 5.2, we can compare $\{(D_{25} - D_{12345})/3\}/\{D_{12345}/17\} = 4868.7/4784.1 = 1.02$ with the upper tail of a $F_{3,17}$ distribution. Since 1.02 is much smaller than 2.96, the upper 5% point of this distribution, we can accept the hypothesis that, for this model, $b_1 = b_3 = b_4 = 0$.

Table 5.15b is concerned with the multiplicative models that assume Poisson errors. The position here is somewhat different: although the fit (in terms of r^2) of the two most complex models is rather better than that of their additive counterparts, the three simpler models fit noticeably less well. Since we no longer have normal errors, the significance of the contributions of individual parameters must be measured using the change in the value of the likelihood ratio, as must the goodness-of-fit of the models.

***Table 5.15** Goodness-of-fit statistics for various normal and Poisson errors models fitted to the Galapagos data*

(a) Normal errors

X variables	$\sum(Y_i - \hat{Y}_i)^2$	r^2
X_1, X_2, X_3, X_4, X_5	81 329	0.76
X_1, X_2, X_4, X_5	81 358	0.76
X_1, X_2, X_5	91 333	0.73
X_2, X_5	95 935	0.72
X_2	170 467	0.50
—	344 051	0

(b) Poisson errors

X variables	$\sum(Y_i - \hat{Y}_i)^2$	r^2	Deviance	Proportion
X_1, X_2, X_3, X_4, X_5	70 643	0.79	603.6	0.79
X_1, X_2, X_4, X_5	81 001	0.76	630.2	0.78
X_1, X_2, X_5	109 640	0.68	762.0	0.74
X_2, X_5	191 090	0.44	1170	0.59
X_2	236 790	0.31	1612	0.44
—	344 051	0	2881	0

Testing the goodness-of-fit of the most complex of the multiplicative models, we compare the observed deviance (equal to twice the log-likelihood ratio) of 603.6 with the upper-tail percentage points of the χ^2_{17} distribution. Of course, such a large value can scarcely be attributed to chance! However, these large deviances are to be taken more as an indicator of the large frequencies involved than an as an indicator that the models are totally inadequate (after all, we would be extremely startled if the model was declared to be a good fit—Darwin's theory of evolution might be at risk!). A large deviance, and large changes in deviance, are inevitable given the approximate nature of any model and the presence of large counts (Bishop *et al.*, 1975). Flowerdew and Aitkin (1982) warn against making too much of the comparative proportion of deviance explained. They argue that a low r^2-value can be attributed either to incorrect $\{\hat{Y}_i\}$ or to large error variance, σ^2 (or both). On the other hand, a high proportion of deviance unexplained by the Poisson errors model can only be due to model misspecification bringing about unrealistic predictions, since λ_i equals both mean and variance for the Poisson distribution. Taken together, however, it is possible to obtain from these statistics an overall evaluation of the performance of the various models.

In order to derive some more informative test statistics, we employ a technique directly analogous to that used in Section 5.2 in connection with the deviances of rival least squares models. To test the joint significance of the parameters X_1, X_3, and X_4, we calculate $\{(1170-603.6)/3\}/\{603.6/17\}=188.8/35.5=5.32$. Since $5.32 \gg 2.96$, the upper 5% point of the $F_{3,17}$ distribution, we reject the null hypothesis that, for this model, $b_1=b_3=b_4=0$. In practice we would normally test just one parameter at a time, and would test in turn, for example, the hypotheses $b_3=0$, $b_4=0$, and $b_1=0$. For this data, only the first of these hypotheses would be accepted.

In summary, therefore, if a simple model is required, then the 'best buy' would appear to be the additive normal errors model involving X_2 and X_5 alone. However, the overall best fitting model is the Poisson errors model involving all five X variables (although X_3 might be omitted). A major advantage of the Poisson errors model is exemplified in Table 5.16, which gives the fitted values of the two most complex models—the normal errors model gives rise to two nonsensical negative estimates for Y. Sadly, neither model is a rival to Darwin.

The estimated values for the normal errors model are derived from the parameter estimates given earlier in Table 5.9a. The parameter estimates for the

Table* 5.16 *Observed and estimated numbers of plant species on the Galapagos Islands

Island (see Table 5.8)	Y	$\hat{Y}$ (Poisson model)	$\hat{Y}$ (normal model)
1	31	28.1	0.4
2	3	31.8	36.9
3	25	27.8	18.5
4	10	9.0	−18.0
5	2	33.7	38.8
6	97	31.7	53.7
7	93	89.2	88.0
8	58	19.2	1.3
9	5	44.8	72.1
10	40	30.9	−0.1
11	347	367.6	383.7
12	51	61.5	92.0
13	2	26.7	12.0
14	104	183.7	218.1
15	108	141.8	160.8
16	70	65.9	83.6
17	280	220.6	209.6
18	237	364.6	288.1
19	444	315.8	271.8
20	62	77.0	94.3
21	285	160.8	204.6
22	16	43.4	60.6
23	21	15.3	20.0

Poisson errors model ($\hat{Y}_i = \exp(\hat{b}_0 + \sum_j \hat{b}_j X_{ji})$) are $\hat{b}_0 = 3.5181$, $\hat{b}_1 = -0.0504$ $\hat{b}_2 = 0.3110$, $\hat{b}_3 = 0.9434$, $\hat{b}_4 = -0.7454$, and $\hat{b}_5 = -0.0575$. Notice that the absolute magnitudes of these $\{\hat{b}_j\}$ reflect not only the relative importance of the X variables, but also the units of measurement of these variables. Direct comparisons are only possible between $\hat{b}_1$ and $\hat{b}_5$ (since X_1 and X_5 are both island areas) and between $\hat{b}_3$ and $\hat{b}_4$ (since X_3 and X_4 are both inter-island distances).

In broad terms, the model states that the number of species increases with increasing elevation (but modified by island area), with increasing isolation from other islands (but modified by its proximity to the group centre), and decreases as the neighbouring island increases in size. There are really two sets of variables that have a bearing on Y. The variables X_1 and X_2 relate to the physical size of the island, while variables X_3, X_4, and X_5 relate to the location of the island. It is not surprising to find that X_1 and X_2 display a very strong non-linear correlation with one another (Spearman's rank correlation = 0.90), reflecting the fact that the larger the island, the greater is the chance of it containing a hill or mountain. From our analysis it is clear that it is elevation rather than area which is the dominant size variable.

The two distance variables, X_3 and X_4, appeared to be of only marginal importance. The positive value for $\hat{b}_3$ implies that there is some advantage in being an island isolate—possibly extra isolation leads to speciation. On the other hand, the negative value for $\hat{b}_4$ implies that there is some advantage in being close to the centre of the archipelago—possibly because of the added chance of species diffusion. Since these two mechanisms work in an opposite fashion to one another, we should not be too surprised at the apparent insignificance of their effects.

The results do make clear, however, that X_5—the area of the adjacent island—is of importance. It may be useful to add that X_5 was selected *a priori* by Hamilton *et al.* (1963) as 'an inverse index of isolation, since a given island with a small neighbouring one might be more isolated than it would be if its neighbour were larger and hence possessed greater numbers of potential dispersers'.

5.5 NON-LINEAR REGRESSION

As mentioned in the opening paragraphs of this chapter, a non-linear structural model may sometimes be a more apt specification of the link between X and Y variables. This realization certainly makes life more difficult for the analyst, for, whilst linear models are all variations on one basic theme, non-linearity introduces infinite variety. We have therefore endeavoured to work with linear models if at all possible.

Where clinging to a linear structural model is stretching the truth too far, non-linearity can be introduced very gently by transforming X (in the simple bivariate case) so that the model takes the form $\mathrm{E}(Y) = b_0 + b_1 U$, where U is some function of X such as $X^{1/2}$ or $\log(X)$. Thus, an element of linearity is preserved, since the transformed model has all the appearance of a conventional regression model: we refer to this as a *linear-in-parameters* model. The advantage of this transformation is that the (linear) least squares methodology remains (potentially) valid and yet non-linearity has been introduced as a more accurate representation of the link between X and Y.

Generally, however, variable transformation has a broader function than was outlined above. This function is to condition a model so that, simultaneously, the residuals correspond to the sampling model idealized in the Gauss–Markov theorem. Thus, transformations have been invoked to stabilize residual variances and to generate that much sought after normal distribution. This can be an alternative to the adoption of unusual sampling models.

Linear-in-parameters models constitute just one of two major groups of non-linear models, the other group being those models that are not linear in the parameters. An example of such a model is provided by

$$Y_i = b_0 X_{1i}^{b_1} X_{2i}^{b_2} \ldots X_{k-1,i}^{b_{k-1}} e_i. \tag{5.60}$$

This second group of models actually contains two subgroups, one of which comprises models readily transformable into the familiar linear regression equation. These models are the so-called *intrinsically-linear* models, as opposed to the more troublesome intrinsically-non-linear variety which refuses to be dealt with by *linear* least squares methodology. Equation (5.60) is an example of the intrinsically-linear model. The linearized version is given by equation (5.61).

$$\log Y_i = \log b_0 + b_1 \log X_{1i} + b_2 \log X_{2i} + \ldots + b_{k-1} \log X_{k-1,i} + \log e_i. \tag{5.61}$$

Note that if we were to write $\log Y_i$ as Z_i, then we could regard this model as being no more than a linear-in-parameters model involving Z_i and the $\{\log e_i\}$.

Intrinsically-non-linear models are exemplified by equation (5.62)

$$Y_i = b_0 X_i^{b_1} + e_i, \tag{5.62}$$

where $\mathbf{e} \sim N(\mathbf{0}, \sigma^2 \mathbf{I})$. Here there are no simple transformations available that will allow us to re-express the model in a linear-in-parameters form. With such a model, one might estimate the $\mathbf{b}$ parameters using maximum likelihood, or, equivalently, by using the so-called method of *non-linear least squares*, which involves minimizing the sum of the squared residuals, where $\hat{e}_i = \hat{Y}_i - \hat{b}_0 X_i^{\hat{b}_1}$. This requires numerical optimization routines such as those described by Mather (1976).

Equation (5.63) is another example of an intrinsically-linear model, as is revealed by the linear-in-parameters version (equation (5.64)) created by taking logarithms:

$$Y_i = e^{b_0} (e^{b_1})^{X_{1i}} (e^{b_2})^{X_{2i}} \ldots (e^{b_{k-1}})^{X_{k-1,i}} u_i; \tag{5.63}$$

$$\log Y_i = b_0 + b_1 X_{1i} + b_2 X_{2i} + \ldots + b_{k-1} X_{k-1,i} + \log u_i. \tag{5.64}$$

In equation (5.63) the e's are, of course, the exponential function and not error terms.

In both equation (5.61) and equation (5.64) the log errors are required to satisfy the distributional assumptions of the Gauss–Markov theorem in order to validate the least squares approach. For inference via the F and t distributions, the log errors must therefore have a joint normal distribution. Note that a model is only specified completely when we have both the form of the model *and* the assumed error distribution. Thus, in Example 5.6 we had an identical structural model to that specified by equation (5.63) (though at that stage we concealed its form!), but we used it in conjunction with Poisson errors.

When we assume that $\log(\mathbf{u}) \sim N(\mathbf{0}, \sigma^2\mathbf{I})$, we are implying that the joint error distribution of the untransformed model (e.g. equation (5.63)), is a joint *log-normal* distribution. The standard work on this distribution is the book by Aitchison and Brown (1957). The log-normal distribution is the expected outcome of the *product* of N independent random variables when N is large, so, by assuming multiplicative errors—which is implied by the adoption of a log-linear equation such as (5.61) or (5.64)—we should be prepared to think that these errors represent 'the net effect of a large number of independent multiplicative factors' (Haworth and Vincent 1979). Essentially, this is an outline of the *multivariate central limit theorem* which suggests that the source of the error term in the model is more than simply error in the measurement of Y.

Note that, unlike the normal distribution, the log-normal distribution is skewed, so that mean and median no longer coincide. In the case of the model specified by equation (5.60), the median of the log-normal distribution of Y is simply $b_0 X_{1i}^{b_1} X_{2i}^{b_2} \ldots X_{k-1,i}^{b_{k-1}}$, whereas E($Y$), the mean of that distribution, is the product of the median with $\exp(\sigma^2/2)$, where $\exp(\sigma^2/2)$ is the mean of the accompanying log-normal error distribution.

Clearly, there are countless other possible transformations, so that the quest for a suitable transformation has considerable potential for degenerating into a trial-and-error procedure in the absence of a theoretical rationale. Taking logarithms is just one common, almost instinctive, reaction when faced with non-linearity. So many different transformations have been advocated to control the residual distribution and generate a tractable structural model, that it is sensible to confine one's search to the members of one specific family of models.

The family of power transformations

To avoid having to resort to a haphazard choice of models, it is natural to consider using some sort of model generalization that includes our earlier models as special cases. Consider the model

$$Y_i^{(\lambda_0)} = b_0 + b_1 X_{1i}^{(\lambda_1)} + b_2 X_{2i}^{(\lambda_2)} + \cdots + b_{k-1} X_{k-1,i}^{(\lambda_{k-1})} + e_i. \tag{5.65}$$

In this model the parentheses around the indices $\{(\lambda_j)\}$ indicate that the corresponding variable (i.e. Y or one of the $\{X_j\}$) is to be transformed by some function that is controlled by the parameter λ_j. Variations in the choice of function lead to alternative families of transformations. Schlesselman (1971) and Haworth and Vincent (1982) mention various alternatives, including the family suggested by Tukey (1957), for which

$$Y_i^{(\lambda_0)} = \begin{cases} Y_i^{\lambda_0} & \lambda_0 \neq 0, \quad (5.66a) \\ \log Y_i, & \lambda_0 = 0. \quad (5.66b) \end{cases}$$

Box and Cox (1964) altered Tukey's family so as to avoid the discontinuity at $\lambda_0 = 0$. their suggestion was

$$Y_i^{(\lambda_0)} = \begin{cases} (Y_i^{\lambda_0} - 1)/\lambda_o, & \lambda_0 \neq 0, \quad (5.67a) \\ \log Y_i, & \lambda_0 = 0. \quad (5.67b) \end{cases}$$

For both the Tukey and Box–Cox families, the logarithm is a natural logarithm and the X variables are to be transformed according to precisely equivalent rules. In what follows we shall concentrate on the Box–Cox family of models.

We stated that a requirement would be that our earlier models could be regarded as special cases of the family. Thus, if $\lambda_j = 1$ for all j, then we have a simple linear model, while if $\lambda_j = 0$ for all j, then we have what is essentially the model specified by equation (5.61).

Were the $\{\lambda_j\}$ known, we could, all being well, proceed by least squares. However, since the $\{\lambda_j\}$ are unknown, we have a parameter estimation problem requiring the method of maximum likelihood. In the multivariate context typified by equation (5.65) this would be an awesome task and we therefore simplify that equation either by setting $\lambda_j = 1$ for all j greater than 0, or by setting $\lambda_j = 0$ for all j greater than 0. This effectively confines the estimation problem to the parameter transforming the Y variable. In matrix terms, with $\lambda_j = 1$ for $j > 0$, equation (5.65) becomes

$$\mathbf{Y}^{(\lambda)} = \mathbf{Xb} + \mathbf{e}, \tag{5.68}$$

where, for simplicity, we replace λ_0 by λ.

Box and Cox (1964) envisaged the existence of a λ-value consistent with both normal errors and a constant variance, thereby tacitly acknowledging the volatility of error distributions subject to transformation. On this basis, we formulate the likelihood function as

$$L = |\mathbf{A}| \frac{1}{\sigma^n (2\pi)^{n/2}} \exp\left(\frac{-\mathbf{e}'\mathbf{e}}{2\sigma^2} \right), \tag{5.69}$$

where

$$\mathbf{e}'\mathbf{e} = (\mathbf{Y}^{(\lambda)} - \mathbf{Xb})'(\mathbf{Y}^{(\lambda)} - \mathbf{Xb}) \tag{5.70}$$

and

$$|\mathbf{A}| = \prod_{i=1}^{n} \left| \frac{\partial Y_i^{(\lambda)}}{\partial Y_i} \right| = \prod_{i=1}^{n} Y_i^{\lambda - 1}. \tag{5.71}$$

Thus, $|\mathbf{A}|$ is the Jacobian of the transformation from $Y_i^{(\lambda)}$ to Y_i. The methodology parallels that of Section 5.4, and a similar quantity appears in (5.28). On taking (natural) logarithms of equation (5.69) we obtain:

$$\log L = \text{constant} - \frac{n}{2}\log \sigma^2 - \frac{1}{2\sigma^2}(\mathbf{e}'\mathbf{e}) + (\lambda - 1)\sum \log Y_i. \tag{5.72}$$

The maximum of the log-likelihood can be obtained, as previously, via an iterative search routine. However, it is worth mentioning the existence of the derivatives of equation (5.72) with respect to λ, for two reasons. First, the fact that a derivative exists at $\lambda = 0$ serves to distinguish the Box–Cox transformation from the Tukey transformation. Second, finding the value of λ for which $\partial(\log \lambda)/\partial\lambda = 0$ is a useful check on the accuracy of the direct search method. Schlesselman (1971) gives the derivative as

$$\frac{\partial(\log \lambda)}{\partial \lambda} = \begin{cases} \dfrac{-n\mathbf{Y}^{(\lambda)'}\mathbf{R}\mathbf{U}^{(\lambda)}}{\mathbf{Y}^{(\lambda)'}\mathbf{R}\mathbf{Y}^{(\lambda)}} + \dfrac{n}{\lambda} + \sum_i \log Y_i, & \lambda \neq 0, \quad (5.73a) \\[2ex] \dfrac{-n\mathbf{V}'\mathbf{R}\mathbf{W}}{2\mathbf{V}'\mathbf{R}\mathbf{V}} + \sum_i \log Y_i, & \lambda = 0, \quad (5.73b) \end{cases}$$

where $\mathbf{R} = \mathbf{I} - \mathbf{X}(\mathbf{X}'\mathbf{X})^{-1}\mathbf{X}'$, $\mathbf{U}^{(\lambda)}$ is the vector whose ith component is $\lambda^{-1} Y_i^{\lambda} \log Y_i$, $\mathbf{V}$ is the vector whose ith component is $\log Y_i$, and $\mathbf{W}$ is the vector whose ith component is $(\log Y_i)^2$.

The maximization problem is eased somewhat if the log-likelihood is written in terms of $Z^{(\lambda)} = Y^{(\lambda)}/|\mathbf{A}|^{1/n}$, since we then have

$$\log L = -\frac{n}{2}\log(\mathbf{Z}^{(\lambda)'}\mathbf{R}\mathbf{Z}^{(\lambda)}/n), \tag{5.74}$$

where $\mathbf{Z}^{(\lambda)'}\mathbf{R}\mathbf{Z}^{(\lambda)}$ is the residual sum of squares of the linear regression of $\mathbf{Z}^{(\lambda)}$ on $\mathbf{X}$. The maximum likelihood estimate $\hat{\lambda}$ is therefore that value of λ which minimizes $\mathbf{Z}^{(\lambda)'}\mathbf{R}\mathbf{Z}^{(\lambda)}$.

***Example* 5.8**

As an example of the simplified Box–Cox transformation, we reconsider the Galapagos Islands data. In order to keep things relatively simple, we take just one of the variables in Table 5.8, namely island area (X_1), as the single explanatory variable that, we hope, will account for the inter-island variation in the number of plant species. In our previous analysis (Example 5.7) we considered that the importance of island area was a by-product of island elevation. Whilst this may be the case, there is, nevertheless, a long tradition of bivariate island area/species modelling (Preston, 1948, 1962). We shall assess the aptness of some previously utilized transformations in the light of the Box–Cox approach and later will discuss the adequacy of the bivariate model.

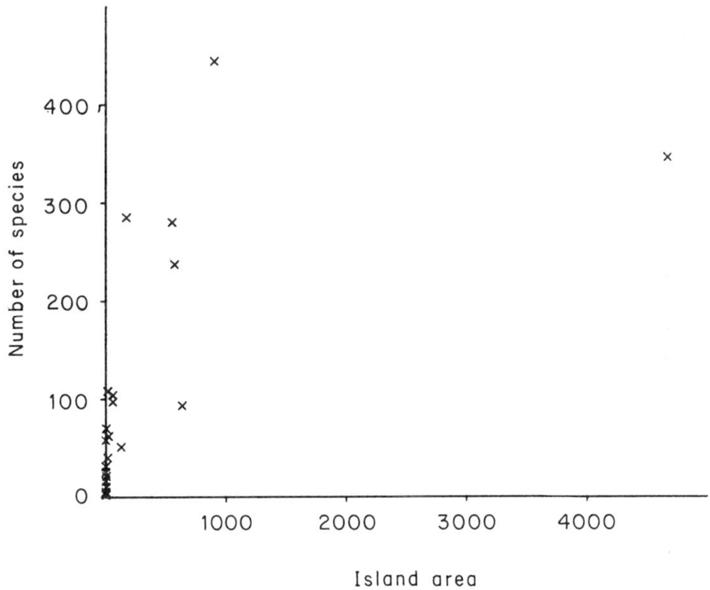

Figure 5.6 The relation between plant species and island area in the Galapagos Islands

A non-linear model is to be expected for the island area/species relationship, since the exclusion of other controlling factors should expose the evident truism that successive increments in area will not produce corresponding increases in the probability of finding a new habitat (Johnson and Raven 1973). The evidence provided by Figure 5.6 certainly indicates that a non-linear link between X and Y is a necessity.

The results of calculating equations (5.72), (5.73a), (5.73b), and the residual sum of squares component of (5.74) are given in Figures 5.7a–c. It is clear from these figures that the maximum likelihood estimate of λ is approximately equal to 0.23. However, as Figure 5.7a demonstrates, the log-likelihood is relatively flat in the neighbourhood of $\lambda = \hat{\lambda}$, and the value of the log-likelihood at $\lambda = 0$ is only marginally less than the maximum.

In order to test the null hypothesis that $\lambda = 0$, we calculate the difference between the value of the log-likelihood at $\lambda = \hat{\lambda}$ and at $\lambda = 0$, recalling that, under the null hypothesis, $2\{\log(L_U) - \log(L_R)\}$ is distributed as χ_1^2 (where L_R and L_U are the natural logarithms of the likelihoods of the restricted and unrestricted models). Since $\log(L_U) = -92.77$ and $\log(L_R) = -94.43$, twice the difference of the logarithms amounts to 3.32 which is less than 3.84, the upper 5% point of a χ_1^2 distribution. In other words, we could not, on the basis of this test, reject the model $E(\log Y_i) = b_0 + b_1 X_i$, though, of course, we are choosing from a very restricted class of models, and this model is disappointingly inadequate. In fact the deviance is equal to 47.09, the total sum of squares is 57.75 and hence r^2 amounts to no more than 0.185. The F test gives a value of 4.75 which is not substantially in excess of 4.325, the upper 5% point of the $F_{1,21}$ distribution. Clearly, this single logarithmic transformation only accounts for a minority of the variation in Y.

The dual logarithmic transformation of equation (5.61) leads to the log-linear model

$$\log Y_i = \log b_0 + b_1 \log X_i + \log e_i, \tag{5.75}$$

which has been widely suggested in the present context, although, as far as we know, it has never before been exposed to the full Box–Cox treatment. We can

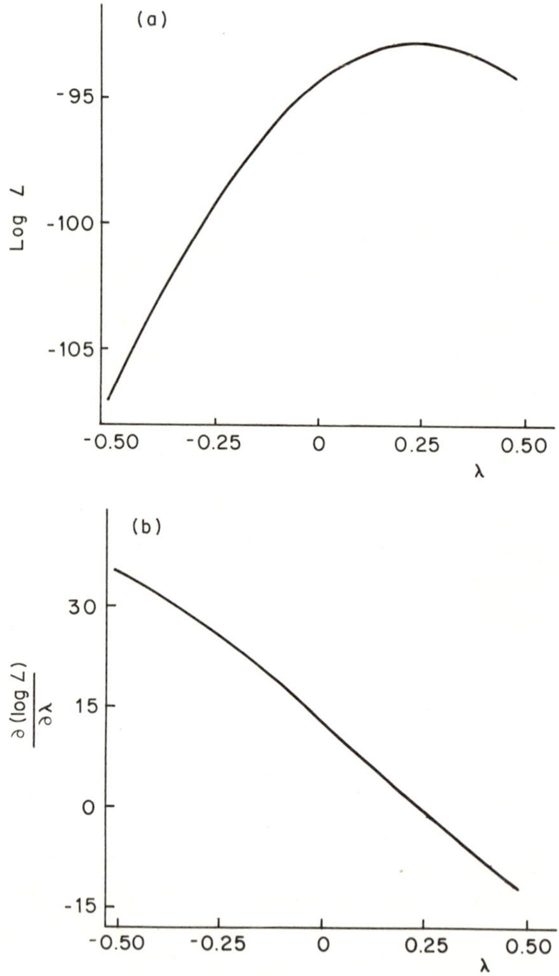

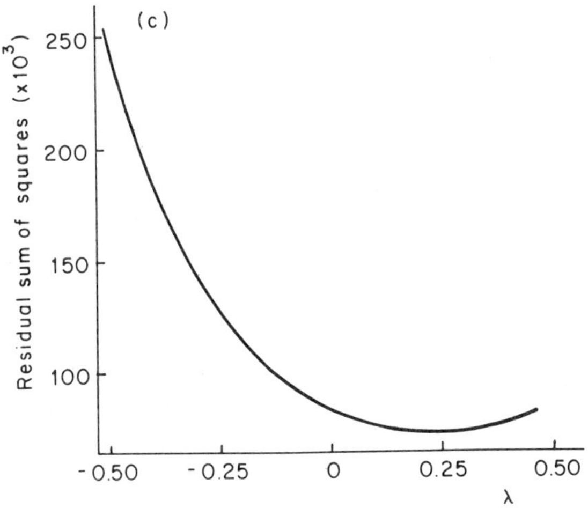

Figure 5.7 Application of the Box–Cox procedure to the Galapagos Islands data. (a) The change in log L with λ. (b) The change in $\partial(\log L)/\partial\lambda$ with λ. (c) The change in the residual sum of squares with λ

anticipate that the log-linear model will be found to be somewhat sub-optimal, but we would be alarmed if we were to discover the difference from the optimum to be significant, since we are aware of its considerable statistical pedigree!

Figures 5.8a–c illustrate the behaviour of the log-likelihood as λ is varied. The maximum value of λ is seen to be about 0.17, while the likelihood-ratio test of the null hypothesis $\lambda = 0$ gives $2\{-78.09 - (-79.96)\} = 3.74$, which, as before, just fails to be significant at the 5% level. So far as the fit of the model is concerned, the improvement over the previous model is startling, since for this model the deviance is just 13.31, which corresponds to a value of 0.77 for r^2 and the immensely significant F-value of 70.06.

The success of the log-linear model, incidentally, estimated as $\widehat{\log Y} = 2.7599 + 0.4153 \log X$, will come as no surprise to biogeographers, for it features prominently in research associated with the MacArthur and Wilson (1967) theory of island biogeography. Vincent (1981), however, tells a cautionary tale. The true processes governing species distributions are undoubtedly too complex to permit full ratification of a simple regression model, no matter how statistically successful. Knowledge of ecological systems demands multivariate models (with due attention to the veracity of the assumed sampling model). Our previous (non-linear, Poisson errors) multivariate model of Example 5.7 fits this prescription nicely; unlike Vincent (1981), who proposes similar models, we are not too dismayed by the large deviance inevitably generated. One cannot become complacent about any model though, for this is a subject-matter rife with alternative points of view. Simpson (1974), for example, suggests that we should be considering the islands as they were in glacial times (when they were bigger and closer together). Box and Cox's (1964) 'proper attitude of sceptical optimism' seems to be the one to nurture!

Parameter bias in log-linear regression

It is important to appreciate that the OLS estimates of the parameters of any regression model are only optimal provided that the Gauss–Markov conditions are satisfied for that model. An added hazard with the log-linear regression model lies in the estimation of b_0. If we work with the model $\log Y_i = \log b_0 + b_1 \log X_i + \log e_i$, then, not unnaturally, our intercept estimate $\widehat{\log b_0}$ is an unbiased estimate of $\log b_0$. Unfortunately, it is *not* the case that $\exp(\widehat{\log b_0})$ is an unbiased

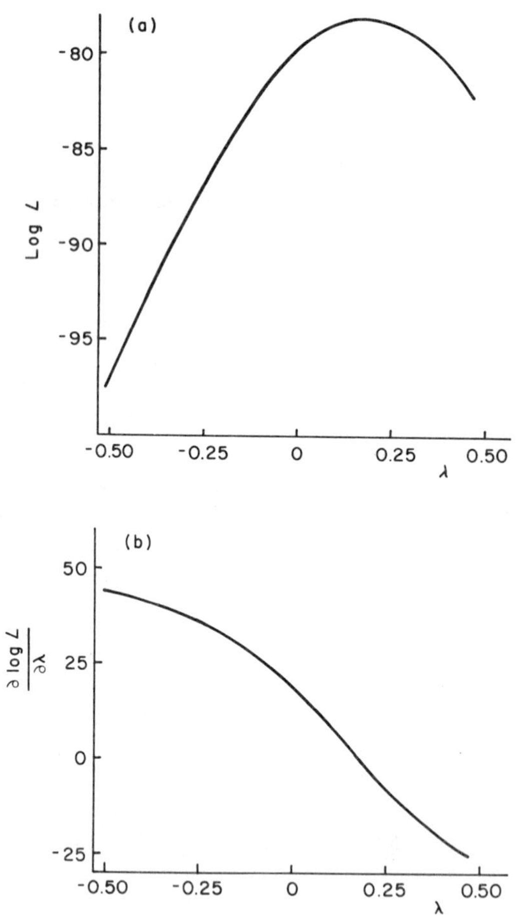

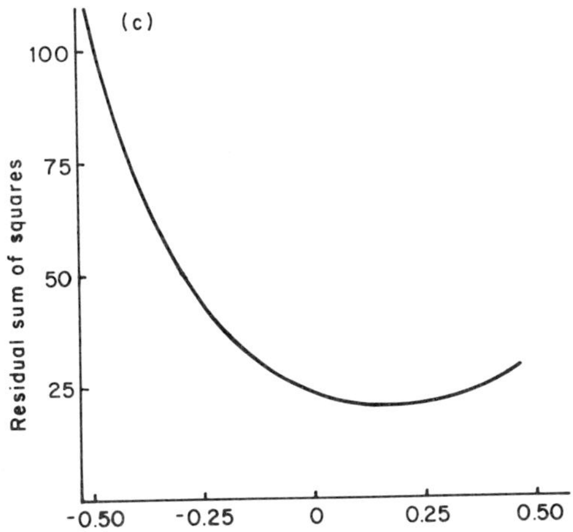

Figure 5.8 Application of the Box–Cox procedure with a log-response. (a) The change in $\log L$ with λ. (b) The change in $\partial(\log L)/\partial\lambda$ with λ. (c) The change in the residual sum of squares with λ

estimate of $\exp(\log b_0) = b_0$. Since b_0 itself is often of considerable interest, it follows that some correction factor will be required.

Goldberger (1968) suggests using $c \cdot \exp(\widehat{\log b_0})$, where c is chosen so that the product is an unbiased estimate of the mean of the distribution of Y. The constant c is calculated from the expression

$$c = \sum_{j=0}^{\infty} f_j d^j / j!, \tag{5.76}$$

where

$$d = \{\hat{\sigma}^2 - \widehat{\text{var}}(\widehat{\log b_0})\}/2, \tag{5.77}$$

$$f_j = \left(\frac{n-k}{2}\right)^j \Gamma\left(\frac{n-k}{2}\right) \Big/ \Gamma\left(\frac{n-k}{2} + j\right), \tag{5.78}$$

and $\Gamma(\cdot)$ is the *gamma function*, defined by

$$\Gamma(m) = (m-1)!, \ \Gamma(m + \tfrac{1}{2}) = \frac{1 \cdot 3 \cdot 5 \cdot \ldots \cdot (2m-1)\sqrt{\pi}}{2^m}. \tag{5.79}$$

It is useful to note that a reasonably accurate first approximation to equation (5.76) is the simple equation $c = 1 + d$.

In equation (5.77), $\hat{\sigma}^2$ is the estimated error variance, while the second term is the estimated variance of the estimate of b_0. From equations (5.9a) and (5.10) we see that

$$\widehat{\text{var}}\,(\widehat{\log b_0}) = \hat{\sigma}^2\{\sum(\log x_i)^2\}/\{\sum(\log x_i)^2 - (\sum \log x_i)^2\}. \tag{5.80}$$

Note that equations (5.76)–(5.80) refer to the correction factor needed in order to obtain an unbiased estimate of the *mean* value of Y. However, since Y has a log-normal distribution, the mean is not equal to the median. In order to obtain an unbiased estimate of the *median* of the distribution it is necessary to replace equation (5.77) by

$$d = -\,\widehat{\text{var}}\,(\widehat{\log b_0})/2, \tag{5.81}$$

where the estimated variance is calculated using equation (5.80). The estimate of the median is then obtained by multiplying the revised value of c by $\exp(\widehat{\log b_0})$.

A further limitation on the complicated formulae given above is that they are applicable only to the case $X = 1$. The full formulae for general X would be a degree more complex! The simplification arises because, if $X = 1$, then $\log X = 0$ and thus b_1 drops out of the estimating equation. For further details and comments see Bradu and Mundlak (1970), Teekens and Koerts (1972), and Vincent and Haworth (1982).

***Example* 5.8** (cont.)

The dual logarithmic transformation of the Galapagos Islands area and species intensity relationship yields $\widehat{\log b_0} = 2.7599$, $\hat{b}_1 = 0.4153$, $\hat{\sigma}^2 = 0.6339$, and $\text{var}\,(\widehat{\log b_0}) = 0.0413$, with $n - k = 21$. Hence, $d = 0.2963$ and therefore $c = 1.2963$ is our first approximation. Pursuing the calculations through equation (5.76) leads to the final result that $c = 1.3396$. Using this value we are able to deduce that the expected number of species on an island having an area of 1 km is $\exp(2.7599)\cdot 1\cdot 3396 \simeq 21$, and that the median number of species on such an island would be approximately 15.

Trend-surface analysis

Consider a linear-in-parameters model $\text{E}(\mathbf{Y}) = \mathbf{zb}$, where $\mathbf{z}$ is an $n \times k$ matrix, the first column of which contains 1's and the remaining $(k-1)$ columns contain the terms of a polynomial power-series expansion:

$$\mathbf{z} = \begin{pmatrix} 1 & X_{11} & X_{21} & X_{11}^2 & X_{11}X_{21} & X_{21}^2 & \cdots \\ 1 & X_{12} & X_{22} & X_{12}^2 & X_{12}X_{22} & X_{22}^2 & \cdots \\ 1 & X_{13} & X_{23} & X_{13}^2 & X_{13}X_{23} & X_{23}^2 & \cdots \\ \cdot & \cdot & \cdot & \cdot & \cdot & \cdot & \cdots \\ \cdot & \cdot & \cdot & \cdot & \cdot & \cdot & \cdots \\ \cdot & \cdot & \cdot & \cdot & \cdot & \cdot & \cdots \\ 1 & X_{1n} & X_{2n} & X_{1n}^2 & X_{1n}X_{2n} & X_{2n}^2 & \cdots \end{pmatrix} \tag{5.82}$$

The model written in non-matrix terms is

$$\mathrm{E}(Y_i) = b_0 + b_1X_{1i} + b_2X_{2i} + b_3X_{1i}^2 + b_4X_{1i}X_{2i} + b_5X_{2i}^2 + \ldots. \quad (5.83)$$

The trend-surface model is the special case of the above model where X_{1i} and X_{2i} are the spatial coordinates of locality i; a more general definition gives us a response surface.

Equation (5.83) is merely a linear regression model with transformed X variables that can be fitted using OLS given the usual provisos. The model works as a highly malleable representation of Y's spatial variation as a function of Y's position. This is a feature which initially attracted application of the model by geophysicists and geologists (Miller, 1956; Krumbein, 1956), partly because of its capacity to accurately interpolate predicted Y-values between very expensive bore-holes! The model has latterly found wider scope among geographers (Tobler, 1964; Chorley and Haggett, 1965) and archaeologists (Hodder and Orton, 1976).

The variants of the trend-surface model come about by curtailing the polynomial expression in equation (5.83) at various points. This creates first-, second-, third-order (and so on) surfaces which vary dramatically in the way $\hat{Y}_i$ responds to location. The first-order, or linear, trend-surface model,

$$\mathrm{E}(Y_i) = b_0 + b_1X_{1i} + b_2X_{2i}, \quad (5.84)$$

is characterized by parallel 'contour' lines connecting equi-valued $\hat{Y}_i$ in two dimensions (see, for example, Figure 5.9 below), and in three dimensions by a sloping 'shed-roof' obtained by infilling $\hat{Y}_i$ at the interstices where no observations are actually made and by representing the entire surface thus created as a block diagram (as in Haggett, Cliff, and Frey, 1977)

The second-order, or quadratic, model is the model given by the six terms specified on the right-hand side of equation (5.83). The contour lines for this model are curves (see, for example, Figure 5.10a). It is obvious that this model is simply the linear model with three additional (quadratic) terms. As a consequence the quadratic model will always fit a set of data at least as well as the linear model fitted that set of data. We can check whether there is a significant improvement in the fit by following the general lines of analysis exemplified earlier in Examples 5.1 (contd.) and 5.2 (contd.)—see also Chayes (1970).

The addition of further terms leads to a cubic surface (such as that illustrated by Figure 5.10b), while still further terms can continue to be added up to the point where the number of model parameters is equal to the number of localities (when, of course, we shall have a perfectly fitting surface). Each increase in order produces extra undulations in the landscape depicted in the block diagram, with the number of inflexions being one less than the order of the surface (p) and with the number of parameters being $k = (p + 1)(p + 2)/2$.

It should be noted that an oscillatory variation in Y is better represented by a model consisting of sin and cos functions of the Fourier family, rather than the polynomial terms that have been discussed so far (see Bassett, 1972; Miesch and Connor, 1968).

Although in every sense merely a regression model, implementation of trend-surface analysis is by no means a routine matter and difficulties are commonly encountered that normally lie dormant in regression analysis in general. A major problem is that of obtaining reliable estimates of the $\{b_j\}$, on account of the ill-conditioning of $\mathbf{z}'\mathbf{z}$ that is due to $\mathbf{z}$ containing a power-series expansion. We briefly mentioned above that numerical matrix inversion procedures are severely challenged by nearly singular matrices: the problem becomes particularly acute when estimating the parameters of cubic and higher-order surfaces. Mather (1976) gives details of alternative numerical approaches that try to avoid the consequences of near-singularity, and some progress may come from relocation of the axes which has the effect of altering $\mathbf{X}$ and hence $\mathbf{z}'\mathbf{z}$. A further problem is that the raising of X_{1i} and X_{2i} to high powers can produce an unnoticed, but important, rounding-error in $\mathbf{z}'\mathbf{z}$, because of the incapacity of computers to distinguish between very large (or very small) numbers. This can be lessened by centring the origin and by scaling the X's to range from -1 to $+1$. In view of these problems one should be somewhat sceptical of the accuracy of computer algorithms (Unwin and Hepple, 1974; Unwin, 1975).

In addition to the purely numerical problems outlined above, there are effects due to the shape of the boundary of the study area and the disposition of sample localities within that boundary. As Ripley (1981) notes, boundary or edge effects are a well-known problem in polynomial regression with 'a tendency to wave the edges to fit points in the centre'; this is best ameliorated by a wide spatial coverage of sample localities to control extrapolated values near the map edge. Mather (1976) cites an example of an edge-affected trend surface.

Highly clustered localities are of little value as a basis for predicting Y at under-represented parts of the study area, such as map edges, since $\hat{Y}_i$ will be weighted towards the area of greatest representation. There is an analogy here with time trends inaccurately projected far into the future. Excessive clustering can also bring on the possibility of spatially autocorrelated residuals, since 'we would expect deviations from a trend surface to be positively correlated over short distances' (Ripley, 1981). As with regression in general, this disrupts normal inference and creates a downwardly biased estimate of σ^2. Haining (1981a) discusses these consequences in (slightly) more detail. Cliff and Ord (1981a) test trend-surface residuals for autocorrelation using methods discussed in a general regression context in the next section.

One way out of the boundary and clustering problem is to relocate the boundary so that the localities no longer have a clustered appearance! The autocorrelation problem is more insidious, demanding physically relocated localities or more sophisticated models. The difficulty arises because often the analyst has no control over sample site location, and may inherit a particularly maldistributed set. Doveton and Parsley (1970), Gray (1972), Robinson (1972), Unwin (1970, 1973a), and Mather (1976) discuss the significance of locality effects. The consensus is (apparently) that whilst mild clustering may often be of little significance, the more ideal and representative locality distributions are random or at rectangular grid intersection points—the latter being particularly useful for simplifying calculations performed manually (Krumbein and Graybill, 1965).

As the concept of the trend surface has diffused, more ambitious applications have been attempted and correspondingly more pressure has been put on the model assumptions. Davis (1973), for example, discusses the problem of the non-replicability of geological (structural) data whereby *all* errors $\{e_i\}$ are supposed to be measurement errors which are usually comparatively negligible. This means that the (autocorrelated) residuals from a given surface mainly reflect the influence of the extra (autocorrelated) terms of the 'true' higher-order surface confounded with (measurement) error. This precludes the usual inferential procedures, although geologists have still found such residuals to be useful as indicators of localities differing from the surface norm.

If replication is possible, then the residual sum of squares, D, can be broken down into two components, that due to pure error $\{e_i\}$, and that due to model misspecification. Given rY measures at locality i, the estimate of the variance due to pure error is $\sum_{j=1}^{r}(Y_{ij}-\bar{Y}_i)^2/(r-1)$. The 'lack of fit' variance estimate is $\{D-\sum_{j=1}^{r}(Y_{ij}-\bar{Y}_i)^2\}/(n-k-r+1)$. If the ratio of the lack-of-fit estimate to the pure-error estimate is significantly greater than the appropriate percentage point of the $F_{(n-k-r+1),(r-1)}$ distribution, then this indicates the need for a model different from the one generating the residuals.

There is also the problem of non-spatially continuous phenomena such as towns or steep mountain hollows (corries) (cf. Norcliffe, 1969; Unwin, 1973b), which cannot in principle be measured at all study area localities. In fact, these phenomena may be so infrequent as to demand measurement at all available localities, thus making the idea that one is working with a (spatial) sample somewhat remote. Great imagination has gone into turning what appears to be a population into a sample, thereby making statistical theory relevant (though we could again appeal to measurement error). Using a surface fitted to all realizations of the process still extant, it is possible to predict, say, the altitudes of corries at impossibly low altitude valley bottoms by claiming that the prediction only holds if and when conditions are identical to those on mountain peaks, albeit subject to some locational shift. One is therefore predicting 'variability in potential for development' (Unwin, 1973b). Wrigley (1977) extends the trend surface idea even further to accommodate a categorical response variable. James (1970) and Attoh and Whitten (1979) consider surfaces dislocated by geological fault-lines.

***Example* 5.9**

Table 5.17 contains data relating to 23 English, Welsh, and Manx weather stations. The Y variable is the mean daily maximum temperature for September 1982. The X variables index the positions of the weather stations using the scale

Table 5.17 The location and temperature of 23 weather stations in England, Wales, and the Isle of Man, together with their altitudes and the fitted values under two alternative models

Location	Coordinates X_{1i}	X_{2i}	Altitude (metres)	Mean September temperature (°C)	Fitted value for linear trend surface	Fitted value for augmented quadratic trend surface
Douglas	− 0.162	0.179	85	15.2	16.29	15.60
Scarborough	0.109	0.190	52	18.1	17.91	18.48
Hull	0.110	0.130	2	19.3	18.21	19.07
Skegness	0.159	0.067	5	18.9	18.82	19.38
Lowestoft	0.258	− 0.004	25	19.3	19.78	19.20
Clacton	0.216	− 0.086	16	19.3	19.92	19.68
Buxton	0.009	0.076	307	16.0	17.85	16.75
Shawbury	− 0.044	0.022	72	18.6	17.78	18.20
Malvern	− 0.020	− 0.056	62	19.3	18.32	18.90
Hurley	0.082	− 0.117	43	19.9	19.25	19.78
Faversham	0.200	− 0.140	48	20.4	20.09	19.69
Margate	0.240	− 0.127	16	19.9	20.27	19.67
Hastings	0.180	− 0.193	45	19.5	20.22	19.95
Squires Gate	− 0.063	0.132	10	17.4	17.13	17.78
Colwyn Bay	− 0.111	0.080	24	17.7	17.09	17.44
Loggerheads	− 0.076	0.068	210	17.3	17.36	16.67
Lake Vyrnwy	− 0.100	0.020	303	16.1	17.45	16.07
Aberporth	− 0.175	− 0.046	134	16.6	17.31	16.68
Swansea	− 0.133	− 0.106	10	18.3	17.86	18.48
Bude	− 0.179	− 0.196	15	18.2	18.02	18.49
Rosewarne	− 0.235	− 0.258	76	18.0	17.97	17.75
Durham	0.026	0.245	102	18.0	17.13	17.38
Poole	0.005	− 0.202	5	19.9	19.18	20.20

Source: *The Monthly Weather Report*, September 1982: Meteorological Office.

indicated on Figure 5.9. A preliminary informal indication of the possibility of a spatial trend is obtained by fitting the linear trend surface model to this data. This model gives rise to the predicted values given in Table 5.17 which are evidently in reasonable agreement with the observed values.

Under the assumption that this linear trend model is an appropriate explanation for the data, together with the assumption that the Gauss–Markov conditions are satisfied, we can use an F test to investigate the significance of the fit. The linear model contains three parameters and has a deviance, D_L, equal to 13.54, while the total sum of squares is $T = 43.01$. Hence $r^2 = 68\%$, while $(n-k)(T-D_L)/(k-1)D_L = 21.75$, which far exceeds the upper 0.1% point (9.95) of the $F_{2,20}$ distribution.

The estimate of the variance due to unexplained factors will incorporate

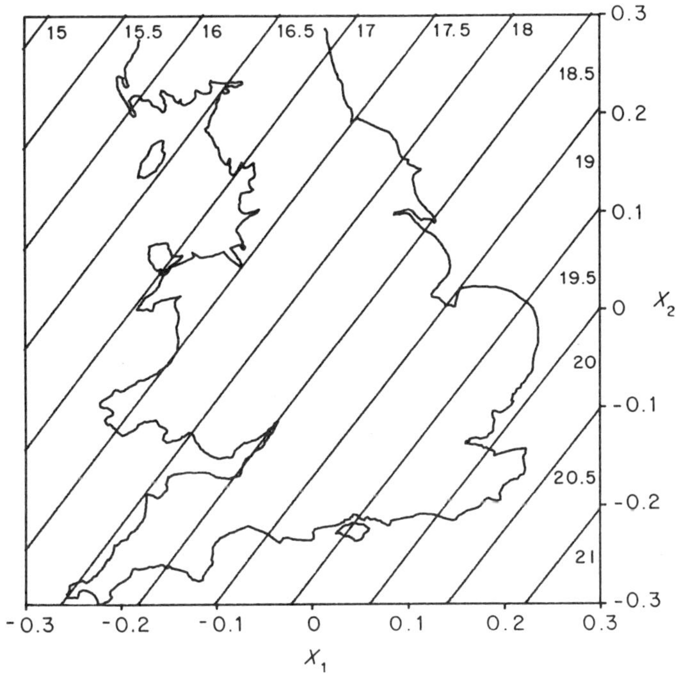

Figure 5.9 Linear trend surface fitted to September temperature data

contributions due to possible non-linear elements of the trend surface as well as any other geographic factors having a bearing on temperature. If any of these have a profound effect on temperature, then the estimate of variance, $\hat{\sigma}^2 = D_L/(n-k)$, will be inflated and a test statistic that incorporates $\hat{\sigma}^2$ in its denominator will be correspondingly depressed. When referring to a 'significant' linear trend, what we really mean is that the linear trend is *significantly large by comparison with all the other influences on temperature.* The equation of the fitted line is estimated as $\bar{Y}_i = 18.16 + 6.193X_{1i} - 4.896X_{2i}$.

Following the success of the fit of the linear trend surface it is natural to consider the possibility that higher-order trend surfaces will produce improved fits. The three extra parameters of the quadratic surface result in a reduction of the deviance to $D_Q = 9.75$, with a further reduction to $D_c = 9.06$ on the addition of the 4 cubic terms. Following our previous arguments and in line with the generally accepted application of regression analysis in this field (Krumbein and Graybill, 1965; Davis, 1973; Cliff and Ord, 1981a), we test the significance of the three quadratic terms by evaluating the ratio of the deviance change per degree of freedom, $(D_Q - D_L)/3$, to the revised estimate of 'experimental error' variance, $D_Q/(n-k)$, where k is now equal to 6. For our data this ratio is 2.2 which does not exceed the upper 5% point of the $F_{3,17}$ distribution. In a similar way, the ratio corresponding to the four cubic terms is $\{(D_c - D_Q)/4\}/(D_c/13) = 0.25$, which is clearly also not significant. On the basis of these tests we can conclude that *if we consider two-dimensional variation alone* then there is no significant evidence of other than a linear trend in temperature.

The procedure adopted above involves a revised estimate of the 'error' variance for each successive F test. If a subsequent F test results in a 'significant' result, then this implies that the estimate of variance used in a previous test was falsely inflated by the setting to zero of parameters now found to be of importance. For this reason Krumbein and Graybill (1965) advocate the use of F-values simply as indices or cut-off points for deciding whether to fit the next higher degree polynomial. Evidently this informal approach has found favour in geological and related studies because of the impracticability of formally correct F tests.

Valid F tests would require the isolation of the deviance due to 'chance' variation, where our definition of 'chance' would be 'all factors of no conceivable interest to the analyst'. This is scarcely feasible in the present context: observations at closely neighbouring locations could be used to obtain an estimate of this 'pure error' component of the variation—but then it is likely that we would be required by a succession of 'highly significant' F-values to fit a trend surface of very high order. Such a surface would be well-nigh uninterpretable and would amount to little more than a recapitulation of the observed values. One possible compromise involves the division of the deviance into major and minor elements, the latter containing a systematic component: Davis (1973) gives details.

Examination of the fitted values for the linear trend model (see Table 5.17) reveals the familiar fact that temperature is a three-dimensional phenomenon, with mountain tops tending to be cooler than surrounding localities! Whilst Buxton (307 m above sea level) and Lake Vyrnwy (303 m) can scarcely be regarded as mountain tops (!) it is nevertheless the case that these locations are associated with the two largest (and negative) residuals. We therefore now deviate from the analysis of 'pure' trend surfaces to create a series of augmented trend surfaces which incorporate altitude as an extra variable. We denote the deviance of the augmented linear trend surface by D_{L+A}, and for our data this has the value 7.47, compared to the deviance of the 'pure' linear model, $D_L = 13.54$. The reduction in the deviance caused by the introduction of a single altitude parameter is striking—the value of r^2 has increased from 68 to 83%. Adopting the previous informal F-value approach, we note that $(13.54 - 7.47)/(7.47/19) = 15.4$, which exceeds 15.08, the upper 0.1% point of the $F_{1,19}$ distribution.

Following our previous investigation, we now consider the augmented quadratic and cubic surfaces, for which we find that $D_{Q+A} = 3.65$ and $D_{C+A} = 2.82$. The informal F-values are $\{(7.47 - 3.65)/3\}/(3.65/16) = 5.58$ and $\{(3.65 - 2.82)/4\}/(2.82/12) = 0.88$, respectively, indicating significant evidence of the need for a quadratic component in the augmented trend surface (since 5.58

exceeds 5.29, the upper 1% point of the $F_{3,16}$ distribution). We conclude that the augmented quadratic trend surface is the most appropriate of the models that we have considered for the September temperature data. This model explains 91.5% of the observed variation in the Y-values and the corresponding fitted values are given as the last column in Table 5.17.

In order to display the dependence of Y on its two-dimensional location alone, as does a 'pure' trend surface, we can utilize all but the last term of the estimated equation $\hat{Y}_i = 19.16 + 6.79X_{1i} - 5.28X_{2i} - 23.32X_{1i}^2 + 10.75X_{1i}X_{2i} - 0.40X_{2i}^2 - 0.00676A_i$, to obtain predicted equivalent sea-level temperature variation over the study area. The resulting sea-level trend surface is given as Figure 5.10a, with, for interest, the corresponding sea-level cubic trend surface being illustrated in Figure 5.10b.

Note that, when working with the augmented models, there is a strong indication of the need to use a non-linear trend surface, whereas, when altitude was omitted, the significance of the quadratic terms was not apparent. The reason for this apparent contradiction is simply that, by taking altitude into account we have accounted for a large chunk of the previously unspecified 'experimental' error and consequently we obtain a much sharper picture of the data.

We could seek to further augment our models by allowing for a possible quadratic altitude effect or by incorporating a 'distance-to-coast' variable. In this way, by introducing more explanatory variables we can attempt to gain a more precise picture of the underlying trend surface—which, we can anticipate, will reflect the influences of latitude and of the position of the continental landmass relative to the region of interest.

5.6 THE ANALYSIS OF RESIDUALS

The form of model that we select for application to a set of data carries with it an implication concerning the anticipated error structure. After fitting a model, it therefore behoves us to examine the estimates of error that we have obtained, in order to confirm that their distribution is in accord with our preconceptions.

Consider the general linear model $\mathbf{Y} = \mathbf{Xb} + \mathbf{e}$, with $\mathrm{E}(\mathbf{e}) = \mathbf{0}$. The estimated errors (i.e. residuals) are given by $\hat{\mathbf{e}} = \mathbf{Y} - \hat{\mathbf{Y}}$, where $\hat{\mathbf{Y}} - \mathbf{X}\hat{\mathbf{b}}$ and $\hat{\mathbf{b}} = (\mathbf{X}'\mathbf{X})^{-1}\mathbf{X}'\mathbf{Y}$ (equation (5.4)). Successive substitutions for $\hat{\mathbf{Y}}$ and for $\hat{\mathbf{b}}$ then give us the result

$$\hat{\mathbf{e}} = \mathbf{Y} - \mathbf{X}(\mathbf{X}'\mathbf{X})^{-1}\mathbf{X}'\mathbf{Y} = \mathbf{MY}, \tag{5.85}$$

where

$$\mathbf{M} = \mathbf{I} - \mathbf{X}(\mathbf{X}'\mathbf{X})^{-1}\mathbf{X}'. \tag{5.86}$$

The matrix $\mathbf{M}$ has some fascinating properties ($\mathbf{M}' = \mathbf{M}$ and $\mathbf{M}'\mathbf{M} = \mathbf{M}$) which we shall use below. We shall also need the result that

$$\mathbf{MXb} = \mathbf{Xb} - \mathbf{X}(\mathbf{X}'\mathbf{X})^{-1}\mathbf{X}'\cdot\mathbf{Xb} = \mathbf{Xb} - \mathbf{Xb} = \mathbf{0}. \tag{5.87}$$

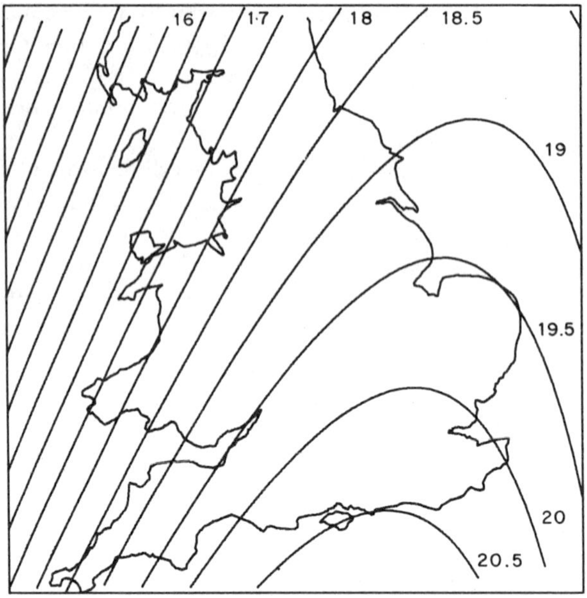

(a)

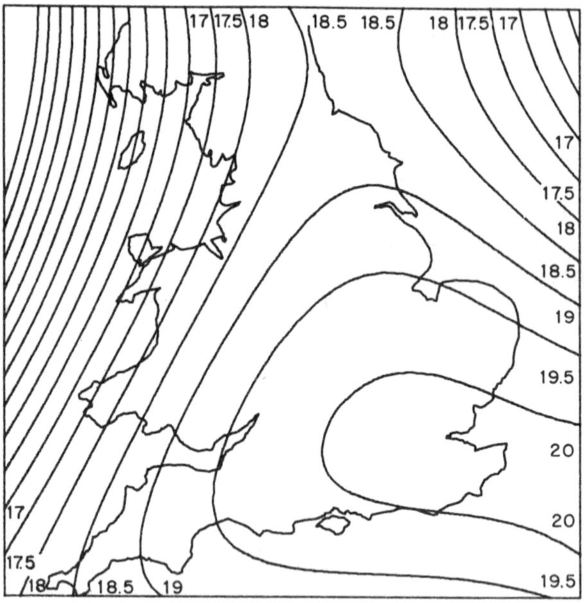

(b)

Figure 5.10(a) Quadratic (sea-level) trend surface fitted to September temperature data. (b) Cubic (sea-level) trend surface fitted to September temperature data

We can now use the equation of the general linear model together with equation (5.85) to obtain an expression for the estimated errors in terms of the true errors:

$$\hat{\mathbf{e}} = \mathbf{Me}. \tag{5.88}$$

Since $\mathbf{M}$ is a constant and $\mathrm{E}(\mathbf{e}) = \mathbf{0}$, it follows that $\mathrm{E}(\hat{\mathbf{e}}) = \mathbf{0}$. If we consider the general case, and presume that $\mathbf{V}(\mathbf{e}) = \sigma^2\mathbf{D}$, then we have that

$$\mathbf{V}(\hat{\mathbf{e}}) = \sigma^2\mathbf{M}'\mathbf{DM} = \sigma^2\mathbf{MDM}. \tag{5.89}$$

Using equation (5.88) in conjunction with these results we can therefore state that the distribution of the $\{\hat{e}_i\}$ will have the same form as that of the $\{e_i\}$, but, while having the same mean ($\mathbf{0}$), it will have a modified variance–covariance structure.

It follows from this that, if the model $\mathbf{Y} = \mathbf{Xb} + \mathbf{e}$ has the correct form, then we can assess the probable form of the distribution of the $\{e_i\}$ by studying the observed distribution of the $\{\hat{e}_i\}$. If the $\{\hat{e}_i\}$ appear to be independent observations from a normal distribution, then it is probable that this was true for the $\{e_i\}$. If the $\{\hat{e}_i\}$ appear to be spatially autocorrelated, then it is probable that the $\{e_i\}$ were spatially autocorrelated.

Of course, if the model that we choose is not appropriate, then the estimated 'errors' will comprise a mixture of experimental error and model error, in which case it may be difficult to make any useful deductions concerning the error distribution.

Sometimes, when we examine the residuals, one or two appear to depart from the general pattern. In such cases it is wise to examine the corresponding data items for the existence of some extenuating circumstance (such as incorrectly recorded data!)

Example 5.10

To illustrate the general techniques of the analysis of residuals, we reconsider the simple normal errors linear model which was applied to the Galapagos Islands data in Example 5.7. Our subsequent analysis in Example 5.8 has suggested that a multiplicative Poisson errors model is a more reasonable model for these data. Consequently, we can expect that the residuals for the normal linear model will reflect this combination of faulty model specifications.

In order to discern a possible Poisson error structure in which variance is proportional to mean, we could plot the estimated errors against the observed Y-values, as shown in Figure 5.11. If there is a Poisson error structure, then we should be able to see a funnelling of the errors towards zero as Y diminishes. In this investigation the actual sign of the residual is of no account and therefore a clearer plot may be obtained by a comparison of $|e_i|$ with Y_i, as shown in Figure 5.12.

From Figure 5.12 there does seem to be some indication of an expansion in the residuals as Y increases, but the effect is evidently somewhat masked by the choice of a linear model.

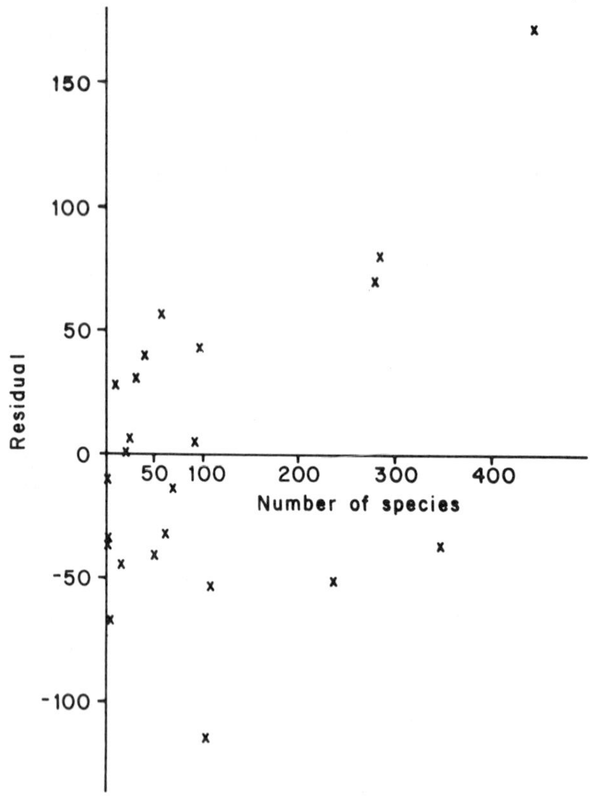

Figure 5.11 Residual size in relation to Galapagos Islands species numbers

Abnormal residuals

Despite our recent obsession with Poisson errors, it remains true that the vast majority of the applications of linear models include the assumption of normally distributed errors. In particular, a common assumption is that $\mathbf{e} \sim N(\mathbf{0},\sigma^2\mathbf{I})$. In this case, from equations (5.88) and (5.89) it follows that $\hat{\mathbf{e}} \sim N(\mathbf{0},\sigma^2\mathbf{M})$. Since σ^2 is unknown, we estimate it by using the 'residual mean square': $\hat{\sigma}^2 = \sum \hat{e}_i^2/(n-k) = D/(n-k)$.

In Example 5.10 we had a definite alternative error structure in mind. This will not generally be the case, so that the plot of $\hat{e}_i$ against Y_i may not be particularly sensible (though, of course, all plots are likely to be informative). Draper and Smith (1981) suggest that a simple linear scatter diagram of the $\{\hat{e}_i\}$ (as illustrated below by Figures 5.13 and 5.14) will often suffice to provide an informal indication of the truth, or otherwise, of the assumption of normality.

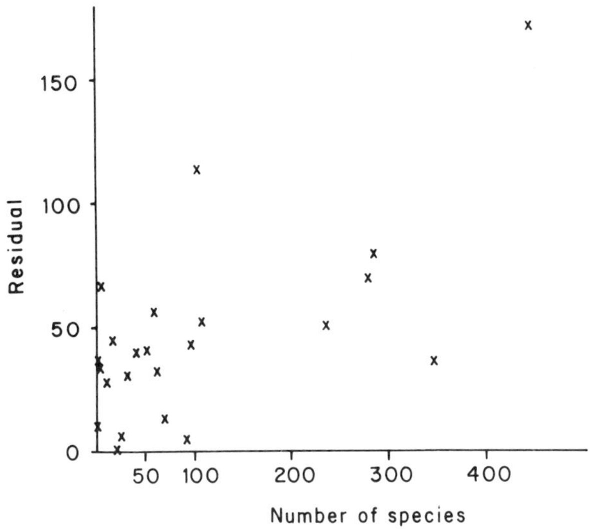

Figure 5.12 Absolute residual size in relation to Galapagos Islands species numbers

A very quick test of approximate normality can also be achieved by calculating the proportions of the $\{\hat{e}_i\}$ that exceed σ or 2σ. This test relies on the matrix $\mathbf{M}$ being not too dissimilar to the identity matrix, $\mathbf{I}$—this will be the case when the number of observations (n) greatly exceeds the number of parameters in the model (k). In this case the two proportions should be about 32% and 5%, respectively.

A much more powerful test of normality is that suggested by Shapiro and Francia (1972). This test makes use of the expected values of the standard normal *order statistics*. Suppose that we take a sample of n independent observations from a $N(0, 1)$ distribution and that we denote by $x_{(i)}$, the value of the ith largest of these n observations. Clearly, $x_{(i)}$ may be regarded as an observation upon a random variable, which we can denote by $X_{(i)}$: this random variable is the order statistic. Repeated samples of size n will lead to different observed values of $X_{(i)}$. Blom (1958) has shown that the expected value of $X_{(i)}$ is closely approximated by

$$m_i = \Phi^{-1}\left(\frac{8i-3}{8n+2}\right), \qquad i = 1, 2, \ldots, n. \tag{5.90}$$

where $\Phi^{-1}(\cdot)$ is the inverse normal distribution function (so that, for example, $\Phi^{-1}(0.975) = 1.96$). Weisberg and Bingham (1975) and Royston (1983) have found this approximation satisfactory, though Neave (1978) has tabulated $E(X_{(i)})$ for small n.

Shapiro and Francia (1972) suggested that a useful test of normality could be based upon the correlation, r_N, between the values of the sample order statistics

$\{x_{(i)}\}$ and the order statistics $\{m_i\}$. Applying this idea to regression residuals, the correlation formula reduces to

$$r_N = \sum m_i \hat{e}_{(i)} \Big/ \left\{ \sum m_i^2 \cdot \sum \hat{e}_{(i)}^2 \right\}. \tag{5.91}$$

Royston (1983) has shown that

$$W = \{(1 - r_N^2)^\lambda - 1\}/\lambda \tag{5.92}$$

has an approximate normal distribution. The parameter λ is given approximately by

$$\lambda = -0.04816 + 0.01972S - 0.01191S^3, \tag{5.93}$$

where

$$S = \log(n) - 5, \tag{5.94}$$

and the logarithm is, as usual, a natural logarithm. Royston shows that the mean and variance of the distribution of W are given by

$$\mathrm{E}(W) = -\exp(1.693 + 0.1442S - 0.01849S^2 + 0.03107S^3 + 0.005572S^4) \tag{5.95}$$

$$\mathrm{var}\,(W) = \{\exp(-0.5107 - 0.1160S - 0.006702S^2 + 0.05447S^3 + 0.008740S^4)\}. \tag{5.96}$$

The coefficients in these formulae, and in the formula for λ, are given here correct to four significant figures and are given to greater accuracy by Royston (1983).

Example 5.11

Figure 5.13 is a plot of the residuals generated by fitting equation (5.11) to the California coast data given in Table 5.2. From Example 5.2 we have that $D = 0.606 \times 10^6$, with $(n-k) = 22$. Thus, $7/26 = 26.9\%$ of the absolute residuals exceed $\{D/(n-k)\}^{1/2}$, and $2/26 = 7.7\%$ exceed $2\{D/(n-k)\}^{1/2}$, compared with the theoretical 32% and 5%, respectively. This suggests that the residuals are likely to be found consistent with the assumption of normality.

Table 5.18 presents the estimated residuals generated from equation (5.11) and the corresponding order statistics (calculated using equation (5.90)). It is evident from the table that there is a high correlation between the two sets of values, and, using equation (5.91), with $\hat{e}_{(i)}$ in place of $x_{(i)}$, we find that $\sum \hat{e}_{(i)}^2 = 605946$, $\sum m_i^2 = 23.5956$, and $\sum m_i \hat{e}_{(i)} = 3754.65$, so that $r_N = 0.9930$. Since $S = -1.7419$, we find that $\lambda = -0.01958$ and so $W = -4.452$, $\mathrm{E}(W) = -3.571$, $\mathrm{var}\,(W) = 0.3421$, and hence $z = \{W - \mathrm{E}(W)\}/\{\mathrm{var}\,(W)\}^{1/2} = -1.506$. Notice that the particularly

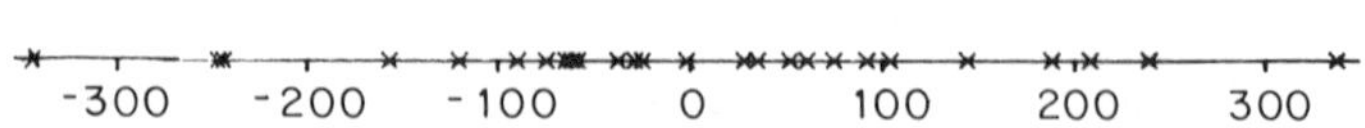

Figure 5.13 The distribution of the Californian residuals

Table* 5.18 *Ordered Californian residuals and standard normal order statistics

Residual, $x_{(i)}$	-341	-247	-243	-156	-120	-90	-75	-65
Order statistic, m_i	-1.982	-1.544	-1.285	-1.091	-0.932	-0.793	-0.668	-0.553
$x_{(i)}$	-62	-61	-40	-28	-26	-1	27	34
m_i	-0.444	-0.341	-0.241	-0.144	-0.048	0.048	0.144	0.241
$x_{(i)}$	52	60	74	92	103	144	188	208
m_i	0.341	0.444	0.553	0.668	0.793	0.932	1.091	1.285
$x_{(i)}$	238	337						
m_i	1.544	1.982						

Table* 5.19 *Ordered Galapagos residuals and standard normal order statistics

Residual, $x_{(i)}$	-114.1	-67.1	-52.8	-51.1	-44.6	-41.0	-36.7	-33.9
Order statistic, m_i	-1.929	-1.481	-1.214	-1.014	-0.847	-0.701	-0.569	-0.446
$x_{(i)}$	-32.3	-30.8	-13.6	-10.0	1.0	5.0	6.5	28.0
m_i	-0.330	-0.218	-0.108	0	0.108	0.218	0.330	0.446
$x_{(i)}$	30.6	40.1	43.3	56.7	70.4	80.4	172.2	
m_i	0.569	0.701	0.847	1.014	1.214	1.481	1.929	

high correlation leads to a negative values of z. The two-sided 90% value of the unit normal distribution is 1.645, which (just) exceeds the observed value of $|z|$. For the present, therefore, we can accept the hypothesis of normally distributed errors. Later, however, we shall discover that there is some evidence that the errors display spatial autocorrelation.

***Example* 5.12**

The linear plot of the residuals that result from the fit of the normal model to the Galapagos Islands data is given as Figure 5.14. The plot is in marked contrast to that of Figure 5.13, since, on this occasion, there is not the same concentration around zero and there appear to be some extreme observations. However, from Example 5.7 we have that $\{D/(n-k)\}^{1/2} = (4784.1)^{1/2} = 69.2$, so that the theoretical two-sided 32% point is exceeded by $5/23 = 22\%$ of the observations, and the theoretical two-sided 5% point is exceeded by 4.3% of the observations. Thus, despite its visual impact, the spread of the residuals seems not too dissimilar to that specified by the normal distribution.

Applying the Shapiro–Francia procedure, the results of Table 5.19 lead to a value of $r_N = 0.9666$. Since $S = -1.8645$, we have $\lambda = -0.00775$ and so $W = -2.7521$, $\mathrm{E}(W) = -3.4072$, $\mathrm{var}(W) = 0.3230$, and hence $z = 1.152$. This value corresponds to a two-tail normal probability of about 25%, which is clearly not unusual! The test does not supply significant evidence of non-normality of the error distribution. Following our analysis of these data in Example 5.8 we can endorse our comment in Example 5.10 that the non-normality of the error distribution has been effectively masked by the poor structure of the chosen model.

Spatially autocorrelated residuals

It is evident that, if we anticipate that the error structure may display spatial influences, then an appropriate plot of the $\{\hat{e}_i\}$ would be a geographical map of the residuals. Often a plot of the signs of the residuals will suffice to draw attention to concentrations of positive or negative residuals which thereby exhibit the influence of spatial autocorrelation. An unusual spatial distribution of residuals from an OLS regression model suggests the existence of missing, but significant, spatial variables (or of higher-order polynomial terms or interaction terms involving the spatial X variables already present); alternatively, there is a need for an autoregressive model of the type discussed elsewhere in this chapter.

In Chapter 3 we examined the general problem of identifying spatial

Figure 5.14 The distribution of the Galapagos Islands residuals

autocorrelation in a set of Y-values, and concluded that, of various alternatives, the most efficient statistic was Moran's I. We now consider the application of I to a set of spatially distributed residuals.

Moran's I for residuals

Following Brandsma and Ketellapper (1979a) we shall choose to work not with I itself, but with the closely related I_k, defined by

$$I_k = \hat{\mathbf{e}}'\mathbf{W}\hat{\mathbf{e}}/\hat{\mathbf{e}}'\hat{\mathbf{e}}. \tag{5.97}$$

In this expression $\mathbf{W}$ is the usual weights matrix involving inter-locational distances.

This definition is directly analagous to that of I given by equation (3.24), although the latter was not expressed in matrix terms. The two differences between the expressions are that I involves a constant multiplier (n/S_0) and the n independent observed values $\{(x_i - \bar{x})\}$, whereas I_k involves the n interdependent residuals $\{\hat{e}_i\}$.

The change of notation serves to remind us of the fundamental distinction between I and I_k, namely that the $\{\hat{e}_i\}$ are subject to k linear constraints brought about by the estimation of the k $\{b_j\}$ terms in the model.

The moments of the original Moran statistic were given by equations (3.28) and (3.29), in the case of randomization, and by equations (3.44a) and (3.44b), in the case of normality. We argued above that, if the number of parameters in the model, k, were much smaller than the number of observations, n, then the matrix $\mathbf{M}$ would be close to the identity matrix $\mathbf{I}$. In plain English, this says that if there are lots of observations for a simple model, then we can overlook the correlations between the residuals and regard the residuals as though they were n independent observations. If this is the case, then it follows that we can use I and the results of Chapter 3, and this does seem to work quite well (Cliff and Ord, 1981a).

An alternative strategy, proposed by Theil (1965), which also seeks to use the original I statistic, involves the discarding of k of the residuals (chosen in some arbitrary fashion). Whilst it is true that the remaining $(n-k)$ residuals may be regarded as being $(n-k)$ independent observations, there is the obvious problem as to which residuals should be discarded, and we shall not pursue this idea. Further discussion is to be found in Brandsma and Ketellapper (1979a) and Cliff and Ord (1973, 1981a).

The I_k distribution under the null hypothesis

We seek to investigate the null hypothesis of no spatial autocorrelation and need to know the sampling distribution of I_k. In view of the interdependencies amongst the residuals, the randomization procedure that we broadly favoured in connection with I may here be less appropriate than the normality approach. The mean and variance of I_k are then found to be

$$E(I_k) = \text{tr}(\mathbf{MW})/(n-k), \tag{5.98}$$

$$\text{var}(I_k) = \frac{\text{tr}(\mathbf{MWMW}') + \text{tr}(\mathbf{MWMW}) + \{\text{tr}(\mathbf{MW})\}^2}{(n-k)(n-k+2)} - E(I_k)^2. \tag{5.99}$$

Here, $\mathbf{M} = \mathbf{I} - \mathbf{X}(\mathbf{X}'\mathbf{X})^{-1}\mathbf{X}$, as before. The sampling distribution of I_k may be assumed to be asymptotically normal. The notation follows that of Brandsma and Ketellapper (1979a) and Ripley (1981). Cliff and Ord (1972, 1981a) provide the equivalent moments for Moran's I.

Example 5.13

In the first example of this chapter we applied the OLS procedure to the Irish blood data in order to examine the separate dependence of Y, the percentage of blood group A, on each of X_2, the ranked town intensity, and X_3, the 'Pale' dummy variable. This led to the estimated relations given by equations (5.7a) and (5.7b). Subsequently, we considered the logit transform $z = \log\{Y/(100 - Y)\}$, and obtained corresponding equations (5.16b) and (5.16d). Throughout the fitting of these models we assumed that the errors were spatially uncorrelated. We now examine estimates of these errors (the residuals) to ascertain whether this was a reasonable assumption.

Using equations (5.92), (5.93), and (5.94) in conjunction with the $\mathbf{W}$ matrix specified in Table 5.3, we obtain the results set out in Table 5.20. The z-values of 2.59 and 3.52 correspond to one-tailed probabilities of 0.48% and 0.02%, respectively, providing unmistakable evidence of some spatial autocorrelation amongst the residuals.

Anticipating the presence of this spatial autocorrelation, in Example 5.3 we introduced models that specified just such a spatial autocorrelation and the results of fitting these models were summarized in Table 5.4. In the earlier subsection entitled 'Inference and maximum likelihood' we suggested a test of the null hypothesis, $\rho = 0$, against the alternative $\rho \neq 0$ (or $\rho > 0$). This consisted of contrasting the ratio of $\hat{\rho}$ to its standard error with the percentage points of the unit normal distribution. The present test, based on I_k, is an alternative to the

***Table* 5.20** *I_k in relation to its normal sampling moments for the Irish blood models*

Model	I_k	$E(I_k)$	$\text{var}(I_k)$	z
(5.7a)	0.4307	−0.0645	0.0198	3.52
(5.7b)	0.3022	−0.0649	0.0201	2.59
(5.16b)	0.4315	−0.0645	0.0198	3.53
(5.16d)	0.3030	−0.0649	0.0201	2.59

Table **5.21** I_k *in relation to its normal sampling moments for the Californian models*

Model	I_k	$E(I_k)$	$var(I_k)$	z
Island-isolation **W**	0.067	−0.033	0.0078	1.13
Two-part distance-based **W**	−0.326	−0.081	0.0179	−1.83

earlier procedure. Brandsma and Ketellapper (1976b) examined the behaviour of these two tests and also of the likelihood-ratio test. On the basis of their results they concluded that the test involving I_k was the most powerful.

Example 5.14

Table 5.21 contains results equivalent to those in Table 5.20, but relating to the Californian regression residuals of Table 5.18. The remarks in the previous example suggest that inferences based on I_k and ρ will be only broadly similar. This is borne out by this example, for in contrast to the highly significant departure of $\hat{\rho}$ from zero (see Example 5.4), there is no strong evidence of positively autocorrelated errors when I_k is used in conjunction with the island-isolation **W** matrix. However, the alternative two-part distance-based **W** matrix does generate significantly negatively autocorrelated residuals and reaffirms our previous conclusion that something was 'going on'. The signal received in Example 5.4 has been amplified by the analysis using I_k.

The **W** matrices referred to above are sparse, especially the island-isolate matrix (as can be seen in Table 5.6). This may cause the I_k sampling distribution to depart somewhat from normality, since it is subject to the same controlling conditions that were outlined in a more general context in Chapter 3. We therefore defer final comments.

Example 5.15

Previous work with the Galapagos Islands data (and the evident insularity of the Californian islands) suggests independent normal errors as a possible alternative to the (more likely) Poisson errors. The residual autocorrelation test offers no evidence to the contrary, since, using residuals derived from Table 5.16, we find that $I_k = 0.0105$, $E(I_k) = -0.1059$, $var(I_k) = 0.0209$, and hence $z = 0.805$.

A randomization approach to autocorrelated residuals

Randomization liberates us from the need to assume that the sampling distribution of I_k is normal and is therefore useful given sparse **W** matrices which may engender non-normality. In any event it provides a supplementary check on our previous inferences.

The procedure commences by obtaining a random sample of n observations from a $N(0, 1)$ error distribution. There are several computer algorithms for the generation of random normal deviates, and there are also many published tables. We can reasonably set the variance to 1 because I_k is scale independent and the value of σ^2 is irrelevant. These random deviates are then assigned in a random fashion to the n locations (so that any apparent spatial autocorrelation is entirely attributable to chance). Having generated these random 'errors' $\mathbf{e}$, we can then use equation (5.88) to obtain the corresponding $\hat{\mathbf{e}}$. However, since the random sample of $\{e_i\}$ need not have a sample mean of 0, it is necessary to adjust the calculated residuals and to work with $\hat{e}_i^* = \hat{e}_i - \bar{\hat{e}}_i$, where $\bar{\hat{e}}_i$ is the mean of the calculated residuals. From these $\{\hat{e}_i^*\}$-values we then calculate the corresponding I_k-value.

Repetition of this procedure to produce a total of, say, 99 I_k-values, then provides us with the yardstick for assessing the significance of the I_k-value that resulted from the observed data.

***Example* 5.13** (cont.)

Application of the randomization procedure for the Irish blood data produces uncontroversial results! For all four models considered, the observed value of I_k is highly significant, being greater than either 98 or 99 of the 99 reference values in every case.

Table 5.22 gives the mean and variance of each of the four sets of 99 reference values. Comparing these actual sample moments with those values given by equations (5.98) and (5.99) which were reported in Table 5.20, it is evident that, while the observed variances were much as expected, the observed means were consistently larger than expected. The increase only amounts to about 0.05 standard deviations, however, which is not enough to worry about.

***Example* 5.14** (cont.)

Table 5.23 contains the I_k randomization distributions for each of the Californian $\mathbf{W}$ matrices. The significant negative residual autocorrelation of the

Table* 5.22** ***I_k and parameters of the I_k randomization distribution for the Irish blood models

Model	I_k	Mean	Variance
(5.7a)	0.4307	−0.0356	0.0193
(5.7b)	0.3022	−0.0557	0.0224
(5.16b)	0.4315	−0.0368	0.0167
(5.16d)	0.3030	−0.0437	0.0240

Table* 5.23 *I_k randomization distribution with Californian residuals

(a) Two-part distance-based proximities (Table 5.6a)

Value of I_k	−0.40	−0.35	−0.30	−0.25	−0.20	−0.15	−0.10	−0.05	0
Freq. $\leq I_k$	0	2	6	19	31	42	56	69	79
	0.05	0.10	0.15	0.20	0.25	0.30			
	87	92	95	98	98	99			

(b) Island isolation proximities (Table 5.6b)

Value of I_k	−0.30	−0.25	−0.20	−0.15	−0.10	−0.05	0	0.05	0.10	0.15
Freq. $\leq I_k$	0	2	7	10	16	36	67	83	94	97
	0.20	0.25	0.30							
	97	98	99							

two-part distance-based **W** matrix remains in evidence, with just five I_k reference values beyond the observed $I_k = -0.326$. The observed $I_k = 0.067$ associated with the island-isolation **W** matrix is exceeded by eleven reference values (Table 5.23b), roughly in accord with the prediction from the assumed normal distribution ($P(\{z > 1.13\}) = 0.129$). Somewhat surprisingly, given our misgivings concerning the structure of these **W** matrices, the normal approximation has performed well.

The above results do little to shake our conviction that islands do not interact in the same way as mainland localities. Clearly, the most appropriate **W** matrix is the island-isolate matrix. The fact that the associated residuals do not display *significant* positive autocorrelation should not cause undue concern, given the volatility of I_k and ρ with changes in the structure of **W**. While we cannot be sure of the precise proximity values to adopt, everything points to a matrix that is broadly of the island-isolate structure.

An alternative to the (island-isolate) autocorrelated errors model might be straightforward multiple regression using distance from the mainland coast as an extra X variable. This variable would take the value 0 for mainland localities. Whilst this variable turns out to make a significant contribution to the fit of the model, its introduction perversely exacerbates the prominent negative residual autocorrelations encountered previously, leading to an I_k-value of -4.065, which is exceeded by all 99 randomly generated I_k-values. Johnson and Raven (1970) try some non-linear models for similar data and discuss relevant processes in the light of biological expertise.

***Example* 5.15** (cont.)

Table 5.24 contains the I_k randomization distribution for the Galapagos Islands data. The observed I_k (0.0105) is exceeded 18 times, which compares with

Table* 5.24 I_k *randomization distribution with Galapagos residuals

Value of I_k	−0.5	−0.4	−0.3	−0.2	−0.1	0	0.1	0.2	0.3	0.4
Freq. $\leq I_k$	0	2	9	31	54	72	94	97	98	99

the forecast of 21 times based on the normal approximation ($z = 0.805$). Once again the normal approximation appears to have been reasonable. Our conclusions concerning the worth of the model are unaltered.

Residual correlograms

Information about the error covariance structure can be gleaned from the variation in residual correlation with inter-locality distance. We refer to a display of the variation as a residual correlogram and construct it in a way that parallels the construction of the correlograms discussed in Chapter 3. A variety of indices could be used to express the amount and direction of residual correlation; here we concentrate on just two, of which one index is $z = \{I_k - \mathrm{E}(I_k)\}/\{\mathrm{var}\,(I_k)\}^{1/2}$, with the terms on the right-hand side being defined in equations (5.97)–(5.99), and the other index is one that will be defined below in equation (5.103). An integral part of the calculations for determining z is the set of $\mathbf{W}$ matrices, $\mathbf{W}_d$ ($d = 1, 2, \ldots, k$), where d indexes the distance class. We take the $\mathbf{W}_d$ proximity values as 1 or 0 depending on whether or not the (i, j) locality pair occupy distance class d. To be consistent with Chapter 3, we could denote the residual autocorrelation at distance d by I_{kd}. We do not formally test for significant residual autocorrelation because of the interdependence of the locality pairs and the sparseness of the $\mathbf{W}_d$ matrices.

***Example* 5.16**

Figure 5.15a depicts the variation of z with d calculated from the Californian residuals originating from equation (5.11) (see Table 5.18). The expected tendency towards close-range negative residual autocorrelation has materialized. We discount suggestions of significance or insignificance in Figure 5.13 for reasons already mentioned. Since there are 26 Californian localities there are 650 distinct inter-locality distances. Somewhat arbitrarily, we have chosen to form these distances into 13 distance classes, each incrementing by 62.5 miles. The numbers of locality pairs per class, $\{n_d\}$, are 86, 70, 66, 62, 90, 84, 48, 52, 18, 24, 16, 16, and 18. The fall-off with increasing distance introduces some extra tail-end instability into the correlogram. This would always be present, even if we chose categories containing equal numbers of pairs, since in that case the final categories would include much larger ranges of inter-locality distances. In the figure, the value of z for each class has been plotted against the mean of the inter-locality distances falling in that class.

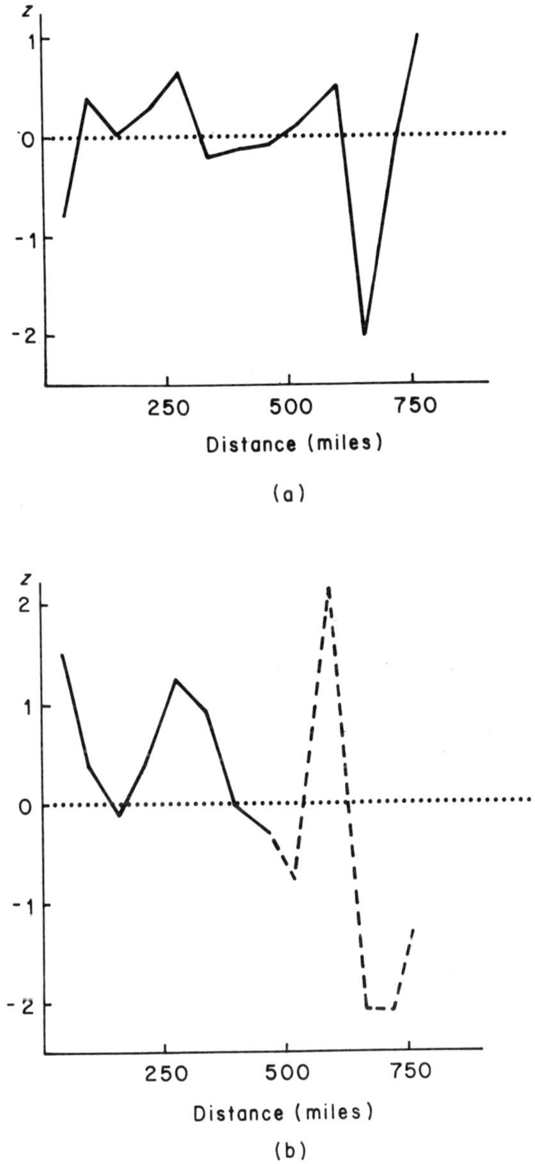

Figure 5.15 Residual correlograms. (a) All Californian localities. (b) Californian Islands only

When we first set up our **W** matrices for these data in Example 5.4, we made the arbitrary assumption that there would be no inter-locality correlations at distances greater than 187.5 miles. However, Figure 5.15a suggests that this cut-off distance may have been too low, since there are suggestions of non-zero autocorrelation for distances up to about 400 miles. The wild values that are noticeable at distances over 600 miles can be partly attributed to the relative dearth of information concerning these distances and partly to the artificial nature of the data, which divorces this section of California from the remainder of the continent.

Figure 5.15b shows what happens when we confine attention to the residuals at the 18 island locations. There is a suggestion of a short-range positive spatial autocorrelation for these localities. However, the model was fitted to all 26 localities and it is possible that Figure 5.15b is simply displaying the side-effect of our failure to include a specific island/mainland dummy variable in the model. The wildly varying values for large d may safely be ignored, since there are only 6, 8, 4, 6, and 4 observations in the final five distance categories.

Following the processes of Chapter 3, we could elaborate on our analysis of Figures 5.15a and b by developing a partial correlogram. Alternatively, we might devise a test of significance by using the randomization principle to generate an 'envelope' of correlograms with which the observed correlogram could be compared. Rather than take up these suggestions, however, we shall investigate the worth of an alternative autocorrelation measure.

Alternative residual correlograms

In Chapter 3 we mentioned the c statistic (equation (3.25)) as being Geary's (1954) alternative to Moran's I. A major distinction between the two measures is the way in which the 'distance' between the locality i- and j-values is measured, a distinction made explicit in the general context by equations (3.26) and (3.27). For residuals, comparable measures of distance are $\hat{e}_i\hat{e}_j$ and $(\hat{e}_i - \hat{e}_j)^2$, respectively.

Pocock *et al.* (1982) and Cook and Pocock (1983) make use of the second distance definition, which is essentially the kernel of Geary's c, to derive an estimate of $\rho(d_{ij})$, the correlation between errors at i and j occurring at distance d_{ij}. The basis of the procedure is the algebraic identity

$$\mathrm{E}\{(e_i - e_j)^2\} = \mathrm{E}(e_i^2) + \mathrm{E}(e_j^2) - 2\mathrm{E}(e_ie_j), \tag{5.100}$$

where $\mathrm{E}(e_i^2) = \mathrm{E}(e_j^2) = \sigma^2$ and $\mathrm{E}(e_ie_j) = \sigma^2\rho(d_{ij})$, so that

$$\mathrm{E}\{(e_i - e_j)^2\} = 2\sigma^2\{1 - \rho(d_{ij})\}. \tag{5.101}$$

A simple rearrangement gives us that

$$\rho(d_{ij}) = 1 - \mathrm{E}\{(e_i - e_j)^2\}/2\sigma^2. \tag{5.102}$$

In order to estimate $\rho(d_{ij})$ we therefore require estimates of σ^2 and of $\mathrm{E}\{(e_i - e_j)^2\}$.

Since the former is independent of distance, our attention focuses on the estimate of the expectation: this estimate is provided by equation (5.103):

$$\mathrm{E}\{(e_i - e_j)^2\} \simeq \frac{1}{n_d} \sum (\hat{e}_i - \hat{e}_j)^2, \tag{5.103}$$

where the summation is taken over all those locality pairs occupying distance class d. It is convenient to denote the sample estimate given on the right-hand side of equation (5.103) as G.

A correlogram is provided by plotting G against $(\sum d_{ij})/n_d$, the mean inter-locality distance of the locality pairs belonging to distance class d. Note that an autocorrelation near to -1 will be represented by a large value of G, while a correlation near $+1$ will be represented by a near 0 value of G. The variation in G with d will provide a clear idea of the extent to which the residuals display autocorrelation, although, without an estimate of σ^2, the precise nature of the autocorrelation will not be known.

***Example* 5.16** (cont.)

Using G for the entire 26 localities of the Californian data we obtain Figure 5.16a, which is essentially the mirror image of Figure 5.15a. There is, perhaps, the suggestion of a horizontal region for values of d between 300 and 600 miles, but the dominant impression is of a gradual rise in G as d increases. A similar remark holds true for the corresponding plot for the 18 islands data, which is given as Figure 5.16b. Here, however, there is a well-marked horizontal step corresponding to values of d between 200 and 500 miles.

Error covariance from correlograms

Suppose that we possessed an accurate estimate of σ^2. It would then be possible to obtain $\rho(d_{ij})$ from equation (5.102) and to show explicitly how $\rho(d_{ij})$ varies with d_{ij}. Pocock *et al.* (1982) and Cook and Pocock (1983) obtain their estimate $\hat{\sigma}^2$ by identifying from the G correlogram the distance u at which errors are evidently uncorrelated. This will correspond to the point at which the correlogram flattens to approximately zero slope. Our experience with the Californian data, where the correlograms clearly do *not* flatten out (though we are suspicious of the values corresponding to large d) suggests that this procedure is likely often to prove difficult.

Assume for the present that the estimation of the value u poses no problem: the next step is to invoke a simple function that represents the G correlogram and thus enables us to provide a succinct explanation of the dependence of autocorrelation on distance. Cook and Pocock (1983), working with a very large data set, observed that their correlogram was an excellent fit to the model $\rho(d_{ij}) = \lambda \exp(-\mu d_{ij})$, for $i \neq j$, and some constant λ and μ. This function was

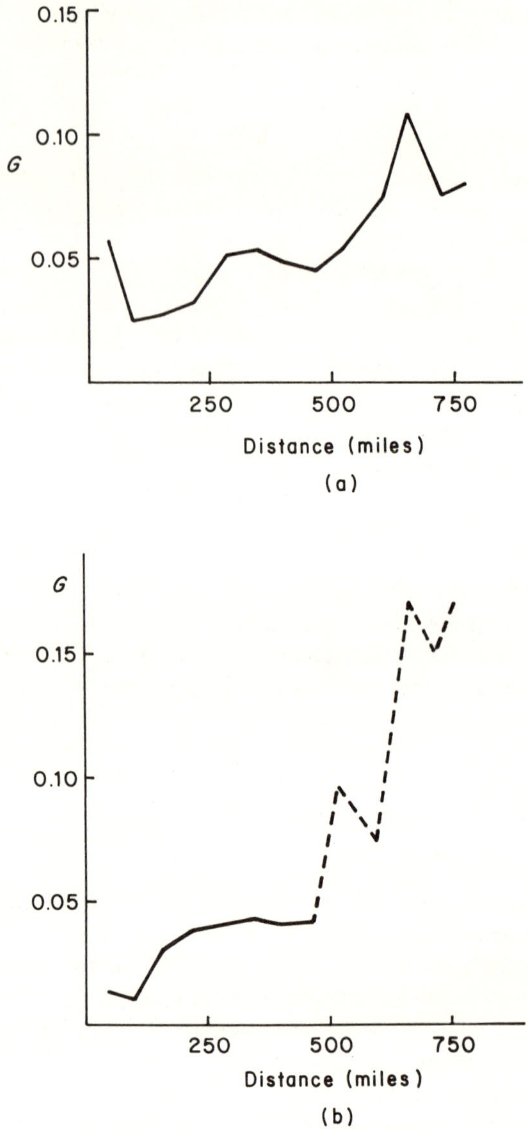

Figure 5.16 Residual correlograms. (a) All Californian localities. (b) Californian Islands only

originally given as equation (5.52), although there its origins were not mentioned and the maximum likelihood procedure to obtain $\hat{\lambda}$, $\hat{\mu}$, and other parameters was barely touched on. In fact, Cook and Pocock's (1983) estimation method is only a minor variation on the methods given earlier for the autocorrelated errors models. Commencing with trial λ- and μ-values, one obtains $\rho(d_{ij})$ for any locality pair. From $\mathbf{V} = \sigma^2\mathbf{D}$, it is apparent that the $\{\rho(d_{ij})\}$ are the elements of $\mathbf{D}$. For given $\mathbf{D}$, it is then possible to obtain $\hat{\mathbf{b}}$ via equation (5.13) and $\hat{\sigma}^2$ via equation (5.15), although because this is maximum likelihood estimation the divisor in equation (5.15) is n rather than $(n-k)$. The variance–covariance matrix of the $\hat{\mathbf{b}}$, $\mathbf{V}(\hat{\mathbf{b}})$, is given by equation (5.14). The log-likelihood is given by equation (5.35), noting that $\mathbf{D} = \mathbf{A}'\mathbf{A}$, and the maximum likelihood estimates are obtained by minimizing, in this case, equation (5.36) by iterating the above through a range of trial λ- and μ-values.

Autoregressive model residuals and residual maps

The foregoing analysis has been on the basis of residuals generated by OLS regression models. Given that spatially autocorrelated residuals are in evidence, one solution is to fit an autocorrelated errors model which incorporates an autoregressive error ($\mathbf{e}$) distribution. An alternative is to fit an autoregressive response model which attributes the autoregressive interaction to the Y variable rather than to the $\{e_i\}$: we discuss this approach in Section 5.7. Having fitted either of these models, the question still remains as to whether any remnant spatial autocorrelation lingers among the $\{u_i\}$. In the case of the autocorrelated errors model, evidence for this would come from mapping the distribution of

$$\hat{\mathbf{u}} = \mathbf{Y} - \hat{\rho}\mathbf{WY} - \mathbf{X}\hat{\mathbf{b}} + \hat{\rho}\mathbf{WX}\hat{\mathbf{b}}, \tag{5.104}$$

which is obtained from equation (5.20). An analogous residual vector can be derived from the autoregressive response model. Unfortunately, as Cliff and Ord (1981a, p. 240) lament, the tests for (residual) spatial autocorrelation that we have worked with thus far 'cease to be valid when the model contains an autoregressive component', and no one has yet come up with a simple, valid test for spatially autocorrelated $\{u_i\}$, though one has been found for the parallel situation in time series analysis (Durbin, 1970). In the absence of a correct procedure, spatial analysts (e.g. Ord, 1975; Burridge, 1981) have resorted to informal approaches such as the use of Moran's I in conjunction with the randomization moments given by equations (3.28) and (3.29). Alternatively, but less informatively, the join-count statistics allied to a map of positive and negative residuals provide a quick appraisal. Even if OLS residuals are under scrutiny, residual maps are of value, for they can reveal residual autocorrelation of a less specific nature than that implied by the particular $\mathbf{W}$ matrix that may happen to have been used in the calculation of equations (5.97), (5.98), and (5.99).

***Example* 5.17**

Figures 5.17a and b display maps of (positive/negative) residuals produced by the (logit) regression model with county ranks as the X variable (equation (5.16b)), and by the equivalent autocorrelated errors model given in Table 5.4. It is not difficult to see that there are regions of positive and negative residuals in Figure 5.17a and that these have been broken up, to a limited extent, by the autocorrelated errors model (Figure 5.17b).

Incidentally, the BW statistic fails to detect significant autocorrelation on either map. This is due to the comparative insensitivity of the BW statistic by comparison with Moran's I, together with the effect of adopting (1, 0) proximity values, as was done with the analysis of Irish blood *per se* in Chapter 3.

***Example* 5.18**

Figures 5.18a and b are the (positive/negative) residual maps emanating from equation (5.16d), the logit regression with dummy X variable, and from the equivalent autocorrelated errors model of Table 5.4. These maps are very similar to their predecessors and the previous remarks hold true once again. A salient feature of Figure 5.18a is the apparent breakdown of the residuals into regions, essentially the inner Pale (positive), outer Pale (negative), inner 'beyond-the-Pale'

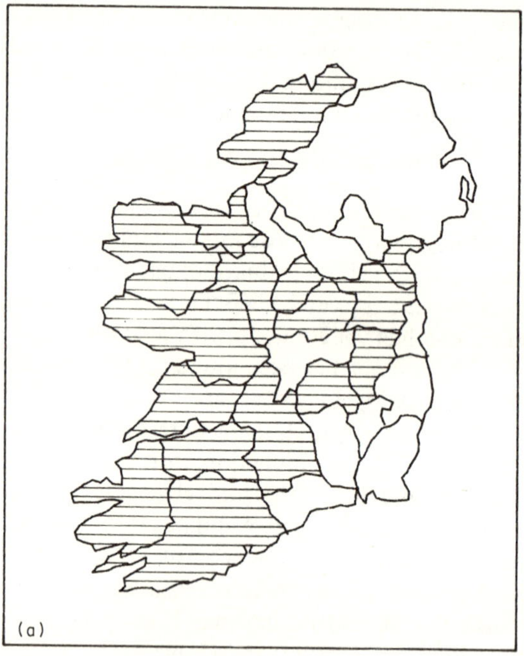

Key: ⊟ = Negative residuals

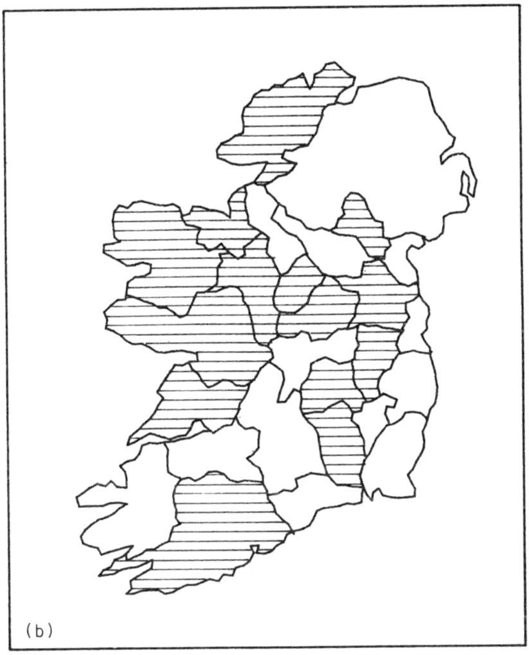

Figure 5.17 Residuals for models involving X_2. (a) From equation (5.16b). (b) From Table 5.4

(positive), and outer 'beyond-the-Pale' (negative). Thus, the model consistently predicts more than the observed log-odds of blood group A in the outer Pale, and less than the observed log-odds in the adjacent counties just beyond the Pale. This suggests that the Pale is not such a significant break-point as the OLS regression model indicates and as would be the case if there was no residual pattern. We should recall that the significance of the Pale becomes considerably watered down on fitting the autocorrelated errors model. Figure 5.18b shows only mild dissipation of the residual pattern, although the residuals are on average smaller, as shown by Table 5.25. The respective estimates of σ^2 are 0.009 and 0.0045.

5.7 AUTOREGRESSIVE RESPONSE MODELS

Previously an autoregressive effect was interwoven into our explanations via the autocorrelated errors model, and in some respects this did rather well (see Table 5.12). It is however apparent from the subsection headed 'The unrestricted autocorrelated errors model' and from equation (5.50) that this model amounts to

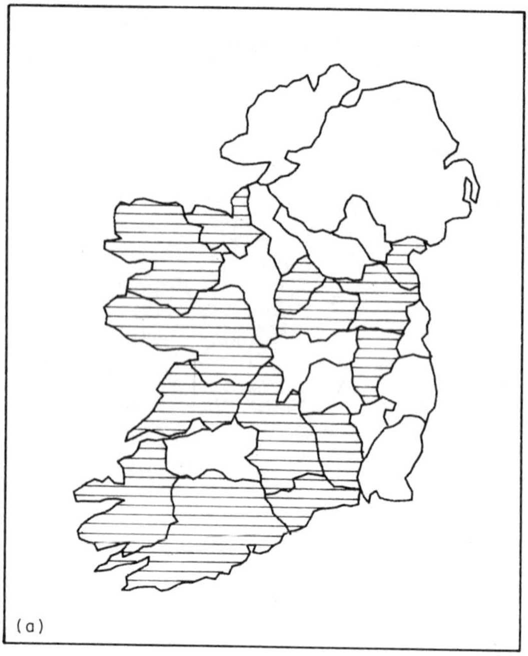

Key: ▤ = Negative residuals

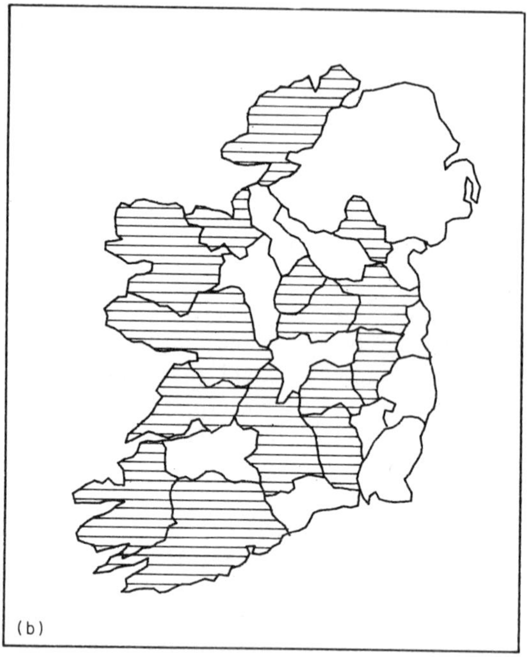

Key: ▤ = Negative residuals

Figure 5.18 Residuals for models involving X_3. (a) From equation (5.16d). (b) From Table 5.4

Table 5.25 Residuals from regression and autoregression models fitted to the Irish blood data

County	Regression residuals (equation (5.16d))	Autoregression residuals (autocorrelated errors)
1 Carlow	0.110	0.063
2 Cavan	0.105	0.057
3 Clare	− 0.050	− 0.051
4 Cork	− 0.189	− 0.171
5 Donegal	0.019	− 0.045
6 Dublin	0.046	0.060
7 Galway	− 0.051	− 0.055
8 Kerry	− 0.123	− 0.026
9 Kildare	− 0.051	− 0.064
10 Kilkenny	− 0.017	− 0.014
11 Laoghis	0.042	− 0.066
12 Leitrim	0.065	0.028
13 Limerick	0.035	0.071
14 Longford	− 0.152	− 0.073
15 Louth	− 0.035	0.030
16 Mayo	− 0.026	− 0.022
17 Meath	− 0.051	− 0.009
18 Monaghan	0.035	− 0.059
19 Offaly	0.129	0.080
20 Roscommon	0.021	0.016
21 Sligo	− 0.011	− 0.035
22 Tipperary	− 0.072	− 0.005
23 Waterford	− 0.034	0.046
24 Westmeath	− 0.092	− 0.011
25 Wexford	0.183	0.126
26 Wicklow	0.166	0.129

a rather complex autoregression mechanism involving two terms, $\rho\mathbf{WY}$ and $\rho\mathbf{WXb}$. Removal of the $\rho\mathbf{WXb}$ term simplifies the autoregression process somewhat and the model then describes a straightforward direct dependence of Y_i on $Y_j\,(i \neq j)$, the extent and nature of the interaction among the Y's being governed by the structure of the $\mathbf{W}$ matrix and by ρ. The simplified model also treats Y_i as a response to X_i, as is usual in ordinary regression models, so that the entire model is

$$\mathbf{Y} = \rho\mathbf{WY} + \mathbf{Xb} + \mathbf{e}, \tag{5.105}$$

where $\mathbf{e} \sim N(\mathbf{0}, \sigma^2\mathbf{I})$ and

$$\mathbf{e} = (\mathbf{I} - \rho\,\mathbf{W})\mathbf{Y} - \mathbf{Xb} = \mathbf{AY} - \mathbf{Xb}. \tag{5.106}$$

The joint likelihood function written in terms of errors is identical to equation (5.24), although $|\mathbf{A}|$ is now the Jacobian of the transformation from $\mathbf{e}$ to $\mathbf{Y}$ rather than from **u** to **e**, as was the case in the section on maximum likelihood with spatially autocorrelated errors. The equivalent of equation (5.28) is therefore

$$L = |\mathbf{A}| \frac{1}{\sigma^n (2\pi)^{n/2}} \exp\left\{ \frac{-(\mathbf{AY}-\mathbf{Xb})'(\mathbf{AY}-\mathbf{Xb})}{2\sigma^2} \right\}. \qquad (5.107)$$

Incorporating the expansion

$$\mathbf{e}'\mathbf{e} = (\mathbf{AY}-\mathbf{Xb})'(\mathbf{AY}-\mathbf{Xb}) = \mathbf{Y}'\mathbf{A}'\mathbf{AY} - 2\mathbf{b}'\mathbf{X}'\mathbf{AY} + \mathbf{b}'\mathbf{X}'\mathbf{Xb}$$

into the logarithm of the terms in equation (5.107) gives:

$$\log L = \text{constant} - \frac{n}{2}\log\sigma^2 - \frac{1}{2\sigma^2}(\mathbf{Y}'\mathbf{A}'\mathbf{AY} - 2\mathbf{b}'\mathbf{X}'\mathbf{AY} + \mathbf{b}'\mathbf{X}'\mathbf{Xb}) + \log|\mathbf{A}|, \qquad (5.108)$$

which is the counterpart of equation (5.30).

The problem, as before, is to find values $\hat{\sigma}^2$, $\hat{\mathbf{b}}$, and $\hat{\rho}$ (which is embodied in **A**) which produce the maximum value of $\log L$. One way to do this is to reduce equation (5.108) to a function of ρ alone by following a process of substitution similar to that outlined for the autocorrelated errors model.

Assume for the moment that ρ is known, so that the maximum likelihood estimates of **b** and of σ^2 are obtainable by setting to zero equations (5.109) and (5.110):

$$\partial(\log L)/\partial\mathbf{b} = -\{-2\mathbf{X}'\mathbf{AY} + 2(\mathbf{X}'\mathbf{X})\mathbf{b}\}/2\sigma^2; \qquad (5.109)$$

$$\partial(\log L)/\partial(\sigma^2) = -n/2\sigma^2 + (\mathbf{e}'\mathbf{e})/2\sigma^4. \qquad (5.110)$$

These provide the estimating equations (5.111) and (5.112) which are equivalent to equations (5.33) and (5.34):

$$\hat{\mathbf{b}} = (\mathbf{X}'\mathbf{X})^{-1}\mathbf{X}'\mathbf{AY}; \qquad (5.111)$$

$$\hat{\sigma}^2 = \{(\mathbf{AY})'\mathbf{R}(\mathbf{AY})\}/n = (\mathbf{Y}'\mathbf{A}'\mathbf{RAY})/n, \qquad (5.112)$$

where $\mathbf{R} = \mathbf{I} - \mathbf{X}(\mathbf{X}'\mathbf{X})^{-1}\mathbf{X}'$.

Substituting these into equation (5.108) leaves us with equation (5.36) for which we need the minimum. This equation is expressed in terms of ρ and σ^2 and so, as before, we replace σ^2 by an equivalent equation in terms of ρ, which may be written as equation (5.113a) or equation (5.113b):

$$\sigma^2 = \{\mathbf{Y}'(\mathbf{I}-\rho\mathbf{W})'\mathbf{R}(\mathbf{I}-\rho\mathbf{W})\mathbf{Y}\}/n, \qquad (5.113a)$$

$$= \{\mathbf{Y}'\mathbf{RY} - 2\rho\mathbf{Y}'\mathbf{RWY} + \rho^2(\mathbf{WY})'\mathbf{RWY}\}/n \qquad (5.113b)$$

thus $\hat{\rho}$ is the value that minimizes

$$M = \log\{\mathbf{Y}'\mathbf{RY} - 2\rho\mathbf{Y}'\mathbf{RWY} + \rho^2(\mathbf{WY})'\mathbf{RWY}\} - \frac{2}{n}\log|\mathbf{A}|. \qquad (5.114)$$

This account follows that of Doreian (1981) who develops Ord's (1975) more concentrated presentation (see also Burt and Doreian 1982). Doreian (1980) compares the autocorrelated errors and autoregressive response models, under the heading 'spatial disturbances or spatial effects'.

As previously, the parameter standard errors are given by the square roots of the main diagonal elements of **V** (equation (5.40)). These we obtain from the expected values of the second partial derivatives obtained from equations (5.109), (5.110), and also from the equation

$$\frac{\partial(\log L)}{\partial \rho} = \frac{\partial \log|\mathbf{A}|}{\partial \rho} - \frac{1}{2\sigma^2}\frac{\partial(\mathbf{e}'\mathbf{e})}{\partial \rho}. \tag{5.115}$$

There are similar terms for the off-diagonal cells.

Since the method is identical to that which connects equations (5.40) and (5.48), we simply give the asymptotic variance–covariance matrix from which the expected values of the second partial derivatives can be extracted if required. The notation follows that of the earlier section:

$$\mathbf{V}(\hat{\sigma}^2,\hat{\rho},\hat{\mathbf{b}}) = \sigma^4 \begin{bmatrix} n/2 & \sigma^2\mathrm{tr}(\mathbf{B}) & \mathbf{0}' \\ \sigma^2\mathrm{tr}(\mathbf{B}) & \sigma^4\mathrm{tr}(\mathbf{B}'\mathbf{B}) - \sigma^4\alpha + \sigma^2\mathbf{b}'\mathbf{X}'\mathbf{B}'\mathbf{B}\mathbf{X}\mathbf{b} & \sigma^2\mathbf{b}'\mathbf{X}'\mathbf{B}\mathbf{X} \\ \mathbf{0} & \sigma^2\mathbf{X}'\mathbf{B}\mathbf{X}\mathbf{b} & \sigma^2\mathbf{X}'\mathbf{X} \end{bmatrix}^{-1} \tag{5.116}$$

***Example* 5.19**

We now fit the autoregressive response model to the Irish blood data of Table 5.1. Since blood type is an inherited characteristic there is some justification for invoking a model with an autoregressive component, although the actual inheritance mechanism is considerably more complex than is suggested by equation (5.105). In the ABO blood group system, a person can be allotted to any one of six genotypes (two of which make up blood group A) and thus there are 36 parent combinations with, at least to the non-biologist, somewhat unpredictable consequences. For example, there is no chance at all that a child whose two parents have blood groups AB and O will inherit either blood type, whereas there is a 50% probability that it will possess blood group A! Our model does not attempt to capture these intricacies, but is more concerned with the *net* outcome of all the combinations.

In fitting the autoregressive response model, use is made of the **W** matrix laid out in Table 5.3, and therefore a (modified) cross-border interdependence among the $\{\mathbf{Y}_i\}$ is assumed. This seems reasonable given what is likely to be a fairly substantial distance component in the marriage-partner choice process, which seems an appropriate facet of social life to consider! Analysis is confined to the logit response $z = \log\{Y/(100 - Y)\}$, for reasons suggested earlier, and X_2 and X_3 are again taken as separate independent variables with the resulting equations being

$$\hat{\mathbf{z}} = \mathbf{c} + \underset{(0.1002)}{0.79}\mathbf{Wz}, \qquad \hat{\sigma}^2 = \underset{(0.0014)}{0.0046}, \tag{5.117a}$$

where the vector $\mathbf{c}$ has elements $\{c_i\}$ given by

$$c_i = \underset{(0.0812)}{-0.1151} - \underset{(0.0020)}{0.00505}X_{2i}, \tag{5.117b}$$

$$\hat{\mathbf{z}} = \mathbf{d} + \underset{(0.1106)}{0.75}\mathbf{Wz}, \qquad \hat{\sigma}^2 = \underset{(0.0014)}{0.0047}, \tag{5.118a}$$

and where the vector $\mathbf{d}$ has elements $\{d_i\}$ given by

$$d_i = \underset{(0.1046)}{-0.2605} + \underset{(0.0305)}{0.0917}X_{3i}. \tag{5.118b}$$

Equations (5.117a) and (5.117b) are the version of the model with county rank as the X variable. The values in parentheses below each parameter estimate are the estimated standard errors taken from the main diagonal of the matrix given in equation (5.116). The autoregressive part of the model is denoted by $\hat{\rho}\mathbf{Wz}$, where $\mathbf{Wz}$ is a vector of spatially lagged $\mathbf{z}$-values. The ratio of $\hat{\rho}$ to its estimated standard error shows that this is a highly significant factor controlling $\mathbf{z}$ and Figure 5.19a illustrates the way in which the log-likelihood reaches a 'maximum' at $\hat{\rho} = 0.79$.

A notable difference between the present model and the autocorrelated errors model lies in the significance presently attributable to X_2, for $\hat{b}_2$ is larger, and its standard error estimate is smaller, than the equivalent values in Table 5.4, giving a highly significant ratio of -2.53 (assuming a $N(0, 1)$ distribution under the null hypothesis).

The explained proportion of the variation of $\{z_i\}$ is given by equation (5.39) as 75.2%, which is on a par with the autocorrelated errors model. Figure 5.19b shows the distribution of the (positive/negative) residuals, $\hat{\mathbf{e}} = \mathbf{z} - \hat{\rho}\mathbf{Wz} - \hat{\mathbf{e}}$, for comparison with Figures 5.17a and 5.17b. These comparisons suggest that the simplification entailed by the autoregressive response model has not been detrimental and that equation (5.117) is an adequate expression of the systematic variation in $\mathbf{z}$.

Equations (5.118a) and (5.118b) utilize the dummy 'Pale' variable in place of X_2. The highly significant autoregressive effect ($\hat{\rho} = 0.75$) is again in evidence (see Figure 5.20a) and the X variable effect again appears enhanced when compared with what was suggested by the equivalent autocorrelated errors model: the current parameter to standard error ratio being 3.01. The model explains 74.6% of the variation in $\mathbf{z}$ (see Table 5.12), and Figure 5.20b shows that there is no apparent loss of residual 'disorder' in transferring to the present model. On the basis of this model, our final interpretation of the effect of the 'Pale' on the odds of encountering blood group A in a randomly chosen individual is that they increase

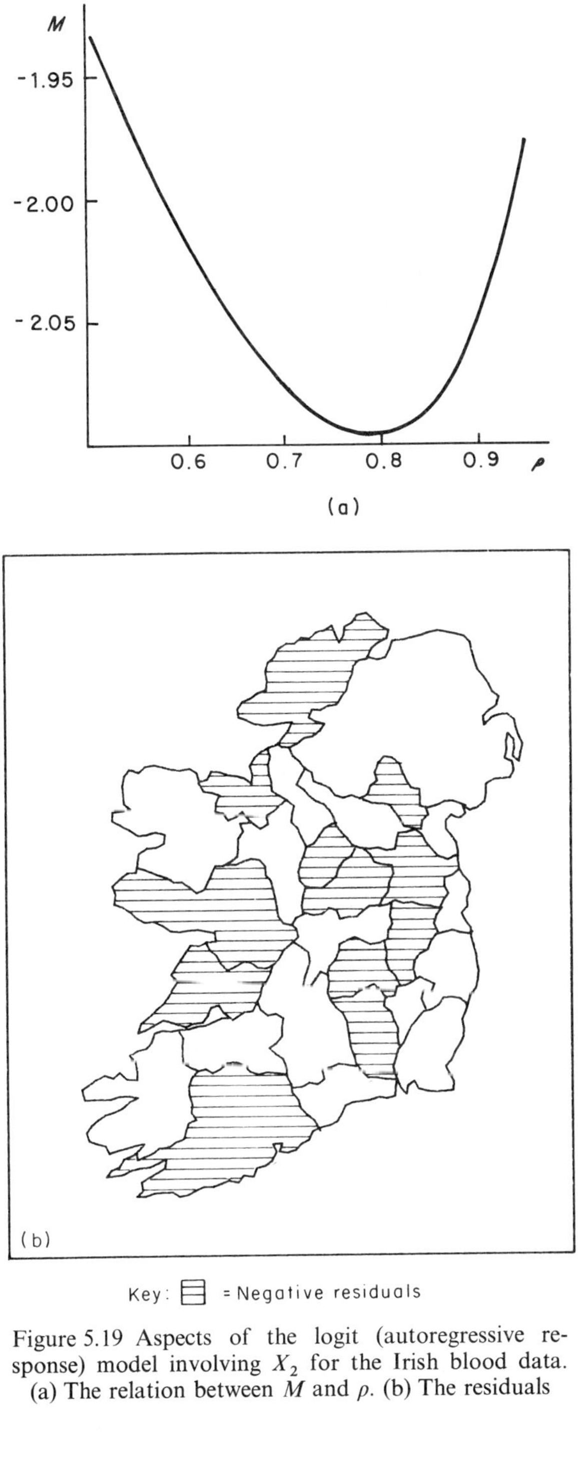

Figure 5.19 Aspects of the logit (autoregressive response) model involving X_2 for the Irish blood data. (a) The relation between M and ρ. (b) The residuals

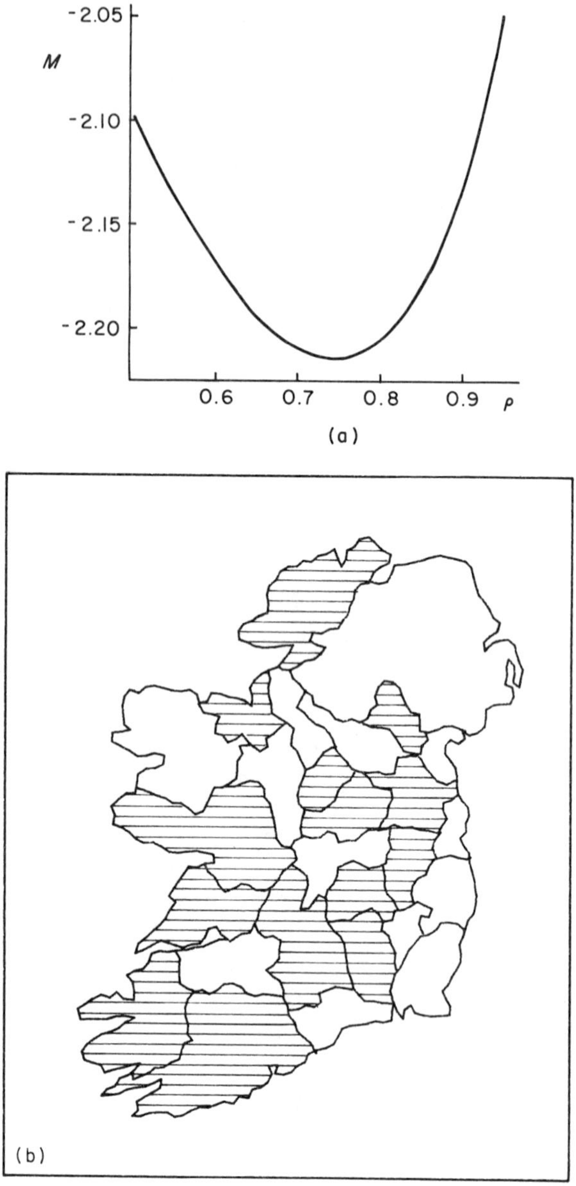

Key: ⊟ = Negative residuals

Figure 5.20 Aspects of the logit (autoregressive response) model involving X_3 for the Irish blood data. (a) The relation between M and ρ. (b) The residuals

by a multiplicative factor of 1.096 (= exp (0.0917)) compared with the factor 1.075 (= exp (0.0722)) suggested earlier via the autocorrelated errors model. Using equation (5.117) an equivalent statement can be made of the effect of a move from one county to the next most 'anglicized'.

***Example* 5.20**

Table 5.26 contains the results of fitting two versions of the autoregressive response model to the Californian data of Table 5.2. The versions differ on account of the difference between the two sets of proximity values given in Table 5.6, which now determine the interactions among the $\{Y_i\}$ directly, rather than among the errors directly as supposed in the autocorrelated errors model. As in Example 5.14, the estimated standard errors are the square roots of the main diagonal elements of the matrix given in equation (5.116). Figures 5.21a and b show the variation in M (equation (5.114)) with ρ and reaffirm the presence of (apparent) negative interaction among Californian localities as a whole. Taken alone, mainland areas interact positively and we again take this as a message that mainland and island localities should be treated differently in our modelling attempts.

The Table 5.26 data bear other similarities to the data in Table 5.7. As before, the strong dependence of the number of different species on the latitude (X_3) is diminished, but not eliminated, by the presumption of an island-isolation **W**

Table* 5.26 *Parameter estimates and estimated standard errors for the autoregressive response model fitted to the Californian data

(a) Using the two-part distance-based **W** matrix of Table 5.6a

Parameter	Estimate	Estimated S.E.	Estimate/S.E.
b_0	− 2347.5	343.2	
b_1	0.1837	0.0394	4.66
b_2	0.0838	0.0289	3.51
b_3	80.0280	11.3772	7.03
ρ	− 0.56	0.1577	− 3.55
σ^2	0.0166	0.0047	

(b) Using the island-isolation **W** matrix of Table 5.6b

Parameter	Estimate	Estimated S.E.	Estimate/S.E.
b_0	− 1046.7	382.3	
b_1	0.1568	0.0446	3.52
b_2	0.0846	0.0277	3.05
b_3	33.6130	11.2786	2.98
ρ	0.34	0.1170	2.91
σ^2	0.0216	0.0060	

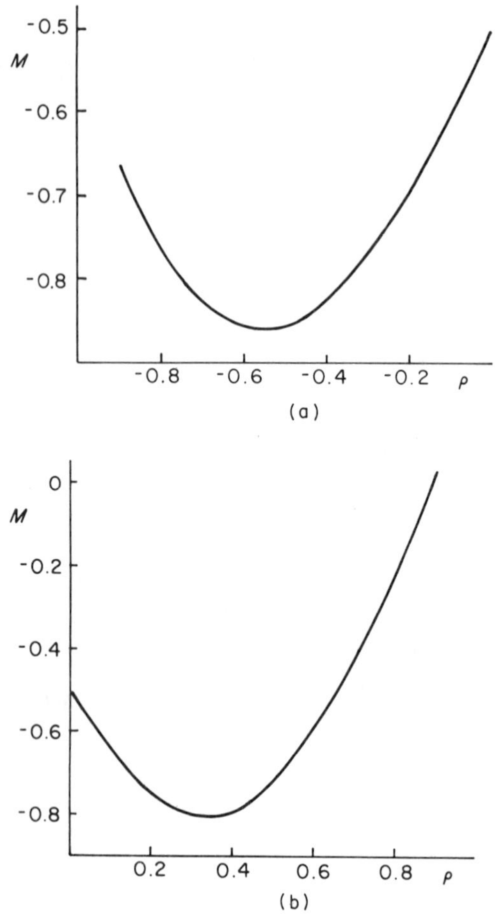

Figure 5.21 The relation between M and ρ for the Californian data using the distance weights. (a) From Table 5.6a. (b) From Table 5.6b

matrix. The other independent variables, elevation (X_2) and area (X_1), remain highly significant according to either version of the model.

Using r^{*2} as a gross measure of model performance we find that the autoregressive response model accounts for 91.1% and 88.5% (island-isolation) of the variation in Y, as opposed to the 91.8% and 90.3% for the equivalent autocorrelated errors models, so that the penalty resulting from the adoption of the simpler model is minimal for this set of data. A more refined appraisal comes from examining the residuals, although we have previously mentioned the absence of any formally correct procedures.

A comparison could be made of the levels of residual autocorrelation among the various sets of residuals, using I as defined in Chapter 3. However, the use of the randomization moments (equations (3.28) and (3.39)) is questionable when k is large compared to n, and so it is somewhat difficult to know whether autocorrelation has been completely eliminated from the autoregressive model residuals. We do know, however, from the empirical work of Cliff and Ord (1981a), that if the autoregressive model residuals are substantially the same as the OLS regression residuals, then they will probably turn out to be significantly negatively autocorrelated as measured by I (randomization), since, contrary to what happens with positive autocorrelation, I is usually liberal as an indicator of OLS regression residual autocorrelation.

Table 5.27 contains the signs of the residuals generated by the OLS regression model (equation (5.11)) and both types of autoregressive model with both versions of **W**. Because this analysis does not purport to be exact, we work with residual signs and search for sign changes as evidence of the dispersal of the negative autocorrelation (though beware that, carried to excess, negative autocorrelation re-emerges!). The maximum number (six) of OLS regression residual sign shifts comes from the autoregressive response model with the two-part distance-based **W** matrix. However, all but one mainland residual generated by this model is positive, and, in any case, we have argued throughout that a negative autoregressive process (i.e. $\rho < 0$) is unreasonable. We therefore opt for the autoregressive response (island-isolation) model as the 'best buy' currently on offer. Not only is $\hat{\rho} > 0$, but also there are five OLS regression residual sign shifts and 11 shifts of the competing autoregressive response ($\hat{\rho} > 0$) residuals. Undoubtedly this model could be improved, perhaps by adjusting **W** so that, following the analysis of Example 5.16, (weak) interaction also takes place when islands are involved.

Pure Autoregression

Elimination of the influence of the X variables from the autoregressive response model produces what we term the pure autoregression model of equation (5.119):

$$\mathbf{Y} = \rho\mathbf{W}\mathbf{Y} + \mathbf{e}, \tag{5.119}$$

in which Y_i responds to 'surrounding' values $\{Y_j, i \neq j\}$ without any modification by $\{X_{ij}, j = 1, 2, \ldots, (k-1)\}$. Writing $\mathbf{e} = (\mathbf{I} - \rho\mathbf{W})\mathbf{Y} = \mathbf{A}\mathbf{Y}$ and assuming that $e \sim N(\mathbf{0}, \sigma^2\mathbf{I})$, the previous arguments can be used to derive the likelihood function and the approximate variance–covariance matrix of the parameter estimates. It follows that the joint likelihood of the $\{Y_i\}$ is

$$\log L = \text{constant} - \frac{n}{2}\log\sigma^2 - \frac{1}{2\sigma^2}(\mathbf{Y}'\mathbf{A}'\mathbf{A}\mathbf{Y}) + \log|\mathbf{A}|, \tag{5.120}$$

Table 5.27 Positive and negative residuals from regression and autoregression models

Locality	Model 1	2	3	4	5
1 Cedros Island	−	+	−	−	−
2 Guadalupe Island	−	−	−	−	−
3 Santa Cruz Island	−	−	−	+	+
4 Santa Rosa Island	+	−	−	+	+
5 Santa Catalina Island	+	+	+	+	+
6 San Clemente Island	−	−	−	−	+
7 San Nicolas Island	−	−	−	−	−
8 San Miguel Island	−	−	−	+	+
9 Natividad Island	+	+	+	+	+
10 Santa Barbara Island	−	−	−	−	−
11 San Martin Island	+	+	−	+	+
12 San Geronimo Island	+	+	+	+	+
13 South Farallon Island	−	−	−	−	−
14 Ano Nuevo	−	−	+	−	−
15 San Benito Island	+	+	+	+	+
16 Anacapa	−	−	−	−	−
17 Los Coronados	−	−	+	−	−
18 Todos Santos Island	+	−	+	+	+
19 San Diego County	−	−	−	−	−
20 Monterey County	−	−	+	−	−
21 Marin County	+	+	+	+	+
22 Santa Cruz Mountains	+	+	+	+	+
23 Santa Monica Mountains	+	+	+	+	−
24 Santa Barbara Area	+	+	+	+	−
25 San Francisco County	+	+	+	+	+
26 Tiburon Peninsula	−	+	+	−	−

Model key: 1. OLS regression.
2. Autocorrelated errors (Table 5.6a).
3. Autoregressive response (Table 5.6a).
4. Autocorrelated errors (Table 5.6b).
5. Autoregressive response (Table 5.6b).

where $|\mathbf{A}|$ is the Jacobian of the transformation from $\mathbf{e}$ to $\mathbf{Y}$ and

$$\hat{\sigma}^2 = (\mathbf{Y}'\mathbf{A}'\mathbf{A}\mathbf{Y})/n, \tag{5.121}$$

so that, written purely as a function of ρ, we have

$$\log L = \text{constant} - \frac{n}{2}\log\left(\frac{\mathbf{Y}'\mathbf{A}'\mathbf{A}\mathbf{Y}}{n}\right) + \log|\mathbf{A}|. \tag{5.122}$$

Again it is practicable to find the value of ρ that minimizes equation (5.36), or, in this case, equation (5.123):

$$M = \log\{\mathbf{Y}'\mathbf{Y} - 2\rho\mathbf{Y}'\mathbf{W}\mathbf{Y} + \rho^2(\mathbf{W}\mathbf{Y})'\mathbf{W}\mathbf{Y}\} - \frac{2}{n}\log|\mathbf{A}|, \tag{5.123}$$

which corresponds to equations (5.114) and (5.38).

On the basis of the first and second partial derivatives we obtain the approximate variance–covariance matrix:

$$V(\hat{\sigma}^2, \hat{\rho}) = \sigma^4 \begin{bmatrix} n/2 & \sigma^2 \mathrm{tr}(\mathbf{B}) \\ \sigma^2 \mathrm{tr}(\mathbf{B}) & \sigma^4 \mathrm{tr}(\mathbf{B}'\mathbf{B}) - \sigma^4 \alpha \end{bmatrix}^{-1}, \tag{5.124}$$

which is identical to the upper-left elements of the corresponding matrix in equation (5.48). As usual, the square roots of the main diagonal element of $\mathbf{V}$ provide the approximate standard errors of $\hat{\sigma}^2$ and of $\hat{\rho}$. Of course, interest lies in establishing whether ρ is significantly different from 0. An alternative approach is to use the likelihood ratio test for which $(n-1)\log\{\exp(M)/(\mathbf{Y}'\mathbf{Y})\} \sim \chi_1^2$ under the null hypothesis that $\rho = 0$ (Whittle, 1954; Haining, 1977, 1981a; Cliff and Ord, 1981a).

Clearly, pure autoregression is an exceedingly simple model which, on account of the stationarity introduced by the elimination of the X variables, cannot hope realistically to fit the non-experimental data that have formed the examples of this chapter. Notwithstanding this, the formative and influential literature of the pure autoregression model predates that relating to the later (and generally more applicable) elaborations, and has a definite vintage quality. Probably the most important early contribution was by Whittle (1954), and for that reason equation (5.119) is sometimes referred to as Whittle's model.

Perusal of the 'pure autoregression' literature shows that it relates very much to the analysis of designed experiments, such as agricultural field trials, in which the analyst is trying to model a partially controllable system unsullied by extraneous factors such as climate and soil variations. In this context, the major problem is (paradoxically) that posed by having an excess of data to hand. For example, while a plot of 30 rows each containing 30 plants does not seem outlandishly large as experiments go (though see Mead, 1974b), it would present some extremely difficult computational problems if the methods of estimation advocated in this chapter were followed to the letter. The principal difficulty is that such an experiment would generate a 900×900 $\mathbf{W}$ matrix, together with an $\mathbf{A}$ matrix of the same size. Obtaining the determinant of $\mathbf{A}$ for use in equation (5.123) (or, for that matter, in either equation (5.114) or equation (5.38)) would be a formidable task, even given the current state of computer technology!

Considerable thought has therefore been put into finding ways around this problem. One solution is to do away entirely with the need to repeatedly evaluate $|\mathbf{A}|$ by specifying a $\mathbf{W}$ matrix so that, whatever the value of ρ, $|\mathbf{A}|$ is equal to unity and hence is redundant as a term in equation (5.123) (or its equivalent). One example of this is what Ripley (1981) calls the TV scan wherein a typical locality (k, l) depends only on the localities in the previous $(k-1)$ rows and the previous $(l-1)$ localities of the present row. The consequence of these restricted dependencies is that the matrix $(\mathbf{I} - \rho\mathbf{W})$ is triangular and therefore has the desired unit determinant (cf. Ord, 1975; Griffith, 1980). Ripley (1981) cites some other examples of 'one-sided' schemes that produce unit determinants for non-triangular matrices. Generally, such processes are probably out of tune with natural spatial processes which will tend to be multilateral, although clever experimental design could make them relevant. One interesting application to Central Place systems is given by Haining (1978b) who exploits the 'rather special way' in which spatial effects transmit through a Central Place system, together with the assumption that the Central Place interactions are one-way rather than two-way (because places are either dominant or subordinate in their relations with other places). This translates into a triangular $\mathbf{A}$ matrix with unit determinant so that the minimizing equation (5.123) reduces to least squares.

An alternative line of attack is to devise methods of simplifying the calculation of $|\mathbf{A}|$ and alternative methods due to Mead (1967), Ord (1975), and Whittle (1954) are described by Cliff and Ord (1981a). The earliest and simplest is relevant when each locality interacts only with its four near neighbours on the lattice (the lattice is, of course, the *sine qua non* of a designed experiment). This simple pattern of interactions implies that the term $-(2/n)\log|\mathbf{A}|$ in equation (5.123) and its counterparts is approximated, for large n, by

$$\sum_j^{\infty} \frac{1}{j} \binom{2j}{j}^2 \rho^{2j}.$$

The most popular method, that of Ord (1975), employs equation (5.42) which requires the calculation of the eigenvalues, $\{\lambda_i\}$, of the matrix $\mathbf{W}$, perhaps using equation (5.49). The subsequent evaluation of $|\mathbf{A}|$ for different values of ρ is then a relatively trivial operation.

Another option, described by Besag (1972, 1974, 1975, 1977b) and by Besag and Moran (1975), is somewhat radical in that is exploits the *conditional* autoregressive scheme that parallels the simultaneous scheme that we have worked with to date: we discuss this at length below.

Conditional autoregression

So far, commencing with the autocorrelated errors model, we have adhered to the so-called simultaneous autoregressive scheme and have only briefly mentioned that a conditional autoregressive scheme has also been advanced (mainly by Besag, but see also Bartlett, 1971; Cliff and Ord, 1981a; Ripley, 1979c, 1981). An essential difference between the two, noted at the outset of the discussion of the autocorrelated errors model, lies in the specification of the covariance matrix, $\mathbf{V}$, which is set equal to $\sigma^2(\mathbf{I}-\rho\mathbf{W})^{-1}$ rather than to $\sigma^2(\mathbf{I}-\rho\mathbf{W})^{-1}(\mathbf{I}-\rho\mathbf{W}')^{-1}$ as has been usual. Hence, $\mathbf{W}$ needs to be symmetrical and $\mathbf{A}=\mathbf{I}-\rho\mathbf{W}$ needs to be strictly positive definite to ensure the existence and symmetry of $\mathbf{V}$ in the conditional scheme.

Written out in conditional expectation form, the model is

$$\mathrm{E}(Y_i|\mathbf{Y}_i^*)=\rho\sum_{j\neq 1}\mathbf{W}_{ij}\mathbf{Y}_j, \tag{5.125}$$

where $\mathbf{Y}_i^*$ is $\mathbf{Y}$ with Y_i, the value at locality i, having been deleted. We also assume that

$$\mathrm{var}(Y_i|\mathbf{Y}_i^*)=\sigma^2. \tag{5.126}$$

For simplicity we assume that the conditional distributions at each locality are normal, so that the joint distribution is given as $\mathbf{Y}\sim N(\mathbf{0},\mathbf{V})$. This relation between the conditional and joint distributions is governed by the Hammersley–Clifford theorem which states in very general terms the conditional distributions that can admissibly be associated with joint distributions. This is discussed by Cliff and Ord (1981a) and, especially, by Besag (1974)—particularly in relation to non-normal distributions (see also Besag, 1972). Bennett (1979) also provides a lengthy interpretation and many relevant references.

The present model specifies the conditional distribution of Y_i given the values of $\{Y_j\}$ at 'neighbouring' localities. If Y_i and Y_j are not neighbours, then, while they are not independent, the conditional distribution of Y_i is not dependent on the value of Y_j.

A consequence of this model specification is that the log-likelihood given globally as equation (5.35) is marginally altered by replacing $\log|\mathbf{A}|$ by $\frac{1}{2}\log|\mathbf{A}|$.

A major advantage claimed for the conditional specification lies in the consistency of the least squares estimators, thus providing a way out of the $|\mathbf{A}|$ problem, on account of the independence of $\{Y_i\}$ and $\{e_j\}$. Furthermore, Cliff and Ord (1981a) give a theorem in which OLS again comes to the rescue to yield consistent estimators for multiparameter conditional schemes. Simultaneous models of this type were discussed above in our subsection on 'more complex autocorrelated errors models'.

The disadvantages are that to facilitate valid least squares estimation, localities are coded so that a subset of them is identified as having conditionally independent $N(\mu,\sigma^2)$ distributions, where μ is a linear function of the unknown parameters. Thus, part of the data is not used for estimation. Also, $\mathbf{I}-\rho\mathbf{W}$ may turn out not to be positive-definite (Ripley, 1981). Bennett (1979) considers advantages and disadvantages in detail and notes particularly that the simultaneous scheme extends directly to space–time models.

It is worth noting that a simultaneous scheme with covariance matrix $\mathbf{V}=\sigma^2(\mathbf{I}-\mathbf{G}_s)^{-1}(\mathbf{I}-\mathbf{G}_s')^{-1}$ has a conditional equivalent with covariance matrix $\mathbf{V}=\sigma^2(\mathbf{I}-\mathbf{G}_c)^{-1}$, where $\mathbf{G}_c=\mathbf{G}_s+\mathbf{G}_s'-\mathbf{G}_s\mathbf{G}_s'$. This implies that two apparently equivalent lattice models such as those specified below in equations (5.127) and (5.128) (which could be written in matrix form with 1/0 **W** matrices) are in fact different, since they amount to Markov models of different orders. Adopting the conditional scheme (equation (5.127)) means that one is invoking a simplification of the conditional scheme deriving from equation (5.128). The equations are:

$$E(Y_{ij}|Y_{ij}^{*}) = \rho(Y_{i-1,j} + Y_{i+1,j} + Y_{i,j-1} + Y_{i,j+1}), \tag{5.127}$$

$$Y_{ij} = \rho(Y_{i-1,j} + Y_{i+1,j} + Y_{i,j-1} + Y_{i,j+1}) + e_{ij}, \tag{5.128}$$

where $e_{ij} \sim N(0, \sigma^2)$ and $|\rho| < 0.25$.

Further discussions of these and related topics are to be found in Cliff and Ord (1981b) and Haining (1979).

Moving averages

Haining (1978a) has advocated the moving average model as an alternative to pure autoregression (see also Cliff and Ord, 1975b). The rationale for this suggestion is the contrasting behaviour of the theoretical correlograms which are associated with the $\{\hat{Y}_i\}$ generated by these models. In the case of pure autoregression, autocorrelation is fairly persistent with increasing distance, whereas the theoretical moving average correlogram decays to near zero quite steeply, and this may be judged to accord more with reality in some applications. For convenience, the mean, $\bar{Y}$, is removed from each of the observations $\{Y_i\}$, so that $E(\mathbf{Y}) = \mathbf{0}$, and in matrix terms the moving average process is

$$\mathbf{Y} = (\mathbf{I} + \rho\mathbf{W})\mathbf{e} = \mathbf{A}\mathbf{e}. \tag{5.129}$$

The non-matrix (lattice) equivalent of equation (5.129) is

$$Y_{ij} = 0.25\rho(e_{i-1,j} + e_{i+1,j} + e_{i,j-1} + e_{i,j+1}) + e_{ij},$$

where $|\rho| < 1$ and each of the $\{e_{ij}\}$ is distributed as $N(0, \sigma^2)$.

The log-likelihood, following the usual arguments, is

$$\log L = \text{constant} - \log|\mathbf{A}| - \frac{n}{2}\log(\sigma^2) - \frac{1}{2\sigma^2}(\mathbf{Y}'(\mathbf{A}\mathbf{A}')^{-1}\mathbf{Y}). \tag{5.130}$$

This log-likelihood clearly resembles its predecessors, though **A** takes on a different definition here and the Jacobian of the transformation from **e** to **Y** is $|\mathbf{A}|^{-1}$. We can reduce the expression to a function of the unknown parameter ρ by substituting $\hat{\sigma}^2 = \mathbf{Y}'(\mathbf{A}\mathbf{A}')^{-1}\mathbf{Y}/n$, which gives us

$$M = \frac{2}{n}\log|\mathbf{A}| + \log(\mathbf{Y}'(\mathbf{A}\mathbf{A}')^{-1}\mathbf{Y}), \tag{5.131}$$

which is to be minimized. Note that this procedure requires the evaluation of both the determinant and the inverse of a large $n \times n$ matrix on each iteration. Haining (1978a) gives some ideas on how this may be accomplished.

The test of the null hypothesis $\rho = 0$ can be performed using the likelihood-ratio test where $\log(L_U)$ is the equation (5.130) with $\mathbf{A} = \mathbf{I} + \hat{\rho}\mathbf{W}$ and $\log(L_R)$ is the equation (5.130) with $\rho = 0$. Haining (1978a) uses Monte-Carlo methods (see below) to examine the behaviour of the ratio (L_R/L_U) in detail and he concludes that it is marginally superior to Moran's I when used as a test of the null hypothesis. Under that null hypothesis, $(n-1)\log\{\exp(M)/(\mathbf{Y}'\mathbf{Y})\}^{-1}$ can again be taken to have a χ_1^2 distribution (Whittle, 1954).

Boundary effects

The boundary effect problem arises only in the situation where Y_i is influenced by Y_j and the locality j is *outside the study area.* In effect we have a model misspecification problem in which the true processes governing Y are not accurately represented. Obviously, since study areas are finite and spatial processes have a tendency to be infinite, this could present an estimation problem, depending on the context. Some explorations of boundary effects on $\hat{\rho}$ have been carried out in the context of stationary lattice processes (i.e. pure autoregression and moving average processes) and hence this research is reported at this juncture. It does, however, have wider ramifications, although apparently little is known about the problem in the context of, say, more complex autoregressions or with areal localities of varying shape and size (such as the Irish counties).

The work of Haining (1977) and of Griffith and Amrhein (1983) shows that $\hat{\rho}$ estimated via equation (5.123) with no correction for an influential boundary effect tends to be an underestimate of the parameter ρ of the pure autoregression process. In contrast, $\hat{\rho}$ is an upwardly biased estimate of the moving average parameter ρ (Haining, 1978a). Fortunately, as n increases the bias decreases, though then we run into our other numerical estimation problems.

There are some correction techniques available which amount to changing the structure of $\mathbf{W}$. For example, torus mapping does away with boundaries altogether, in which case extra non-zero elements appear in $\mathbf{W}$. Alternatively, a buffer zone can be introduced, thereby increasing the dimension of $\mathbf{W}$.

Unfortunately, the work of Griffith and Amrhein (1983) suggests that these 'corrections' provide scarcely any improvement on doing nothing!

Projecting these findings onto our autoregressive response models of Ireland and California, we conjecture that boundary effects will be operating in both cases, presumably downwardly biasing $\hat{\rho}$. Otherwise, in the Irish case, we are assuming in our $\mathbf{W}$ matrix (Table 5.3) that the Irish border is roughly equivalent to the Atlantic Ocean in its effect on social intercourse! Obviously an analysis combining the counties of the Republic and Northern Ireland is one way to refine our estimate of ρ. A similar problem exists with the Californian analysis, where Y is undoubtedly influenced by the Y-values in mainland localities excluded from the analysis. All this amounts to a call for a reanalysis with more data, which, as is almost always the case, helps to refine one's ideas on the operation of the system of interest. Certainly there appears to be no magic trick to resolve boundary effect bias; as Griffith and Amrhein (1983) observe, 'current techniques are inadequate' (even for stationary lattice data). This impasse has led Ord (1981) to suggest that attention should be switched to the conditional scheme, which holds out more prospect of a satisfactory solution. Alternatively, one should strive for a study area with natural boundaries that halt the process 'in its tracks'. Griffith (1983) gives an overview of the boundary effect problem and surveys currently available methods of handling it.

Much information about boundary effects and the behaviour of statistics relating to the pure autoregression model and the moving average process has been derived from lattice simulation experiments. Haining *et al.* (1983) remark that further development 'will depend at least to some extent on the ability to simulate appropriate data', and they give details of important factors to consider and references to much relevant literature (see also Griffith and Amrhein 1983).

5.8 ALTERNATIVE APPROACHES

Prediction methods related to Kriging

The approach generally referred to as Kriging (on account of the work of the geostatistician D. G. Krige, 1966; see also Krige *et al.*, 1969) aims to produce accurate predictions by interpolating from localities with known Y-values to 'prediction localities' with missing values. Many applications of these methods have been in the mining industry, particularly South African gold mines, where accurate prediction is essential and where a smooth function such as a trend surface is an inadequate representation of the variation typically associated with, say, an ore-body. The Y surface formed by ore-grade variations around a mine tends to be continuous and intensely locally variable, so that trend surfaces *per se* provide, via the equation $\hat{\mathbf{Y}} = \mathbf{z}\hat{\mathbf{b}}$, predictions that are only broadly accurate, since the estimates $\hat{\mathbf{b}}$ are the consequence of fitting the surface to the entire data set and assuming normal errors. Although an ore-body or similar surface may actually contain such a large-scale component, a more effective prediction technique is to combine knowledge of the trend at the prediction locality ($\hat{T}_i$) with a local component ($\hat{L}_i$) derived from the covariance structure of the residuals from that trend, so that $\hat{Y}_i = \hat{T}_i + \hat{L}_i$. By definition, the trend will describe the large-scale fluctuation of the surface and the local or residual component will represent small-scale fluctuation with a tendency to peter out.

In its 'pure' form, Kriging is concerned with the local component of the surface partition, although no partition is envisaged and the entire surface is treated as the outcome of a localized spatial interdependence process, without any initial extraction of trend. Here, however, we adopt the approach of Ripley (1981) who combines local and trend components (see also Matheron, 1969). This serves to unite Kriging with other topics in this chapter, a unity which unfortunately has been largely absent from the Kriging literature which, possibly due to translation difficulties, has developed a somewhat separate existence and nomenclature and which has been considered more as a competitor to trend surface analysis than as an integral component (see, for example, Davis, 1973; Watson, 1971; Whitten, 1966; Krige, 1966; Matheron, 1967).

It is evident that the residuals exposed by fitting a trend to the surface will, of necessity, contain an autocorrelation structure, and so our OLS method of fitting trend surfaces is somewhat suboptimal. A much more appropriate method is generalized least squares, because it downweights intercorrelated adjacent observations which merely replicate the information yielded by their neighbours, although, of course, this requires the error variance–covariance matrix $\mathbf{V}$ to be known. This suggests replacing $\mathbf{V}$ by a simple parameterization, $\mathbf{V} = \sigma^2 ((\mathbf{I} - \rho\mathbf{W}')(\mathbf{I} - \rho\mathbf{W}))^{-1}$, say, and estimating ρ, σ^2, and $\mathbf{b}$ using maximum likelihood, as was done generally for autocorrelated errors models. However, as we have mentioned, this is difficult with a more general multiparameter specification of $\mathbf{V}$. The generalized least squares approach can be pursued by, initially, fitting a trend surface by OLS to expose the residuals. These could provide the input to a (somewhat) approximate error variance–covariance matrix $\mathbf{V}$ using the methods described above (see the subsection 'Error covariance from correlograms'). This $\mathbf{V}$ matrix could then be used to refit the trend surface by generalized least squares, thereby producing 'refined' residuals $\hat{\mathbf{e}} = \mathbf{Y} - \mathbf{z}\hat{\mathbf{b}}$, and a more accurate approximation to $\mathbf{V}$. Ripley (1981, p. 58) suggests that this procedure could be repeated, and that $\mathbf{V}$ should then form a biased, but reasonably accurate, approximation.

We have mentioned that the estimates $\mathbf{z}\hat{\mathbf{b}}$ will provide a smooth and, depending on the context, inadequate summary of the variation in Y. We denoted the local component of the prediction for locality i by $\hat{L}_i$. This relates to the residuals obtained above as in the equation

$$\hat{L}_i = \boldsymbol{\lambda}'\hat{\mathbf{e}}. \tag{5.132}$$

The term $\boldsymbol{\lambda}$ is a vector of weights that also links the error variance–covariance matrix $\mathbf{V}$ to a column vector $\mathbf{c}$ containing the residual convariance of prediction locality i with the other localities:

$$\mathbf{V}\boldsymbol{\lambda} = \mathbf{c}, \tag{5.133}$$

and

$$\boldsymbol{\lambda} = \mathbf{V}^{-1}\mathbf{c}, \tag{5.134}$$

hence

$$\hat{L}_i = (\hat{\mathbf{e}}'\mathbf{V}^{-1})\,\mathbf{c}. \tag{5.135}$$

We assume that $\mathbf{c}$ is a copy of what is known to happen generally and can be simply extracted from the smoothed variogram (see below). Ripley (1981) also gives formulae for estimating the error of the prediction $\hat{Y}_i$. Clearly, much rests on finding a suitable function giving the elements of **V** and **c**. Within the Kriging literature, there is reference to the variogram as a source of this function (Matheron, 1971; Kooijman, 1976). However, this is no more than the equivalent of what we referred to as the 'alternative residual correlogram' (see equation (5.111)), although the concept has been developed further by Matheron (1971). This development introduces the option of a discontinuity at distance zero so that $\rho(d_{ij})$ does not tend to 1 as d_{ij} tends to 0 and hence the variogram does not tend smoothly to 0 (see also Delfiner and Delhomme, 1975; Huijbrechts, 1975). This is designed to handle what Matheron terms the 'nugget effect', a synonym for localized variation on the surface such that it fails to settle down to near perfect correlation at very short distances.

One way to generate the 'nugget effect' is to adopt equation (5.136) as the smoothed variogram:

$$\mathrm{E}\{(e_i - e_j)^2\} = \theta_0 - \theta_1 \exp(-\theta_2 d_{ij}). \tag{5.136}$$

Apart from θ_0, this is the same function as was used by Cook and Pocock (1983) for the case $i \neq j$, and, at the origin ($d_{ij} = 0$) this could generate a non-zero value depending on $\{\theta_k\}$. Cliff and Ord (1981a) suggest obtaining estimates of the $\{\theta_k\}$ by non-linear least squares. Eagleson (1967) suggests an alternative (restricted) cubic polynomial function, evidently having experienced rough (residual) covariance functions akin to Figure 5.19. Ripley (1981) also provides examples of smoothing functions that are continuous at the origin, though he resorts to fitting these by eye.

Extensions to the methodology include allowing for changes in the covariance function according to direction (geometric anisotropy), an important consideration in mining applications, and predicting at block rather than point localities. These and other topics have generated an extensive literature which we do not attempt to review. The aim here is to give an outline that particularly emphasizes the links that exist between Kriging and associated methods and the other methods of this chapter. Ripley (1981) should be consulted for a comprehensive mathematical treatment that pursues the integrative theme and collates much of the literature (see also Craig and Labovitz, 1981). Earlier accessible accounts are provided by Blais and Carlier (1967), Matheron (1963, 1970), and Agterburg (1970).

Space–time models

By consciously restricting the models of this chapter to purely spatial processes, we have left unopened the Pandora's box of space–time models. In this subsection we lift the lid slightly to provide a brief outline of their shape and size; analysts seriously intending to use these models should regard the book by Bennett (1979) as compulsory reading. Such models, of course, can only be fitted given access to spatio-temporal data, which might consist of observations at T separate points in time and at each of n localities. We briefly considered the analysis of such data from an exploratory rather than a modelling standpoint in Chapter 3.

We commence our discussion by considering the following model:

$$\mathbf{Y}_t = \rho \mathbf{W}\mathbf{Y}_t + \mathbf{X}\mathbf{b} + \mathbf{e}_t, \qquad t = 1, \ldots, T. \tag{5.137}$$

This model displays a close affinity to the autoregressive response model described by equation (5.105). The two models differ on account of the subscript t, which indicates that the observations are time indexed. Parameter estimation could be achieved by the usual maximum likelihood procedure, although, as noted by Cliff and Ord (1975b), the determinant in the likelihood function now appears T times!

Despite its apparent simplicity, equation (5.137) is a somewhat troublesome formulation, the difficulty arising from the suggestion, provoked by the subscript t (when interpreted as a precise point in time), that the autoregressive Y interaction is instantaneous. Clearly, there is a degree of philosophical intractability associated with this idea, and this has troubled some analysts (see Bennett, 1979). If the time at which an observation is made does have a bearing on its value, then it seems unreasonable to suggest a model in which interaction time is disallowed.

The literal impossibility of instantaneous interaction is acknowledged in the version of the model given by equation (5.138), in which the autoregressive response is allotted a period in which to take effect:

$$\mathbf{Y}_t = \rho \mathbf{W}\mathbf{Y}_{t-1} + \mathbf{X}\mathbf{b} + \mathbf{e}_t, \qquad t = 1, \ldots, T. \tag{5.138}$$

This specification also resolves another difficulty, for, as Ripley (1981) observes, in this model 'we do not encounter any of the problems posed by the determinant'. A moment's reflection and reference back to the start of Section 5.7 reveals why this is true. From equation (5.138) we have:

$$\mathbf{e}_t = \mathbf{Y}_t - \rho \mathbf{W}\mathbf{Y}_{t-1} - \mathbf{X}\mathbf{b} \tag{5.139}$$

and, if we follow the previous practice and introduce $\mathbf{A}$, we have:

$$\mathbf{e}_t = \mathbf{A}\mathbf{Y}_t - \rho \mathbf{W}\mathbf{Y}_{t-1} - \mathbf{X}\mathbf{b}, \tag{5.140}$$

although, of course, in this case $\mathbf{A} = \mathbf{I}$ and $|\mathbf{A}| = 1$, so that $|\mathbf{A}|$ drops out of the likelihood function and the maximum likelihood procedure amounts to OLS.

Very many models of this genre have been proposed by Box and Jenkins (1970), although it was up to geographers, mainly, to explore the spatial aspects and to develop the spatio-temporal analogues of the Box–Jenkins time series models. These developments are comprehensively summarized by Cliff *et al.* (1975) and by Haggett *et al.* (1977), who follow Box and Jenkins in using a shorthand to describe the models. For example, we can refer to equation (5.138) as a STARR model, where the letters are an acronym for Space–Time AutoRegressive model with additional regression components (the $\mathbf{X}\mathbf{b}$).

The equations (5.141a) and (5.141b) constitute a STARIMAR model (Space–Time AutoRegression Integrated Moving Average model with additional Regression terms!) which captures a representative collection of the various effects in a single model:

$$\mathbf{Y}_t = \mathbf{X}_{t-1}\mathbf{b} + \beta\, \mathbf{Y}_{t-1} + \rho_1 \mathbf{W}_1 \mathbf{Y}_{t-1} + \mathbf{u}_t; \tag{5.141a}$$

$$\mathbf{u}_t = \gamma \mathbf{e}_{t-1} + \rho_2 \mathbf{W}_2 \mathbf{e}_{t-1} + \mathbf{e}_t. \tag{5.141b}$$

Setting $\beta = \gamma = \rho_2 = 0$ reduces the STARIMAR model to the STARR model, and setting other subsets of parameters to zero provides other options. Note that one immediate difficulty is the likely temporal trend in Y. A simple solution to this non-stationarity problem is to replace Y_{it} by the temporal differences $\Delta Y_{it} = Y_{it} - Y_{i,t-1}$ for reasons akin to those outlined in the subsection on spatial differencing, although repeated differencing may be required to induce stationarity (see Cliff *et al.* 1975). Another difficulty is, as ever, the arbitrariness that may enter into the choice of the **W** matrices. More generality can be achieved from multiparameter covariance structures, although this adds to an already immense estimation problem. With some justification, the STARIMAR model has been described as a 'monster' (Cliff and Ord, 1975c) which is occasionally wheeled out for demonstration purposes only. For those utterly determined to prove that it works, estimation methods are suggested by Box and Jenkins (1970) and Martin (1975).

Further extensions to the family of space–time models are possible by making adjustments to the space and time lags and by allowing parameter variation according to location and occasion. The latter is explored in the context of linear regression and analysis of variance by Cliff *et al.* (1975). Bennett (1975a, 1975b, 1975c, 1975d, 1975e, 1976), Martin (1978), and Hepple (1979b) use Kalman (1960) filters for similar purposes. The reviews by Hepple (1981) and Bennett (1981) provide additional references.

A second model of interest is the SARMAR (Spatial AutoRegressive Moving Average model with additional Regression components) model

$$\mathbf{Y} = \rho_1 \mathbf{W}_1 \mathbf{Y} + \mathbf{X}\mathbf{b} + \rho_2 \mathbf{W}_2 \mathbf{e} + \mathbf{e}. \tag{5.142}$$

Notice that this model is the STARIMAR denuded of time indexing and temporal autoregressive terms. Hence, the SARMAR model simply combines the autoregressive response and moving average models of Section 5.7, and could be estimated in a similar manner. Having affirmed that instantaneous interaction is impossible, we take the view with this and the other purely spatial models that the absence of the time subscript reflects a belief in the irrelevance of the time attribute which could be assigned to observations. In many applications, it is likely that time is an unknowable attribute of the data, or one that is forcibly constant, rather than irrelevant, in which case considerable care should be exercised in drawing conclusions about a dynamic system from an essentially static picture. We shall again take up the important issue of time-dependent observations when we discuss categorical data in Volume 2.

5.9 A RECAPITULATION OF SPATIAL REGRESSION MODELS

Regression and autoregression are terms encompassing such a diverse set of models that some recapitulation is called for. In this, to encourage succinctness, derivations and details of the methodology of model fitting are omitted.

We commence with the simplest approach, in which we use OLS to fit a *linear regression model*, as described in Section 5.2. As we noted at the outset, the Gauss–Markov assumptions associated with this model are rather tough. For example, a principal reason for progressing to other models was the requirement that errors should be independent, which, as we discovered, is a difficult condition to meet given spatial data. The penalty for ignoring the existence of (what appeared to be) positively spatially autocorrelated errors was a downwardly biased variance estimate, $\hat{\sigma}^2$, leading to correspondingly depressed estimates of the terms within the variance–covariance matrix, $\mathbf{V(b)}$, and hence to inflated values of t, F, and r^2 which overstated the apparent significance of the X variables. The reverse effect, which happens with negative autocorrelation, was also demonstrated.

The Gauss–Markov condition that the error variance is constant can also, if violated, provide potentially misleading conclusions, although this is more easily circumvented than error autocorrelation—the solution being to use *Weighted Least Squares* (WLS). Methods for assessing whether or nor the errors conform to the conditions required by OLS and inference under Gauss–Markov conditions were described in Section 5.6.

It is highly likely that, whatever preconceptions the analyst may possess concerning the 'final' model, he or she will want to fit a regression using OLS as a preliminary exploration of the data. This is comparatively simple to do, and may expose residuals suggestive of more apt models (see Section 5.6). These need not necessarily involve a reformulation of the error–covariance structure, but could involve some *non-linear regression model*, of the type detailed in Section 5.5.

It is convenient at this point to recount the *logit transformation* which permeated the chapter during our attempts to handle the technical problem presented by the case where the Y variable was a percentage. Without the transformation, any model stands a risk of producing estimates for Y that lie outside the feasible range (0–100). By invoking the transformation $z = \log\{Y/(100 - Y)\}$, and fitting models to the variation in z, we obviate this difficulty. The penalty incurred by this procedure is that the resulting models may be somewhat difficult to interpret.

A similar problem arises when Y consists of small counts forming a lumpy distribution close to a lower bound of zero. In this case the *Poisson distribution* may be more appropriate than the normal distribution as a model of the sampling variation in Y. In this case, model fitting is by ML (maximum likelihood), although Poisson distributions (and, indeed, many other distributions) are a standard facility in modern software packages, so the mention of ML need not invoke trauma. Inference is based on likelihood-ratio criteria, which we found to be the logical counterpart of inference under Gauss–Markov conditions. Poisson and normal response models were compared in Example 5.7.

Notwithstanding the facility one has to invoke alternative error distributions like the Poisson, regression models fitted by OLS and assuming normal errors remain the main vehicle in the environmental, biological, behavioural, and social sciences for portraying the dependence of Y on X variables. We have stressed continually, however, that when the data possesses a spatial dimension, the independent errors (and hence independent Y-values) assumption is either apparently or actually in jeopardy, whilst it is often the case that there is systematic variation in the accuracy with which Y is measured.

It is possible to avert both these dangers by invoking the *Generalized Least Squares* (GLS) formulation, which was shown in Section 5.3 to be only a few steps removed from that of OLS. One form of GLS, namely WLS, was used as part of the ongoing discussion of Example 5.1. In place of the assumption of a constant error structure, we allowed for variations in accuracy from place to place. Estimation of the *autocorrelated errors model* using GLS is problematical, since it entails knowledge of the error variance–covariance structure and requires the use of ML estimation, as described below. *Spatial differencing*, which assumes a particular structure for the variances and covariances, helps to a degree to get around the more substantial computational problems associated with ML in this context, but unsubstantiated assumptions are not entirely satisfactory.

In this situation, therefore, we have used the *autocorrelated errors model estimated by ML* as, it turned out, something of a 'catch-all' to diminish the adverse effect that autocorrelated residuals have on the variance estimates, and thus to permit valid inferences concerning the dependence of Y on the X variables, since it combines non-zero covariances with a structural model akin to the OLS regression model.

There are several disadvantages with this model which should be noted. First, fitting the model requires some facility with a programming language, as it is unlikely to be a standard option of a multiple regression package. Also, to specify the error interaction either requires a degree of *a priori* information in the form of a $\mathbf{W}$ matrix, in order to render parameter estimation comparatively straightforward, or it requires the estimation of a more complex multiparameter error covariance structure (see the subsections on 'Error covariance from correlograms', 'More complex autocorrelated error models', and 'Prediction methods related to Kriging').

An alternative to the autocorrelated errors model is the *autoregressive response model* of Section 5.7, which also specifies autoregressive interaction, but among the $\{Y_i\}$ rather than the $\{e_i\}$, so that Y_i depends on 'neighbouring' values

$\{Y_j\}$. What we term *pure autoregression* and *moving averages* are unlikely to provide viable models in a non-experimental setting, unless supplemented by the addition of X variables as utilized in Examples 5.19 and 5.20. We preferred not to invoke *conditional autoregression* models because, despite some attractive features, they did not appeal to us as practicable when compared with the simultaneous scheme with which we are familiar. Here, we could be accused of possessing mild personal foibles in this regard!

In order to fit the autoregressive response model, maximum likelihood is again appropriate and is carried out using a procedure very similar to that used with the autocorrelated errors model. The need to use maximum likelihood must count as a disadvantage, as must the need to specify an appropriate **W** matrix. Our experience with both types of autoregressive model, related in Examples 5.19 and 5.20, was that essentially the same amount of information was yielded by both types, although there were differences of detail. Generally, in such a situation it is preferable to choose the simplest model, and in this case this can be shown to be the autoregressive response model, which is essentially a simplified version of the autocorrelated errors model (see the subsection entitled 'The unrestricted autocorrelated errors model'). It may be conceptually difficult to envisage direct autoregressive Y interaction in some applications, as for example in the mortality study by Cook and Pocock (1983): in their study it is hard to see how the mortality rate at location i can, *per se*, directly influence the mortality rate at location j, and this type of imponderable should override considerations of statistical convenience when it comes to the choice of a model.

In striving for a 'final' model, a simple measure of model performance is provided by r^2 and its analogues, although taken alone this is an inadequate statistic and should be supplemented by the diagnostic evidence contained in the residual distribution. Broadly, one would expect a satisfactory model to produce residuals that are spatially random, although we have noted in Section 5.6 that this is a difficult-to-measure attribute of spatial autoregressions. For example, we cannot be entirely sure that we would ever arrive at random residuals for the Irish blood data, since there is every reason to believe that there are many other factors that contribute to the overall process deriving from the socio-economic structure of Irish society.

If one admits at this juncture the possibility of extending the list of X variables, then some of the above-mentioned 'slack' may be taken out of the residuals, even if these new variables amount to no more than dummy variables used to label clusters of like-valued residuals (cf. Diggle, 1982). Further analysis could proceed by examining each individual observation, since just one or two may exert an exorbitant influence on the form of the model finally chosen. The examination of residuals serves to identify incorrectly recorded data and to point out atypical observations. While a 'high leverage' point will not necessarily reveal itself as a salient residual, exploration and detection are comparatively straightforward (see, for example, Atkinson, 1981, and, for a review, Wrigley, 1983).

References

Aboav, D. A. (1970). The arrangement of grains in a polycrystal. *Metallography*, **3**, 383–390. [1.7]

Aboav, D. A. (1980). The arrangement of cells in a net. *Metallography*, **13**, 43–58. [1.7]

Agterburg, F. P. (1970). Autocorrelation functions in geology, pp. 113–141 in *Geostatistics: A Colloquium* (Merriam, D. F., Ed.). Plenum: New York. [5.8]

Aherne, W. A. and Diggle, P. J. (1978). The estimation of neuronal population density by a robust distance method. *Journal of Microscopy*, **114**, 285–293. [1.1]

Aitchison, J. and Brown, J. A. C. (1957). *The Lognormal Distribution*. Cambridge University Press: Cambridge. [5.5]

Anderson, D. R. and Pospahala, R. S. (1970) Correction of bias in belt transect studies of immotile objects. *Journal of Wildlife Management*, **34**, 141–146. [2.3]

Appel, K. and Haken, W. (1977). Every map is four-colorable, I, II. *Illinois Journal of Mathematics*, 429–527. [1.1]

Arora, S. S. and Brown, M. (1977). Alternative approaches to spatial autocorrelation: an improvement over current practice. *International Regional Science Review*, **2**, 67–78. [3.4]

Askew, G. R. (1983). Two methods for evaluating spatial pattern in naturally regenerated loblolly pine stands. *Forest Science*, **29**, 514–518. [1.5]

Atkinson, A. C. (1981). Two graphical displays for outlying and influential observations in regression. *Biometrika*, **68**, 13–20. [5.9]

Attoh, R. and Whitten, E. H. T. (1979). Computer program for regression model for discontinuous structural surfaces. *Computers and Geosciences*, **5**, 47–71. [5.5]

Baker, L. (1971). A selection of geographical computer programs, Geographical Papers No. 6, London School of Economics and Political Science. [3.6]

Barlow, R. E., Bartholomew, D. J., Bremner, J. M., and Brunk, H. D. (1972) *Statistical Inference Under Order Restrictions*. Wiley: New York. [2.3]

Barnard, G. A. (1963). Discussion of the paper by Prof. M. S. Bartlett. *Journal of the Royal Statistical Society, Series B*, **25**, 294. [3.1]

Bartholomew, D. J. (1981). *Mathematical Methods in Social Science*. Wiley: London. [5.2, 5.4]

Bartlett, M. S. (1936). Some notes on insecticide tests in the laboratory and in the field. *Supplement to the Journal of the Royal Statistical Society*, **3**, 185–194. [1.2]

Bartlett, M. S. (1948). Determination of plant densities. *Nature*, **162**, 621–624. [1.2]

Bartlett, M. S. (1964). The spectral analysis of two-dimensional point processes. *Biometrika*, **51**, 299–311. [1.1, 1.5]

Bartlett, M. S. (1971). Physical nearest-neighbour models and non-linear time series. *Journal of Applied Probability*, **8**, 222–232. [5.7]

Bartlett, M. S. (1974). The statistical analysis of spatial pattern. *Advances in Applied Probability*, **6**, 336–358. [1.1]

Barton, D. E. and David, F. N. (1966). The random intersection of two graphs, pp. 455–459 in *Research Papers in Statistics: Festchrift for J. Neyman* (David, F. N., Ed.). Wiley: New York. [3.2]

Bassett, K. A. (1972). Numerical methods for map analysis. *Progress in Human Geography*, **4**, 217–254. [5.5]

Batcheler, C. L. (1971). Estimation of density from a sample of joint point and nearest neighbour distances. *Ecology*, **3**, 703–709. [2.4]

Batcheler, C. L. (1973). Estimating density and dispersion from truncated or unrestricted joint point-distance nearest-neighbour distances. *Proceedings of the New Zealand Ecological Society*, **20**, 131–147. [2.4]

Batcheler, C. L. (1975a). Probable limits of error of the point distance-neighbour distance estimate of density. *Proceedings of the New Zealand Ecological Society*, **22**, 28–33. [2.4]

Batcheler, C. L. (1975b). Development of a distance method of deer census from pellet groups. *Journal of Wildife Management*, **39**, 641–652. [2.2, 2.4]

Batcheler, C. L. and Bell, D. J. (1970). Experiments in estimating density from point- and nearest neighbour distance samples. *Proceedings of the New Zealand Ecological Society*, **17**, 111–117. [2.4]

Batcheler, C. L. and Hodder, R. A. C. (1975). Tests of a distance technique for inventory of pine plantations. *New Zealand Journal of Forest Science*, **5**, 3–17. [2.4]

Beavon, K. S. O. (1977). *Central Place Theory: A Reinterpretation* Longman: London. [1.1]

Bennett, R. J. (1975a). The representation and identification of spatio-temporal systems: an example of population diffusion in North-West England. *Transactions of the Institute of British Geographers*, **66**, 73–94. [3.5, 5.8]

Bennett, R. J. (1975b). Dynamic systems modelling of the north-west region. 1. Spatio-temporal representation and identification. *Environment and Planning*, **7**, 525–538. [5.8]

Bennett, R. J. (1975c). Dynamic systems modelling of the north-west region. 2. Estimation of the spatio-temporal policy model. *Environment and Planning*, **7**, 539–566. [5.8]

Bennett, R. J. (1975d). Dynamic systems modelling of the north-west region. 3. Adaptive parameter policy model. *Environment and Planning*, **7**, 617–636. [5.8]

Bennett, R. J. (1975e). Dynamic systems modelling of the north-west region. 4. Adaptive spatio-temporal forecasts. *Environment and Planning*, **7**, 887–898. [5.8]

Bennett, R. J. (1976). Non-stationary parameter estimation for small sample situations: a comparison of methods. *International Journal of System Science*, **7**, 257–275. [5.8]

Bennett, R. J. (1979). *Spatial Time Series*. Pion: London. [3.5, 5.7, 5.8]

Bennett, R. J. (1981). Spatial and temporal analysis: spatial time series, pp. 96–103 in *Quantitative Geography: a British view* (Wrigley, N. and Bennett, R. J., Eds.). Routledge and Kegan Paul: London. [5.8]

Bergmann, J. (1970). *Die altere Bronzezeit Nordwestdeutschlands*. Elwert: Marburg. [1.1]

Berry, B. J. L. (1967). *The Geography of Market Centres and Retail Distribution*. Prentice-Hall: Englewood Cliffs. [1.1]

Besag, J. E. (1972). Nearest neighbour systems and the auto-logistic model for binary data, *Journal of the Royal Statistical Society, Series B*, **34**, 75–83. [5.7]

Besag, J. E. (1974). Spatial interaction and the statistical analysis of lattice systems. *Journal of the Royal Statistical Society, Series B*, **36**, 192–236. [5.7]

Besag, J. E. (1975). Statistical analysis of non-lattice data. *The Statistician*, **24**, 179–195. [5.7]

Besag, J. E. (1977a). Some methods of statistical analysis for spatial data. *Bulletin of the International Statistical Institute*, **47**, 77–92. [5.7]

Besag, J. E. (1977b). Efficiency of pseudo-likelihood estimators for simple Gaussian fields. *Biometrika*, **64**, 616–618. [5.7]

Besag, J. E. and Diggle, P. J. (1977). Simple Monte Carlo tests for spatial pattern. *Applied Statistics*, **26**, 327–333. [1.5, 3.1, 3.6]

Besag, J. E. and Gleaves, J. T. (1973). On the detection of spatial pattern in plant communities. *Bulletin of the International Statistical Institute*, **45**, 153–158. [1.4]

Besag, J. E. and Moran, P. A. P. (1975). On the estimation and testing of spatial interaction in Gaussian lattice processes. *Biometrika*, **62**, 555–562. [5.7]

Bishop, Y. M. M., Fienberg, S. E., and Holland, P. W. (1975). *Discrete Multivariate Analysis: Theory and Practice*. MIT Press: Cambridge, Mass. [5.4]

Bitterlich, W. (1948). Die Winkelzählprobe. *Allgemeine forst-und Holzwirtschaftliche Zeitung*, **59**, 4–5. [2.6]

Bivand, R. (1980). A Monte Carlo study of correlation coefficient estimation with spatially autocorrelated observations. *Quaestiones Geographicae*, **6**, 5–10. [4.2]

Blais, R. A. and Carlier, P. A. (1967). Applications of geostatistics in ore evaluation. *Ore Reserve Estimation and Grade Control*, **9**, 41–68. [5.8]

Bliss, C. I. and Fisher, R. A. (1953). Fitting the negative binomial distribution to biological data and a note on the efficient fitting of the negative binomial. *Biometrics*, **9**, 176–200. [1.3]

Blom, G. (1958). *Statistical Estimates and Transformed Beta-variables*. Wiley: New York. [5.6]

Boots, B. N. (1973). Some models of the random subdivision of space. *Geografiska Annaler*, **55B**, 34–48. [1.7]

Boots, B. N. (1974). Delaunay triangles: an alternative approach to point pattern analysis. *Proceedings of the Association of American Geographers*, **6**, 26–29. [1.7]

Boots, B. N. (1975a). Some observations on the structure of socioeconomic cellular networks. *Canadian Geographer*, **19**, 107–120. [1.7]

Boots, B. N. (1975b). Patterns of urban settlements revisited. *The Professional Geographer*, **27**, 426–431. [1.7]

Boots, B. N. (1977). Contact number properties in the study of cellular networks. *Geographical Analysis*, **9**, 379–387. [1.7]

Boots, B. N. (1980). Packing polygons: some empirical evidence. *Canadian Geographer*, **24**, 406–411. [1.7]

Boots, B. N. (1981). Weighting Theissen polygons. *Economic Geography*, **57**, 248–259. [1.7]

Boots, B. N. (1982). The arrangement of cells in 'random' networks. *Metallography*, **15**, 53–62. [1.7]

Boots, B. N. and Murdoch, D. J. (1983). The spatial arrangement of random Voronoi polygons. *Computers and Geosciences*, **9**, 351–365. [1.7]

Box, G. E. P. and Cox, D. R. (1964). An analysis of transformations. *Journal of the Royal Statistical Society, Series B*, **26**, 211–252. [5.5]

Box, G. E. P. and Jenkins, G. M. (1970). *Time Series Analysis: Forecasting and Control*. Holden-Day: San Francisco. [3.5, 5.8]

Bradu, D. and Mundlak, V. (1970). Estimation in lognormal linear models. *Journal of the American Statistical Association*, **65**, 198–211. [5.5]

Brandsma, A. S. and Ketellapper, R. H. (1979a). Further evidence on alternative procedures for testing of spatial autocorrelation amongst regression disturbances, pp. 113–136 in *Exploratory and Explanatory Statistical Analysis of Spatial Data* (Bartels, C. P. A. and Ketellapper, R. H., Eds.) Martinus Nijhoff: Boston, Mass. [5.4, 5.6]

Brandsma, A. S. and Ketellapper, R. H. (1979b). A biparametric approach to spatial autocorrelation. *Environment and Planning, Series A*, **11**, 51–58. [5.4]

Bray, J. R. (1956). A study of mutal occurrence of plant species. *Ecology*, **37**, 21–28. [4.2]

Bromley, K. J. (1980). Trader mobility in systems of periodic and daily markets, pp. 133–174 in *Geography and the Urban Environment*, (Herbert, D. T. and Johnston, R. J., Eds.), Vol. 3. Wiley: Chichester. [1.1]

Brown, D. (1975). A test of randomness of nest spacing. *Wildfowl*, **26**, 102–103. [1.5]

Brown, D. and Rothery, P. (1978). Randomness and local regularity of points in a plane. *Biometrika*, **65**, 115–122. [1.5]

Brown, G. S. (1965). Point density in stems per acre. *New Zealand Forestry Service Research Notes*, **38**, 1–11. [2.5]

Brown, S. and Holgate, P. (1974). The thinned plantation. *Biometrika*, **61**, 253–261. [1.4]

Brown, T. C. (1979). Position dependent and stochastic thinning of point processes. *Stochastic Processes and their Applications*, **9**, 189–193. [4.1]

Brown, T. C., Silverman, B. W., and Milne, R. K. (1981). A class of two-type point processes. *Zeitschrift für Wahrscheinlichkeitstheorie und Verwandte Gebiete*, **58**, 299–308. [4.1]

Buckland, S. T. (1982). A note on the Fourier series model for analysing line transect data. *Biometrics*, **38**, 469–477. [2.3]

Burnham, K. P. (1979). A parametric generalization of the Hayne estimator for line transect sampling. *Biometrics*, **35**, 587–596. [2.3]

Burnham, K. P. and Anderson, D. R. (1976). Mathematical models for nonparametric inferences from line transect data. *Biometrics*, **32**, 325–336. [2.3]

Burnham, K. P., Anderson, D. R., and Laake, J. L. (1980). Estimation of density from line transect sampling of biological populations. Wildlife monograph no. 72. Supplement to *Journal of Wildlife Management*, **44**. [2.3]

Burridge, P. (1981). Testing for a common factor in a spatial autoregression model. *Environment and Planning, Series A*, **13**, 795–800. [5.4, 5.6]

Burt, R. S. and Doreian, P. (1982). Testing a structural model of perception: conformity and deviance with respect to journal norms in elite sociological methodology. *Quality and Quantity*, **16**, 109–150. [5.7]

Byth, K. (1981). θ-stationary point processes and their second-order analysis. *Journal of Applied Probability*, **18**, 864–878. [1.5]

Byth, K. (1982). On robust distance-based intensity estimators. *Biometrics*, **38**, 127–135. [2.4]

Byth, K. and Ripley, B. D. (1980). On sampling spatial patterns by distance methods. *Biometrics*, **36**, 279–284. [1.4, 2.4]

Carpenter, S. R. and Cheney, J. E. (1983). Scale of spatial pattern: four methods compared. *Vegetatio*, **53**, 153–160. [1.3]

Carter, H. (1972). *The Study of Urban Geography*. Arnold: London. [1.1]

Chapman, G. P. (1970). The application of information theory to the analysis of population distributions in space. *Economic Geography*, **46**, 317–331. [1.7]

Chatterjee, S. and Price, B. (1977). *Regression Analysis by Example*. Wiley: New York. [5.1]

Chayes, F. (1970). On deciding whether trend surfaces of progressively higher order are meaningful. *Bulletin of the Geological Society of America*, **81**, 1273–1278. [5.5]

Chorley, R. J. and Haggett, P. (1965). Trend surface mapping in geographical research. *Transactions of the Institute of British Geographers*, **37**, 47–67. [5.5]

Christaller, W. (1933). *Die zentralen Orte in Suddeutschland*, translated as *Central Places in Southern Germany* by Baskin, C. W. (1966). Prentice-Hall: Englewood Cliffs, New Jersey. [1.1]

Clapham, A. R. (1936). Over-dispersion in grassland communities and the use of statistical methods in plant ecology. *Journal of Ecology*, **24**, 232–251. [1.2]

Clark, P. J. (1956). Grouping in spatial distributions. *Science*, **123**, 373–374. [1.5]

Clark, P. J. and Evans, F. C. (1954). Distance to nearest neighbour as a measure of spatial relationships in populations. *Ecology*, **35**, 445–453. [1.5, 1.6]

Clark, P. J. and Evans, F. C. (1955). On some aspects of spatial pattern in biological populations. *Science*, **121**, 397–398. [1.5]

Cliff, A. D. (1970). Computing the spatial correspondence between geographical patterns. *Transactions of the Institute of British Geographers*, **50**, 143–154. [3.6]

Cliff, A. D., Haggett, P., Ord, J. K., Bassett, K. A., and Davies, R. B. (1975). *Elements of Spatial Structure: A Quantitative Approach*. Cambridge University Press: Cambridge. [3.1, 3.5, 5.8]

Cliff, A. D. and Ord, J. K. (1969). The problem of spatial correlation, pp. 25–55 in *London Papers in Regional Science 1: Studies in Regional Science* (Scott, A. J., Ed.). Pion: London. [3.4]

Cliff, A. D. and Ord, J. K. (1971). Evaluating the percentage points of a spatial autocorrelation coefficient. *Geographical Analysis*, **3**, 5162. [3.3]

Cliff, A. D. and Ord, J. K. (1972). Testing for spatial autocorrelation among regression residuals. *Geographical Analysis*, **4**, 267–284. [5.6]

Cliff, A. D. and Ord, J. K. (1973). *Spatial Autocorrelation*. Pion: London. [3.1, 3.2, 3.3, 5.2, 5.4, 5.6]

Cliff, A. D. and Ord, J. K. (1975a). The choice of a test for spatial autocorrelation, pp. 54–77 in *Display and Analysis of Spatial Data* (Davis, J. C. and McCullagh, M. J., Eds.). Wiley: Chichester. [3.3, 3.6, 5.7]

Cliff, A. D. and Ord, J. K. (1975b). Model building and the analysis of spatial pattern in human geography. *Journal of the Royal Statistical Society, Series B*, **37**, 297–348. [5.8]

Cliff, A. D. and Ord, J. K. (1975c). Space–time modelling with an application to regional forecasting. *Transactions of the Institute of British Geographers*, **64**, 119–128. [5.8]

Cliff, A. D. and Ord, J. K. (1981a). *Spatial Processes: Models and Applications*. Pion: London. [1.1, 1.5, 3.1, 3.2, 3.3, 3.4, 3.5, 3.6, 5.3, 5.4, 5.5, 5.6, 5.7]

Cliff, A. D. and Ord, J. K. (1981b). Spatial and temporal analysis: autocorrelation in space and time, pp. 104–110 in *Quantitative Geography: A British View* (Wrigley, N. and Bennett, R. J., Eds.). Routledge and Kegan Paul: London. [5.7]

Collins, R. (1972). Melting and statistical geometry of simple liquids, in *Phase Transitions and Critical Phenomena* (Domb, C. and Green, M. S., Eds.), Vol. 2. Academic Press: New York. [1.7]

Cook, D. G. and Pocock, S. J. (1983). Multiple regression in geographic mortality studies with spatially correlated errors. *Biometrics*, **39**, 361–371. [5.4, 5.6, 5.8, 5.9]

Cormack, R. M. (1977). The invariance of Cox and Lewis's statistic for the analysis of spatial patterns. *Biometrika*, **64**, 143–144. [1.4]

Cormack, R. M. (1979). Spatial aspects of competition between individuals, pp. 151–212 in *Spatial and Temporal Analysis in Ecology* (Cormack, R. M. and Ord, J. K., Eds.). International Co-operative Publishing House: Fairland, Maryland. [1.1, 1.3, 1.4]

Cox, D. R. (1955). Some statistical methods related with series of events (with discussion). *Journal of the Royal Statistical Society, Series B*, **17**, 129–164. [4.1]

Cox, D. R. (1970). *The Analysis of Binary Data*. Methuen: London. [5.3]

Cox, T. F. (1976). The robust estimation of the density of a forest stand using a new conditioned distance method. *Biometrika*, **63**, 493–499. [2.4]

Cox, T. F. (1981). Reflexive nearest neighbours. *Biometrics*, **37**, 367–370. [1.5]

Cox, T. F. and Lewis, T. (1976). A conditioned distance ratio method for analyzing spatial patterns. *Biometrika*, **63**, 483–491. [1.4, 2.4]

Craig, R. G. and Labovitz, M. L. (Eds.) (1981). *Future Trends in Geomathematics*. Pion: London. [5.8]

Crain, B. R., Burnham, K. P., Anderson, D. R., and Laake, J. L. (1979). Nonparametric estimation of population density using Fourier series. *Biometrical Journal*, **21**, 731–748. [2.3]

Crain, I. K. (1972). Monte Carlo simulation of random Voronoi polygons: preliminary results. *Search*, **3**, 220–221. [1.7]

Crain, I. K. (1978). The Monte-Carlo generation of random polygons. *Computers and Geosciences*, **4**, 131–141. [1.7]

Crain, I. K. and Miles, R. E. (1976). Monte Carlo estimates of the distributions of the random polygons determined by random lines in a plane. *Journal of Statistical Computation and Simulation*, **4**, 293–325. [1.7]

Cresswell, W. L. and Froggart, P. (1963). *The Causation of Bus Driver Accidents—An Epidemiological Study*. Oxford University Press: London. [1.1]

Curry, L. (1964). The random spatial economy: an exploration in settlement theory. *Annals of the Association of American Geographers*, **54**, 138–146. [1.1]

Curry, L. (1967). Central places in the random spatial economy. *Journal of Regional Science*, **7**, 217–238. [1.1]

Curtis, J. T. and McIntosh, R. P. (1950). The interrelations of certain analytic and synthetic phytosociological characters. *Ecology*, **31**, 434–455. [1.2]

Dacey, M. F. (1960). The spacing of river towns. *Annals of the Association of American Geographers*, **50**, 59–61. [1.5]

Dacey, M. F. (1964). Modified Poisson probability law for point patterns more regular than random. *Annals of the Association of American Geographers*, **54**, 559–565. [1.1]

Dacey, M. F. (1965). The geometry of central place theory. *Geografiska Annaler*, **47B**, 111–124. [1.1, 3.2]

Dacey, M. F. (1966a). A county seat model for the areal pattern of an urban system. *Geographical Review*, **56**, 527–542. [1.1]

Dacey, M. F. (1966b). A compound probability law for a pattern more dispersed than random with areal inhomogeneity. *Economic Geography*, **42**, 172–179. [1.1]

Dacey, M. F. (1968a). An empirical study of the areal distribution of houses in Puerto Rico. *Transactions of the Institute of British Geographers*, **45**, 51–70. [1.2, 3.2]

Dacey, M. F. (1968b). A review of measures of contiguity for two and k-color maps, pp. 479–495 in *Spatial Analysis: A Reader in Statistical Geography* (Berry, B. J. L. and Marble, D. F., Eds.). Prentice-Hall: Englewood Cliffs, New Jersey. [3.2]

Dacey, M. F. (1969). Proportion of reflexive nth order neighbours in spatial distribution. *Geographical Analysis*, **1**, 385–388. [1.5]

David, F. N. and Moore, P. G. (1954). Notes on contagious distributions in plant populations. *Annals of Botany of London*, **18**, 47–53. [1.2]

Davis, J. C. (1973). *Statistics and Data Analysis in Geology*. Wiley: New York. [1.1, 5.5, 5.8]

Delfiner, P. and Delhomme, J. P. (1975). Optimum interpolation by Kriging, pp. 96–114 in *Display and Analysis of Spatial Data* (Davis, J. C. and McCullagh, M. J., Eds.). Wiley: Chichester. [5.8]

De Vos, S. (1973). The use of nearest neighbour methods. *Tijdschrift voor Economische en Sociale Geografie*, **64**, 307–319. [1.5]

Diggle, P. J. (1975). Robust density estimation using distance methods. *Biometrika*, **62**, 39–48. [1.1, 2.4]

Diggle, P. J. (1977a). The detection of random heterogeneity in plant populations. *Biometrics*, **33**, 390–394. [1.4]

Diggle, P. J. (1977b). A note on robust density estimation for spatial point patterns. *Biometrika*, **64**, 91–96. [2.4]

Diggle, P. J. (1979a). On parameter estimation and goodness-of-fit testing for spatial point patterns. *Biometrics*, **35**, 87–101. [1.5]

Diggle, P. J. (1979b). Statistical methods for spatial point patterns in ecology, pp. 95–150 in *Spatial and Temporal Analysis in Ecology* (Cormack, R. M. and Ord. J. K., Eds.). International Co-operative Publishing House: Fairland, Maryland. [1.1, 1.5, 3.4, 4.2]

Diggle, P. J. (1981a). Some graphical methods in the analysis of spatial point patterns, pp. 55–73 in *Interpreting Multivariate Data* (Barnett, V. D., Ed.). Wiley: Chichester. [1.5]

Diggle, P. J. (1981b). Binary mosaics and the spatial pattern of heather. *Biometrics*, **37**, 531–539. [2.7]

Diggle, P. J. (1982). Contribution to the discussion of the paper by Pocock, S. J., Cook, D. G. and Shaper, A. G. *Journal of the Royal Statistical Society, Series A*, **145**, 332–333. [5.4, 5.9]

Diggle, P. J., Besag, J., and Gleaves, J. T. (1976). Statistical analysis of spatial point patterns by means of distance methods. *Biometrics*, **32**, 659–667. [1.1, 1.4]

Diggle, P. J. and Cox, T. F. (1983). Some distance-based tests of independence for sparsely-sampled multivariate spatial point patterns. *International Statistical Review*, **51**, 11–23. [1.4, 4.1, 4.2, 4.4]

Diggle, P. J. and Matérn, B. (1980). On sampling designs for the study of point–event nearest neighbour distributions in R^2. *Scandinavian Journal of Statistics*, **7**, 80–84. [1.5]

Diggle, P. J. and Milne, R. K. (1983). Bivariate Cox processes: some models for bivariate spatial point patterns. *Journal of the Royal Statistical Society, Series B*, **45**, 11–21. [4.1, 4.2, 4.3]

Dirichlet, G. L. (1850). Uber die Preduction der positiven quadritschen Formen mit drei unbestimmten ganzen Zahlen. *Journal für die reine und angewandte Mathematik*, **40**, 209–234. [1.7]

Donnelly, K. (1978). Simulation to determine the variance and edge-effect of total nearest neighbour distance, pp. 91–95 in *Simulation Methods in Archaeology* (Hodder, I. R., Ed.). Cambridge University Press: London. [1.5]

Doreian, P. (1980). Linear models with spatially distributed data. *Sociological Methods and Research*, **9**, 29–60. [5.4, 5.7]

Doreian, P. (1981). Estimating linear models with spatially distributed data, pp. 359–388 in *Sociological Methodology 1980* (Leinhardt, S., Ed.) Jossey-Bass: San Francisco. [5.4, 5.7]

Douglas, J. B. (1975). Clustering and aggregation. *Sankhya, Series B*, **37**, 398–417. [1.1, 1.2]

Doveton, J. H. and Parsley, A. J. (1970). Experimental evaluation of trend surface distortions induced by inadequate data point distributions. *Transactions of the Institute of Mining and Metallurgy*, B, **79**, B197–B207. [5.5]

Draper, N. R. and Smith, H. (1981). *Applied Regression Analysis*, 2nd edition. Wiley: London. [5.1, 5.5]

Durbin, J. (1960). Estimation of parameters in time series regression models. *Journal of the Royal Statistical Society, Series B*, **22**, 139–153. [3.5]

Durbin, J. (1970). Testing for serial correlation in least squares regression when some of the regressors are lagged dependent variables, *Econometrica*, **38**, 410–421. [5.6]

Eagleson, P. S. (1967). Optimum density of rainfall networks. *Water Resources Research*, **3**, 1021–1033. [5.8]

Eberhardt, L. L. (1967). Some developments in 'distance sampling'. *Biometrics*, **23**, 207–216. [1.4]

Eberhardt, L. L. (1968). A preliminary appraisal of line transects. *Journal of Wildlife Management*, **32**, 82–88. [2.3]

Eberhardt, L. L. (1978). Transect methods for population studies. *Journal of Wildlife Management*, **42**, 1–31. [2.3, 2.7]

Eberhardt, L. L. (1979). Line transects based on right-angle distances. *Journal of Wildlife Management*, **43**, 768–774. [2.3]

Edgington, E. S. (1969). Randomization tests. *Journal of Psychology*, **57**, 445–449. [3.1]

Errington, J.C. (1973). The effect of regular and random distributions on the analysis of pattern. *Journal of Ecology*, **61**, 99–105. [1.3]

Erschbamer, B., Grabherr, G., and Reisigl, H. (1983). Spatial pattern in dry grassland communities of the Central Alps and its ecophysiological significance. *Vegetatio*, **54**, 143–154. [I]

Evans, U. R. (1945). The laws of expanding circles and spheres in relation to the lateral growth of surface films and the grain-size of metals. *Transactions of the Faraday Society*, **41**, 365–374. [1.7]

Feller, W. (1943). On a general class of contagious distributions. *Annals of Mathematical Statistics*, **14**, 389–400. [1.1]

Ferns, P. N. and Mudge, G. P. (1981). Accuracy of counts at a mixed colony of herring and lesser black-backed gulls. *Brid Study*, **28**, 244–245. [2.2]

Fingleton, B. (1978). Factors affecting optimum locations. *Area*, **10**, 96–101. [1.1]

Fingleton, B. (1983a). Independence, stationarity, categorical spatial data and the chisquare test. *Environment and Planning, Series A*, **15**, 483–499. [4.2]

Fingleton, B. (1983b). Log-linear models with dependent spatial data. *Environment and Planning, Series A*, **15**, 801–814. [4.2]

Flowerdew, R. and Aitkin, M. (1982). A method of fitting the gravity model based on the Poisson distribution. *Journal of Regional Science*, **22**, 191–202. [5.4]

Ford, E. D. (1976). The canopy of a Scots Pine forest: description of a surface of complex roughness. *Agricultural Meteorology*, **17**, 9–32. [3.5]

Fraser, A. R. (1977). Triangle based probability polygons for forest sampling. *Forest Science*, **23**, 111–121. [2.5]

Fraser, A. R. and van den Driessche, P. (1972). Triangles, density and pattern in point populations, pp. 277–286 in *Proceedings of the 3rd Conference of the Advisory Group of Forest Statisticians*. International Union for Research organization, Institut National de la Recherche Agronomique: Jouy-en-Josas, France. [1.5, 1.7, 2.5]

Freeman, G. H. (1953). Spread of diseases in a rectangular plantation with vacancies. *Biometrika*, **40**, 287–296. [[5.4]

Freeman, T. W. (1969). *Ireland. A General and Regional Geography*. Methuen: London. [5.2]

Galiano, A. F. (1982) Pattern detection in plant populations through the analysis of plant-to-all-plants distances. *Vegetatio*, **49**, 39–43. [1.5]

Gardner, M. J. (1973). Using the environment to explain and predict mortality. *Journal of the Royal Statistical Society, Series A*, **136**, 421–440. [5.3]

Gates, C. E. (1969). Simulation study of estimators for the line transect sampling method. *Biometrics*, **25**, 317–328. [2.3]

Gates, C. E., Marshall, H., and Olson, D. P. (1968). Line transect method of estimating grouse population. *Biometrics*, **24**, 135–145. [2.3]

Gatrell, A. C. (1979). Autocorrelation in spaces. *Environment and Planning Series A*, **11**, 507–516. [3.4]

Geary, R. C. (1954). The contiguity ratio and statistical mapping. *The Incorporated Statistician*, **5**, 115–145. [3.3, 5.6]

Ghent, A. W. (1963). Studies of regeneration of forest stands devastated by the spruce budworm. *Forest Science*, **9**, 295–310. [2.2]

Gilbert, E. N. (1962). Random subdivisions of space into crystals. *Annals of Mathematical Statistics*, **33**, 958–972. [1.7]

Glass, L. and Tobler, W. R. (1971). Uniform distribution of objects in a homogenous field: cities on a plain. *Nature*, **233**, 67–68. [1.1, 3.6]

Glick, B. J. (1979). Tests for space–time clustering used in cancer research. *Geographical Analysis*, **11**, 202–208. [3.6]

Glick, B. J. (1982). A spatial rank-order correlation measure. *Geographical Analysis*, **14**, 177–181. [4.5]

Goldberger, A. S. (1968). The interpretation and estimation of Cobb–Douglas functions. *Econometrica*, **35**, 464–472. [5.5]

Goodall, D. W. (1961). Objective methods for the classification of vegetation IV. Pattern and minimal area. *Australian Journal of Botany*, **9**, 162–196. [1.3]

Goodall, D. W. (1965). Plot-less tests of inter-specific association. *Journal of Ecology*, **53**, 197–210. [4.4]

Goodall, D. W. (1974). A new method for the analysis of spatial pattern by random pairing of quadrats. *Vegetatio*, **29**, 135–146. [1.3]

Goudsmit, S. (1945). Random distribution of lines in a plane. *Review of Modern Physics*, **17**, 321–322. [1.7]

Gray, J. M. (1972). Trend surface analysis: trends through clusters. *Area*, **4**, 275–279. [5.5]

Green, P. J. and Sibson, R. (1978). Computing Dirichlet tesselations in the plane. *Computer Journal*, **21**, 168–173. [1.5]

Green, R. F. (1979). A graphical test to test interference in selecting nest-sites, pp. 439–452 in *Contemporary Quantitative Ecology and Related Econometrics* (Patil, G. P. and Rosenzweig, M. L., Eds.). International Co-operative Publishing House: Fairland, Maryland. [3.6]

Greig-Smith, P. (1952). The use of random and contiguous quadrats in the study of the structure of plant communities. *Annals of Botany*, **16**, 293–316. [1.3]

Greig-Smith, P. (1964). *Quantitative Plant Ecology*, 2nd edition. Butterworth: London. [1.3]

Greig-Smith, P. (1979). Pattern in vegetation. *Journal of Ecology*, **67**, 755–779. [1.1]

Griffith, D. A. (1980). Towards a theory of spatial statistics. *Geographical Analysis*, **12**, 325–339. [I, 5.4, 5.7]

Griffith, D. A. (1981). Towards a theory of spatial statistics: a rejoinder. *Geographical Analysis*, **13**, 91–93. [I, 5.4]

Griffith, D. A. (1983). The boundary value problem in spatial statistical analysis. *Journal of Regional Science*, **23**, 377–387. [5.7]

Griffith, D. A. and Amrhein, C. G. (1983). An evaluation of correction techniques for boundary effects in spatial statistical analysis: traditional method. *Geographical Analysis*, **15**, 352–360. [5.7]

Gulmon, S. L. and Mooney, H. A. (1977). Spatial and temporal relationships between two desert shrubs *Atriplex hymnelytra* and *Tidestromia oblongifolia* in Death Valley, California. *Journal of Ecology*, **65**, 831–838. [1.1, 1.2, 3.6, 4.4]

Gustavsson, R. (1974). Type stands for cleaning. *Research notes No. 70*. Department of Operational Efficiency, Royal College of Forestry: Garpenberg, Sweden. [4.2]

Hackett, W. E. R. and Dawson, G. W. P. (1958). The distribution of the ABO and simple 'Rhesus' (D) blood groups in the Republic of Ireland from a sample of 1 in 37 of the adult population. *Irish Journal of Medical Science*, **6**, 387, 99–109. [5.2]

Hagerstrand, T. (1953). On Monte Carlo simulation of diffusion, reprinted as pp. 1–32 in *Quantitative Geography, Part I: Economic and Cultural Topics* (Garrison, W. L. and Marble, D. F., Eds., 1967: Studies in Geography, 13). Northwestern University: Evanston, Illinois. [3.5, 3.6]

Haggett, P. (1976). Hybridizing alternative models of an epidemic diffusion process, *Economic Geography*, **52**, 136–146. [3.4]

Haggett, P. (1980). Boundary problems in statistical geography, in *Die Bedeutung von Grenzen in der Geographie* (Kishimoto, H., Ed.). Kummerley and Frey: Zurich. [1.5]

Haggett, P., Cliff, A. D. and Frey, A. (1977). *Locational Analysis in Human Geography* Arnold: London. [I, 1.1, 1.5, 3.6, 5.3, 5.5, 5.8]
Haining, R. P. (1977). Model specification in stationary random fields. *Geographical Analysis*, **9**, 107–129. [3.3, 5.7]
Haining, R. P. (1978a). The moving average model for spatial interaction. *Transactions of the Institute of British Geographers*, **3**, 202–225. [3.3, 5.7]
Haining, R. P. (1978b). Interaction modelling on central place lattices. *Journal of Regional Science*, **18**, 217–228. [5.7]
Haining, R. P. (1979). Statistical tests and process generators for random field models. *Geographical Analysis*, **11**, 45–64. [5.7]
Haining, R. P. (1980). Spatial autocorrelation problems, pp. 1–44 in *Geography and the Urban Environment* (Herbert, D. T. and Johnston, R. J., Eds.). Wiley: Chichester. [3.1]
Haining, R. P. (1981a). Spatial interdependencies in population distributions: a study in univariate map analysis. 1. Rural population desnities. *Environment and Planning, Series A*, **13**, 65–84. [5.4, 5.5]
Haining, R. P. (1981b). Spatial interdependencies in population distributions: a study in univariate map analysis. 2. Urban population distributions. *Environment and Planning, Series A*, **13**, 85–96. [5.4]
Haining, R. P. (1981c). Analysing univariate maps. *Progress in Human Geography*, **5**, 58–78. [3.4]
Haining, R. P., Griffith, D. A., and Bennett, R. J. (1983). Simulating two-dimensional autocorrelated surfaces. *Geographical Analysis*, **15**, 247–255. [5.7]
Hamilton, T. H., Rubinoff, I., Barth, R. H. Jr., and Bush, G. L. (1963). Species abundance: natural regulation of insular variation. *Science*, **142**, 1575–1577. [5.2, 5.4]
Hanisch, K.-H. and Stoyan, D. (1979). Formulas for the second-order analysis of marked point processes. *Mathematische Operationsforschung und Statistik; Series Statistics*, **10**, 555–560. [4.1, 4.2]
Hansel, B. (1968). *Beitrage zur Chronologie der mittleren Bronzezeit im Karpathenbecken.* Beitrage zur ur-und Fruhgeschischtlichen Archaoligie des Mittelmeer-Kulturraumes, 7 and 8. Rudolf Habelt: Bonn. [4.4]
Harkness, R. D. and Isham, V. (1984). A bivariate spatial point pattern of ants' nests. *Applied Statistics*, **32**, 293–303. [4.4]
Haworth, J. M. and Vincent, P. J. (1979). The stochastic disturbance specification and its implications for log-linear regression. *Environment and Planning, Series A*, **11**, 781–790. [5.5]
Haworth, J. M. and Vincent, P. J. (1982). Power transformations and linear regression models in geography. *Area*, **14**, 137–143. [5.5]
Hayes, R. J. and Buckland, S. T. (1983). Radial-distance models for the line-transect method. *Biometrics*, **39**, 29–42. [2.3]
Hayne, D. W. (1949). An examination of the strip census method for estimating animal populations. *Journal of Wildlife Management*, **13**, 145–157. [2.3]
Hemingway, P. (1971). Field trials of the line transect method of sampling large populations of herbivores, pp. 405–411 in *The Scientific Management of Animal and Plant Communities for Conservation* (Duffey, E. and Watt, A. S., Eds.). Blackwell: Oxford. [2.3]
Hepple, L. W. (1976). A maximum likelihood model for econometric estimation with spatial series, pp. 90–104 in *London Paper in Regional Science*, 6. *Theory and Practice in Regional Science* (Masser, I., Ed.). Pion: London. [5.3, 5.4]
Hepple, L. W. (1979a). Bayesian analysis of the linear model with spatial dependence, pp. 179–199 in *Exploratory and Explanatory Statistical Analysis of Spatial Data* (Bartels, C. P. A. and Ketellapper, R. H., Eds.). Martinus Nijhoff: Boston. [5.4]

Hepple, L. W. (1979b). Regional dynamics in British unemployment, and the impact of structural change, pp. 45–63 in *Statistical Applications in the Spatial Sciences* (Wrigley, N., Ed.). Pion: London. [5.8]

Hepple, L. W. (1981). Spatial and temporal analysis: time series analysis, pp. 92–96 in *Quantitative Geography: A British View* (Wrigley, N. and Bennett, R. J., Eds.). Routledge and Kegan Paul: London. [5.8]

Hill, A. R. (1973). The distribution of drumlins in County Down, Ireland. *Annals of the Association of American Geographers*, **63**, 229–241. [1.4]

Hill, M. O. (1973). The intensity of spatial pattern in plant communities. *Journal of Ecology*, **61**, 225–236. [1.1, 1.3]

Hinde, A. L. and Miles, R. E. (1980). Monte Carlo estimates of the distributions of the random polygons of the Voronoi tessellation with respect to a Poisson process. *Journal of Statistical Computation and Simulation*, **10**, 205–223. [1.7]

Hines, W. G. S. and Hines, R. J. O'H. (1979). The Eberhardt statistic, and the detection of non-randomness of spatial point distributions. *Biometrika*, **66**, 73–79. [1.4]

Hodder, I. R. (1975). Some new directions in the spatial analysis of archaeological data, in *New Directions in Archaeology* (Clarke, D. L., Ed). Methuen: London. [1.4]

Hodder, I. R. and Hassall, M. (1971). The non-random spacing of Romano-British walled towns. *Man*, **6**, 391–407. [1.1]

Hodder, I. R. and Orton, C. (1976). *Spatial Analysis in Archaeology*. Cambridge University Press: Cambridge. [1.1, 1.4, 4.4, 5.5]

Hoel, P. G. (1943). On indices of dispersion. *Annals of Mathematical Statistics*, **14**, 155–162. [1.2]

Hoel, P. G. (1971). *Introduction to Mathematical Statistics*. Wiley: New York. [5.4]

Holgate, P. (1965). Some new tests of randomness. *Journal of Ecology*, **53**, 261–266. [1.4]

Holgate, P. (1967). The angle-count method. *Biometrika*, **54**, 615–623. [2.6]

Hooper, I. (1947). Blood group distributions in Ireland. *Irish Journal of Medical Science*, **6**, 259, 471–479. [5.2]

Hooper, P. M. and Hewings, G. J. D. (1981). Some properties of space–time processes. *Geographical Analysis*, **13**, 203–223. [5.2]

Hopkins, B. (1954). A new method for determining the type of distribution of plant individuals. *Annals of Botany*, **18**, 213–227. [1.4]

Hopkins, B. (1965). Observations on savanna burning in the Olokemeji Forest Reserve, Nigeria. *Journal of Applied Ecology*, **2**, 367–381. [1.3, 3.2]

Hubert, L. J. (1978). Nonparametric tests for patterns in geographic variation: possible generalizations. *Geographical Analysis*, **10**, 86–88. [3.6]

Hubert, L. J. and Golledge, R. G. (1982a). Measuring association between spatially defined variables: Tjostheim's index and some extensions. *Geographical Analysis*, **14**, 273–278. [4.5]

Hubert, L. J. and Golledge, R. G. (1982b). Comparing rectangular data matrices. *Environment and Planning, Series A*, **14**, 1087–1095. [4.5]

Hubert, L. J., Golledge, R. G., and Costanzo, C. M. (1981). Generalized procedures for evaluating spatial autocorrelation. *Geographical Analysis*, **13**, 224–233. [3.1, 3.3, 3.6]

Hubert, L. J., Golledge, R. G., Costanzo, C. M., Gale, N. D., and Halperin, W. C. (1983). Nonparametric tests for directional data, in *Recent Developments in Spatial Analysis: Methodology, Measurement, Models* (Bahrenberg, G., Fischer, M., and Nijkamp, P., Eds.). Gower: Aldershot. [3.1]

Hudson, J. C. and Fowler, P. M. (1966). *The concept of Pattern in Geography*. Department of Geography, University of Iowa, Discussion paper: Series 1 (mimeo). [1.1]

Huijbrechts, C. (1975). Regionalised variables and quantitative analysis of spatial data, pp. 38–53 in *Display and Analysis of Spatial Data* (Davis, J. C. and McCullagh, M. J., Eds.). Wiley: Chichester. [5.8]

Hutchings, M. J. (1979). Standing crop and pattern in pure stands of *Merculialis perennis* and *Rubus fruticosus* in mixed deciduous woodland. *Oikos*, **31**, 351–357. [4.1]

Huxley, J. S. (1939). Clines: an auxiliary method of taxonomy. *Bijdragen tot de Dierkunde*, **27**, 491–520. [3.3]

Isard, W. (1956). *Location and Space Economy*. Wiley: New York. [1.1]

Iwao, S. (1968). A new regression method for analysing the aggregation pattern in animal populations. *Research in Population Ecology*, **10**, 1–20. [1.2]

Jacob-Friesen, K. K. (1967). *Bronzezeitliche Lanzenspitzen Norddeutschlands und Skandinaviens*, Veröffentlichungen der urgeschichtlichen Sammlungen des Landesmuseums zu Hannover, 17, Hildesheim. [I]

James, W. R. (1970). Regression models for faulted structural surfaces. *Bulletin of the American Association of Petroleum Geologists*, **54**, 638–646. [5.5]

Johnson, M. P., Mason, L. G., and Raven, P. H. (1968). Ecological parameters and plant species diversity. *American Naturalist*, **102**, 297–306. [5.2]

Johnson, M. P. and Raven, P. H. (1970). Natural regulation of plant species diversity. *Evolutionary Biology*, **4**, 127–162. [5.2, 5.4, 5.6]

Johnson, M. P. and Raven, P. H. (1973). Species number and endemism: the Galapagos revisited. *Science*, **179**, 893–895. [5.4, 5.5]

Johnson, W. A. and Mehl, R. F. (1939). Reaction kinetics in processes of nucleation and growth. *Transactions of the American Institute of Mining, Metallurgical and Petroleum Engineers*, **135**, 416–456. [1.7]

Johnston, R. J. and Wrigley, N. (1981). Urban geography, pp. 335–351 in *Quantitative Geography: A British View* (Wrigley, N. and Bennett, R. J., Eds.). Routledge and Kegan Paul: London. [1.1]

Jolivet, E. (1978). Caractérisation et test du caractère agrégatif des processus ponctuels stationnaires sur R^2. *Lecture notes in Mathematics*, **636**, 1–25. [1.5]

Jolivet, E. (1980). Théorème centrale limite et convergence de processus empiriques pour des processus ponctuels stationnaires. *Point Processes and Queueing Systems* (Bártfai, P. and Tomkó, J., Eds.). North-Holland: Amsterdam. [1.5]

Jones Hughes, T. (1970). Town and baile in Irish place names, pp. 244–258 in *Irish Geographical Studies in Honour of E. Estyn Evans* (Stephens, N. and Glasscock, R. E., Eds.). Department of Geography, Queen's University of Belfast: Belfast. [5.2]

Jumars, P. A. (1978). Spatial autocorrelation with RUM (Remote Underwater Manipulator): vertical and horizontal structure of a bathyal benthic community. *Deep Sea Research*, **25**, 589–604. [3.4, 3.5]

Jumars, P. A., Thistle, D., and Jones, M. L. (1977). Detecting two-dimensional spatial structure in biological data. *Oecologia*, **28**, 109–123. [3.3]

Kagan, Y. Y. and Knopoff, L. (1976a). Statistical search for non-random features of the seismicity of strong earthquakes. *Physics of the Earth and Planetary Interiors*, **12**, 291–318. [3.6]

Kagan, Y. Y. and Knopoff, L. (1976b). Earthquake risk prediction as a stochastic process. *Physics of the Earth and Planetary Interiors*, **14**, 97–108. [3.6]

Kalman, R. E. (1960). A new approach to linear filtering and prediction problems. *Transactions of the American Society of Mechanical Engineers, Journal of Basic Engineering, Series D*, **82**, 35–45. [5.8]

Kathirgamatamby, N. (1953). Note on the Poisson index of dispersion. *Biometrika*, **40**, 225–228. [1.3]

Kelker, G. H. (1945). Measurement and interpretation of forces that determine populations of managed deer. Ph.D. thesis, University of Michigan: Ann Arbor. [2.3]

Kelly, F. P. and Ripley, B. D. (1976). A note on Strauss's model for clustering. *Biometrika*, **63**, 357–360. [1.1]

Kemp, C. D. (1967). On a contagious distribution suggested for accident data. *Biometrics*, **23**, 241–255. [1.1]

Kendall, M. G. and Moran, P. A. P. (1963). *Geometrical Probability*. Griffin: London. [1.4]

Kendall, M. G. and Stuart, A. (1979). *The Advanced Theory of Statistics*, Vol. 2, 3rd edition. Griffin: London. [5.4]

Kershaw, K. A. (1957). The use of cover and frequency in the detection of pattern in plant communities. *Ecology*, **38**, 291–299. [1.3]

Keuls, M., Over, H. J., and de Wit, C. T. (1963). The distance method for estimating densities. *Statistica Neerlandica*, **17**, 71–91. [1.1]

Knox, E. G. (1964). The detection of space–time interactions. *Applied Statistics*, **13**, 25–29. [3.2, 3.6]

Kooijman, S. A. L. M. (1976). Some remarks on the statistical analysis of grids especially with respect to ecology. *Annals of Systems Research*, **5**, 113–132. [3.4, 5.8]

Kooijman, S. A. L. M. (1979). The description of point patterns, pp. 305–332 in *Spatial and Temporal Analysis in Ecology* (Cormack, R. M. and Ord, J. K., Eds.). International Co-operative Publishing House: Fairland, Maryland. [1.1, 1.4, 2.4]

Krige, D. G. (1966). Two-dimensional weighted moving average trend surfaces for ore evaluation, pp. 13–38 in *Proceedings of the Symposium on Mathematical Statistics and Computer Applications in Ore Valuation, Johannesburg*. [5.8]

Krige, D. G., Watson, M. I., Oberholzer, W. J., and du Toit, S. R. (1969). The use of contour surfaces as predictive models for ore values, pp. 127–161 in *A Decade of Digital Computing in the Mineral Industry* (Weiss, A., Ed.). American Institute of Mining, Metallurgical and Petroleum Engineers: New York. [5.8]

Krishna Iyer, P. V. A. (1949). The first and second moments of some probability distributions arising from points on a lattice, and their applications. *Biometrika*, **36**, 135–141. [3.2]

Kruk, J. (1973). *Studia osadnicze nad neolitem wyzyn lessowych*. Polska akademia nauk insytut historii kultury materialnej. [1.4]

Krumbein, W. C. (1956). Regional and local components in facies maps. *Bulletin of the American Association of Petroleum Geologists*, **40**, 2163–2194. [5.5]

Krumbein, W. C. and Graybill, F. A. (1965). *An Introduction to Statistical Models in Geology*. McGraw-Hill: New York. [5.5]

Lackritz, J. R. and Scheaffer, R. L. (1983). Asymptotic properties of estimators derived from line transect sampling. *Communications in Statistics, Theory and Methods*, **12**, 849–864. [2.7]

Lebart, L. (1966). *Les Variables Socio-economiques Departmentales ou Regionales—I Methodes Statistiques d'etude*. Institut d'Etude du Developpement Economique et Social. Université de Paris: Paris. [3.5]

Lebart, L. (1969). Analyse statistique de la contiguité. *Publications de l'Institut Statistique de l'Université de Paris*, **18**, 81–112. [3.5, 5.3]

Lee, Y. (1979). A nearest-neighbour spatial association measure for the analysis of firm interdependence. *Environment and Planning, Series A*, **11**, 169–176. [4.1, 4.4]

Lewis, S. M. (1975). Robust estimation of density for a two-dimensional point process. *Biometrika*, **62**, 519–521. [2.4]

Liebetrau, A. M. (1977). Tests for randomness in two dimensions. *Communications in Statistics, Theory and Methods, Series A*, **6**, 1367–1383. [1.5]

Liebetrau, A. M. (1978). The weak convergence of a class of estimators of the variance function of a two-dimensional Poisson process. *Journal of Applied Probability*, **15**, 433–439. [1.5]

Liebetrau, A. M. and Rothman, E. D. (1977). A classification of spatial distributions based upon several cell sizes. *Geographical Analysis*, **9**, 14–28. [1.5]

Lloyd, M. (1967). Mean crowding. *Journal of Animal Ecology*, **36**, 1–30. [1.2]

Losch, A. (1954). *The Economics of Location*. Yale University Press: New Haven. [1.1]

Lotwick, H. W. and Silverman, B. W. (1982). Methods of analysing spatial processes of several types of points. *Journal of the Royal Statistical Society, Series B*, **44**, 406–413. [4.4]

Ludwig, J. A. (1979). A test of different quadrat variance methods for the analysis of spatial pattern, pp. 289–304 in *Spatial and Temporal Analysis in Ecology* (Cormack, R. M. and Ord, J. K., Eds.) International Co-operative Publishing House: Fairland, Maryland. [1.3]

Ludwig, J. A. and Goodall, D. W. (1978). A comparison of paired- with blocked-quadrat variance methods for the analysis of spatial pattern. *Vegetatio*, **38**, 49–59. [1.3]

Lyon, L. J. (1968). An evaluation of density sampling methods in a shrub community. *Journal of Range Management*, **21**, 16–20. [I]

MacArthur, R. H. and Wilson, E. O. (1967). *The Theory of Island biogeography*. Princeton University Press: Princeton, New Jersey. [5.5]

Mantel, N. (1967). The detection of disease clustering and a generalized regression approach. *Cancer Research*, **27**, 209–220. [3.6]

Marble, D. F. (1967). *Some Computer Programs for Geographical Research*. Dept of Geography. Northwestern University: Evanston, Illinois. [3.6]

Mardia, K. V. Edwards, R., and Puri, M. L. (1977). Analysis of central place theory. *Bulletin of the International Statistical Institute*, **47**, 93–110. [1.7]

Marriott, F. H. C. (1981). Letter to the editor. *Biometrics*, **37**, 178. [1.4]

Marshall, R. J. (1982). Discussion of the paper by Dr Pocock *et al*. *Journal of the Royal Statistical Society, Series A*, **145**, 337. [5.4]

Martin, R. L. (1974). On spatial dependence, bias and the use of first spatial differences in regression analysis. *Area*, **6**, 185–194. [5.3]

Martin, R. L. (1975). Identification and estimation of dynamic space-time forecasting models for geographical data. *Advances in Applied Probability*, **7**, 455–456. [5.8]

Martin, R. L. (1978). Kalman filter modelling of time-varying processes in urban and regional analysis, pp. 104–126 in *Dynamic Modelling for Space–time Systems* (Martin, R. L., Thrift, N. J., and Bennett, R. J., Eds.). Pion: London. [5.8]

Martin, R. L. and Oeppen, J. E. (1975). The identification of regional forecasting models using space–time correlation functions. *Transactions of the Institute of British Geographers*, **66**, 95–118. [3.5]

Matérn, B. (1960). Spatial variation. *Meddelanden Statens från Skogsforskningsinstitut Stockholm*, **49**(5), 1–144. [1.1]

Matérn, B. (1971). Doubly stochastic Poisson process in the plane, pp 195–213 in *Statistical Ecology* (Patil, G. P., Pielou, E. C., and Waters, W. E., Eds.), Vol. 1. Pennsylvania State University Press: College Town, Pennsylvania. [1.1]

Matérn, B. (1972). Poisson processes in the plane and related models for clumping and heterogeneity. *NATO Advanced Study Institute in Statistical Ecology*. Pennsylvania State University Press: College Town, Pennsylvania. [1.5]

Mather, P. M. (1976). *Computational Methods of Multivariate Analysis in Physical Geography*. Wiley: London. [5.5]

Matheron, G. (1963). Principles of geostatistics, *Economic Geology*, **58**, 1246–1266. [5.8]

Matheron, G. (1967). Kriging or polynomial interpolation procedures. *Canadian Mining and Metallurgical Bulletin*, **60**, 1041–1045. [5.8]

Matheron, G. (1969). *Le Krigeage Universal*. Centre de Morphologie Mathematique: Fontainebleau. [5.8]

Matheron, G. (1970). Structures aléatoires et géologie mathématique *International*

Statistical Review, **38**, 1–11. [5.8]

Matheron, G. (1971). *The Theory of Regionalised Variables.* Centre de morphologie mathematique: Fontainebleau. [5.8]

Mayr, E. (1942). *Systematics and the Origin of Species.* Columbia University Press: N.Y. [3.4]

Mayr, E. (1963). *Animal Species and Evolution.* Belknap: Cambridge. [3.4]

Mayr, E. and Vaurie, C. (1948). Evolution of the family Dicruridae (birds), *Evolution*, **4**, 238–265. [3.3]

Mead, R. (1966). A relationship between individual plant spacing and yield. *Annals of Botany*, **30**, 301–309. [1.7]

Mead, R. (1967). A mathematical model for the estimation of inter-plant competition. *Biometrics*, **23**, 189–205. [5.7]

Mead, R. (1974a). A test for spatial pattern at several scales using data from a grid of contiguous quadrats. *Biometrics*, **30**, 295–307. [1.3]

Mead, R. (1974b). Discussion of the paper by Besag. *Journal of the Royal Statistical Society, Series B*, **36**, 192–236. [5.7]

Meijering, J. L. (1953). Interface area, edge length and number of vertices in crystal aggregates with random nucleation. *Philips Research Reports*, **8**, 270–290. [1.7]

Mielke, P. W. (1978). Clarification and appropriate inferences for Mantel and Valand's nonparametric multivariate analysis technique. *Biometrics*, **34**, 277–282. [3.1]

Miesch, A. T. and Connor, J. J. (1968). Stepwise regression and non-polynomial models in trend analysis. *Computer Contribution*, **27**, *State Geological Survey, University of Kansas*. [5.5]

Miles, R. E. (1964a). Random polygons determined by random lines in a plane. I. *Proceedings of the National Academy of Sciences*, **52**, 901–907. [1.7]

Miles, R. E. (1964b). Random polygons determined by random lines in a plane. II. *Proceedings of the National Academy of Sciences*, **52**, 1157–1160. [1.7]

Miles, R. E. (1970). On the homogeneous planar Poisson point process. *Mathematical Bioscience*, **6**, 85–127. [1.7]

Miller, R. L. (1956). Trend surfaces: their application to analysis and description of environments of sedimentation. *Journal of Geology*, **64**, 425–446. [5.5]

Moellering, H. and Tobler, W. R. (1972). Geographical variances. *Geographical Analysis*, **4**, 34–50. [1.3]

Mollison, D. (1977). Spatial contact models for ecological and epidemic spread. *Journal of the Royal Statistical Society, Series B*, **39**, 283–326. [4.0]

Moore, P. G. (1954). Spacing in plant populations. *Ecology*, **35**, 222–227. [1.4, 2.4]

Moran, P. A. P. (1948). The interpretation of statistical maps. *Journal of the Royal Statistical Society, Series B*, **10**, 243–251. [3.2]

Moran, P. A. P. (1950). Notes on continuous stochastic phenomena. *Biometrika*, **37**, 17–23. [3.3]

Moran, P. A. P. (1951). The random division of an interval, II. *Journal of the Royal Statistical Society, Series B*, **13**, 147–150. [1.5]

Morisita, M. (1954). Estimation of population density by a spacing method. *Memoirs of the Faculty of Science, Kyushu University, Series E, Biology*, **1**, 187–197. [1.4]

Morisita, M. (1957). A new method for the estimation of density by the spacing method applicable to non-randomly distributed populations. *Physiological Ecology, Kyoto*, **7**, 134–144. [2.4]

Morisita, M. (1959). Measuring of the dispersion and analysis of distribution patterns. *Memoires of the Faculty of Science, Kyushu University, Series E. Biology*, **2**(4), 215–235. [1.2]

Mosby, H. S. (1959). Reconnaissance mapping and map use, pp. 119–134 in *Wildlife Management Technique* (Giles, R. H. Jr, Ed.) Edward Bros.: Ann Arbor. [1.5]

Mountford, M. D. (1961). On E. C. Pielou's index of non-randomness. *Journal of Ecology*, **49**, 271–275. [1.4]

Mourant, A. E., Kopec, A. C., and Bomoniewska-Solczak, K. (1958). *The ABO Blood Group*. Thomas: Springfield, Ill. [3.2]

Neave, H. R. (1978). *Statistics Tables*. Allen and Unwin: London. [5.6]

Nelder, J. A. and Wedderburn, R. W. M. (1972). Generalized linear models. *Journal of the Royal Statistical Society, Series A*, **135**, 370–384. [5.1, 5.2, 5.4]

Newton, I., Marquiss, M., Weir, D. N., and Moss, D. (1977). Spacing of sparrowhawk nesting territories. *Journal of Animal Ecology*, **46**, 425–441. [1.1]

Neyman, J. (1939). On a new class of contagious distributions applicable in entomology and bacteriology. *Annals of Mathematical Statistics*, **10**, 35–57. [1.1]

Neyman, J. and Scott, E. L. (1952). A theory of the spatial distribution of the galaxies. *Astrophysics Journal*, **116**, 144–163. [1.1]

Neyman, J. and Scott, E. L. (1958). A statistical approach to problems of cosmology. *Journal of the Royal Statistical Society, Series B*, **20**, 1–43. [1.1]

Norcliffe, G. B. (1969). The uses and limitations of trend surface models. *Canadian Geographer*, **13**, 338–348. [5.5]

Norden, R. H. (1972). A survey of maximum likelihood estimation. *Review of the International Statistics Institute*, 329–354. [5.4]

Numata, M. (1961). Forest vegetation in the vicinity of Choshi, Coastal flora and vegetation at Choshi, Chiba Prefecture IV. *Bulletin of the Choshi Marine Laboratory, Chiba University*, **3**, 28–48. [1.5]

Openshaw, S. (1977). A geographical solution to scale and aggregation problems in region-building, partitioning, and spatial modelling. *Transactions of the Institute of British Geographers*, **2**, 459–472. [4.2]

Openshaw, S. and Taylor, P. J. (1979). A million or so correlation coefficients: three experiments on the modifiable areal unit problem, pp. 127–144 in *Statistical Applications in the Spatial Sciences* (Wrigley, N., Ed.) Pion: London. [4.2]

Ord, J. K. (1972). *Families of Frequency Distributions*. Griffin: London. [1.1]

Ord, J. K. (1975). Estimation methods for models of spatial interaction. *Journal of the American Statistical Association*, **70**, 120–126. [5.3, 5.4, 5.6, 5.7]

Ord, J. K. (1978). How many trees in a forest? *Mathematical Scientist*, **3**, 23–33. [2.5, 2.6]

Ord, J. K. (1980). Test of significance using non-normal data. *Geographical Analysis*, **12**, 387–392. [3.1, 5.4]

Ord, J. K (1981). Towards a theory of spatial statistics: a comment. *Geographical Analysis*, **13**, 86–91. [I]

Paloheimo, J. E. (1971). On a theory of search. *Biometrika*, **58**, 61–75. [1.1]

Patil, S. A., Burnham, K. P., and Kovner, J. L. (1979). Nonparametric estimation of plant density by the distance method. *Biometrics*, **35**, 597–604. [2.3]

Patil, S. A., Kovner, J. L., and Burnham, K. P. (1982). Optimum nonparametric estimation of population density based on ordered distances. *Biometrics*, **38**, 243–248. [2.3]

Perry, J. N. and Mead, R. (1979). On the power of the index of dispersion test to detect spatial pattern. *Biometrics*, **35**, 613–622. [1.2]

Persson, O. (1964). Distance methods. *Studia Forestalia Suecica*, **15**, 1–68. [1.4]

Pfeifer, P. E. and Deutsch, S. J. (1980). A STARIMA model-building procedure with application to description and regional forecasting. *Transactions of the Institute of British Geographers*, **5**, 330–349. [3.5]

Pickard, D. K. (1982). Isolated nearest neighbours. *Journal of Applied Probability*, **19**, 444–449. [1.5]

Pielou, E. C. (1959). The use of point-to-plant distances in the study of the pattern of plant populations. *Journal of Ecology*, **47**, 607–613. [1.4]

Pielou, E. C. (1961). Segregation and symmetry in two species populations as studied by nearest neighbour relations. *Journal of Ecology*, **49**, 255–269. [4.4]

Pielou, E. C. (1963). Runs of healthy and diseased trees in transects through an infected forest. *Biometrics*, **19**, 603–614. [4.3]

Pielou, E. C. (1977). *Mathematical Ecology*. Wiley: New York. [I, 1.1, 1.3, 1.5, 1.7, 2.1, 3.5, 4.3]

Pinder, D. A. and Witherick, M. E. (1972). The principles, practice and pitfalls of nearest-neighbour analysis. *Geography*, **57**, 277–288. [1.5]

Pirie, H. R. and Hamdan, M. A. (1972). Some revised continuity corrections for discrete distributions. *Biometrics*,, **28**, 693–701. [4.2]

Plackett, R. L. (1960). *Regression Analysis*. Clarendon: Oxford. [5.1]

Pocock, S. J., Cook, D. G., and Beresford, S. A. A. (1981). Regression of area mortality rates on explanatory variables: what weighting is appropriate? *Applied Statistics*, **30**, 286–295. [5.3]

Pocock, S. J., Cook, D. G., and Shaper, A. G. (1982). Analysing geographic variation in cardiovascular mortality: methods and results. *Journal of the Royal Statistical Society, Series A*, **145**, 313–341. [5.4, 5.6]

Pollard, J. H. (1971). On distance estimators of density in randomly distributed forests. *Biometrics*, **27**, 991–1002. [1.4, 2.4]

Pollock, K. H. (1978). A family of density estimators for line transect sampling. *Biometrics*, **34**, 475–478. [2.3]

Porter, P. W. (1960). Earnest and the Orephagians: A fable for the instruction of young geographers. *Annals of the Association of American Geographers*, **50**, 297–299. [1.5]

Preston, F. W. (1948). The commonness, and rarity, of species. *Ecology*, **29**, 254–283. [5.5]

Preston, F. W. (1962). The canonical distribution of commonness and rarity. I. *Ecology*, **43**, 186–215. [5.5]

Puri, M. L. and Sen, P. K. (1971). *Nonparametric methods in Multivariate Analysis*. Wiley: New York. [4.5]

Ramsey, F. L. (1979). Parametric models for line transect surveys. *Biometrika*, **66**, 505–512. [2.3]

Ripley, B. D. (1976). The second-order analysis of stationary point processes. *Journal of Applied Probability*, **13**, 255–266. [1.5]

Ripley, B. D. (1977). Modelling spatial patterns. *Journal of the Royal Statistical Society, Series B*, **39**, 172–212. [1.1, 1.5]

Ripley, B. D. (1978). Spectral analysis and the analysis of pattern in plant communities. *Journal of Ecology*, **66**, 965–981. [1.5]

Ripley, B. D. (1979a). Tests of 'randomness' for spatial point patterns. *Journal of the Royal Statistical Society, Series B*, **41**, 368–374. [1.5]

Ripley, B. D. (1979b). Simulating spatial patterns: dependent samples from a multivariate density: Algorithm AS137. *Applied Statistics*, **28**, 109–112. [1.1]

Ripley, B. D. (1979c). The analysis of geographical maps, pp. 53–72 in *Exploratory and Explanatory Statistical Analysis of Spatial Data* (Bartels, C. P. A. and Ketellapper, R. H., Eds.). Nijhoff: Boston. [5.7]

Ripley, B. D. (1981). *Spatial Statistics*. Wiley: New York. [I, 1.1, 1.3, 1.5, 3.3, 5.3, 5.4, 5.5, 5.6, 5.7]

Ripley, B. D. and Silverman, B. W. (1978). Quick tests for spatial interaction. *Biometrika*, **65**, 641–642. [1.5]

Robinette, W. L., Loveless, C. M., and Jones, D. A. (1974). Field tests of strip census methods. *Journal of Wildlife Management*, **38**, 81–96. [2.3]

Robinson, G. (1972). Trials on trends through clusters of cirques. *Area*, **4**, 104–113. [5.5]

Roder, W. (1975). A procedure for assessing point patterns without reference to area or density. *The Professional Geographer*, **27**, 432–440. [1.5]

Rohlf, F. J. and Archie, J. W. (1978). Least-squares mapping using interpoint distances. *Ecology*, **59**, 126–132. [1.5]

Romanes, G. J. (1897). *Darwin, and after Darwin*. Open Court: Chicago. [3.4]

Royaltey, H. H., Astrachan, E., and Sokal, R. R. (1975). Tests for patterns in geographic variation, *Geographical Analysis*, **7**, 369–394. [3.3]

Royston, J. P. (1983). A simple method for evaluating the Shapiro–Francia *W* test of non-normality. *The Statistician*, **32**, 297–300. [5.6]

Rushton, G. (1972). Map transformations of point patterns: central place patterns in areas of variable population density. *Papers of the Regional Science Association*, **28**, 111–129. [1.1]

Satyamurthi, K. R. (1979). Density, derived from measured distances, for studying the spatial patterns. *Sankhya, Series B*, **40**, 197–203. [1.4, 2.4]

Saunders, R., Kryscio, R. J., and Funk, G. M. (1982). Poisson limits for a hardcore clustering model. *Stochastic Processes and their Applications*, **12**, 97–106. [1.5]

Schlesselman, J. (1971). Power families: a note on the Box and Cox transformations. *Journal of the Royal Statistical Society, Series B*, **33**, 307–311. [5.5]

Schweder, T. (1977). Point process models for line transect experiments, pp. 221–242 in *Recent Developments in Statistics* (Barra, J. R., *et al*., Eds.). North-Holland: Amsterdam. [2.3]

Seber, G. A. F. (1977). *Linear Regression Analysis*. Wiley: New York. [5.1]

Shapiro, S. S. and Francia, R. S. (1972). An approximate analysis of variance test for normality. *Journal of the American Statistical Association*, **67**, 215. [5.6]

Siegel, S. (1956). *Nonparametric Statistics for the Behavioural Sciences*. McGraw-Hill: New York. [4.4, 4.5]

Siemiatycki, J. (1978). Mantel's space–time clustering statistic: computing higher moments and a comparison of various data transforms. *Journal of Statistical Computation and Simulation*, **7**, 13–31. [3.1]

Silk, J. (1979). *Statistical Concepts in Geography*. Allen & Unwin: London. [3.2]

Silverman, B. W. and Brown, T. C. (1978). Short distances, flat triangles and Poisson limits. *Journal of Applied Probability*, **15**, 815–825. [1.5]

Simberloff, D. (1979). Nearest neighbour assessments of spatial configurations of circles rather than points. *Ecology*, **60**, 679–685. [1.6]

Simpson, B. B. (1974). Glacial migrations of plants: island biogeographical evidence. *Science*, **182**, 698–700. [5.5]

Skellam, J. G. (1952). Studies in statistical ecology: I Spatial pattern, *Biometrika*, **39**, 346–362. [1.1, 1.4]

Smith, G. E. J. (1979). Some aspects of line transect sampling when the target population moves. *Biometrics*, **35**, 323–330. [2.3]

Sokal, R. R. (1979). Ecological parameters inferred from spatial correlograms, pp. 167–196 in *Contemporary Quantitative Ecology and Related Ecometrics* (Patil, G. P. and Rosenzweig, M. L., Eds.) International Co-operative Publishing House; Fairland, Maryland. [3.5]

Sokal, R. R. and Oden, N. L. (1978a). Spatial autocorrelation in biology, 1. Methodology. *Biological Journal of the Linnean Society*, **10**, 199–228. [3.1, 3.3, 3.4, 3.5]

Sokal, R. R. and Oden, N. L. (1978b). Spatial autocorrelation in biology, 2. Some biological applications of evolutionary and ecological interest. *Biological Journal of the Linnean Society*, **10**, 229–249. [3.2, 3.5, 5.2]

Stine, J. H. (1962). Temporal aspects of tertiary production elements in Korea, pp. 68–88 in *Urban Systems and Economic Development* (Pitts, F. R., Ed.). University of Oregon Press: Eugene, Oregon. [1.1]

Stoyan, D. (1979). Interrupted point processes. *Biometrical Journal*, **21**, 607–610. [4.1]

Strand, L. (1972). A model for stand growth, pp. 207–216 in *Proceedings of the 3rd*

Conference of the Advisory Group of Forest Statisticians, International Union for Research Organization, Institut National de la Recherche Agronomique, Jouy-en-Josas, France. [I]

Strauss, D. J. (1975). A model for clustering, *Biometrika*, **62**, 467–475. [1.1, 1.5]

Swed, F. S. and Eisenhart, C. (1943). Tables for testing randomness of grouping in a sequence of alternatives. *Annals of Mathematical Statistics*, **14**, 66–85. [4.3]

Tanner, J. C. (1983). The proportion of quadrilaterals formed by random lines in a plane. *Journal of Applied Probability*, **20**, 400–404. [1.7]

Taylor, L. R. (1961). Aggregation, variance and the mean. *Nature*, **189**, 732–735. [1.2]

Taylor, L. R., Woiwod, I. P., and Perry, J. N. (1978). The density-dependence of spatial behaviour and the rarity of randomness. *Journal of Animal Ecology*, **47**, 383–406. [1.2]

Teekens, R. and Koerts, J. (1972). Some statistical implications of the log transformation of multiplicative models. *Econometrica*, **40**, 793–819. [5.5]

Theil, H. (1965). The analysis of disturbances in regression residuals. *Journal of the American Statistical Association*, **60**, 1067–1079. [5.6]

Theil, H. (1969). A multinomial extension of the linear logit model. *International Economic Review*, **10**, 251–259. [5.3]

Theil, H. (1970). On the estimation of relationships involving qualitative variables. *American Journal of Sociology*, **76**, 103–154. [5.3]

Thiessen, A. W. (1911). Precipitation averages for large areas. *Monthly Weather Review*, **39**, 1082–1084. [1.7]

Thomas, M. (1949). A generalisation of Poisson's binomial limit for use in ecology. *Biometrika*, **36**, 18–25. [1.1]

Thompson, H. R. (1955). Spatial point processes with applications to ecology. *Biometrika*, **42**, 102–115. [1.3]

Thompson, H. R. (1956). Distribution of distance to nth nearest neighbour in a population of randomly distributed individuals. *Ecology*, **37**, 391–394. [1.4]

Thompson, H. R. (1958). The statistical study of plant distribution patterns using a grid of quadrats. *Australian Journal of Botany*, **6**, 322–342. [1.3]

Tinkler, K. J. (1973). The topology of rural periodic market systems, *Geografiska Annaler*, B, **55**, 121–133. [1.1]

Tinline, R. R. (1971). Linear operators in diffusion research, pp. 71–91 in *Regional Forecasting* (Chisholm, M., Frey A., and Haggett, P., Eds.) Butterworths: London. [3.5]

Tjostheim, D. (1978). A measure of association for spatial variables. *Biometrika*, **65**, 109–114. [4.5]

Tobler, W. R. (1963). Geographic area and map projections. *Geographical Review*, **53** 59–78. [1.1]

Tobler, W. R. (1964). A polynomial representation of the Michigan population. *Papers of the Michigan Academy of Science, Arts and Letters*, **49**, 445–452. [5.5]

Tobler, W. R. (1970a). A computer movie simulating urban growth in the Detroit Region. *Economic Geography*, **46**, 234–240. [3.5]

Tobler, W. R. (1970b). *Selected Computer Programs*. University of Michigan: Ann Arbor, Michigan. [3.6]

Tobler, W. R. (1975). Linear operators applied to areal data, pp. 14–37 in *Display and Analysis of Spatial Data* (Davis, J. C. and McCullagh, M. J., Eds.). Wiley: London. [3.5]

Tornquist, G. Nordbeck, S., Rystedt, B., and Gould, P. R. (1972). Multiple Location Analysis. *Lund Studies in Geography, Series C.*, **12**. [1.1]

Tukey, J. W. (1957). On the comparative anatomy of transformations. *Annals of Mathematical Statistics*, **28**, 602–632. [5.5]

Unwin, D. J. (1970). Percentage RSS in trend surface analysis, *Area*, **2**, 25–28. [5.5]

Unwin, D. J. (1973a). Trials on trends. *Area*, **5**, 31–33. [5.5]

Unwin, D. J. (1973b). The distribution and orientation of corries in northern Snowdonia, Wales. *Transactions of the Institute of British Geographers*, **58**, 85–97. [5.5]

Unwin, D. J. (1975). Numerical error in a familiar technique: a case study of polynomial trend surface analysis. *Geographical Analysis*, **7**, 197–203. [5.5]

Unwin, D. J. and Hepple, L. W. (1974). The statistical analysis of spatial series. *The Statistician*, **23**, 211–227. [5.5]

Upton, G. J. G. (1982). A comparison of alternative tests for the 2 × 2 comparative trial. *Journal of the Royal Statistical Society, Series A*, **145**, 86–105. [4.2, 4.4]

Upton, G. J. G. (1984). On Mead's test for pattern. *Biometrics*, **40**. [1.3]

Usher, M. B. (1969). The relation between mean square and block size in the analysis of similar patterns. *Journal of Ecology*, **57**, 505–514. [1.3]

Usher, M. B. (1975). Analysis of pattern in real and artificial plant populations. *Journal of Ecology*, **63**, 569–586. [1.3]

Vere-Jones, D. (1978). Space time correlations for microearthquakes—a pilot study. *Supplement Advances in Applied Probability*, **10**, 73–87. [3.6]

Vincent, P. J. (1981). From theory into practice—a cautionary tale of island biogeography. *Area*, **13**, 115–118. [5.5]

Vincent, P. J. and Haworth, J. M. (1982). Statistical bias in species–area regression models. *Area*, **14**, 301–303. [5.5]

Vincent, P. J., Haworth, J. M., Griffiths, J. G., and Collins, R. (1976). The detection of randomness in plant patterns. *Journal of Biogeography*, **3**, 373–380. [1.7]

Vithayasai, C. (1971). Runs of healthy and diseased trees in transects through an infected forest: statistical and computational methods. *Biometrics*, **27**, 1003–1016. [4.3]

Voronoi, G. F. (1909). Recherches sur les paralleloèdres primitifs. *Journal für die reine und angewandte Mathematik*, **136**, 67–179. [1.7]

Walker, G. (1931). On periodicity in a series of related terms. *Proceedings of the Royal Society, Series A*, 518–532. [3.5]

Warren, W. G. (1962). *Contributions to the Study of Spatial Point Processes*. University of North Carolina, Institute of Statistics, Series No. 337 (mimeo). [1.1]

Warren, W. G. (1971). The center–satellite concept as a basis for ecological sampling, in *Statistical Ecology* (Patil, G. P., Pielou, E. C., and Waters, W. E., Eds.) Vol. 2. Pennsylvania State University Press: College Town, Pennsylvania. [1.1]

Warren, W. G. and Batcheler, C. L. (1979). The density of spatial patterns: robust estimation through distance methods, pp. 247–269 in *Spatial and Temporal Analysis in Ecology* (Cormack, R. M. and Ord, J. K., Eds.) International Co-operative Publishing House: Fairland, Maryland. [2.2, 2.4]

Waters, W. E. (1959). A quantitative measure of aggregation in insects. *Journal of Economics in Entomology*, **52**, 1180–1184. [1.2]

Watson, G. S. (1971). Trend surface analysis. *Journal of the International Association of Mathematical Geology*, **3**, 215–226. [5.8]

Watt, A. S. (1947). Pattern and process in the plant community. *Journal of Ecology*, **35**, 1–22. [1.3]

Weaire, D. (1974). Some remarks on the arrangement of grains in a polycrystal. *Metallography*, **7**, 157–160. [1.7]

Weisberg, S. and Bingham, C. (1975). An approximate analysis of variance test for non-normality suitable for machine calculation. *Technometrics*, **17**, 133–134. [5.6]

Westcott, M. (1982). Approximations to hard-core models and their application to statistical analysis. *Journal of Applied Probability*, **19A**, 281–292. [1.5]

White, K. W. (1977). Dynamic central place theory: results of a simulation approach. *Geographical Analysis*, **9**, 226–243. [1.1]

Whitten, E. H. T. (1966). The general linear equation in prediction of gold content in Witwatersrand rocks, South Africa, pp. 124–141 in *Proceedings of the Symposium on Mathematical Statistics and Computer Applications in Ore Valuation, Johannesburg.* [5.8]

Whittle, P. (1954). On stationary processes in the plane. *Biometrika*, **41**, 434–449. [5.4, 5.7]

Wiggins, I. L. and Porter, D. M. (1971). *Flora of the Galapagos Islands*, Stanford University Press: Stanford, U.S.A. [5.4]

Wigner, E. and Seitz, F. (1933). On the constitution of metallic sodium. *Physics Review*, **43**, 804–810. [1.7]

Williams, I. N. (1976). Optimistic theory validation from spatially grouped regression: theoretical aspects. *Transactions of the Martin Centre*, **1**, 113–145. [4.2]

Wrigley, N. (1977). Probability surface mapping: a new approach to trend surface mapping. *Transactions of the Institute of British Geographers*, **2**, 129–140. [5.5]

Wrigley, N. (1983). Quantitative methods: on data and diagnostics. *Progress in Human Geography*, **7**, 567–577. [5.9]

Yates, F. (1934). Contingency tables involving small numbers and the χ^2 test. *Journal of The Royal Statistical Society, Supplement*, **1**, 217–235. [4.2]

Yates, F. (1984). Tests of significance for 2×2 tables. *Journal of the Royal Statistical society Series A*, **147**. [4.2, 4.4]

Yeates, M. H. and Garner, B. J. (1971). *The North American City*. Harper and Row: New York. [1.1]

Yule, G. U. (1927). On a method of investigating periodicities in a disturbed series, with special reference to Wolfer's sunspot numbers. *Philosophical Transactions of the Royal Society of London, Series A*, **226**, 267–298. [3.5]

Yule, G. U. and Kendall, M. G. (1965). *An Introduction to the Theory of Statistics*. 14th edition. Griffin: London. [4.2]

Zahl, S. (1974). Application of the *S*-method to the analysis of spatial pattern. *Biometrics*, **30**, 513–524. [1.3]

Zahl, S. (1977). A comparison of three methods for the analysis of spatial pattern. *Biometrics*. **33**, 681–692. [1.3]

Zellner, A. (1983). Applications of Bayesian analysis in econometrics. *The Statistician*, **32**, 23–34. [5.4]

Appendices

APPENDIX A: THE NORMAL DISTRIBUTION FUNCTION

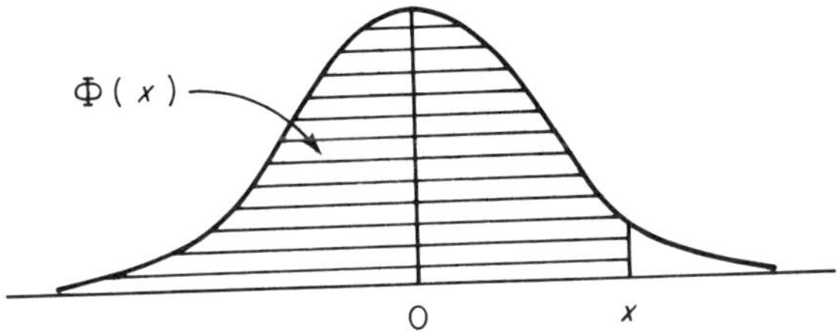

$\Phi(x) = P[X \leqslant x]$, where X has a normal distribution with mean 0 and variance 1. (Note that if Z has a normal distribution with mean μ and variance σ^2, then $P[Z \leq z] = \Phi[(z-\mu)/\sigma]$.)

x	$\Phi(x)$	x	$\Phi(x)$
0.0	0.500	1.5	0.933
0.1	0.540	1.6	0.945
0.2	0.579	1.7	0.955
0.3	0.618	1.8	0.964
0.4	0.655	1.9	0.971
0.5	0.692	2.0	0.977
0.6	0.726	2.1	0.982
0.7	0.758	2.2	0.986
0.8	0.788	2.3	0.989
0.9	0.816	2.4	0.9918
1.0	0.841	2.5	0.9938
1.1	0.864	2.6	0.9953
1.2	0.885	2.7	0.9965
1.3	0.903	2.8	0.9974
1.4	0.919	2.9	0.9981

$\Phi(-x) = 1 - \Phi(x)$

APPENDIX B: PERCENTAGE POINTS OF THE χ^2 DISTRIBUTION

The figure in the table corresponding to the *d* row (*d* degrees of freedom) is that value, *T*, for which $P[\chi_d^2 > T] = P$, where *P* is the probability specified by the column heading.

Degrees of freedom	Tail probability *P*			
d	0.10	0.05	0.01	0.001
1	2.71	3.84	6.63	10.83
2	4.61	5.99	9.21	13.81
3	6.25	7.81	11.34	16.27
4	7.78	9.49	13.28	18.47
5	9.24	11.07	15.09	20.52
6	10.64	12.59	16.81	22.46
7	12.02	14.07	18.48	24.32
8	13.36	15.51	20.09	26.12
9	14.68	16.92	21.67	27.88
10	15.99	18.31	23.21	29.59
12	18.55	21.03	26.22	32.91
14	21.06	23.68	29.14	36.12
16	23.54	26.30	32.00	39.25
18	25.99	28.87	34.81	42.31
20	28.41	31.41	37.57	45.31
25	34.38	37.65	44.31	52.62
30	40.26	43.77	50.89	59.70
40	51.81	55.76	63.69	73.40
50	63.17	67.50	76.15	86.66
60	74.40	79.08	88.38	99.61
70	85.53	90.53	100.4	112.3
80	96.58	101.9	112.3	124.8
90	107.6	113.1	124.1	137.2
100	118.5	124.3	135.8	149.4
	10%	5%	1%	0.1%

APPENDIX C: PERCENTAGE POINTS OF THE F DISTRIBUTION

$$\Pr\{F(n_1, n_2)\text{-variable} \geq F_{n_1,n_2,\alpha}\} = \alpha = 0.05$$

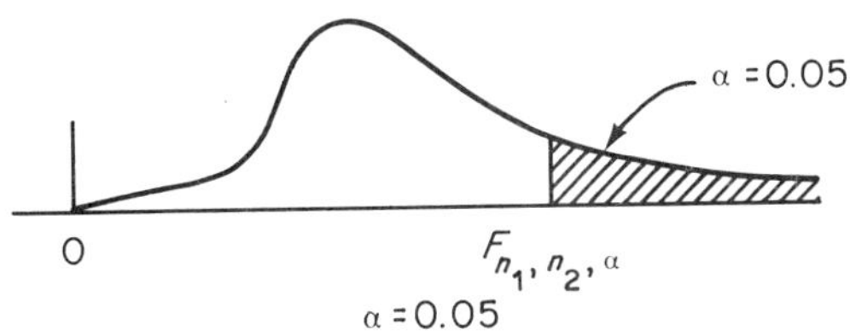

n_2 (denom. d.f.)	n_1 (numerator d.f.) 1	2	4	6	8	10	12	24	∞
	Values of $F_{n_1,n_2,\alpha}$								
1	161.4	199.5	224.6	234.0	238.9	241.9	243.9	249.1	254.3
2	18.51	19.00	19.25	19.33	19.37	19.40	19.41	19.45	19.50
3	10.13	9.55	9.12	8.94	8.85	8.79	8.74	8.64	8.53
4	7.71	6.94	6.39	6.16	6.04	5.96	5.91	5.77	5.63
5	6.61	5.79	5.19	4.95	4.82	4.74	4.68	4.53	4.36
6	5.99	5.14	4.53	4.28	4.15	4.06	4.00	3.84	3.67
7	5.59	4.74	4.12	3.87	3.73	3.64	3.57	3.41	3.23
8	5.32	4.46	3.84	3.58	3.44	3.35	3.28	3.12	2.93
9	5.12	4.26	3.63	3.37	3.23	3.14	3.07	2.90	2.71
10	4.96	4.10	3.48	3.22	3.07	2.98	2.91	2.74	2.54
11	4.84	3.98	3.36	3.09	2.95	2.85	2.79	2.61	2.40
12	4.75	3.89	3.26	3.00	2.85	2.75	2.69	2.51	2.30
13	4.67	3.81	3.18	2.92	2.77	2.67	2.60	2.42	2.21
14	4.60	3.74	3.11	2.85	2.70	2.60	2.53	2.35	2.13
15	4.54	3.68	3.06	2.79	2.64	2.54	2.48	2.29	2.07
20	4.35	3.49	2.87	2.60	2.45	2.35	2.28	2.08	1.84
25	4.24	3.39	2.76	2.49	2.34	2.24	2.16	1.96	1.71
30	4.17	3.32	2.69	2.42	2.27	2.16	2.09	1.89	1.62
40	4.08	3.23	2.61	2.34	2.18	2.08	2.00	1.79	1.51
60	4.00	3.15	2.53	2.25	2.10	1.99	1.92	1.70	1.39
120	3.92	3.07	2.45	2.17	2.02	1.91	1.83	1.61	1.25
∞	3.84	3.00	2.37	2.10	1.94	1.83	1.75	1.52	1.00

Source: Abridged from Table 18 of Pearson and Hartley (1954), *Biometrika Tables for Statisticians, Volume I*, published at the Cambridge University Press for *Biometrika* Trustees, with kind permission of the authors and publishers.

The diagram, and this presentation of the table, are taken from Table 4a of S. R. Searle (1971) *Linear Models*, published by John Wiley and Sons.

Appendix C (cont.)

$$\Pr\{F(n_1, n_2)\text{-variable} \geq F_{n_1,n_2\alpha}\} = \alpha = 0.01$$

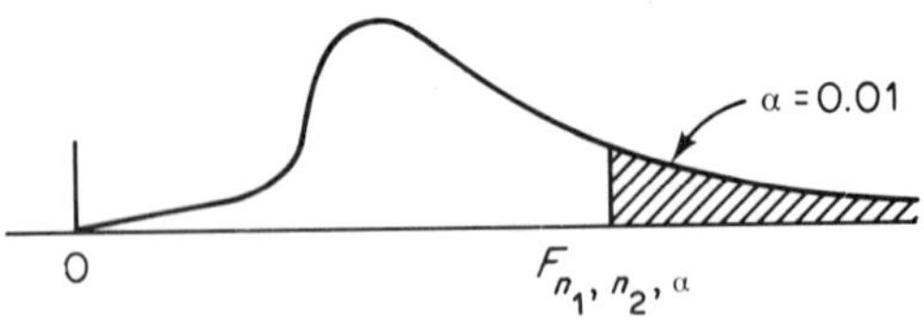

α = 0.01

n_2 (denom. d.f.)	n_1 (numerator d.f.) 1	2	4	6	8	10	12	24	∞
				Values of $F_{n_1,n_2,\alpha}$					
1	4052	5000	5625	5859	5982	6056	6106	6235	6366
2	98.50	99.00	99.25	99.33	99.37	99.40	99.42	99.46	99.50
3	34.12	30.82	28.71	27.91	27.49	27.23	27.05	26.60	26.13
4	21.20	18.00	15.98	15.21	14.80	14.55	14.37	13.93	13.46
5	16.26	13.27	11.39	10.67	10.29	10.05	9.89	9.47	9.02
6	13.75	10.92	9.15	8.47	8.10	7.87	7.72	7.31	6.88
7	12.25	9.55	7.85	7.19	6.84	6.62	6.47	6.07	5.65
8	11.26	8.65	7.01	6.37	6.03	5.81	5.67	5.28	4.86
9	10.56	8.02	6.42	5.80	5.47	5.26	5.11	4.73	4.31
10	10.04	7.56	5.99	5.39	5.06	4.85	4.71	4.33	3.91
11	9.65	7.21	5.67	5.07	4.74	4.54	4.40	4.02	3.60
12	9.33	6.93	5.41	4.82	4.50	4.30	4.16	3.78	3.36
13	9.07	6.70	5.21	4.62	4.30	4.10	3.96	3.59	3.17
14	8.86	6.51	5.04	4.46	4.14	3.94	3.80	3.43	3.00
15	8.68	6.36	4.89	4.32	4.00	3.80	3.67	3.29	2.87
20	8.10	5.85	4.43	3.87	3.56	3.37	3.23	2.86	2.42
25	7.77	5.57	4.18	3.63	3.32	3.13	2.99	2.62	2.17
30	7.56	5.39	4.02	3.47	3.17	2.98	2.84	2.47	2.01
40	7.31	5.18	3.83	3.29	2.99	2.80	2.66	2.29	1.80
60	7.08	4.98	3.65	3.12	2.82	2.63	2.50	2.12	1.60
120	6.85	4.79	3.48	2.96	2.66	2.47	2.34	1.95	1.38
∞	6.63	4.61	3.32	2.80	2.51	2.32	2.18	1.79	1.00

Source: Abridged from Table 18 of Pearson and Hartley (1954), *Biometrika Tables for Statisticians, Volume I*, published at the Cambridge University Press for the *Biometrika* Trustees, with kind permission of the authors and publishers.

The diagram, and this presentation of the table, are taken from Table 4b of S. R. Searle (1971) *Linear Models*, published by John Wiley and Sons.

APPENDIX D: PERCENTAGE POINTS OF THE EBERHARDT STATISTIC

Critical values for $E = m\sum X_i^2/(\sum X_i)^2$, *where m is the sample size*

	Regular alternative				Aggregate alternative			
m \ α	0.005	0.01	0.025	0.05	0.05	0.025	0.01	0.005
10	1.0340	1.0488	1.0719	1.0932	1.4593	1.5211	1.6054	1.6727
12	1.0501	1.0644	1.0865	1.1069	1.4472	1.5025	1.5769	1.6354
14	1.0632	1.0769	1.0983	1.1178	1.4368	1.4872	1.5540	1.6060
16	1.0740	1.0873	1.1080	1.1268	1.4280	1.4743	1.5352	1.5821
18	1.0832	1.0962	1.1162	1.1344	1.4203	1.4633	1.5195	1.5623
20	1.0912	1.1038	1.1232	1.1409	1.4136	1.4539	1.5061	1.5456
22	1.0982	1.1105	1.1293	1.1465	1.4078	1.4456	1.4945	1.5313
24	1.1044	1.1164	1.1348	1.1515	1.4025	1.4384	1.4844	1.5189
26	1.1099	1.1216	1.1396	1.1559	1.3978	1.4319	1.4755	1.5080
28	1.1149	1.1264	1.1439	1.1598	1.3936	1.4261	1.4675	1.4983
30	1.1195	1.1307	1.1479	1.1634	1.3898	1.4209	1.4604	1.4897
35	1.1292	1.1399	1.1563	1.1710	1.3815	1.4098	1.4454	1.4715
40	1.1372	1.1475	1.1631	1.1772	1.3748	1.4008	1.4333	1.4571
50	1.1498	1.1593	1.1738	1.1868	1.3644	1.3870	1.4151	1.4354
60	1.1593	1.1682	1.1818	1.1940	1.3565	1.3768	1.4017	1.4197
70	1.1668	1.1753	1.1882	1.1996	1.3504	1.3689	1.3915	1.4077
80	1.1730	1.1811	1.1933	1.2042	1.3455	1.3625	1.3833	1.3981
90	1.1782	1.1859	1.1976	1.2080	1.3414	1.3572	1.3765	1.3903
100	1.1826	1.1900	1.2013	1.2112	1.3379	1.3528	1.3709	1.3837
150	1.1979	1.2043	1.2139	1.2223	1.3260	1.3377	1.3519	1.3619
200	1.2073	1.2130	1.2215	1.2290	1.3189	1.3289	1.3408	1.3492
300	1.2187	1.2235	1.2307	1.2369	1.3105	1.3184	1.3279	1.3344
400	1.2257	1.2299	1.2362	1.2417	1.3055	1.3122	1.3203	1.3258
600	1.2341	1.2376	1.2429	1.2474	1.2995	1.3049	1.3113	1.3158
800	1.2391	1.2422	1.2468	1.2509	1.2960	1.3006	1.3061	1.3099
1000	1.2426	1.2454	1.2496	1.2532	1.2936	1.2977	1.3025	1.3059

Source: Reproduced from Table 1 of Hines and Hines (1979) by kind permission of the trustees of the *Biometrika Trust*.

Author Index

Subject Index

Example Index

Applied Probability and Statistics (Continued)

DRAPER and SMITH • Applied Regression Analysis, *Second Edition*

DUNN • Basic Statistics: A Primer for the Biomedical Sciences, *Second Edition*

DUNN and CLARK • Applied Statistics: Analysis of Variance and Regression

ELANDT-JOHNSON and JOHNSON • Survival Models and Data Analysis

FLEISS • Statistical Methods for Rates and Proportions, *Second Edition*

FOX • Linear Statistical Models and Related Methods

FRANKEN, KÖNIG, ARNDT, and SCHMIDT • Queues and Point Processes

GALAMBOS • The Asymptotic Theory of Extreme Order Statistics

GIBBONS, OLKIN, and SOBEL • Selecting and Ordering Populations: A New Statistical Methodology

GNANADESIKAN • Methods for Statistical Data Analysis of Multivariate Observations

GOLDBERGER • Econometric Theory

GOLDSTEIN and DILLON • Discrete Discriminant Analysis

GREENBERG and WEBSTER • Advanced Econometrics: A Bridge to the Literature

GROSS and CLARK • Survival Distributions: Reliability Applications in the Biomedical Sciences

GROSS and HARRIS • Fundamentals of Queueing Theory

GUPTA and PANCHAPAKESAN • Multiple Decision Procedures: Theory and Methodology of Selecting and Ranking Populations

GUTTMAN, WILKS, and HUNTER • Introductory Engineering Statistics, *Third Edition*

HAHN and SHAPIRO • Statistical Models in Engineering

HALD • Statistical Tables and Formulas

HALD • Statistical Theory with Engineering Applications

HAND • Discrimination and Classification

HILDEBRAND, LAING, and ROSENTHAL • Prediction Analysis of Cross Classifications

HOAGLIN, MOSTELLER, and TUKEY • Understanding Robust and Exploratory Data Analysis

HOEL • Elementary Statistics, *Fourth Edition*

HOEL and JESSEN • Basic Statistics for Business and Economics, *Third Edition*

HOGG and KLUGMAN • Loss Distributions

HOLLANDER and WOLFE • Nonparametric Statistical Methods

IMAN and CONOVER • Modern Business Statistics

JAGERS • Branching Processes with Biological Applications

JESSEN • Statistical Survey Techniques

JOHNSON and KOTZ • Distributions in Statistics
Discrete Distributions
Continuous Univariate Distributions—1
Continuous Univariate Distributions—2
Continuous Multivariate Distributions

JOHNSON and KOTZ • Urn Models and Their Application: An Approach to Modern Discrete Probability Theory

JOHNSON and LEONE • Statistics and Experimental Design in Engineering and the Physical Sciences, Volumes I and II, *Second Edition*

JUDGE, HILL, GRIFFITHS, LÜTKEPOHL and LEE • Introduction to the Theory and Practice of Econometrics

JUDGE, GRIFFITHS, HILL, LÜTKEPOHL and LEE • The Theory and Practice of Econometrics, *Second Edition*

KALBFLEISCH and PRENTICE • The Statistical Analysis of Failure Time Data

KEENEY and RAIFFA • Decisions with Multiple Objectives

LAWLESS • Statistical Models and Methods for Lifetime Data

LEAMER • Specification Searches: Ad Hoc Inference with Nonexperimental Data

LEBART, MORINEAU, and WARWICK • Multivariate Descriptive Statistical Analysis: Correspondence Analysis and Related Techniques for Large Matrices